Advanced Applications in Acoustics, Noise and Vibration

T0174000

Advanced Applications in Acoustics, Noise and Vibration

Edited by
Frank Fahy and John Walker

CRC Press
Taylor & Francis Group
Boca Raton London New York

CRC Press is an imprint of the
Taylor & Francis Group, an **informa** business

A SPON PRESS BOOK

CRC Press
Taylor & Francis Group
6000 Broken Sound Parkway NW, Suite 300
Boca Raton, FL 33487-2742

First issued in paperback 2019

© 2004 by Taylor & Francis Group, LLC
CRC Press is an imprint of Taylor & Francis Group, an Informa business

Typeset in Sabon by
Integra Software Services Pvt. Ltd, Pondicherry, India

No claim to original U.S. Government works

ISBN-13: 978-0-415-23791-1 (hbk)
ISBN-13: 978-0-367-39388-5 (pbk)

British Library Cataloguing in Publication Data
A catalogue record for this book is available from the
British Library

Library of Congress Cataloging in Publication Data
A catalog record for this book has been requested

Visit the Taylor & Francis Web site at
http://www.taylorandfrancis.com

and the CRC Press Web site at
http://www.crcpress.com

Contents

Preface

F.J. Fahy and J.G. Walker

This book is a companion volume to *Fundamentals of Noise and Vibration* published in 1998, which was based upon material presented in the first semester course of the one-year ISVR postgraduate Masters programme 'Sound and Vibration Studies' and was essentially pedagogic in form, content and purpose. This successor is based partly on material covered in a selection of elective modules in the second semester of the Masters programme and partly on material presented in the annual ISVR short course 'Advanced Course in Acoustics, Noise and Vibration'. The principal aim is to provide comprehensive and up-to-date overviews of knowledge, applications and research activities in a range of topics that are of current interest in the practice of Engineering Acoustics and Vibration Technology. Naturally, the selection of topics reflects particular academic interests and expertise of members of the ISVR, together with those of some external associates.

As the title of the book indicates, the authors have, in general, attempted to emphasise applied aspects of the subjects of their chapters. However, the focus of those chapters that represent modules of the Masters course is biased towards instruction in the concepts, principles and techniques that underlie the technology of the topics addressed. Although this is not a textbook, its origins naturally lend it a tutorial element, albeit of an advanced nature, that is largely absent from technical handbooks, journal papers and the proceedings of specialist conferences. Consequently, it is anticipated that academics will find selected chapters suitable to support programmes of advanced instruction in various topics that are represented. The combination of expositions of underlying principles, phenomenological aspects, theoretical models and methods of analysis, and experimental methodology, together with comprehensive reference to current research and recent published material, should also appeal to the reader who is seeking an entrée into unfamiliar areas of sound and vibration technology. This is not a handbook in that it does not provide the engineer with readily applied sets of data and recipes for 'instant' solutions to problems.

A multi-authored book inevitably lacks the uniformity of style of a single author volume. We have endeavoured to promote coherence and consistency by cross-referencing between chapters where appropriate. This editorial task was considerably easier in the case of the first book, since the form and content of the first semester introductory element of the Masters course has evolved over 50 years of presentation to, and feedback from, students who were mostly new to the subject. The matter of mathematical notation has proved to be particularly problematic. Complete uniformity has proved to be impracticable on account of the diversity of conventions and symbols employed in the various specialisations represented in the book. For this reason, symbols specific to each chapter are defined at the appropriate place in the text. However, the complex phasor representation of time-harmonic variation $e^{j\omega t}$ is used throughout the book. Readers should note this convention in relation to time derivatives, mobilities, impedances and other functions of complex quantities. Bold type is used to indicate vector quantities.

Contributors

M.J. Brennan holds a personal Chair in Engineering Dynamics in the ISVR and is Chairman of the Dynamics Research Group. He joined the ISVR in 1995 after spending 23 years in the Royal Navy. His work at ISVR has centred on the use of smart structures for active vibration control, active control of structurally radiated sound and the condition monitoring of gear boxes by the analysis of vibration data and rotor dynamics. More recent work has involved the design of active tunable resonant devices that can deliver large forces and displacements for active control systems and the use of tuned systems to control sound transmission through structures.

J. Dixon is Technical Manager (Automotive and Marine) of ISVR Consulting. He joined the ISVR in 1971 and has researched many aspects of automotive noise and vibration, primarily in a commercial context. His special interests are in low noise IC engine design, vehicle noise path analysis and exterior noise optimisation. He and his team have gained a reputation throughout the industry for their wide knowledge of automotive NVH and their often novel and straightforward engineering approach to supporting industrial issues.

S.J. Elliott has been Professor of Adaptive Systems in the ISVR since 1994. His research interests have been concerned with the connection between the physical world and digital signal processing, specifically in relation to the modelling and synthesis of speech and to the active control of sound and vibration. His current work includes the active control of structural waves and active vibration isolation, as well as research into biomedical signal processing and control. He has authored or co-authored three books and over 120 papers in refereed journals.

F.J. Fahy is Emeritus Professor of Engineering Acoustics in the ISVR where he joined at its inception in 1963. He has a wide spectrum of interests in acoustics and vibration, ranging from statistical energy analysis to sound intensity measurement. He is a recipient of the Tyndall Silver and Rayleigh Gold Medals of the Institute of Acoustics and is an Honorary Fellow of the IOA. He has written three widely read text books, most recently *Foundations of Engineering Acoustics*, published in 2000, and was joint editor with John Walker of *Fundamentals of Noise and Vibration*.

N.S. Ferguson is Senior Lecturer in Structural Dynamics. He joined the ISVR in 1985 and his research projects are primarily analytical/numerical and include vibration transmission, shock and non-linear response, acoustic fatigue, SEA and variability problems. He has over 40 papers in conference and refereed journal publications. In addition to his academic responsibilities, he has worked closely with ISVR Consulting and is undergraduate programme co-ordinator for the engineering programmes of the ISVR.

I.H. Flindell joined the ISVR as a Lecturer in 1986. His research interest centres on environmental noise assessment and related issues. He has served on numerous standards committees, and has managed the ISVR involvement in large-scale research contracts with EPSRC, the European Commission and UK Government Departments, as well as with the USAF and the Canadian Department of Defense. He has recently chaired the EC Working Group on Noise Indicators, advising the Commission on the 2002 European Directive on the assessment and management of environmental noise. He has consultancy interests in environmental noise assessment, audio public address systems and acoustic design work at airports. He has also written over 40 technical papers and articles in conference proceedings.

P. Gardonio is a Senior Lecturer having joined the ISVR in 1995. His main research involves active isolation of vibration transmission, the active control of structure-borne and airborne sound transmission through single and double partitions. He has also worked on the development of smart structures for the implementation of active structural acoustic control. He has written 80 journal and conference papers. In parallel with the research activity he is the coordinator of the European Doctorate in Sound and Vibration Studies programme.

N.A. Halliwell is Professor of Optical Engineering at Loughborough University. His research develops optical diagnostic techniques for application in fluid and solid mechanics. He pioneered the development of Particle Image Velocimetry (PIV) in the UK and invented the Laser Torsional or Rotational Vibrometer at ISVR in 1983. In 1991, he was awarded a DSc by Southampton University for published work in laser technology applications to engineering and was elected Fellow of the Royal Academy of Engineering (FREng) in 1996. He is the author of more than

150 publications and is a member of the editorial board of *Optics and Lasers in Engineering, The Journal of Laser Applications* and the *Journal of Engineering Science and Technology*. The International Society for Optical Engineers in the USA elected him as Fellow (FSPIE) in 1997.

J.K. Hammond is Professor of Signal Analysis in the ISVR and was formerly Director of the ISVR before his appointment as Dean of the School of Engineering, Science and Mathematics in the University of Southampton. His particular research interests are in the area of applied digital signal processing, and theory and application of time series analysis with particular reference to problems in acoustics and dynamics. He is a member of the editorial boards of the *Journal of Mechanical Systems and Signal Processing* and of *Acta Acustica*.

C.J.C. Jones is Lecturer in Railway Dynamics in the ISVR. He joined the ISVR from British Rail Research in 1997 where his main interest was in the development and application of numerical models for noise and ground vibration from trains. He continues to work for the railway industry internationally through research and consultancy. He teaches numerical methods in acoustics on the BEng and MSc programmes and has authored over 50 publications of which about 20 are academic journal papers.

P.F. Joseph is Senior Lecturer in Acoustics in the ISVR. After gaining his doctorate in 1990 he spent 4 years in Australia working on low frequency active sonar. Since joining the ISVR in 1994 his research interests included broad band fan noise, shallow water acoustics, active noise control and duct acoustics. He is currently a member of the Rolls Royce University Technology Centre in Gas Turbine Noise which is undertaking research that includes broad band fan noise, advanced measurement techniques and active noise control.

R.S. Langley is Professor of Mechanical Engineering at the University of Cambridge, a post he has held since 1998. Previous appointments include Professor of Structural Dynamics in the Department of Aeronautics and Astronautics at the University of Southampton and Lecturer in Structural Dynamics in the College of Aeronautics at the Cranfield Institute of Technology. He undertakes research and consultancy on various aspects of structural dynamics, including the dynamics of offshore structures and the analysis of noise and vibration.

P.A. Nelson is Professor of Acoustics and Director of the ISVR. His main research interests are in the fields of acoustics, vibrations, fluid dynamics and signal processing, with a particular interest in the active control of noise and vibration. He has authored or co-authored two books, over 90 papers in refereed journals and 12 patents. He is a Fellow of the Royal Academy of Engineering, a Fellow of the Acoustical Society of

America and a Member of the International Commission of Acoustics. He is a recipient of the Rayleigh and Tyndall Medals of the Institute of Acoustics, the latter jointly with S.J. Elliott.

M. Petyt is Emeritus Professor of Structural Dynamics in the ISVR. He has also held appointments as a Research Professor at George Washington University, USA, and as Visiting Professor at the Université d'Aix-Marseille, France, and at the National University of Singapore. He has a special interest in Finite Element Methods and other techniques for structural vibration analysis. He is the author of more than 150 publications and of the book *Introduction to Finite Element Vibration Analysis*. He is also Editor Emeritus of the *Journal of Sound and Vibration*.

J.N. Petzing joined the Department of Mechanical Engineering at Loughborough University in 1991 and is now the Senior Lecturer in Metrology within the Wolfson School of Mechanical & Manufacturing Engineering. After completing his PhD, he developed research interests within optical metrology and more recently has been working in other areas of dimensional metrology. He is the author of more than 70 publications, detailing aspects of metrology development and application.

M.G. Smith is Technical Manager (Vibroacoustics and Modelling) of ISVR Consulting. His interests cover a wide range of problems in acoustics and structural dynamics including low noise design, silencer design, SEA methods, ultrasonic flow measurement, etc. He has particular links to the aircraft industry, first through Rolls Royce Aeroengines Ltd, where he worked between 1978 and 1983 and more recently as a member of the Rolls Royce University Technology Centre at the ISVR, and second through recent collaborative studies with Airbus (UK) Ltd on EC-funded projects on the modelling and control of noise from landing gears.

D.J. Thompson is Professor of Railway Noise and Vibration in the ISVR. Prior to his appointment in the ISVR in as Lecturer 1996, he spent nearly 10 years with British Rail Research, followed by over 6 years in TNO Institute of Applied Physics in Delft. Whilst at TNO, he continued to work on railway noise and vibration, developing and validating the TWINS computer model for rolling noise generation on behalf of the European Rail Research Institute, and developing a prediction model of the noise radiated by steel bridges. His research interests continue to be dominated by railway noise and vibration although he is also becoming more involved in automotive noise research.

J.G. Walker is a Visiting Senior Lecturer in the ISVR. His research activities have been centred on environmental noise research, with a particular interest in transportation noise issues. He was deeply involved in establishing environmental noise standards for railway noise exposure in the

UK as well as participating in several multi-national studies on aircraft noise exposure. He has authored or co-authored over 50 scientific papers as well as co-editing *Fundamentals of Acoustics* with Frank Fahy.

P.R. White joined the ISVR in 1988 and is Senior Lecturer in Underwater Systems. After gaining his PhD, he developed special interests in sonar signal processing, including array processing and transient signal analysis, time frequency and time scale methods, condition monitoring, biomedical signal processing and image processing. He has published numerous scientific papers and is a member of the editorial board of the *Journal of Condition Monitoring and Diagnostic Engineering Management*.

Acknowledgements

The editors wish to acknowledge the substantial effort and time expended on the preparation of this book by all our present and former colleagues in the Institute of Sound and Vibration Research. We include in this acknowledgement not only the academic and research staff, but also the technical and secretarial support staff. Without the contributions of all our colleagues over many years in support of research, teaching and consultancy, this book would not have become a reality.

We also wish to place on record our thanks to the staff of the publishers, Spon Press, who have provided so much helpful advice and encouragement to us and, equally importantly, have shown much patience.

Part 1

Signal processing

Chapter 1

Signal processing techniques

P.R. White and J.K. Hammond

1.1 Introduction

Signal processing plays an important role in acoustics and vibration technology; for example, in the fields of SONAR, condition monitoring, reproduced sound fields and seismic analysis. The scope of methodologies available is enormous, ranging from filtering to higher order spectra, Bayesian analysis, non-linear dynamics and many more.

This chapter is an intermediate-level review, moving from the foundations of digital signal processing (DSP), presented in the earlier work (Hammond 1998), to an introduction to digital filter design, high-resolution spectral estimation and time–frequency methods. The intention is to provide the reader with a flavour of the area rather than attempt to review the full gamut of DSP techniques available. Section 1.2 provides a basic background to digital systems theory, whilst Sections 1.3 and 1.4 outline some of the methods that are used in designing digital filters. We then consider parametric spectral estimation methods in Section 1.5; many of these techniques employ an underlying model based on filtering operations. The final section discusses the problem of estimating the spectrum of a non-stationary signal using time–frequency analysis.

1.2 Digital filters

Digital filters fulfil the same tasks as analogue filters do in circuit design. Digital filters are based on the implementation of *difference equations* acting upon sampled data, as opposed to analogue filters which can be regarded as realising *differential equations* (implemented via electronic circuits) that manipulate continuous time signals. The main advantage offered by digital filters is that one can achieve great accuracy of design implementation; the design of analogue filters can be compromised by variabilities in component values. This precision allows one to implement filter designs that achieve much greater performance than can be reasonably achieved using analogue circuits. A further consequence of the control that can be exhibited over digital filters is that one can design them with more complex responses than can be achieved in analogue form. Analogue filters do offer some advantages over digital implementations: their power consumption is generally much lower than digital filters, they are normally very much cheaper and they are not band-limited as digital implementations necessarily are.

1.2.1   Linear time invariant (LTI) digital filters

The most widely employed class of digital filters is the LTI form. Such filters are characterised by the following two properties:

(i) If $y_1(n)$ and $y_2(n)$ are the responses of the filter to input signals $x_1(n)$ and $x_2(n)$, respectively, then the filter's response to $c_1 x_1(n) + c_2 x_2(n)$ is given by $c_1 y_1(n) + c_2 y_2(n)$ where c_1 and c_2 are scalar constants.
(ii) If $y(n)$ is the filter's response to $x(n)$, then $y(n-m)$ is the filter's response to $x(n-m)$, where m is an integer.

Any LTI digital filter is completely described by its impulse response function $h(n)$, defined as the filter's response to an input, which is a Kronecker delta $\delta(n)$:

$$\delta(n) = 0: \quad n \neq 0$$
$$= 1: \quad n = 0 \tag{1.1}$$

One can write any sequence $x(n)$ as a sum of scaled Kronecker delta functions thus,

$$x(n) = \sum_{m=-\infty}^{\infty} \delta(m) x(n - m) \tag{1.2}$$

Using the stated properties of an LTI digital filter one can show that the output $y(n)$ of a digital filter in response to an arbitrary input $x(n)$ is given by

$$y(n) = \sum_{m=-\infty}^{\infty} h(m) x(n-m) = \sum_{m=-\infty}^{\infty} h(n-m) x(m) = h(n) * x(n) \tag{1.3}$$

where $*$ denotes digital convolution. If such a filter is to be realisable then the output $y(n)$ should not depend upon input samples occurring after time n (one cannot compute an output if it depends on data that has not yet been received). This constraint is called 'causality', and for a filter to be causal, its impulse response must be zero for negative n. In this case, the input/output relation (the convolution sum) can be written as:

$$y(n) = \sum_{m=0}^{\infty} h(m) x(n-m) = \sum_{m=-\infty}^{n} h(n-m) x(m) \tag{1.4}$$

To be practical, the filter should also be stable. Whilst there are several definitions of stability, we shall illustrate the principles by restricting our attention to bounded input bound output (BIBO) stability (Rabiner and Gold 1975). BIBO stability requires that if the input sequence $x(n)$ is bounded such that $|x(n)| < C_1$, then the output is also bounded $|y(n)| < C_2$, for C_1 and C_2 finite. A necessary and sufficient condition for BIBO stability is that the impulse response is absolutely summable, meaning that for a causal system (Rabiner and Gold 1975; Oppenheim *et al.* 1997),

$$\sum_{n=0}^{\infty} |h(n)| < \infty \tag{1.5}$$

1.2.2 Difference equations

Digital filters are implemented through the application of difference equations. The general form of a (causal) difference equation is

$$y(n) = b_0 x(n) + b_1 x(n-1) + \cdots + b_M x(n-M) - a_1 y(n-1)$$
$$-a_2 y(n-2) - \cdots - a_N y(n-N) \tag{1.6}$$
$$= \sum_{p=0}^{M} b_p x(n-p) - \sum_{q=1}^{N} a_q y(n-q)$$

The output of a difference equation $y(n)$ is constructed by summing a linear combination of previous input samples with a linear combination of past output samples. The values M and N control the number of terms in the two summations. The coefficients a_q and b_p are weighting factors; the coefficient a_q is defined in the negative for later convenience. Difference equations are very suitable for implementation on a digital computer. It can be shown that any impulse response $h(n)$ can be synthesised using such difference equations. However, to achieve such a synthesis one may be required to use a difference equation of infinite order, so that M and N may not be finite.

Two subclasses of difference equations are commonly defined. Those difference equations in which the output is constructed through combining only passed inputs, so that $a_q = 0$ for all q, are called moving average (MA) systems, with the difference equation

$$y(n) = b_0 x(n) + b_1 x(n-1) + \cdots + b_M x(n-M) = \sum_{p=0}^{M} b_p x(n-p) \tag{1.7}$$

The impulse response of an MA system is given by

$$h(n) = b_n : \quad 0 \le n \le M$$
$$= 0 : \quad \text{Otherwise} \tag{1.8}$$

A second subclass of difference equation is obtained by assuming that the output only depends upon the current input and passed outputs. These systems are called autoregressive (AR). The form of difference equation is

$$y(n) = b_0 x(n) - a_1 y(n-1) - a_2 y(n-2) - \cdots - a_N y(n-N)$$
$$= b_0 x(n) - \sum_{q=1}^{N} a_q y(n-q) \tag{1.9}$$

These AR systems are widely used in signal modelling applications, particularly in the area of speech processing (Rabiner and Schafer 1978). Since the general difference equation (1.6) consists of a mixture of AR and MA elements, it is referred to as an ARMA system.

1.2.3 Analysis of digital filters/systems

The tools used to analyse digital systems (specifically digital filters) are analogous to those used to analyse continuous time systems (differential equations). If one considers the Fourier transform of the convolution sum equation (1.8), it can be shown that

$$Y_s(f) = H_s(f) X_s(f) \tag{1.10}$$

where $Y_s(f)$, $X_s(f)$ and $H_s(f)$ are the Fourier transforms of the sequences $y(n)$, $x(n)$ and $h(n)$ respectively. The Fourier transform of $x(n)$ is defined as

$$X_s(f) = F\{x(n)\} = \sum_{n=-\infty}^{\infty} x(n)e^{-2\pi jfn\Delta} \tag{1.11}$$

where Δ is the sampling interval and $F\{.\}$ represents the Fourier operator. $H_s(f)$ is the frequency response of the digital system and is equal to the Fourier transform of the impulse response $h(n)$; or by re-arranging equation (1.10) it can be computed as the ratio between the Fourier transforms of the input and the output. If one takes the Fourier transform of equation (1.6) and notes that

$$F\{x(n-p)\} = e^{-2\pi jpf\Delta} X_s(f) \tag{1.12}$$

then,

$$H_s(f) = \frac{Y_s(s)}{X_s(s)} = \frac{b_0 + b_1 e^{-2\pi jf\Delta} + \cdots + b_M e^{-2\pi jfM\Delta}}{1 + a_1 e^{-2\pi jf\Delta} + \cdots + a_N e^{-2\pi jfN\Delta}} \tag{1.13}$$

The frequency response has a physical interpretation; specifically if the input signal is a complex exponential $x(n) = Ae^{2\pi j f_0 n \Delta}$, then the system's output is

$$y(n) = H(f_0) x(n) = A |H(f_0)| e^{2\pi j f_0 n \Delta + j\text{Arg}\{H(f_0)\}} \tag{1.14}$$

Hence, the magnitude of $H(f_0)$ dictates the change in amplitude experienced by a complex exponential as it passes through the system, whereas the phase of $H(f_0)$ defines the phase change experienced by such a complex sinusoid. The phase response of a filter at any one frequency controls the delay experienced by frequency components close to that frequency. The delay experienced by a narrow band of frequencies centred on f_0 is called the group delay denoted by $\tau_g(f_0)$, which can be determined from the phase response via

$$\tau_g(f_0) = -\frac{1}{2\pi} \frac{d}{df} \text{Arg}\{H(f)\}\bigg|_{f=f_0} \tag{1.15}$$

If the group delay varies with frequency, different delays are experienced by components in different frequency bands; such a system introduces phase distortion. To avoid such distortion, one seeks to design a filter for which the group delay is independent of frequency. Thus,

$$\tau_g(f) - \alpha \text{Arg}\{H(f)\} = -2\pi\alpha f \tag{1.16}$$

where α is a constant. Systems that introduce no phase distortion must have a phase response which is a linear function of frequency and are therefore referred to as linear phase filters. One can show that a linear phase filter must have an impulse response that is symmetric (Rabiner and Gold 1975). This observation precludes causal filters with infinitely long impulse responses from being in the linear phase because a function that is non-zero over the interval $(0, \infty)$ cannot be symmetric. Hence, the only causal filters that can have linear phase are those with an impulse response that has a finite duration.

In addition to the frequency response, a digital system can be described by its transfer functions (sometimes referred to as the 'characteristic function'). It is defined in terms of a z-transform (in contrast to analogue systems theory in which one employs the Laplace transform to specify a transfer function). A key property of the z-transform, that is widely used in digital linear systems theory, is

$$Z\{x(n-m)\} = z^{-m} X(z) \tag{1.17}$$

where $X(z)$ is the z-transform of $x(n)$. Applying this to the definition of an ARMA system, equation (1.6), one obtains the transfer function $H(z)$:

$$H(z) = \frac{Y(z)}{X(z)} = \frac{b_0 + b_1 z^{-1} + \cdots + b_M z^{-M}}{1 + a_1 z^{-1} + \cdots + a_N z^{-N}} \tag{1.18}$$

It can also be shown that this transfer function satisfies

$$H(z) = \mathbf{Z}\{h(n)\} \tag{1.19}$$

The frequency response function $H(f)$ can be regarded as a special case of the transfer function $H(z)$. Specifically, one can show that $H(f)$ is obtained by evaluating $H(z)$ on the unit circle using $z = e^{2\pi i f\Delta}$. For a finite order ARMA system, the transfer function $H(z)$ always takes the form of the ratio of two polynomials, in which case $H(z)$ is completely determined from knowledge of the roots of numerator and denominator along with an overall gain g. Under these circumstances, one can write

$$H(z) = g \frac{\prod\limits_{p=1}^{M}\left(1 - z_p z^{-1}\right)}{\prod\limits_{q=1}^{N}\left(1 - p_q z^{-1}\right)} \tag{1.20}$$

where z_p are the roots of the numerator and are simply termed the 'zeros', whilst p_q are the roots of the denominator and are termed the 'poles'. If a pole and a zero coincide, then, their effects cancel out and an equivalent ARMA system of lower order can be found.

The position of the poles of $H(z)$ indicates the stability of the system. Assuming that there are no coincident poles and zeros, then if all the poles of $H(z)$ have magnitude less than 1, the digital system is stable. Since it is only the poles that dictate a system's stability, a finite order MA system contains no poles and is always stable.

1.2.4   General comments on design of digital filters

There are two classes of digital filters: finite impulse response (FIR) filters and infinite impulse response (IIR) filters. There is a strong connection between this classification and that of digital systems into AR, MA and ARMA. A filter is FIR if its impulse response is identically zero for all $n > M$; whereas IIR filters are filters whose impulse response is such that there is no finite value K for which $h(n)$ is zero for all $n > K$. All (finite order) MA systems necessarily define an FIR filter, but, because of the pole–zero cancellation, one can also construct ARMA systems that are FIR. However, it is usually the case that AR and ARMA systems are IIR filters.

The filters we shall seek to design are of a simple generic form. The ideal filters to be approximated are assumed to perfectly partition the frequency axis, so that some bands of frequencies are passed unattenuated (the pass bands), whilst the gain in adjacent bands is zero (the stop bands). This class of problems we shall refer to as *standard filter design* problems. The four most common examples of the standard filter design problem are: low-pass, high-pass, band-pass and band-stop filters. For example, the ideal low-pass filter has the form

$$|H_{\text{ideal}}(f)| = 1 : \quad f < f_c$$
$$= 0 : \quad f > f_c \tag{1.21}$$

where f_c is the cut-off frequency. Note that the gain at the point $f = f_c$ is undefined and that the ideal frequency response is discontinuous at this frequency. We shall concentrate on the case of low-pass filter design since this encapsulates all of the design issues faced when considering any standard filter design problem.

It is anticipated that any practical filter will not exactly achieve the ideal response. There are two aspects of the differences between the practical and the ideal filters that are usually specified; these are illustrated, in the context of a low-pass filter, in Figure 1.1. The frequency band close to $f = f_c$ is termed the transition band (see Figure 1.1). In any practical filter, the transition band has a non-zero width but its breadth should be minimised. The second class of departures from the ideal filter is oscillations of the gain in the pass and the stop bands. The objectives of keeping the transition band narrow and minimising the size of the oscillations work in opposition; a filter with a broad transition zone may have small oscillations in the pass and the stop bands, whilst if the oscillations are reduced the transition zone necessarily broadens.

The procedures used to design digital filters are very different depending upon whether the filter is to be FIR or IIR. We shall first consider the design of FIR filters, followed by that of IIR filters.

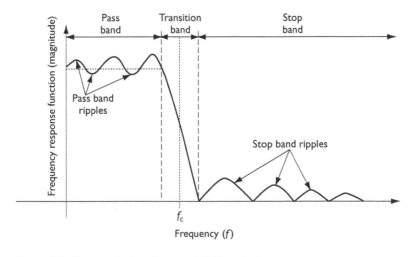

Figure 1.1 Characteristics of a practical filter design.

1.3 The design of FIR filters

The design of FIR filters is made relatively straightforward by virtue of equation (1.8), which informs us that the impulse response of an FIR filter is identical to the coefficients b_p. Hence the frequency response of the filter is given by

$$H(f) = \sum_{n=0}^{M} h(n) e^{-2\pi j f n \Delta} = \sum_{p=0}^{M} b_p e^{-2\pi j f p \Delta} \tag{1.22}$$

The objective of the design procedure is to select the coefficients b_p so that $H(f)$ closely approximates the ideal filter. The FIR filter coefficients and the filter's frequency response are linearly related, and it is this fact that facilitates the use of direct design procedures for FIR filters.

1.3.1 The windowing design technique

The magnitude of the frequency response function of the ideal filter is specified. If one assumes a structure for the phase, then the complete frequency response is known. The simplest assumption that can be made regarding the phase of a filter is that it has zero phase. A zero-phase filter introduces no phase distortion, and there is no group delay through the filter, implying that the input propagates through to the output instantaneously (a concept that is intuitively unreasonable and, as we shall see, leads to impractical filters). Under the assumption of a zero phase, the frequency response of the ideal low-pass filter is

$$\begin{aligned} H_{\text{ideal}}(f) &= 1: \quad f < f_c \\ &= 0: \quad f > f_c \end{aligned} \tag{1.23}$$

Note that this defines the complete frequency response and not only its magnitude, as specified in equation (1.21). The ideal filter's impulse response $h_{\text{ideal}}(n)$ can be computed through the use of the inverse Fourier transform thus,

$$h_{\text{ideal}}(n) = \Delta \int_{-f_s/2}^{f_s/2} H_{\text{ideal}}(f)\, e^{2\pi j f n \Delta}\, \mathrm{d}f \tag{1.24}$$

from which

$$h_{\text{ideal}}(n) = \frac{\sin(2\pi f_c n \Delta)}{\pi n}: \quad -\infty < n < \infty \tag{1.25}$$

This impulse response is plotted in Figure 1.2 for the case of $f_c = 100$ Hz, assuming a sampling frequency $f_s(= 1/\Delta)$, of 1000 Hz. This 'ideal' impulse response is not causal, the impulse response is non-zero for negative n, and is also of infinite duration.

It is rather surprising to note that this 'ideal' low-pass filter is not BIBO-stable (the impulse response is not absolutely summable). This lack of stability is a consequence of the slow decay of the impulse response, which asymptotically decays like $1/n$. This slow decay is a consequence of the discontinuity in $H_{\text{ideal}}(f)$. Since the frequency responses of all the filters considered in the standard filter design problem possess such discontinuities, all such 'ideal' filters will not be BIBO-stable. This lack of stability is only associated with the 'ideal' filters and, with care, can be avoided in practical filter designs. Stability is never an issue in the case of FIR filters, but when we consider IIR filters we must take precautions to ensure that the designed filters are more stable than the filters they seek to approximate.

The windowing design method proceeds by truncating the ideal impulse response by multiplying $h_{\text{ideal}}(n)$ with a spectral windowing function $w(n)$ to give the truncated impulse response $h_w(n)$.

$$h_w(n) = w(n)\, h_{\text{ideal}}(n) \qquad (1.26)$$

The windowing function is assumed to have finite support. Specifically,

$$w(n) = 0 \quad |n| > \frac{M}{2} \qquad (1.27)$$

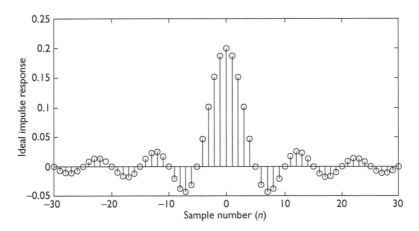

Figure 1.2 Ideal impulse response for a low-pass filter cut-off frequency of 100 Hz: sampling frequency 1000 Hz.

where it has been assumed that M is even (so that there is an odd number of filter coefficients, b_p). The response $h_w(n)$ is only non-zero for $-M/2 \leq n \leq M/2$. If M is sufficiently large, one intuitively anticipates that the frequency response $h_w(n)$ will converge to the ideal response. If a rectangular window is used, then,

$$
\begin{aligned}
h_w(n) &= h_{\text{ideal}}(n): \quad |n| \leq \frac{M}{2} \\
&= 0: \qquad\qquad |n| > \frac{M}{2}
\end{aligned}
\tag{1.28}
$$

Figure 1.3 shows the result of applying a rectangular window to the impulse response shown in Figure 1.2. The impulse response is 25 samples long ($M = 24$).

The frequency response corresponding to the windowed impulse response can be related to the frequency response of the windowed frequency response through

$$
H_w(f) = W(f) * H_{\text{ideal}}(f) = \int_{-\infty}^{\infty} H_{\text{ideal}}(v)\, W(f - v)\, \mathrm{d}v
\tag{1.29}
$$

where $H_w(f)$ and $W(f)$ are the Fourier transforms of $h_w(n)$ and $w(n)$ respectively.

The filter defined by the impulse response $h_w(n)$ has the advantage over $h_{\text{ideal}}(n)$ of having finite duration; however, it remains non-causal. From Figure 1.3 it is evident that the impulse response is still non-zero for $n < 0$;

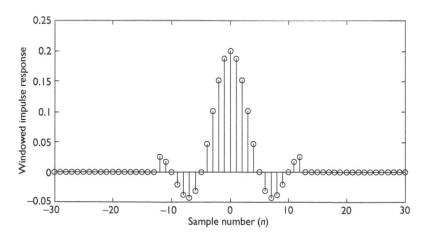

Figure 1.3 Rectangular windowed impulse response for a low-pass filter cut-off frequency of 100 Hz: sampling frequency 1000 Hz, $M = 24$.

that is to say it is non-causal. To resolve this problem, the impulse response is shifted to the right by $M/2$ samples. The impulse response of the resulting filter $h(n)$ is written as:

$$h(n) = h_w\left(n - \frac{M}{2}\right): \quad 0 \leq n \leq M$$

$$= 0: \qquad\qquad \text{Otherwise}$$

(1.30)

Figure 1.4 shows the result of this operation applied to the low-pass filter example. It is evident that this filter has an impulse response that is both causal and finite in duration, and so represents a practical FIR filter. Recall that the coefficients of the filter are the same as the impulse response, so that $b_p = h(p)$, and in this manner the design is complete.

The shifting operation in the final stage of the design method only affects the phase of the frequency response, since

$$H(f) = F\{h(n)\} = F\left\{h_w\left(n - \frac{M}{2}\right)\right\} = H_w(f)\,e^{-2\pi j\left(\frac{M}{2}\right)f\Delta}$$

(1.31)

The frequency response $H_w(f)$ is real (zero phase) so that the phase of $H(f)$ is given by $-\pi f M\Delta$; this is a linear phase filter with delay $M/2$ samples and is free of phase distortion.

When applying this design method the only degrees of freedom available to the user are the number of filter coefficients and the choice of spectral window used to truncate the impulse response. Increasing the number of

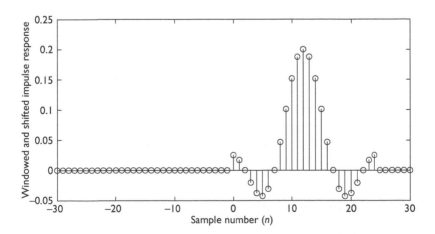

Figure 1.4 Windowed and shifted impulse response for a low-pass filter cut-off frequency of 100 Hz: sampling frequency 1000 Hz, $M = 24$.

coefficients will tend to improve the characteristics of the filter. However, longer filters require more computation to implement and introduce greater delays. In practice, the filter length is commonly determined by practical considerations; for example, available time in which to compute the filter output. The choice of spectral window controls the trade-off between the width of the transition band and the size of oscillations in the pass and the stop bands.

The effect of the choice of spectral window is most clearly illustrated through considering the windowing operation as a convolution in the frequency domain. The Fourier transform of the windowing function $W(f)$ is characterised by two measures: first the width of the main lobe and second the size of its sidelobes. The design of a spectral window involves a trade-off between these two quantities: windows with narrow main lobes have high sidelobes, whilst windows with broad main lobes can be constructed with small sidelobes (Hammond 1998). Consideration of the convolution equation (1.29) leads one to realise that the broadening of the transition zone relates to the width of the main lobe of $W(f)$, and the size of the oscillations in the pass and the stop bands is controlled by the size of the sidelobes of $W(f)$. This leads one to conclude that an FIR filter designed using a rectangular window will have a narrow transition zone but high sidelobes.

Figure 1.5 illustrates the frequency response of the low-pass filter example considered previously, which employed a rectangular window. The two plots show the same response on a linear and a logarithmic (dB) scale. The use of a linear scale tends to highlight oscillations in the pass band, whilst the use of a logarithmic scale emphasises stop-band oscillations. The transition band for this filter is relatively narrow, by virtue of the narrow main lobe of the rectangular window, but the oscillations, particularly in the stop band, are generally considered as unacceptable.

The effect of different choices of windowing function is shown in Figure 1.6. Two different windows (Hamming and Blackman) are shown. The Blackman window has a broader main lobe than the Hamming window, which in turn has a broader main lobe than the rectangular window, with the result that the transition bands of the Blackman and the Hamming filters are consequently broader. However, the oscillations in the stop band of these filters are very much smaller; in this case, around 20 dB and 40 dB lower than those obtained using a rectangular window.

1.3.2   The frequency sampling method

An alternative approach to designing an FIR digital filter is offered by the frequency sampling approach. This technique is based upon application of the inverse discrete Fourier transform (IDFT), rather than the inverse

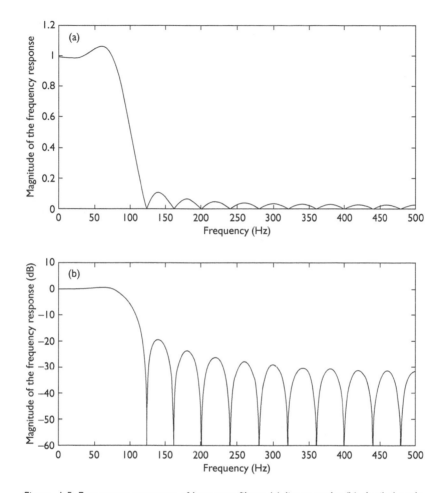

Figure 1.5 Frequency response of low-pass filter: (a) linear scale; (b) decibel scale.

Fourier transform of a sequence. The basic principle is to first define an ideal frequency response, in which a linear phase is imposed, so that

$$H_{\text{ideal}}(f) = |H_{\text{ideal}}(f)| e^{-\pi j f M \Delta} \tag{1.32}$$

where, as before, it is assumed that M is even. A sampled response can be obtained by evaluating this function at the frequency point f_k such that

$$H(k) = H_{\text{ideal}}(f_k) \tag{1.33}$$

where $f_k = k f_s / M = k / M \Delta \quad k = 0, 1, 2, \ldots, M - 1$

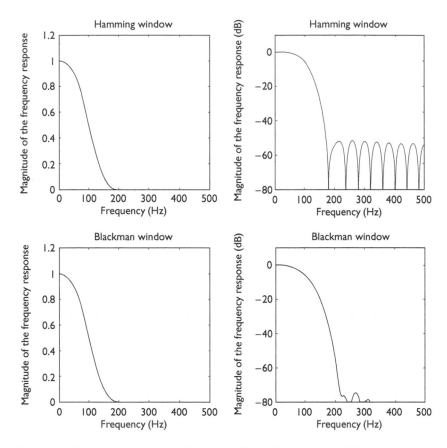

Figure 1.6 Frequency responses of low-pass filters designed with different spectral windows.

The impulse response (and hence filter coefficients) is obtained simply by computing the IDFT of $H(k)$ as

$$h(n) = \frac{1}{M} \sum_{k=0}^{M-1} H(k)\, e^{\frac{2\pi j n k}{M}} \tag{1.34}$$

If this approach is applied directly to a problem, the results are identical to those obtained using the windowing method with a rectangular window. The frequency response function of the resulting filter is obtained by Fourier transforming $h(n)$ (see equation (1.22)). One can show that

$$H(f) = \frac{1 - e^{-2\pi j f M \Delta}}{M} \sum_{k=0}^{M-1} \frac{H(k)}{1 - e^{2\pi j \left(\frac{k}{M} - f\Delta\right)}} \tag{1.35}$$

This equation relates the frequency response $H(f)$ for any frequency f to the sampled frequency point $H(k)$. One can readily show that for $f = k/M\Delta$, $H(f) = H(k)$ $(k = 0, 1, \ldots, M-1)$, so that the frequency response of the filter passes through the sampling points; this is illustrated in Figure 1.7. Note how the solid line (describing $H(f)$) is constrained to pass through each of the circles indicating $H(k)$. We can regard the frequency design method as constructing a filter whose frequency response is obtained by interpolating between the points $H(k)$. The precise manner of this interpolation is prescribed by equation (1.35). Filters designed in this fashion have a response that makes the transition from a gain of unity to a gain of zero between two samples in frequency, so that the width of transition zone can be said to be $1/M\Delta$ Hz.

The essential trade-off in the design of a digital filter is that between the width of the transition band and the size of oscillations in the pass and the stop bands. The windowing design method manipulates this trade-off through the choice of windowing function and as such provides indirect control of the relevant performance measures. In the frequency design method one can control the width of the transition zone directly. This is achieved by relaxing the constraint that all of the sampled points must lie on the ideal frequency response. For example, one can define a low-pass response of the form defined by

$$
\begin{aligned}
H(k) &= 1: \quad k = 0, 1, \ldots, k_{\text{cut}} - 1 \\
&= a: \quad k = k_{\text{cut}} \\
&= 0: \quad k_{\text{cut}} < k \le \frac{M}{2}
\end{aligned}
\tag{1.36}
$$

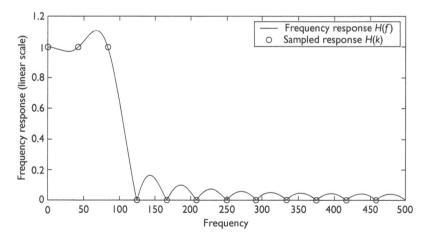

Figure 1.7 Frequency response of a low-pass filter designed using frequency sampling.

where k_{cut} is the index of the sampling point closest to the desired cut-off frequency. The above only defines $H(k)$ for samples below the Nyquist frequency; the points above Nyquist can be inferred from the symmetry condition $H(M - k) = H(k)^*$. The value of a can be selected by the user and should satisfy $0 < a < 1$. This point lies within the transition zone whose width is $2/M\Delta$ Hz. The size of the oscillations induced depends upon the choice of value for a. Figures 1.8 and 1.9 illustrate the effect of different choices for a. From these plots, one can see that the attenuation in the stop band critically depends on the value of a. The value of 0.3904 has been included since this can be shown to minimise the size of the oscillations in the stop band (Oppenheim and Schafer 1975). This can be most clearly seen in Figure 1.9, where the maximum value in the stop band is about -40 dB.

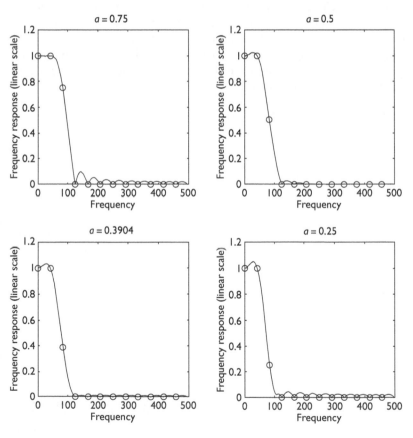

Figure 1.8 Frequency sampling designs using different values of a displayed on a linear scale.

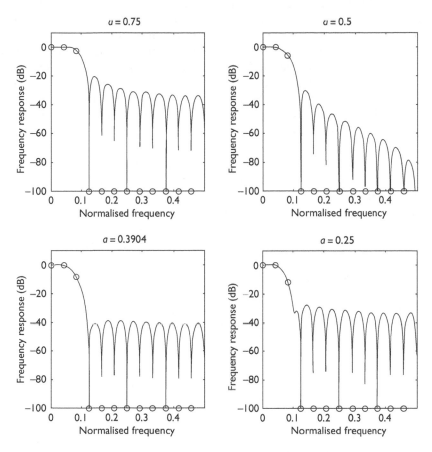

Figure 1.9 Frequency sampling designs using different values of *a* displayed on a decibel scale.

1.3.3 Optimised FIR filter design

An extension of the above is to relax the constraint that the points $H(k)$ are samples from the ideal response and to let them become free parameters. These values can then be selected so that the resulting response is optimised through some criterion. This is the principle behind the methods discussed in this is sub-section. However, one need not define the filter points in terms of their DFT values $H(k)$, rather one can directly optimise the response with respect to the filter coefficient b_n.

The simplest approach is to solve the following problem: find the set of coefficient b_n that minimise the least squares cost function Ψ_{LS} defined as

$$\Psi_{LS} = \int_0^{\frac{f_s}{2}} \Omega(f) \, |H_{ideal}(f) - H(f)|^2 \, df \tag{1.37}$$

where $H(f)$ is defined as the frequency response of a zero-phase filter with coefficient b_n and $\Omega(f)$ is a weighting function. Once a zero-phase filter has been designed, a linear phase filter is simply obtained by shifting the coefficients (see Section 1.2.1). The cost function Ψ_{LS} is a quadratic function of the filter coefficient b_n, and so this minimisation requires one to solve a linear system of equations which can be accomplished in a non-iterative fashion. The weighting function $\Omega(f)$ is used to emphasise (and de-emphasise) various frequency bands. In the transition band one usually sets $\Omega(f) = 0$ since differences between the designed and the ideal filters can be tolerated there.

This least squares problem can be solved analytically for simple choices of the weighting function. If $\Omega(f) = 1$ for all f, the filter that results is in fact identical to that obtained using the windowing design technique with a rectangular window, an example of which is shown in Figure 1.5. As previously discussed, this filter has poor properties, mainly relating to the large oscillations that occur close to the cut-off frequency. By careful selection of the weighting function, these effects can be mitigated. But it remains a characteristic of least squares FIR filters that their response has oscillations that increase in size as the cut-off frequency is approached.

An alternative optimisation solution is obtained if one changes the cost function. A common choice is to use the minimax cost function Ψ_m:

$$\Psi_m = \min\{\max[\Omega(f)\,|H_{ideal}(f) - H(f)|]\} \tag{1.38}$$

This cost function aims to minimise the largest absolute deviation of $H(f)$ from the ideal frequency response. The weighting function is once again used to define the transition band where errors are tolerated. Solution of this optimisation task is more difficult to achieve, requiring iterative procedures. However reliable, computationally efficient and numerically stable, algorithms have been developed (Rabiner and Gold 1975). Minimax problem solutions have some desirable properties. Principally, the oscillations in the pass and the stop bands have constant amplitude, leading to these filters sometimes being called equi-ripple filters. The relative size of the oscillations in the pass and the stop bands can also be controlled. This form of filter is probably the most widely used FIR filter because of its well-defined characteristics. An example of a low-pass equi-ripple filter is shown in Figure 1.10. This filter has a higher cut-off frequency than previous examples so that the oscillations in the pass band can be more easily seen.

1.4 IIR filter design

The design of IIR filters exploits a very different principle from that used when designing FIR filters. The reason for this is that the direct approach adopted during FIR filter design relies heavily upon the fact that an FIR

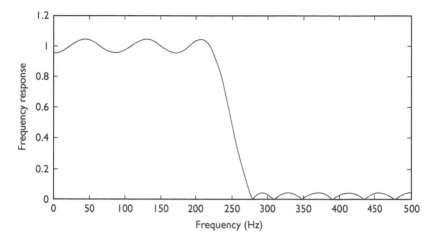

Figure 1.10 Design of a 24-coefficient equi-ripple filter with cut-off frequency of 250 Hz.

filter's response is obtained by Fourier transforming its coefficients (see equation (1.22)). The frequency response function of an IIR filter is related to the filter's coefficients through equation (1.13), which is a more involved relation.

The traditional approach to design digital IIR filters is to use existing designs for analogue filters and convert them into digital filters. This is made possible by the existence of established, familiar designs for analogue filters. Examples of analogue filters that are used for this purpose include: Butterworth filters, Elliptic filters and Bessel filters. The matter of designing a digital IIR filter is essentially that of mimicking analogue filters in the digital domain. Since the goal of these methods is to design filters, one seeks a digital system whose frequency response approximates that of the analogue filter. We shall not dwell on the form of the analogue filters but will concentrate our attention on how the transformation from analogue to digital filter is constructed.

The three basic steps involved in IIR filter design are:

(i) Select the order and type of analogue filter to be used; for example, a fourth order Butterworth filter. This specification defines an analogue low-pass filter with a cut-off frequency at 1 radian, a unit low-pass filter, with transfer function $H_u(s)$.

(ii) Transform the unit low-pass filter into a new analogue filter of the correct form with the required cut-off and cut-on frequencies, with transfer function $H_a(s)$.

(iii) Transform this analogue filter into digital form.

Table 1.1 Mapping for a unit low-pass filter (all frequencies are expressed in radians)

Output filter type	Mapping
Low-pass filter Cut-off frequency ω_c	$s \rightarrow \dfrac{s}{\omega_c}$
High-pass filter Cut-on frequency ω_c	$s \rightarrow \dfrac{\omega_c}{s}$
Band-pass filter Cut-on frequency ω_0 Cut-off frequency ω_1	$s \rightarrow \dfrac{s^2 + \omega_0 \omega_1}{s(\omega_1 - \omega_0)}$
Band-stop filter Cut-off frequency ω_0 Cut-on frequency ω_1	$s \rightarrow \dfrac{s(\omega_1 - \omega_0)}{s^2 + \omega_0 \omega_1}$

The transformation implemented in step (ii) is relatively simple mapping $s \rightarrow s$. Table 1.1 details the mappings from a unit low-pass filter to an arbitrary low-pass filter, a high-pass filter and band-pass/band-stop filters. To use these mappings, one replaces each occurrence of s in $H_u(s)$ with the functional form given in the right-hand column of Table 1.1.

Once the suitable analogue filter has been determined, the remaining task is to perform the mapping from the analogue filter to the digital filter. Three methods will be examined to perform this task. Before detailing each mapping method in detail we shall consider what is required from a general mapping. There are three conditions that we require our mapping to satisfy:

(i) If the analogue system is stable then we require that the digital system is stable. This requires that if s is such that $\text{Re}\{s\} < 0$ then $|z| < 1$, so that all points in the left-half plane of the s-plane are mapped to the interior of the unit disc in the z-plane.

(ii) The mapping should be one-to-one, so that for a given value of s there is a unique value z, and *vice versa*. This ensures that aliasing does not occur.

(iii) The frequency axis in the s-plane maps to the frequency axis in the z-plane, so that if $s = j\omega$, then $z = e^{j\theta}$. By ensuring that this is the case, the frequency response of the analogue filter is reproduced in the digital filter, albeit in a distorted form. This distortion must occur since one seeks to map an infinite line in the s-plane to a circle in the z-plane.

1.4.1 The method of mapping differentials

This is essentially Euler's method for solving differential equations (Press *et al.* 1993). It is based on the finite difference approximation to the differential operator

$$\frac{dx(t)}{dt} \approx \frac{x(t) - x(t - \delta t)}{\delta t} \tag{1.39}$$

If this is evaluated at a sample point $t = n\Delta$, and δt is selected as the sampling interval Δ, then one can write

$$\frac{dx(t)}{dt} \approx \frac{x(n) - x(n-1)}{\Delta} \tag{1.40}$$

The differential operator s is associated with the derivative on the left-hand side, and the right-hand side can be expressed in operator form as $(1 - z^{-1})/\Delta$, and one can write

$$s \approx \frac{1 - z^{-1}}{\Delta} \tag{1.41}$$

This is the basis of the method of mapping differentials, specifically

$$H(z) = H(s) \Big|_{s = \frac{1 - z^{-1}}{\Delta}} \tag{1.42}$$

Unfortunately, this mapping has severe limitations. The most apparent of which occurs when one considers how the frequency axis in the s-plane maps to the z-plane. Specifically, if $s = j\omega$ then

$$z = \frac{1}{2} \left(1 + e^{j2\tan^{-1}(\omega\Delta)} \right) \tag{1.43}$$

which evidently does not lie on the frequency axis in the z-plane. In fact, the locus of the points described by z is a circle of radius 1/2 centred at $z = 1/2$ (see Figure 1.11). In the low-frequency region the locus lies close to the digital frequency axis, so that in this band the design maybe reasonable. The shortcomings of this method mean that it is of little practical use for the design of digital filters.

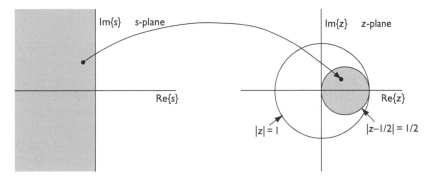

Figure 1.11 Mapping of points from the s-plane to the z-plane using the method of mapping differentials.

1.4.2   The impulse-invariant method

The methodology adopted by the impulse-invariant method is dissimilar to that of the other two approaches described above. The principle is to construct the digital system as follows:

(i) Compute the impulse response of the analogue system $h_a(t)$ by inverse Laplace transforming $H_a(s)$.

(ii) The impulse response $h(n)$ of the associated digital system is defined as a sampled version of the analogue system's impulse response. Thus,

$$h(n) = h_a(t)|_{t=n\Delta} \tag{1.44}$$

(iii) The transfer function $H(z)$ of the digital system is then obtained by z-transforming the digital impulse response.

The success of this process depends upon whether the sampling process step (ii) introduces distortions due to aliasing. Specifically, if the frequency response $H_a(f)$ is negligible for $f > 1/(2\Delta)$ then the sampling is performed without aliasing, and the resulting filter will be an accurate mimic of the analogue system. This condition precludes the use of impulse-invariant filter design for high-pass and band-stop filters, since they necessarily violate the condition that $H_a(f)$ is negligible for $f > 1/(2\Delta)$. However, for low-pass and band-pass filters, assuming that $1/(2\Delta) > f_c$, where f_c is the cut-off frequency, then good designs may be achieved.

The frequency responses of the impulse-invariant digital filter $H(f)$ is related to the frequency response of the analogue filter $H_a(f)$, through the Poisson sum formula

$$H(f) = \frac{1}{\Delta} \sum_{k=-\infty}^{\infty} H_a\left(f - \frac{k}{\Delta}\right) \tag{1.45}$$

The impulse-invariant method has the property that stable analogue filters map to stable digital filters and the analogue frequency axis maps to the digital frequency axis. However, the mapping is not one-to-one, since multiple points in the analogue domain are mapped to the same point in the digital domain. Indeed, there are infinitely many analogue filters that generate the same digital filter using the impulse-invariant mapping.

The impulse-invariant method has obvious advantages if one seeks to create a digital system with an impulse response that matches that of a given analogue system. The technique can be extended so that other system responses are matched; that is to say that the method can be modified so that the system's step responses are matched.

1.4.3 The bilinear transform

The most widely used method for mapping analogue filters to digital filters is the bilinear transform. This transform is based on the substitution

$$s \rightarrow \frac{2}{\Delta} \left[\frac{1 - z^{-1}}{1 + z^{-1}} \right] \qquad (1.46)$$

This can be viewed as a corrected version of the method of mapping differentials, since it is that substitution multiplied by a factor $2(1 + z^{-1})^{-1}$. The use of the bilinear transform can be justified by consideration of the first-order analogue system

$$c_1 \frac{dy(t)}{dt} + c_0 y(t) = d_0 x(n)$$

$$H(s) = \frac{d_0}{c_1 s + c_0} \qquad (1.47)$$

If one notes that

$$\int_{(n-1)\Delta}^{n\Delta} \frac{dy(t)}{dt} dt = y(n) - y(n-1) \approx \frac{\Delta}{2} \left[\left. \frac{dy(t)}{dt} \right|_{t=n\Delta} + \left. \frac{dy(t)}{dt} \right|_{t=(n-1)\Delta} \right]$$

$$(1.48)$$

where the right-hand approximation is obtained by use of the Trapezoidal rule to approximate the integral, combining equations (1.47) and (1.48) yields

$$y(n) - y(n-1) \approx \frac{\Delta}{2} \left[\frac{d_0}{c_1} (x(n) + x(n-1)) + \frac{c_0}{c_1} (y(n) + y(n-1)) \right] \qquad (1.49)$$

Accepting the approximation and z-transforming leads one to the digital transfer function

$$H(z) = \frac{d_0}{c_1 \dfrac{2}{\Delta} \left\{ \dfrac{1 - z^{-1}}{1 + z^{-1}} \right\} + c_0} \qquad (1.50)$$

If this is compared with the transfer function of the analogue system, as specified in equation (1.47), then evidently one can be obtain from the other by using the substitution given by equation (1.46).

The bilinear transform maps the analogue frequency axis to the frequency axis in the digital domain. To see this, consider the points on the digital frequency axis $z = e^{j\omega_d \Delta}$. One can show that under the bilinear transform

$$s = \frac{2}{\Delta} j \tan \left(\frac{\omega_d \Delta}{2} \right) \qquad (1.51)$$

Defining ω_a as the angular frequency in the analogue domain,

$$\frac{\omega_a \Delta}{2} = \tan\left(\frac{\omega_d \Delta}{2}\right) \tag{1.52}$$

This relation describes which frequencies in the analogue domain relate to which frequencies in the digital domain. The 'tan' function serves to compress the complete frequency axis in the analogue domain onto the circumference of the unit disc in the z-plane. This compression distorts the frequency axis in a non-linear manner. An important consequence of this distortion is that if an analogue filter with a transition frequency at $\omega_a = \omega_c$ is used as the basis for a filter design, then after application of the bilinear transform the transition point will have moved to a frequency ω_d, as defined by equation (1.52). Hence, to design a digital filter with a transition frequency ω_d one needs to ensure that the analogue filter is specified with a transition frequency ω_a given by equation (1.52) with $\omega_d = \omega_c$. This process adjusts the transition points in a filter to allow for the distortion that is introduced by the bilinear transform and is called 'pre-warping'.

To illustrate the need for pre-warping, consider the problem of designing a digital filter consisting of four pass bands, each of width 100 Hz, centred on 100, 300, 500 and 700 Hz and using a sampling rate of 1600 Hz. The bottom frame in Figure 1.12 shows what happens if one naively bases the design on an analogue filter, specified without pre-warping, and applies the bilinear transform. The lower frame of Figure 1.12 shows the resulting digital filter. The analogue design which contains four equal band-pass regions has been distorted so that in the digital domain the pass bands all have different widths; the high-frequency bands being narrower than the lower-frequency bands. Note that the 'tan' form of distortion means that in the low-frequency region the mapping is nearly linear, but at high frequencies a much greater compression of the frequency axis is effected. Figure 1.13 illustrates the correct manner in which to design such a filter. The original analogue filter is pre-warped so that the high-frequency band-pass regions are broader than those in the low-frequency region, anticipating the manner in which the tan function will distort the frequency axis. The frequency axis in the upper frame of Figure 1.13 is extended to include the wider range of frequencies needed to encompass the design. After application of the bilinear transformation, the resulting digital filter has the required equally spaced pass bands, each with bandwidths of 100 Hz.

The bilinear transform satisfies all of the stated requirements for a mapping from analogue to digital systems. It maintains stability of a system, it is one-to-one, and the analogue frequency axis maps to the digital

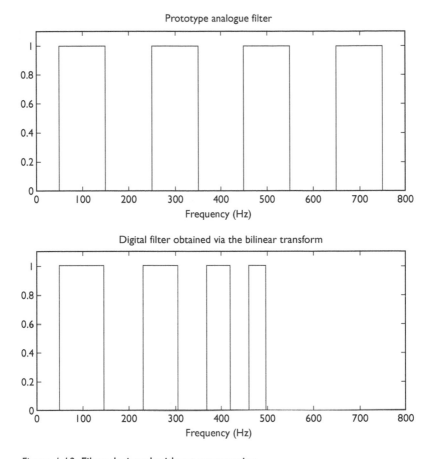

Figure 1.12 Filter designed without pre-warping.

frequency axis. It is for these reasons that the bilinear transform is the widely employed mapping technique.

1.5 Parametric spectral analysis

Conventional methods of spectral estimation use narrow-band filtering, Fourier transformation of correlation functions and, most commonly, the 'segment averaging' approach based on averaging the periodogram (the squared magnitude of the Fourier transform of segments of the time history) (Hammond 1998). These methods, sometimes referred to as non-parametric methods, are based on representing a time history as the sum of sinusoids and cosinusoids. The Fourier transform-based methods involve the use of

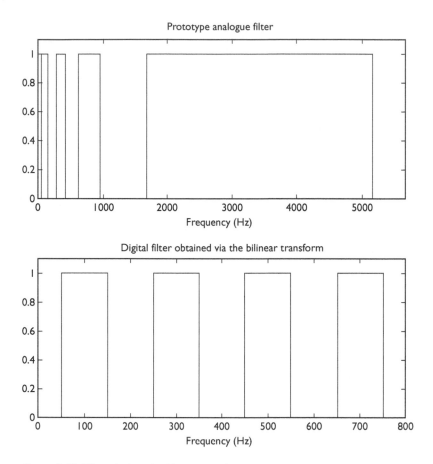

Figure 1.13 Filter designed with pre-warping.

window functions, referred to as quadratic tapering when applied to the correlation function and linear tapering when applied directly to the time history in the segment averaging approach. The windowing operations can distort the results owing to a loss of resolution that must be set against the reduction in the fluctuations of the estimates (variance reduction).

There are alternative approaches to spectral analysis that can offer advantages in many circumstances. They rely on incorporating models of the data generation process, and procedures arise that turn out to be 'data dependent', and in this sense are sometimes termed 'non-linear' methods of spectral analysis (Haykin 1983). These methods side-step the windowing problem. This section describes some of these methods which are variously described as 'model based' or 'high resolution'.

1.5.1   Time series models (Gutowski et al. 1978; Priestley 1988)

It seems self-evident that the procedure used to estimate a spectrum of a time series should depend upon the process by which the data are generated. If the mechanism of data generation is unknown, then there is no one 'correct' method of spectral estimation.

A time series $x(n)$ will have some structure reflected in its mean, covariance, etc., and various models have been proposed to account for the behaviour. A common approach is to regard $x(n)$ as having arisen from appropriate operations on white noise $u(n)$ (defined by $E[u(n)u(k)] = 0$, $n \neq k$ and $E[u(n)^2] = \sigma^2$). We shall restrict our discussions to *linear* operations and *stationary* time series.

A general linear model for $x(n)$ can be written as

$$\sum_{k=-\infty}^{\infty} g(k)\,x(n-k) = u(n) \tag{1.53}$$

When expressed in this form, it is perhaps not apparent that $u(n)$ is an 'input' and $x(n)$ an 'output'. The reason for choosing this structure for the model is that we are trying to find some linear relationship between past, present and future values of the time series $x(n)$ which reduces it to white noise. The model, as posed, is non-causal (anticipative), and to ensure that $x(n)$ depends only on past values, we set $g(n) = 0$ for $n < 0$, so that the model becomes

$$\sum_{k=0}^{\infty} g(k)\,x(n-k) = u(n) \tag{1.54}$$

Assuming that we can 'invert' this equation, we write

$$\sum_{k=0}^{\infty} h(k)\,w(n-k) = x(n) \tag{1.55}$$

This is an alternative form for the general causal linear model, where $x(n)$ is regarded as a sum of weighted values of the past white noise. It is in this form that $h(n)$ can be regarded as a system impulse response sequence, $u(n)$ a white noise input sequence and $x(n)$ the output sequence.

The models in equations (1.54) and (1.55) are characterised by the *infinite* set of parameters $g(n)$ or $h(n)$. The calculation of these parameters by fitting such models to observed data would provide the required spectrum because the spectrum of $x(n)$ following from equation (1.55) is $|H(f)|^2\sigma^2$. However, for practical purposes, it is necessary to reduce the models to those containing only a *finite* number of parameters; typically AR, MA or ARMA models are used to achieve this. As noted at the beginning of this section, if we wish to estimate the power spectral density from a finite length of data, it is only sensible to choose an estimation procedure appropriate to the data model. That is, we must know whether the data are best described by AR, MA or ARMA process, though in a practical situation we may have few clues as to which model is the most appropriate. The MA model can be treated by the classical windowed correlation method (Gutowski *et al.* 1978), and the AR model arises naturally out of the maximum entropy method described in Section 1.5.2.

1.5.2   Maximum entropy spectral analysis

The conventional approach to spectral estimation matches the finite MA model since the autocovariance function is of finite length and so windowing effects might be discounted. But if the MA model is of infinite order (equation (1.15)) then so is the autocovariance function, and so windowing results in an information loss resulting in bias/resolution problems. If such an infinite order MA process could be described by a finite order AR or ARMA process, this might be a useful model on which to base the estimation procedure. We might consider some method of estimating the (say) AR parameters directly or 'suitably' extending the autocovariance function in some logical manner rather than arbitrarily truncating it after 'running out of data'. It turns out that these procedures are equivalent (for AR processes) and constitute the maximum entropy method (MEM). However, we have yet to account for why the concept of 'entropy' arises in the description, and this we now do rather briefly. (The references Burg 1968; Ulrych and Bishop 1975; Kaveh and Cooper 1976; Kay 1988, and particularly, Ables 1974 are illuminating and serve as the basis of this section.)

The problem of spectral analysis is one of data reduction and analysis and it seems self-evident that a basic principle of data reduction should be that any transformation imposed on experimental data shall incorporate and be consistent with all relevant data and be maximally non-committal with regard to unavailable data. The act of setting estimates of the auto-correlation function to zero outside a particular range is hardly being non-committal. It would seem more appropriate to extend the autocovariance function in some 'reasonable' way, but whilst still being non-committal with regard to unavailable data. It is appropriate to introduce the concept of entropy at this stage. The 'unavailable data' are referred to as a lack of information, and information and entropy are related. In fact, entropy is a measure of uncertainty, or disorder, of a system. To see how the notion of maximising entropy enters the picture, we attempt to state the basic principle of data reduction in rather more specific terms.

Suppose we possess some data about a signal $x(n)$ in the form of values of the autocovariance function, $R(0), R(\pm 1) \ldots R(\pm p)$, etc., and we wish to produce from this finite length sequence a spectral density of a signal in the range $-f_s/2 < f < f_s/2$, we know that the (theoretical) autocovariance sequence $R(n)$ and spectral density $S(f)$ are related by

$$R(n) = \Delta \int_{-\frac{f_s}{2}}^{\frac{f_s}{2}} S(f)\, e^{j2\pi f n\Delta}\, df \qquad (1.56a)$$

and

$$S(f) = \sum_{n=-\infty}^{\infty} R(n)\, e^{-j2\pi f n\Delta} \qquad (1.56b)$$

When dealing with estimates $\tilde{R}(n)$ and $\tilde{S}(f)$, the relationships in equations (1.56a, b) do not provide a unique way of obtaining an estimate of $S(f)$ from the estimate of $R(n)$. The traditional way is to choose a lag window $w(n)$ from an infinite set in the inversion and form

$$\tilde{S}(f) = \sum_{n=-p}^{p} w(n)\, \tilde{R}(n)\, e^{-j2\pi f n\Delta} \qquad (1.57)$$

thus contravening the basic principle of data reduction by assuming $\tilde{R}(n)$ is zero for $|n| > p$, and further, 'rejecting' good data in any tapering process imposed by $w(n)$. Since we should be non-committal with regard to the unavailable covariance function estimates (or our lack of information about them), it is desirable to be able to quantify our ignorance of the original signal as expressed by the spectral density $S(f)$, and maximise this, but still take full account of the measured data as constraints. What we still need, however, is an expression for the ignorance (entropy) given in terms of the spectrum $S(f)$. The usual definition of entropy (strictly the relative entropy) is given by

$$H = -\int_{-\infty}^{\infty} p(u) \log p(u)\, du \qquad (1.58)$$

where $p(u)$ is the signal amplitude probability density function. This is not immediately useful for spectral analysis, but Bartlett (1966) has argued that the (Gaussian) signal with spectral density $S(f)$ can be regarded as having been created by passing white noise through a linear filter, and the entropy difference between the input process (which has the maximum entropy of all processes with the same variance) and the output is the entropy loss δE, where

$$\delta E = -\Delta \int_{-\frac{f_s}{2}}^{\frac{f_s}{2}} \log(S(f))\, df \qquad (1.59)$$

From equation (1.56b) the equation (1.59) is rewritten as

$$\delta E = -\Delta \int_{-\frac{f_s}{2}}^{\frac{f_s}{2}} \log \left(\sum_{n=-\infty}^{\infty} R(n) e^{-j2\pi f n \Delta} \right) df \qquad (1.60)$$

Now, minimising δE (or maximising the entropy) with respect to the unknown $R(n)$, so that $|n| > p$, subject to the constraint that the spectrum is consistent with the known $R(n)$, or $|n| < p$, leads to the maximum entropy spectrum and is the spectral estimate that arises from making the least number of assumptions about the unmeasured data, whilst remaining consistent with the known information.

Thus, the maximum entropy method finally reduces to a calculus of variations problem, solved using Lagrange multipliers, and the resulting MEM spectrum is (Ulrych and Bishop 1975) of the form

$$\hat{S}_{\text{MEM}}(f) = \frac{K}{\left| 1 + \sum_{n=1}^{p} a_n e^{j2\pi f n \Delta} \right|^2} \qquad (1.61)$$

where K is a constant.

We see from equation (1.61) that the spectral estimate has a form similar to the transfer function of an all-pole digital filter, being in fact the squared magnitude of the frequency response of the all-pole filter excited by white noise. This shows the equivalence of the MEM approach and AR processes.

1.5.3 MEM and AR processes

The MEM spectrum is equivalent to the spectrum of a white excited AR filter. This correspondence between MEM and AR processes was formalised by van den Bos (1971) for Gaussian processes where the starting point for the AR form is

$$x(n) = -a_1 x(n-1) - a_2 x(n-2) - \cdots - a_p x(n-p) + u(n) \qquad (1.62)$$

and determining the extension to the autocovariance function is equivalent to calculating the parameters a_i of the AR model in equation (1.62). Indeed, knowledge of the coefficients and $R(n)$ for $n = 0, \ldots, p$ implies knowledge of $R(n)$ for all n if $x(n)$ is described by equation (1.62). This follows from equation (1.62) by multiplying by $x(n-k)$ for $k > 0$ and taking expectations. This yields

$$R(k) = -a_1 R(k-1) - a_2 R(k-2) - \cdots - a_p R(k-p), \quad k > 0 \qquad (1.63)$$

Knowledge of $R(0), \ldots, R(p-1)$ and the a_i is sufficient to go on generating $R(n)$ for $n > 0$. We also note that when $k = 0$ we can show

$$R(0) = -a_1 R(1) - a_2 R(2) - \cdots - a_p R(p) + \sigma^2 \tag{1.64}$$

In the estimation problem we only know the autocovariance function estimates and not the a_i, which must be determined to calculate the spectrum, which was considered in Section 1.4.5.

Before considering the problem of estimation of the a_i it is worth noting the interpretation of the AR process of equation (1.62) as one that identifies $x(n)$ with a value that is predicted from the previous values of the process with an error $e(n) = u(n)$, and the coefficients $a_i, \ldots, a_p$ define a p-point prediction filter with $u(n) = x(n) + a_1 x(n-1) + \cdots + a_p x(n-p)$, so that the prediction error filter has coefficients $(1, a_1, \ldots, a_p)$ (Makhoul 1975).

1.5.4 Estimation of AR coefficients

We shall now consider how the MEM spectrum may be computed using equation (1.61). It is necessary to determine (i) the length p of the prediction filter and, logically, then (ii) the coefficients a_i. However, we shall see in Section 1.5.5 that the choice of p requires a knowledge of the a_i and so we shall assume here that p is fixed and, subject to that, calculate the estimates $\hat{a}_i$ of a_i.

We emphasise that the discussions so far have related to the definition of the MEM spectrum using exact values of the autocovariance function. No mention has been made of the use of *estimates* of the autocovariance function to produce estimates of the MEM spectra. The estimates of the autocovariance function contain the effects of additive noise and data truncation. The first problem was considered by Ables (1974) who suggested the modification that the autocorrelation be constrained to pass close to measured values. The problem of data truncation amounts to a violation of the basic principle of data reduction.

1.5.4.1 The Yule Walker (YW) method

A widely used method of obtaining estimates of a_i follows directly from equation (1.63) which can be written out in matrix form for $k = 1, \ldots, p$ as

$$
\begin{aligned}
a_1 R(0) + a_2 R(1) + a_3 R(2) + \cdots + a_p R(p-1) &= -R(1) \\
a_1 R(1) + a_2 R(0) + a_3 R(1) + \cdots + a_p R(p-2) &= -R(2) \\
a_1 R(2) + a_2 R(1) + a_3 R(0) + \cdots + a_p R(p-3) &= -R(3) \\
\vdots \qquad \vdots \qquad \vdots \qquad\qquad \vdots \qquad\quad &\ \ \vdots \\
a_1 R(p-1) + a_2 R(p-2) + a_3 R(p-3) + \cdots + a_p R(0) &= -R(p)
\end{aligned}
\tag{1.65}
$$

or

$$
\begin{bmatrix}
R(0) & R(1) & R(2) & \cdots & R(p-1) \\
R(1) & R(0) & R(1) & \cdots & R(p-2) \\
R(2) & R(1) & R(0) & \cdots & R(p-3) \\
\vdots & \vdots & \vdots & \ddots & \vdots \\
R(p-1) & R(p-2) & R(p-3) & \cdots & R(0)
\end{bmatrix}
\begin{bmatrix}
a_1 \\ a_2 \\ a_3 \\ \vdots \\ a_p
\end{bmatrix}
=
\begin{bmatrix}
-R(1) \\ -R(2) \\ -R(3) \\ \vdots \\ -R(p)
\end{bmatrix}
$$

$$(1.66)$$

From equations (1.65) and (1.66) we note that knowledge of $R(0), \ldots, R(p)$ gives enough information to obtain the a_i. Any usual technique (for example, Gauss elimination) may be used to solve equation (1.66) but note that the coefficient matrix is symmetrical and has a diagonal symmetry of the coefficients (elements on leading diagonals are equal). Such matrices are called Toeplitz matrices, and their special structure can be used to speed the solution using a recursive procedure originated by Levinson and improved by Durbin (Ulrych and Bishop 1975).

To write down the spectral estimate, equation (1.61) requires not just the coefficients a_i but also the scale factor σ^2, and this follows from equation (1.65), which when this final equation is added to the set of equations in equation (1.66) results in the alternative form

$$
\begin{bmatrix}
R(0) & R(1) & R(2) & \cdots & R(p) \\
R(1) & R(0) & R(1) & \cdots & R(p-1) \\
R(2) & R(1) & R(0) & \cdots & R(p-2) \\
\vdots & \vdots & \vdots & \ddots & \vdots \\
R(p) & R(p-1) & R(p-2) & \cdots & R(0)
\end{bmatrix}
\begin{bmatrix}
1 \\ a_1 \\ a_2 \\ \vdots \\ a_p
\end{bmatrix}
=
\begin{bmatrix}
\sigma^2 \\ 0 \\ 0 \\ \vdots \\ 0
\end{bmatrix}
\qquad (1.67)
$$

The correspondence between the AR process and the prediction of $x(n)$ from past data means that σ^2 may be identified with the prediction error variance E_p. We emphasise that this formulation has used ideal values for $R(n)$. In any practical estimation problem, we only have the measured data $x(n)$ from which we might make estimates $\hat{R}(n)$, and from the $(p+1)$ equation (1.67) we solve for the $(p+1)$ unknowns $a_1, a_2, \ldots, a_p$ and σ^2.

The YW estimates of the AR coefficients may be criticised since (i) the AR coefficients should be estimated so as to be non-committal with regard to unavailable information. However, the usual methods of estimation of $\hat{R}(n)$ imply $\hat{R}(n) = 0$ outside the available data length, thus contradicting the principle of maximum entropy. (ii) If $\hat{R}(n)$ is estimated using the unbiased estimator it is sometimes possible to obtain estimates that do not constitute a valid autocovariance matrix (Claerbout 1976). (iii) The YW estimates are sensitive to rounding errors (Ulrych and Bishop 1975) when poles of the linear system lie near $|z| = 1$.

A method for the computation of the $\hat{a}_i$ proposed by Burg (1968, 1972) avoids problems (i) and (ii). The need for calculation of the autocovariance of the data is side-stepped altogether, and the scheme essentially uses the interpretation of the MEM spectrum as the spectrum of the process, which is the output of an AR filter. In fact, the method obtains the prediction error filter coefficients directly using least squares methods.

1.5.5 Choice of filter length

The MEM spectral density estimate requires the length p (order of the AR process) to be known. Without prior information, this is a difficult task. Underestimates of p tend to lead to 'oversmoothed' spectra, and overestimates should be avoided since (i) computation should be minimised, (ii) ill-conditioning of the equations increases with numbers of poles and (iii) spurious ripples occur in the spectrum.

Several authors have considered the problem of order determination, and the most successful and widely used method of order determination can be attributed to Akaike (1974) (though there are others (like Gustaffson and Hjalmarsson 1995)). The basic idea on which the criteria are based use the 'error variance' which is monitored at each iteration. As the order of the filter increases, the error decreases until it flattens out. The idea is to choose p so that an increase in order results only in a small reduction in error. Akaike's approach is to combine into a single function both the error and the number of parameters, which should demonstrate a minimum beyond which point an increase in order is not appropriate. Two such functions are given in equations (1.68) and (1.69). The terms FPE and AIC stand for final prediction error and Akaike's information criterion, respectively.

$$FPE_p = \frac{N+p+1}{N-p-1} E_p \tag{1.68}$$

$$AIC_p = N\log(E_p) + 2p \tag{1.69}$$

where N is the number of samples in the time series.

Gustaffson and Hjalmarsson (1995) have reconsidered the problem of model selection as an estimation task where model structure is the unknown parameter and treats the parameters a_i as stochastic variables with some 'prior' distribution. A comprehensive description of 21 maximum likelihood estimators is presented and related to a Bayesian approach for maximum a posteriori (MAP) estimators.

1.5.6 High-resolution spectral analysis (Capon's method)

An approach often referred to as 'Capon's method' (Capon 1969; Kay 1988) will be described in this section. This is sometimes called the 'minimum variance' or 'maximum likelihood' approach. The principle is based on the analysis of a signal using a filter bank, wherein a stationary random signal $x(n)$, when passed through a filter, produces an output $y(n)$. The output power spectral density of $y(n)$ is

$$S_{yy}(f) = |H(f)|^2 S_{xx}(f) \tag{1.70}$$

So, if $|H(f)| = 1$ at frequency f, the output power spectral density is the same as the input power spectral density. If the filter $H(f)$ is a narrow-band filter, then the output power is the power of $x(n)$ in a narrow band. So, narrow-band filtering is an approach to spectral analysis.

'Ordinary'-band-pass filtering usually sets up a fixed filter structure (equal bandwidths or in 1/3 octaves, etc.). Capon's method sets up filters that are 'data dependent'.

1.5.6.1 The basis of the method

The data $x(n)$ is analysed for its behaviour at frequency f Hz (normal narrow-band filter behaviour) but with the constraint that the total output power of the filter is minimised, such that, extraneous power from adjacent frequencies does not contribute to the filter output.

The power spectral density of the output $y(n)$ of filter operating on a signal $x(n)$ is

$$S_{yy}(f) = |H(f)|^2 S_{xx}(f) \tag{1.71}$$

and the total output power is

$$E\left[y(n)^2\right] = \int_{-\frac{1}{2}}^{\frac{1}{2}} S_{yy}(f)\, \mathrm{d}f \tag{1.72}$$

For the purposes of high-resolution spectral analysis, we want the filter $H(f)$ to give us a focus on the spectrum of $x(n)$ at frequency f, so

$$H(f) = 1 \quad \text{at frequency } f \tag{1.73}$$

and we want to reject effects from frequencies other than f Hz. So we wish to minimise the output power, that is, $E[y(n)^2]$ subject to the constraint that $H(f) = 1$. If we can find the filter $H(f)$ that achieves this, then the output power of the filter is the power of the signal in the 'close' vicinity of f. Note that only by division by a bandwidth is it a density function. To develop this further, we will assume the filter to be an L-point FIR filter with impulse response $h(n)$ which can be expressed in vector form as

$$\mathbf{h} = [h(0) \quad h(1) \quad \cdots \quad h(L-1)]^t \tag{1.74}$$

and the input is the time sequence $x(n)$, which can also be expressed through the vector

$$\mathbf{x}_n = [x(n) \quad x(n-1) \quad x(n-2) \quad \cdots \quad x(n-L+1)]^t \tag{1.75}$$

in which case the output of the filter can be expressed as a vector inner product

$$y(n) = \mathbf{h}^t \mathbf{x}_n \tag{1.76}$$

The problem is to find $\mathbf{h}$ such that the output variance of the filter $E[y(n)^2]$ is minimised, subject to the constraint that the transfer function of the filter is unity at frequency f Hz, that is, $H(f) = 1$. This constraint is written out fully as

$$\sum_{n=0}^{L-1} h(n) e^{-j2\pi fn} = H(f) = 1 \tag{1.77}$$

or expressed in vector notation as

$$\mathbf{e}_f^{\dagger} \mathbf{h} = \mathbf{h}^{\dagger} \mathbf{e}_f = 1 \quad \text{at frequency } f \tag{1.78}$$

where $\mathbf{e}_f = \begin{bmatrix} 1 & e^{j2\pi f} & e^{j2\pi 2f} & \cdots & e^{j2\pi(L-1)f} \end{bmatrix}^t$ and $\dagger$ denotes the conjugate transpose. This is a constrained optimisation problem that can be solved by the method of Lagrange multipliers as explained in Section 1.5.6.2.

1.5.6.2 The method of Lagrange multipliers applied to spectral analysis

The filter variance can be written as

$$E\left[|y(n)|^2\right] = E\left[\mathbf{h}^{\dagger} \mathbf{x}_n \mathbf{x}_n^{\dagger} \mathbf{h}\right] = \mathbf{h}^{\dagger} E\left[\mathbf{x}_n \mathbf{x}_n^{\dagger}\right] \mathbf{h} = \mathbf{h}^{\dagger} \mathbf{R} \mathbf{h} \tag{1.79}$$

where $\mathbf{R}$ is the covariance matrix of the data $x(n)$. This definition allows for the possibility that the data are complex-valued.

We want to find $\mathbf{h}$ that minimises the output power subject to constraint $\mathbf{e}_f^{\dagger} \mathbf{h} = 1$. This is achieved by using the method of Lagrange multipliers; we introduce the (as yet) undetermined multiplier λ and create the augmented cost function J' as

$$J' = \mathbf{h}^{\dagger} \mathbf{R} \mathbf{h} + \lambda \left(1 - \mathbf{h}^{\dagger} \mathbf{e}_f\right) \tag{1.80}$$

Minimising J' with respect to $\mathbf{h}$ gives

$$\mathbf{h}_{\text{opt}} = \frac{\lambda \mathbf{R}^{-1} \mathbf{e}_f}{2} \tag{1.81}$$

Combining this with the constraint equation gives

$$\lambda = \frac{2}{e_f{}^\dagger R^{-1} e_f} \qquad (1.82)$$

and

$$h_{\text{opt}} = \frac{R^{-1} e_f}{e_f{}^\dagger R^{-1} e_f} \qquad (1.83)$$

The variance of the output of the optimal filter output is

$$E\left[|y(n)|^2\right] = h_{\text{opt}}{}^\dagger R h_{\text{opt}} = \frac{1}{e_f{}^\dagger R^{-1} e_f} = S_{\text{Capon}}(f) \qquad (1.84)$$

This is the estimate of the power of the signal in the 'close' vicinity of f Hz and constitutes the Capon spectral estimate.

1.5.7 Eigen-based high-resolution methods

The problem of detecting and estimating sinusoidal signals in noise has received enormous attention, and the so-called eigen-based methods have evolved around this problem (see, for example, Vaseghi 1996). The eigenvalues and eigenvectors are computed from the autocorrelation matrix of the measured signal (sinusoid plus noise) and allow the decomposition of the measured signal into components lying in the (true) signal subspace and the noise subspace (orthogonal to the signal subspace). This decomposition leads to a spectral estimate that is sharply peaked at the sinusoidal component frequencies. A widely used method is the so-called MUSIC (multiple signal classification) algorithm briefly described in this section following Vaseghi (1996).

Suppose $x(n)$ denotes the 'true' signal which is the sum of P sinusoids, so that,

$$x(n) = \sum_{k=1}^{P} A_k e^{j(2\pi f_k n \Delta + \phi_k)} \qquad (1.85)$$

and let $y(n)$ denote a measurement which is contaminated by a noise $v(n)$, so that

$$y(n) = x(n) + v(n) \qquad (1.86)$$

Suppose N values of $y(n)$ are known and the column vector y_n contains the previous $L(> p)$ samples of the sequence. Then,

$$y_n = x_n + v_n = Sa + v_n \qquad (1.87)$$

where $\mathbf{a} = [A_1 e^{j2\pi\phi_1} \quad A_2 e^{j2\pi\phi_2} \quad \cdots \quad A_P e^{j2\pi\phi_P}]^t$ and S is an $L \times P$ matrix of complex exponentials involving the frequencies $f_1, \ldots, f_P$, specifically

$$
S = [\mathbf{e}_{f_1} \quad \mathbf{e}_{f_2} \quad \cdots \quad \mathbf{e}_{f_P}]
$$

$$
= \begin{bmatrix}
1 & 1 & 1 & \cdots & 1 \\
e^{j2\pi f_1 \Delta} & e^{j2\pi f_2 \Delta} & e^{j2\pi f_3 \Delta} & \cdots & e^{j2\pi f_P \Delta} \\
e^{j2\pi 2 f_1 \Delta} & e^{j2\pi 2 f_2 \Delta} & e^{j2\pi 2 f_3 \Delta} & \cdots & e^{j2\pi 2 f_P \Delta} \\
\vdots & \vdots & \vdots & \ddots & \vdots \\
e^{j2\pi(L-1)f_1 \Delta} & e^{j2\pi(L-1)f_2 \Delta} & e^{j2\pi(L-1)f_3 \Delta} & \cdots & e^{j2\pi(L-1)f_P \Delta}
\end{bmatrix} \tag{1.88}
$$

The P-dimensional vector space defined by the columns of S is referred to as the signal subspace. Any noise-free measurements $\mathbf{x}_n$ span this subspace, which means that $\mathbf{x}_n$ is formed as linear combination of the columns of S. Assuming signal and noise are uncorrelated, then

$$
\mathbf{R}_{yy} = \mathbf{R}_{xx} + \mathbf{R}_{vv} \tag{1.89}
$$

where $\mathbf{R}_{yy}, \mathbf{R}_{xx}$ and $\mathbf{R}_{vv}$ are the $(L \times L)$ correlation matrices of $y(n), x(n)$ and $v(n)$, respectively. If $v(n)$ is 'white' (temporally uncorrelated), with variance σ_v^2, then,

$$
\mathbf{R}_{yy} = \mathbf{S}\mathbf{Q}\mathbf{S}^\dagger + \sigma_v^2 I \tag{1.90}
$$

Matrix Q is diagonal with elements A_i^2. It can be shown (Vaseghi 1996) that the correlation matrix $\mathbf{R}_{yy}$ can be expressed as

$$
\mathbf{R}_{yy} = \sum_{k=1}^{P} \left(\lambda_k + \sigma_v^2\right) \mathbf{u}_k \mathbf{u}_k^\dagger + \sum_{k=P+1}^{L} \sigma_v^2 \mathbf{u}_k \mathbf{u}_k^\dagger \tag{1.91}
$$

where the λ_k are the eigenvalues of $\mathbf{R}_{xx}$ and the $\mathbf{u}_k$ are eigenvectors of $\mathbf{R}_{yy}$.

The eigenvalues of $\mathbf{R}_{yy}$ (the noisy signal) decompose into the principal eigenvalues $\{(\lambda_k + \sigma_v^2), \ k = 1, 2, \ldots, P\}$ and the noise eigenvalues σ_v^2. The eigenvectors $(\mathbf{u}_1, \ldots, \mathbf{u}_P)$ span the signal subspace (the column space of the matrix S). The vectors $(\mathbf{u}_{P+1}, \ldots, \mathbf{u}_L)$ span a vector space usually referred to as the noise subspace. Since the eigenvectors of a correlation matrix are orthogonal, the signal and noise subspaces are orthogonal. From this, the frequencies can be deduced and the MUSIC spectrum is computed as

$$
S_{\mathrm{MUSIC}}(f) = \frac{1}{\left| \sum_{k=P+1}^{L} \mathbf{e}_f^\dagger \mathbf{u}_k \right|^2} \tag{1.92}
$$

The vector $\mathbf{e}_f$, in general, lies partly in the signal subspace and partly in the noise subspace. However, when the value of f coincides with one of the frequencies $f_1, \ldots, f_P$ then $\mathbf{e}_f$ is completely contained within the signal subspace and as such is orthogonal to the noise subspace, meaning that it is orthogonal to $\mathbf{u}_{P+1}, \ldots, \mathbf{u}_N$. Consequently, the denominator in the MUSIC spectrum becomes zero, leading to a singularity. In practice, the eigenvectors $\mathbf{u}_k$ are calculated from an *estimate* of correlation matrix $\mathbf{R}_{yy}$, so that only approximate orthogonality is attained due to the finite data length, leading to peaks, rather than singularities, appearing in the MUSIC spectrum.

The MUSIC spectrum peaks at frequencies where there are likely to be sinusoidal components and is not a measure of power within a filter band. So it is inappropriate to regard it as a true spectral estimate, but rather as a method for identifying frequencies of sinusoids in noise.

1.5.8 Performance example

This section considers the performance of the spectral estimators discussed in the preceding sections. A simple test signal is used as the basis of these comparisons. This test signal is selected so that it contains features commonly observed in acoustic and vibration measurements. It is constructed by passing Gaussian white noise (with a variance of 1) through a linear digital system with the transfer function

$$H(z) = \frac{1 + 1.41z^{-1} + 0.944z^{-2}}{1 + 0.6z^{-1} + 0.944z^{-2}} \tag{1.93}$$

The system exhibits a resonance at a normalised frequency of 0.6 and an anti-resonance at 0.76. There are also three sinusoidal signals added to the signal at normalised frequencies 0.05, 0.06 and 0.15; each sinusoid has unit amplitude. Note that the two lowest frequency sinusoids have very similar frequencies and are something of a challenge to resolve. The time series analysed contains 512 samples.

In addition to the methods discussed herein, a classical non-parametric method is also used to compute a spectral estimate. Specifically, a segment averaging approach is employed (for details see Hammond 1998). The estimated spectra are shown in Figure 1.14. In each frame, the spectrum of the filtered noise is shown as a dotted curve and the frequencies of the sinusoidal components are indicated by vertical dotted lines. Figure 1.14a,b shows the results obtained from segment averaging. Use of the shorter window (Figure 1.14a) results in a heavily smoothed spectral estimate in which the two closely spaced sinusoids are not resolved and the heights and the depths of the resonance and anti-resonance are underestimated. The result of using a longer window in segment averaging is to improve the resolution (the closely spaced sinusoids are resolved) but there is a significant loss of smoothness. Figure 1.14c,d depicts the results of applying an AR spectral estimation scheme (Burg 1968) with different model orders

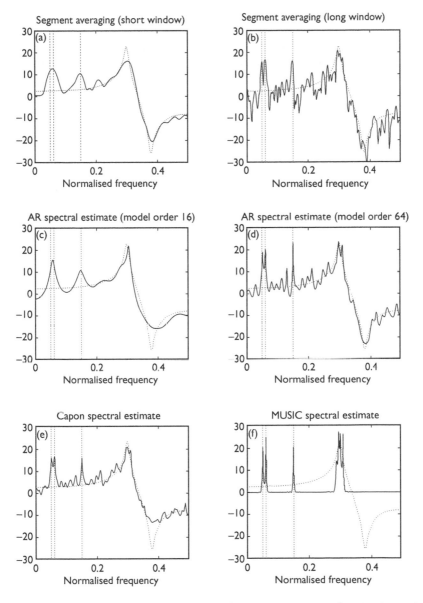

Figure 1.14 Spectra of test signal computed using various spectral estimation techniques: (a) and (b) segment averaging using a Hanning window of 64 and 256 samples respectively; (c) and (d) AR spectral estimates using model orders of 16 and 64; (e) Capon spectrum; (f) MUSIC spectrum.

(16 and 64 respectively). The spectrum computed using a model order 16 shows distinct signs of excessive smoothing; the two closest sinusoids are not resolved. One should note that the peak at the resonance is well resolved in comparison to the anti-resonance. This behaviour is to be expected since the AR method is based on modelling data using an all-pole model and, since the presence of the resonance corresponds to a pole, this feature is readily reproduced in the AR spectrum. However the anti-resonance, resulting from the presence of a zero in the transfer function, is poorly modelled using only poles (although in principle a model of sufficient order is able to capture such behaviour). The higher order AR spectrum (Figure 1.14d) has a significantly greater resolution, the two low-frequency components are clearly resolved and the isolated frequency component is well localised. The anti-resonance is reproduced with some accuracy, but in this case there is a double peak close to the resonance which suggests the presence of two poles (rather than the one which is actually present). The Capon spectrum (Figure 1.14e) shares many of the features of the higher-order AR spectrum; it is somewhat smoother, with a consequent loss of resolution and has failed to reproduce the anti-resonance. Finally, the MUSIC spectrum (Figure 1.14f) has resolved the sinusoidal components better than any of the previous methods, but has clearly failed to model the rest of the spectrum with any accuracy. This highlights the fact that the MUSIC spectrum cannot be interpreted as a measure of signal energy in a band; it should be interpreted as indicating the likelihood that a sinusoid exists at that frequency.

1.6   Time–frequency analysis

The spectrum $S(f)$ of a signal $x(t)$, describes how the power of the signal is distributed in frequency. The spectrum is only an appropriate characterisation of the signal, if the signal is stationary. If the signal is non-stationary then, generally, the spectrum of the signal changes as the signal evolves in time. This leads one to the intuitive concept of a time-dependent spectrum; it is the goal of time–frequency analysis to estimate such spectra. A time–frequency representation aims to describe how the power of a signal is distributed as a function of both time and frequency. In this Section we shall consider the case of continuous time signals; the extension to digital signals is straightforward (Qian and Chen 1996).

1.6.1   The spectrogram

Historically, the first form of time–frequency analysis considered was the spectrogram (Gabor 1946). The spectrogram $S(t, f)$ is derived from the short-time Fourier transform (STFT) $X(t, f)$ which is defined as

$$X(t, f) = \int_{-\infty}^{\infty} w(t - \tau)x(\tau)e^{-j2\pi f\tau}\,\mathrm{d}\tau \tag{1.94}$$

The STFT is complex-valued and represents the Fourier transform of the windowed signal, the window being centred at time t. The spectrogram is defined as the squared magnitude of the STFT

$$S(t,f) = |X(t,f)|^2 = \left| \int_{-\infty}^{\infty} w(t-\tau)x(\tau)e^{-j2\pi f\tau}d\tau \right|^2 \tag{1.95}$$

The spectrogram is a powerful tool for analysing non-stationary signals and is widely employed in a variety of applications including speech processing (Rabiner and Schafer 1978), passive SONAR analysis (Knight *et al.* 1981) and structural analysis (Hammond and White 1996). It has the attractions of being efficient to compute (in digital form it can be computed using fast Fourier transforms) and is simple to interpret. Figure 1.15 shows an example of a spectrogram computed for part of the bird call of a wren (*Troglodytes troglodytes*), along with the time series data. The spectrogram

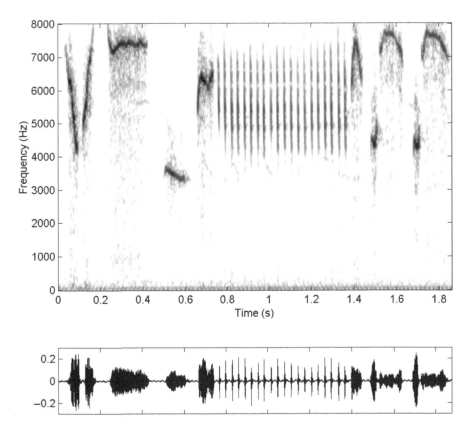

Figure 1.15 Analysis of a wren's call. Upper frame is the spectrogram, lower frame is the time series data.

is plotted using a logarithmic grey scale, with dark shades representing high energy. The time series highlights the fact that this call contains discrete events (chirps), and each chirp can be associated with a curve in the spectrogram. This curve gives information about the frequency behaviour within the chirp. For example, the first chirp can be seen to consist of a downward-sweeping frequency component and is followed by a second chirp which is an upward sweep. The spectrogram gives a good visualisation of this signal, presenting data to the analyst in an intuitive manner.

The spectrogam suffers from severe limitations, the most important of which relates to its lack of resolution. At a particular time, the spectrogram corresponds to the magnitude of a windowed Fourier transform; the frequency resolution of this Fourier transform improves as the duration of the window increases. Conversely, the temporal resolution of the spectrogram improves as the window duration reduces. The structure of the spectrogram strongly depends upon the duration of the window employed, and one cannot achieve good time and frequency resolution simultaneously. This is illustrated in Figure 1.16 that shows the spectrogram of the signal defined in equation (1.96).

$$x(t) = \sin(2\pi 101t): \quad t < 0.98$$
$$= 0: \qquad\qquad 0.98 \leq t < 1.02 \tag{1.96}$$
$$= \sin(2\pi 105t): \quad t \geq 1.02$$

Two spectrograms computed from this signal are shown in Figure 1.16. Figure 1.16a shows a spectrogram computed using a long duration window and, as a consequence of the resulting good frequency resolution, the difference in frequencies between the two signal halves can clearly be seen. However, this long window also obscures the 40 ms gap in the time series, whereas in Figure 1.16b the gap between the two signal halves is evident, because of the good temporal resolution, but the difference in frequencies is concealed, because of the poor frequency resolution.

1.6.2 The Wigner-Ville distribution (WVD)

The spectrogram is not a fundamental signal representation since its characteristics depend upon the windowing function chosen by the user. A more fundamental time–frequency representation is provided by WVD $W(t, f)$ defined as

$$W(t, f) = \int_{-\infty}^{\infty} x\left(t - \frac{\tau}{2}\right)^{*} x\left(t + \frac{\tau}{2}\right) e^{-j2\pi f\tau} d\tau \tag{1.97}$$

This can be recognised as the Fourier transform of the instantaneous autocorrelation function $r(t, \tau)$.

$$r(t, \tau) = x\left(t - \frac{\tau}{2}\right)^{*} x\left(t + \frac{\tau}{2}\right) \tag{1.98}$$

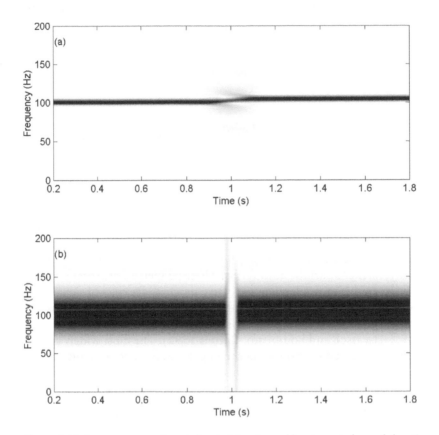

Figure 1.16 Spectrograms of test signal: (a) using a Hanning window of duration 320 ms; (b) using a Hanning window of duration 40 ms.

which is the product of the signal values separated by τ seconds, centred at time t. This interpretation of the WVD, as a Fourier transform of time-dependent correlation function, is analogous to the definition of stationary spectrum as the Fourier transform of a stationary correlation function.

The WVD is capable of offering greater resolution than that available using the spectrogram, but this performance is achieved at a price. There are two major problems associated with WVD: (i) for almost all signals the WVD contains negative values, which is inconsistent with its interpretation as decomposition of a signal's power; (ii) if a signal contains two distinct signal components, for example, two frequencies, the WVD will consist of three elements, one due to each of the components and a third (spurious) element that is a result of the interaction of the two components. These additional elements are variously called cross-terms, interference terms or even 'junk' terms. The presence of these terms can make interpretation of the WVD difficult. They arise because of the quadratic nature of the WVD. This can be illustrated through a simple example; consider the signal

$$x(t) = A_1 e^{j2\pi f_1 t} + A_2 e^{j2\pi f_2 t} \tag{1.99}$$

so that

$$r(t, \tau) = |A_1|^2 e^{j2\pi f_1 \tau} + |A_2|^2 e^{j2\pi f_2 \tau} + 2\mathrm{Re}\{A_1{}^* A_2 e^{j2\pi(f_2 - f_1)t}\} e^{j2\pi(f_2 + f_1)\frac{\tau}{2}} \tag{1.100}$$

$$W(t, f) = |A_1|^2 \delta(f_1) + |A_2|^2 \delta(f_2) + 2\mathrm{Re}\{A_1{}^* A_2 e^{j2\pi(f_2 - f_1)t}\} \delta(f_m) \tag{1.101}$$

where f_m is the mean of frequencies f_1 and f_2. The third term in this WVD is the cross-term. It should be noted that this cross-term lies midway between the two auto-terms and it oscillates. The magnitude of the cross-term is twice that of the auto-terms (assuming $|A_1| = |A_2|$). The cross-term oscillates more rapidly, further apart the auto-terms. These properties of cross-terms are not exclusive to this example; they apply to a much wider class of signals.

Much research effort in the field of time–frequency analysis has centred on the problem of reducing the cross-terms introduced by the WVD (Cohen 1989; Hlawatsch and Boudreaux-Bartels 1992). The most widely considered technique is based on exploiting the key difference between auto-terms and cross-terms, namely, that cross-terms oscillate and auto-terms commonly have slowly varying amplitudes. Alternatively, if the WVD is viewed as an image, the cross-terms are associated with high spatial frequencies and the auto-terms occupy the low spatial frequency. By applying a two-dimensional low-pass filter to the WVD one can reduce the cross-terms, but in doing so one also smears sharp lines resulting in a loss of resolution. The class of time–frequency methods based on two-dimensional smoothing of the WVD is referred to as Cohen's class. A general Cohen's class (Cohen 1989, 1995) time–frequency representation $C(t, f)$ can be expressed as

$$C(t, f) = \Phi(t, f) ** W(t, f) \tag{1.102}$$

where $**$ is used to denote two-dimensional convolution and the function $\Phi(t, f)$ is the low-pass filter, the form of which distinguishes the different members of Cohen's class.

The design of suitable kernel functions $\Phi(t, f)$ involves a trade-off between resolution and the degree to which the cross-terms are suppressed. Although not obvious, the spectrogram is in fact a member of Cohen's class, and so it can be obtained by suitably smoothing the WVD. Additionally, one may design the kernel functions so that $C(t, f)$ retains some of the desirable theoretical properties of the WVD.

1.6.3   Example: analysis of beam response data

To demonstrate the effectiveness of time–frequency representations, we shall consider the analysis of the data shown in Figure 1.17. These data were collected from an accelerometer attached to a beam. The beam had one end free and one end embedded in sand; the sand provides a near-anechoic termination. The response was obtained by exciting the beam with an impact at a short distance from the measurement point. The response is dominated by two components: the first corresponding to the arrival of the direct wave, the second the same wave after reflection from the free end. Remaining echoes are strongly attenuated by the sandbox termination. As the waves travel in the beam, the low-frequency components travel more slowly than the high frequencies (due to dispersion). Consequently, both of the observed wave packets initially contain the high-frequency components with the low-frequency components arriving later (see Figure 1.17). This effect is more pronounced in the second wave packet that has travelled a greater distance so that the disparity in the speeds of frequency components produces greater time differences.

Figure 1.18 shows the results of applying various time–frequency analysis tools to this data. The same conventions are used in this plot, as were used in Figure 1.15. Figure 1.18a shows the results of applying a spectrogram with a relatively long window. The result is a distribution in which the temporal resolution is poor and the structure of the signal is not clear. In Figure 1.18b a shorter window is used which allows one to observe the underlying structure of this signal. The two curves describing the arrival times of frequency components in each of the wave packets can be seen. In principle,

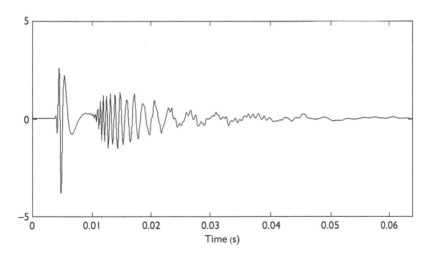

Figure 1.17 Beam response measurement.

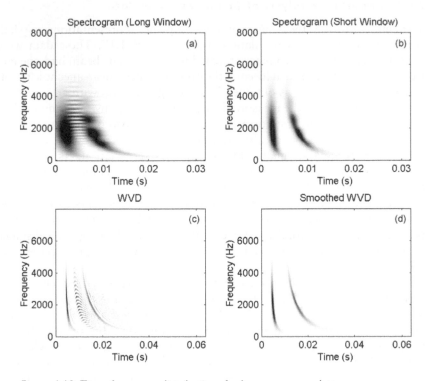

Figure 1.18 Time–frequency distributions for beam response data.

measurements from these curves, allied to knowledge of the spacing between the site of excitation and measurement, allow one to estimate the dispersion laws for this beam.

Figure 1.18c depicts the WVD for this data. Two curves, defining the dispersion curves for the two wave packets, can be identified. In this case, these curves are drawn as sharper lines, implying that one can identify the arrival time of a specific frequency component with greater accuracy, suggesting that using the WVD may allow one to estimate dispersion laws more accurately than can be achieved using the spectrogram. However, the cross-terms in the WVD are also evident, most prominently as a chevron pattern between the two dispersion curves. In this case, the presence of these interference terms slightly impedes the identification of the dispersion curves. When analysing more complex signals, the presence of these cross-terms can be far more troublesome. Figure 1.18d shows an example of a smoothed WVD distribution; note how the cross-terms have been largely attenuated and the resolution of the dispersion curves somewhat reduced. This distribution has a resolution that is greater than that which can be achieved using a spectrogram, and the effect of the cross-terms has been greatly reduced relatively to the WVD.

1.7 Concluding remarks

Advanced signal processing methods are currently very much in evidence in acoustics and vibration applications. This chapter provides a few 'stepping stones' amongst an array of exciting possibilities, with the areas of non-linearity, non-Gaussianity and non-stationarity set within a probabilistic framework offering the greatest scope.

References

Ables, J.G. (1974) Maximum entropy spectral analysis. *Astronomy Astrophysics Supplement Series*, **15**, 383–393.

Akaike, H. (1974) A new look at statistical model identification. *IEEE Transactions on Automatic Control*, **AC-19**, 716–723.

Bartlett, M.S. (1966) *Introduction to Stochastic Processes* (Cambridge: Cambridge University Press).

Burg, J.P. (1968) A new analysis technique for time series data. *NATO Advanced Study Institute on Signal Processing with Emphasis on Underwater Acoustics*, Enschede, The Netherlands.

Burg, J.P. (1972) The relationship between maximum entropy spectra and maximum likelihood. *Geophysics*, **37**, 375–376.

Capon, J. (1969) High resolution frequency wavenumber spectrum analysis. *Proceedings of the IEEE*, **57**, 1408–1418.

Claerbout, J.F. (1976) *Fundamentals of Geophysical Data Processing* (New York: McGraw-Hill).

Cohen, L. (1989) Time–frequency distributions – A review. *Proceedings of the IEEE*, **77**, 941–981.

Cohen, L. (1995) *Time–Frequency Analysis* (Englewood Cliffs: Prentice-Hall).

Gabor, D. (1946) Theory of communications. *Journal of IEE*, **93**, 429–457.

Gustaffson, F. and Hjalmarsson, H. (1995) Twenty-one ML estimators for model selection. *Automatica*, **31**, 1377–1392.

Gutowski, P.R., Robinson, E.A. and Treitel, S. (1978) Spectral estimation: fact or fiction. *IEEE Transactions on Geoscience Electronics*, **GE-16**, 80–84.

Hammond, J.K. (1998) Fundamentals of signal processing. In F.J. Fahy and J.G. Walker (eds) *Fundamentals of Noise and Vibration* (London, New York: E & FN Spon), pp. 311–372.

Hammond, J.K. and White, P.R. (1996) The analysis of nonstationary signals using time frequency methods. *Journal of Sound and Vibration*, **190**, 419–447.

Haykin, S. (ed.) (1983) *Nonlinear methods of spectral analysis*. Vol. 34, Topics in Applied Physics. (Berlin: Springer-Verlag).

Hlawatsch, F. and Boudreaux-Bartels, G. (1992) Linear and quadratic time–frequency representations. *IEEE Signal Processing Magazine*, 21–66.

Kaveh, M. and Cooper, G.R. (1976) An empirical investigation of the properties of the autoregressive spectral estimator. *IEEE Transactions on Information Theory*, **IT-22**, 313–323.

Kay, S.M. (1988) *Modern Spectral Estimation: Theory and Application* (Englewood Cliffs: Prentice-Hall).

Knight, W.C., Pridham, R.G. and Kay, S.M. (1981) Digital signal processing for sonar. *Proceedings of the IEEE*, **69**, 1451–1506.

Makhoul, J. (1975) Linear prediction: A tutorial review. *Proceedings of the IEEE*, **63**, 561–580.

Oppenheim, A.V. and Schafer, R.W. (1975) *Digital Signal Processing* (Englewood Cliffs: Prentice-Hall).

Oppenheim, A.V., Wilsky A.S. and Nawab, S.H. (1997) *Signals and Systems* (Englewood Cliffs: Prentice-Hall).

Press, W.H., Teukolsky, S.A., Vetterling, W.T. and Flannery, B.P. (1993) *Numerical Recipes in C: The Art of Scientific Computing* (Cambridge: Cambridge University Press).

Priestley, M.P. (1988) *Non-linear and Non-stationary Time Series* (New York: Academic Press).

Qian, S. and Chen, D. (1996) *Joint Time–Frequency Analysis* (Englewood Cliffs: Prentice-Hall).

Rabiner, L.R. and Gold, B. (1975) *Theory and Application of Digital Signal Processing* (Englewood Cliffs: Prentice-Hall).

Rabiner, L.R. and Schafer, R.W. (1978) *Digital Processing of Speech* (Englewood Cliffs: Prentice-Hall).

Ulrych, T.J. and Bishop, T.N. (1975) Maximum entropy spectral analysis and autoregressive decomposition. *Reviews of Geophysics and Space Physics*, **13**, 183–200.

van den Bos, A. (1971) Alternative interpretation of maximum entropy spectral analysis. *IEEE Transactions on Information Theory*, **IT-17**, 493–494.

Vaseghi, S.V. (1996) *Advanced Signal Processing and Digital Noise Reduction* (New York: Wiley and Teubner).

Part II

Acoustic modelling

Chapter 2

Numerical methods in acoustics

M. Petyt and C.J.C. Jones

2.1 Introduction

Analytical solutions of the governing differential equations in acoustics can only be obtained if the physical boundaries can be described simply in mathematical terms (Nelson 1998). This is rarely the case in engineering and so it is generally necessary to employ approximate numerical methods. The two most widely used are the finite element method (FEM) and the boundary element method (BEM).

Finite element methods were first developed for analysing complex engineering structures. Once the method had been given a firm mathematical foundation, it was only natural that it should be used for analysing other physical problems which could be represented by partial differential equations. The field of acoustics has been no exception. The BEM was first developed to predict the noise radiated from vibrating structures which are immersed in an infinite acoustic medium. Subsequently, it was applied to interior problems. Vibroacoustic problems can be analysed by combining the equations of motion of the vibrating structure, obtained using finite element techniques, with the equations of motion of the acoustic medium, which can be obtained using either FEM or BEM.

Both FEM and BEM represent a continuous system, which has an infinite number of degrees of freedom, by a discrete system having a finite number of degrees of freedom. The accuracy of the solution depends upon the number of degrees of freedom used. The higher the frequency of interest the more degrees of freedom that are required. Therefore, at high frequencies such methods become inefficient. Consequently, they are only used at low frequencies where wavelengths are of similar order of magnitude to the defined geometry. High-frequency acoustic problems are analysed using geometrical or ray-tracing techniques (Pierce 1989), and vibroacoustic problems are handled by means of statistical energy analysis (Lyon and DeJong 1995; Craik 1996). Further details are given in Chapter 11. This chapter is confined to low-frequency analysis in acoustics and vibroacoustics.

Section 2.2 deals with the derivation of the finite element formulation for the acoustic field in an irregularly shaped closed cavity, with Section 2.3 presenting examples of the analysis of acoustic modes for one-, two- and three-dimensional and axisymmetric cases. Section 2.4 uses the axisymmetric formulation to demonstrate the application of the method to the calculation of the transmission loss of a simple duct system. Sections 2.5 and 2.6 show briefly how the finite element method can be extended to apply to externally radiated acoustic fields and the acoustics of porous media. The direct and indirect BEMs are described in Section 2.7, with examples of the use of the direct BEM. Section 2.8 briefly indicates how the methods are used in coupled structural-fluid models.

2.2   Finite element methods for interior problems

2.2.1   Statement of the problem

Consider a cavity of volume V enclosed by a non-rigid surface, as shown in Figure 2.1. Part of the surface S_0 is acoustically rigid, the part S_1 is flexible and vibrating with a harmonic normal velocity $\mathbf{v}$, and the part S_2 is covered with a sound-absorbing material. If this material is assumed to be locally reacting, then its properties can be represented by a specific acoustic impedance z_s.

Within V, the acoustic pressure p, must satisfy the Helmholtz equation

$$\nabla^2 p + \left(\frac{\omega}{c_0}\right)^2 p = 0 \tag{2.1}$$

where ∇^2 is Laplacian operator, ω the circular frequency and c_0 the speed of sound. Over the boundary surface, the fluid particle velocity normal to the surface is equal to the normal velocity of the surface. This gives rise to the following boundary conditions:

$$\mathbf{n} \cdot \nabla p = 0 \quad \text{on } S_0 \tag{2.2a}$$

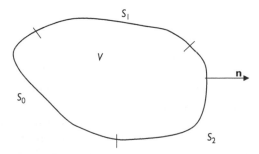

Figure 2.1 Irregular-shaped cavity with non-rigid walls.

where n is the outward-directed unit vector normal to the surface and ∇ is the gradient operator,

$$\mathbf{n} \cdot \nabla p = -j\rho_0 \omega \mathbf{n} \cdot \mathbf{v} \quad \text{on } S_1 \tag{2.2b}$$

where ρ_0 is the mean density of the acoustic medium and

$$\mathbf{n} \cdot \nabla p = -j\rho_0 \omega \frac{p}{z_s} \quad \text{on } S_2 \tag{2.2c}$$

Analytical solutions to equation (2.1) subject to the boundary conditions of equation (2.2) can be found for simple geometries. However, for fluid-filled spaces of arbitrary shape this is not possible. It is therefore necessary to obtain an approximate solution.

2.2.2 Variational formulation

In order to obtain an approximate solution, the requirement that the differential equation (2.1) and boundary conditions in equation (2.2) be satisfied at every point is relaxed. Instead, a solution is sought which satisfies them in some average sense. This can be obtained using a weighted residual technique (Finlayson 1972).

If $\bar{p}$ is an approximate solution then, from equation (2.1)

$$\nabla^2 \bar{p} + \left(\frac{\omega}{c_0}\right)^2 \bar{p} = \varepsilon_1 \neq 0 \tag{2.3}$$

where ε_1 is a non-zero error. A solution is obtained by requiring the weighted average of this error to be zero. Multiplying equation (2.3) by the weight $\bar{w}$, and integrating over the volume gives

$$\int_V \bar{w} \nabla^2 \bar{p} \, dV + \omega^2 \int_V \frac{1}{c_0^2} \bar{w} \cdot \bar{p} \, dV = 0 \tag{2.4}$$

Various weights can be used which give rise to various weighted residual methods.

Applying Green's theorem in the form

$$\int_V \bar{w} \nabla^2 \bar{p} \, dV = \int_S \mathbf{n} \cdot (\bar{w} \nabla \bar{p}) \, dS - \int_V \nabla \bar{w} \cdot \nabla \bar{p} \, dV \tag{2.5}$$

to the first term in equation (2.4) gives

$$\int_V \nabla \bar{w} \cdot \nabla \bar{p} \, dV - \omega^2 \int_V \frac{1}{c_0^2} \bar{w} \cdot \bar{p} \, dV = \int_S \bar{w} \mathbf{n} \cdot \nabla \bar{p} \, dS \tag{2.6}$$

where $S = S_0 + S_1 + S_2$.

Again, since $\bar{p}$ is an approximate solution, from boundary conditions in equation (2.2)

$$\mathbf{n} \cdot \nabla \bar{p} = \varepsilon_2 \neq 0 \quad \text{on } S_0 \tag{2.7a}$$

$$\mathbf{n} \cdot \nabla \bar{p} + j\rho_0 \omega \mathbf{n} \cdot \mathbf{v} = \varepsilon_3 \neq 0 \quad \text{on } S_1 \tag{2.7b}$$

and

$$\mathbf{n} \cdot \nabla \bar{p} + j\rho_0 \omega \frac{\bar{p}}{z_s} = \varepsilon_4 \neq 0 \quad \text{on } S_2 \tag{2.7c}$$

Multiplying each of the equations (2.7) by the weight $\bar{w}$, integrating over the appropriate part of the surface, adding and equating the weighted average of each error to zero gives

$$\int_S \bar{w} \mathbf{n} \cdot \nabla \bar{p} \, dS = -j\omega \int_{S_1} \rho_0 \bar{w} \mathbf{n} \cdot \mathbf{v} \, dS - j\omega \int_{S_2} \frac{\rho_0}{z_s} \bar{w} \bar{p} \, dS \tag{2.8}$$

Substituting equation (2.8) into equation (2.6) gives

$$\int_V \nabla \bar{w} \cdot \nabla \bar{p} \, dV - \omega^2 \int_V \frac{1}{c_0^2} \bar{w} \bar{p} \, dV + j\omega \int_{S_2} \frac{\rho_0}{z_s} \bar{w} \bar{p} \, dS = -j\omega \int_{S_1} \rho_0 \bar{w} \mathbf{n} \cdot \mathbf{v} \, dS \tag{2.9}$$

All that remains now is to construct an assumed form of an approximate solution $\bar{p}$ to substitute into this equation. This is normally done by assuming $\bar{p}$ to have the form

$$\bar{p} = \sum_{i=1}^n \phi_i(\mathbf{r}) q_i \tag{2.10}$$

where $\phi_i(\mathbf{r})$ are prescribed functions of position $\mathbf{r}$ and q_i are unknown parameters to be determined. The functions $\phi_i(\mathbf{r})$ should be linearly independent and satisfy the essential boundary conditions ($\bar{p}$ prescribed on the boundary). In addition, they should be continuous and have finite first derivatives (see equation (2.9)). Finally, they should form a complete series such as polynomials or trigonometric functions. These conditions ensure that the approximate solution converges as the number of terms n in expression (2.10) increases. For the present application, the weight $\bar{w}$ is taken to be of the form

$$\bar{w} = \sum_{i=1}^{n} \phi_i(\mathbf{r}) w_i \qquad (2.11)$$

Use of this particular form of weighting function is known as the Galerkin method. Substituting expressions (2.10) and (2.11) into equation (2.9) results in a set of equations for the parameters q_i. The most convenient way of constructing the functions $\phi_i(\mathbf{r})$ for practical problems is the finite element method. For convenience, the overbar on the pressure p and weight w will be excluded from the remainder of this chapter.

2.2.3 Finite element formulation

The finite element method is an automatic procedure for constructing the functions $\phi_i(\mathbf{r})$ in the expression (2.10). It also ensures that the parameters q_i represents physical quantities. The volume V is divided into a number of sub-volumes called *elements*, as illustrated in Figure 2.2. Adjacent elements are assumed to be connected at a discrete set of points, usually the vertices of the element. These points are refered to as *nodes*.

The function $\phi_i(\mathbf{r})$ is constructed in a piece-wise manner by prescribing a variation of pressure over each element. The condition that the approximating functions should be continuous and have finite first derivatives can be satisfied by constructing a continuous function over each element and ensuring that there is continuity of pressure between elements. Therefore, the pressure distribution within an element can be approximated by an expression of the form

$$p = \lfloor \mathbf{N}_a(x, y, z) \rfloor_e \{\mathbf{p}\}_e \qquad (2.12)$$

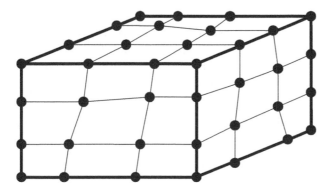

Figure 2.2 Finite element idealisation.

where $\lfloor N_a(x, y, z) \rfloor_e$ is a row matrix of assumed 'shape' functions and $\{p\}_e$ is a column matrix of nodal pressure values for element e. The weight function is expressed in the form

$$w = \lfloor N_a(x, y, z) \rfloor_e \{w\}_e \tag{2.13}$$

The expressions (2.12) and (2.13) are substituted into equation (2.9) and the integrations carried out as the sum of integrations over individual elements. This gives

$$\int_{V_e} \nabla w \cdot \nabla p \, dV = \{w\}_e^T [k_a]_e \{p\}_e \tag{2.14}$$

where V_e is the volume of element e and

$$[k_a]_e = \int_{V_e} [B]_e^T [B]_e \, dV \tag{2.15}$$

and

$$[B]_e = \left\lfloor \frac{\partial}{\partial x} \frac{\partial}{\partial y} \frac{\partial}{\partial z} \right\rfloor^T \lfloor N_a \rfloor_e \tag{2.16}$$

Also

$$\left(\frac{1}{c_0^2}\right) \int_{V_e} wp \, dV = \{w\}_e^T [m_a]_e \{p\}_e \tag{2.17}$$

where

$$[m_a]_e = \left(\frac{1}{c_0^2}\right) \int_{V_e} \lfloor N_a \rfloor_e^T \lfloor N_a \rfloor_e \, dV \tag{2.18}$$

If an element is in contact with either S_1 or S_2, it will contribute to one of the surface integrals in equation (2.9). The variation of pressure over one surface of an element will be denoted by $\lfloor N_a(S) \rfloor_e$. This is obtained from the volume shape function $\lfloor N_a \rfloor_e$. Therefore,

$$\int_{S_e} \left(\frac{\rho_0}{Z_s}\right) w \cdot p \, dS = \{w\}_e^T [c_a]_e \{p\}_e \tag{2.19}$$

where

$$[c_a]_e = \int_{S_e} \left(\frac{\rho_0}{Z_s}\right) \lfloor N_a(S) \rfloor_e^T \lfloor N_a(S) \rfloor_e \, dS \tag{2.20}$$

The normal velocity distribution over an element of the surface S_1, assumed to be identical to one face of an acoustic element, is approximated by

$$\mathbf{n} \cdot \mathbf{v} = \lfloor \mathbf{N}_S \rfloor_e \{\mathbf{v}\}_e \qquad (2.21)$$

where $\{\mathbf{v}\}_e$ is a column matrix of normal velocities at the nodes. Therefore, the final integral is expressed as

$$\int_{S_e} \rho_0 w \mathbf{n} \cdot \mathbf{v} \, dS = \{\mathbf{w}\}_e^{\mathrm{T}} [\mathbf{s}]_e \{\mathbf{v}\}_e \qquad (2.22)$$

where

$$[\mathbf{s}]_e = \int_{S_e} \rho_0 \lfloor \mathbf{N}_a(S) \rfloor_c^{\mathrm{T}} [\mathbf{N}_S]_e \, dS \qquad (2.23)$$

Adding the contributions from each element together and substituting the results into equation (2.9) gives the following equation in terms of global matrices:

$$\{\mathbf{w}\}^{\mathrm{T}} \left[\mathbf{K}_a - \omega^2 \mathbf{M}_a + j\omega \mathbf{C}_a \right] \{\mathbf{p}\} = -j\omega \{\mathbf{w}\}^{\mathrm{T}} [\mathbf{S}] \{\mathbf{v}\} \qquad (2.24)$$

where $\{\mathbf{p}\}$ is a column matrix of nodal pressures for the complete volume. Since $\{\mathbf{w}\}$ is arbitrary, then

$$\left[\mathbf{K}_a - \omega^2 \mathbf{M}_a + j\omega \mathbf{C}_a \right] \{\mathbf{p}\} = -j\omega [\mathbf{S}] \{\mathbf{v}\} \qquad (2.25)$$

This equation can be solved for the nodal pressures for a specified normal velocity distribution. Note that $\mathbf{K}_a, \mathbf{M}_a$ and $\mathbf{S}$ are real matrices, whilst $\mathbf{C}_a$ will be complex if z_s is complex.

2.3 Analysis of cavities with rigid walls

When the complete wall of a cavity is rigid, the portions S_1 and S_2 in Figure 2.1 do not exist. In this case, equation (2.25) reduces to

$$\left[\mathbf{K}_a - \omega^2 \mathbf{M}_a \right] \{\mathbf{p}\} = 0 \qquad (2.26)$$

This equation is a linear eigenvalue problem which can be solved by one of the methods described in Chapter 10. The eigenvalues represent the square of the natural frequencies and the eigenvectors represent the pressure distribution of the acoustic modes of the cavity.

2.3.1   A one-dimensional example: modes of a
closed tube

In order to illustrate the process of performing a finite element calculation, a one-dimensional acoustic cavity is now considered. Such a cavity may be thought of as a narrow tube with the 'rigid wall' boundary condition, meaning that it is closed at both ends. Although the example problem is not one of very great interest, it does serve to demonstrate many aspects of the solution process for more general two- and three-dimensional cases. The results from the finite element model are compared below with the simple sinusoidal solutions for the standing waves in the tube.

An element type with two nodes and linear shape functions is chosen. A diagram of the element is shown in Figure 2.3. The matrix $\lfloor N_a \rfloor$ is therefore given by

$$\lfloor N_a(\xi) \rfloor = \left[\frac{1}{2}(1-\xi) \quad \frac{1}{2}(1+\xi) \right] \tag{2.27}$$

where ξ is a co-ordinate parameter defined only within the element. The element has a length $2a$ and a cross-sectional area A.

For the specific element and choice of shape functions, equation (2.16) gives

$$[B]_e = \frac{\partial}{\partial x} \lfloor N_a \rfloor = \frac{1}{a} \left[\frac{\partial}{\partial \xi} N_{a_i}(\xi) \quad \frac{\partial}{\partial \xi} N_{a_j}(\xi) \right] = \frac{1}{a} \left[-\frac{1}{2} \quad \frac{1}{2} \right] \tag{2.28}$$

In this case, the integration over the element is simple enough in form to be evaluated analytically, so that the element matrices $[k]_e$ and $[m]_e$ are given by equations (2.15) and (2.18) as

$$[k]_e = \int_{-1}^{+1} \frac{A}{a} \begin{bmatrix} -\frac{1}{2} \\ \frac{1}{2} \end{bmatrix} \left[-\frac{1}{2} \quad \frac{1}{2} \right] d\xi = \frac{A}{2a} \begin{bmatrix} 1 & -1 \\ -1 & 1 \end{bmatrix} \tag{2.29}$$

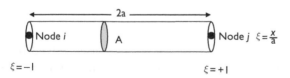

Figure 2.3 Diagram of a two-noded, one-dimensional, acoustic element.

and

$$[m]_e = \int_{-1}^{+1} \frac{Aa}{c_0^2} \begin{bmatrix} \frac{1}{2}(1-\xi) \\ \frac{1}{2}(1+\xi) \end{bmatrix} \begin{bmatrix} \frac{1}{2}(1-\xi) & \frac{1}{2}(1+\xi) \end{bmatrix} d\xi = \frac{Aa}{3c_0^2} \begin{bmatrix} 2 & 1 \\ 1 & 2 \end{bmatrix}$$

(2.30)

Note that the integration over the element has been performed over the local element co-ordinate. The factor of a scales the result of the integration for the actual size of the element; a is found for each element from the nodal co-ordinates. The wave speed c_0 defines the property of the material of the element. It can be specified to have different values in different elements to represent, for example, a variation in temperature. In the one-dimensional acoustic formulation, the cross-sectional area A must be defined. This, too, may be set as an element parameter.

The next step is to relate the pressures at the nodes of a single element to the pressures at the set of nodes for the complete tube. Formally, this may be represented by the application of element transformation matrices (see Chapter 10). However, in practice, for each element, the terms of $[k]_e$ are merely added into the corresponding positions in a *global system matrix* $[K_a]$; that is, a matrix representing the whole cavity with matrix element indices corresponding to the node numbers (for the acoustic analysis case where the field variable, pressure, is a scalar value at each node). This process is called finite element assembly; the terms of $[k]_e$ must be added into the row and column positions to which the element nodes correspond in the global node numbering system. The global matrix $[M_a]$ is constructed from the element matrix $[m]_e$ by the same process.

Figure 2.4 shows a 1 m long tube modelled using four elements. The numbering system for the nodes is shown in the figure. In defining the model of the tube, the node numbers in the global system have to be related to the element-local numbers i and j.

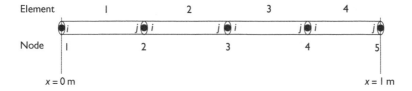

Figure 2.4 Finite element model of a one-dimensional cavity (tube) using four two-noded elements.

This numbering leads to the global matrices

$$[K_a] = \frac{A}{2a} \begin{bmatrix} 1 & -1 & & & \\ -1 & 2 & -1 & & \\ & -1 & 2 & -1 & \\ & & -1 & 2 & -1 \\ & & & -1 & 1 \end{bmatrix}, \quad [M_a] = \frac{Aa}{3c_0^2} \begin{bmatrix} 2 & 1 & & & \\ 1 & 4 & 1 & & \\ & 1 & 4 & 1 & \\ & & 1 & 4 & 1 \\ & & & 1 & 2 \end{bmatrix}$$

(2.31)

The matrices calculated for this simple example display the properties general to finite element system matrices of being symmetric and sparse. With the particular node numbering system used, the matrices have a banded structure (i.e. there are a limited number of off-diagonals of the matrix with non-zero terms beyond which all matrix elements are zero). These properties allow the global matrices of a finite element model to be stored in a fraction of the computer memory space that would be required to store full square matrices of the same order. $[M_a]$ is positive definite and $[K_a]$ is positive semi-definite, so that they are well-conditioned and amenable to standard solution methods.

Eigenvalue and eigenvector solution of the equation

$$([K_a] - \omega^2 [M_a]) \{p\} = 0 \tag{2.32}$$

yields the modal frequencies $f_i = \omega_i/2\pi$ and the mode shapes. The results for the four-element model for $c_0 = 340 \, ms^{-1}$ are shown in Figure 2.5. The analytical solution both for the mode shapes (standing waves) and their corresponding frequencies are shown along with the finite element model results. The lowest eigenvalue to emerge from the model corresponds to a uniform pressure distribution in the tube. The corresponding frequency is zero (allowing for the numerical error). The uniform pressure solution should always be expected from such a model. The mode calculated at 174 Hz is the first mode of the cavity. The exact frequency of this standing wave is 170 Hz. The error in the approximate finite element calculation can be seen to increase with mode order, as the linear shape function approximation to the mode shape also becomes poorer. Since there are five nodes, the global matrices are of order five, and the eigensolution produces five modes. With more elements, the model produces smoother mode shapes that approximate better to the sinusoidal solutions. In this case, the matrices would be of higher order, allowing, in principle, more modes to be found and the analysis to be used to higher frequencies. In practice, only a fraction of the eigenvectors produced by a large finite element model approximate the mode shapes of the physical system closely.

Figure 2.6 shows the percentage error in the calculated modal frequencies for the 1 m tube, as the finite element model is refined by using an increasing

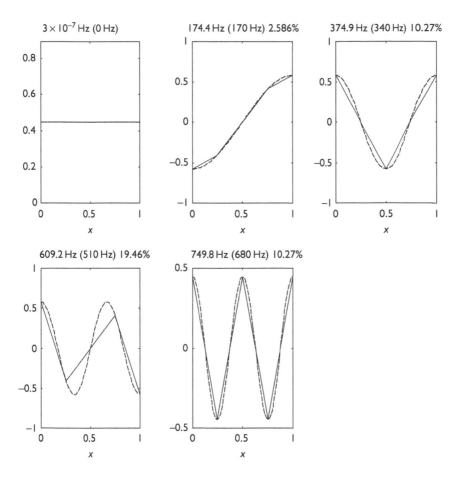

Figure 2.5 Mode shape, and modal frequencies produced by the four-element model compared with the exact results.

number of elements. For any particular mode, the calculated frequency converges as more elements are used. It is a feature of finite element models that the error in the modal frequency is always positive. As a 'rule of thumb' it may be said that for reasonable accuracy, in the finite element approximation method, at least six linear elements per wavelength are required to obtain reasonable accuracy. Fewer elements are not adequate to approximate a mode shape.

2.3.2 A two-dimensional example: modes of a reverberation chamber

Although most analyses in acoustics are for three-dimensional cases, a two-dimensional model can often be used to provide insight into the physical

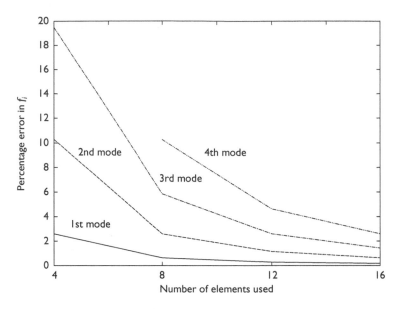

Figure 2.6 Error incurred in the evaluation of the modal frequencies of the 1 m tube for different numbers of elements.

situation. The relatively small amount of work to set up and solve simple models and the shorter time needed to study the results allow the researcher to carry out many analyses with variations of parameters with less effort than that required to set up, run and analyse the results of a single, detailed, three-dimensional model requiring many finite elements.

In particular, for calculating the modes, a two-dimensional model may be used to represent a three-dimensional cavity, if two opposite sides are parallel. At a modal frequency, the pressure distribution along the axis normal to the two parallel sides of the cavity constitutes a standing wave, i.e. if this is the z-direction,

$$p(x, y, z) = p(x, y) \cos\left(\frac{n\pi z}{L_z}\right), \quad n = 0, 1, \ldots \tag{2.33}$$

where L_z is the distance between the parallel sides. Using this to perform the differentiation and the integration with respect to dz, equation (2.9), for a hard walled cavity, becomes,

$$\int_A \nabla_2 w \cdot \nabla_2 p \, dA - \bar{\omega}^2 \int_A \frac{1}{c_0^2} wp \, dA = 0 \tag{2.34}$$

where $\bar{\omega}^2 = \omega^2 - (n\pi c_0/L_z)^2$, $\nabla_2 = (\partial/\partial x, \partial/\partial y)$, n is the order of the standing wave in the z-direction and A is the area in the x–y plane.

An example of a two-dimensional analysis is provided by the case of a reverberation chamber that was built at the Philips Research Laboratories in Eindhoven (van Nieuwland and Weber 1979). For simplicity, and comparability with the original reference, the analysis presented considers only modes where $n = 0$, i.e. the two-dimensional case. In designing a reverberation chamber to promote a diffuse field, the acoustician must examine the frequency distribution of the modes of the room. Examination of the spatial distribution of pressure will warn of the coincidence of possible measurement positions in the room with nodes or anti-nodes of the room modes. To enable such an analysis to be made, a finite element model generating the mode shapes of modes up to a high order may be required.

Figure 2.7 shows the floor plan of the reverberation room of van Nieuwland and Weber. A finite element model has been constructed using two-dimensional quadrilateral elements having four nodes, as illustrated in Figure 2.8. For this element, as is the general case, the analytical integration of the element matrix expressions (equations (2.15) and (2.18)) is not possible and the integrations over the element area are carried out numerically.

The element is mapped onto a *generic element* with a local co-ordinate system (ξ, η). For this purpose, functions are required to describe the variation of the geometry of the true element with the local co-ordinates, i.e. so that the integrand f over the element is

$$f(x, y) = f(x(\xi, \eta), y(\xi, \eta)) = \hat{f}(\xi, \eta) \tag{2.35}$$

In the finite element approach it is convenient to choose the same method of approximating the geometry as that used for the approximation of the

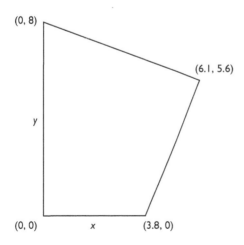

Figure 2.7 Geometry of reverberation room (co-ordinates in metres).

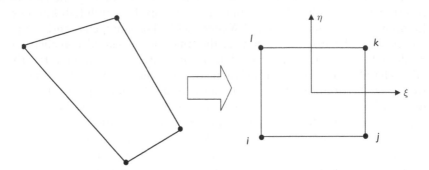

Figure 2.8 Four-noded quadrilateral element.

pressure. In the present case, therefore, the same shape functions are used, i.e. linear shape functions are chosen to approximate the pressure and the value of x and y over these elements, i.e. the elements of the row matrix $\lfloor N_a \rfloor$ (see equation (2.12)) are

$$N_{a_i}(\xi, \eta) = \frac{1}{4}(1 - \xi)(1 - \eta)$$

$$N_{a_j}(\xi, \eta) = \frac{1}{4}(1 + \xi)(1 - \eta)$$

$$N_{a_k}(\xi, \eta) = \frac{1}{4}(1 + \xi)(1 + \eta)$$

$$N_{a_l}(\xi, \eta) = \frac{1}{4}(1 - \xi)(1 + \eta)$$

(2.36)

and

$$x = \lfloor N_a(x, y) \rfloor \{x\}_e$$
$$y = \lfloor N_a(x, y) \rfloor \{y\}_e$$
$$p = \lfloor N_a(x, y) \rfloor \{p\}_e$$

(2.37)

where $\{x\}_e$ is the vector of nodal x-ordinates, etc. An element which makes use of the same shape functions for the geometry and the state variable (here pressure) is said to be *isoparametric*. It is possible, and sometimes more efficient, to use a lower-order shape function for the geometry than for the pressure. In that case the element would be said to be *sub-parametric*.

The integration of a function $f(x, y)$ over an element in local co-ordinates is

$$\int_A f(x, y)\, dx\, dy = \int_A \hat{f}(\xi, \eta)\, |J(\xi, \eta)|\, d\xi\, d\eta$$

(2.38)

where

$$[\mathbf{J}] = \begin{bmatrix} \dfrac{\partial x}{\partial \xi} & \dfrac{\partial y}{\partial \xi} \\[2mm] \dfrac{\partial x}{\partial \eta} & \dfrac{\partial y}{\partial \eta} \end{bmatrix} \tag{2.39}$$

is called the *Jacobian* matrix of the *mapping* or *transformation*. Its terms can be calculated inside a finite element program, simply from the derivatives of the shape functions in a manner similar to equation (2.16), for the derivative terms of $[\mathbf{k}]_e$. The role of $|\mathbf{J}(\xi, \eta)|$ in the two-dimensional analysis is similar to that of the factor a in the one-dimensional example, in that it provides the factor between the integration over the generic element and the actual element.

With this, the calculation for the element matrix (equation (2.18))

$$[\mathbf{m}]_e = \left(\dfrac{1}{c_0^2}\right) \int_{V_e} \lfloor \mathbf{N}_{\mathrm{a}} \rfloor_e^{T} \lfloor \mathbf{N}_{\mathrm{a}} \rfloor_e \mathrm{d}V_e \tag{2.40}$$

becomes

$$[\mathbf{m}]_e = \left(\dfrac{1}{c_0^2}\right) \int_{-1}^{+1} \int_{-1}^{+1} \lfloor \mathbf{N}(\xi, \eta) \rfloor^{T} \lfloor \mathbf{N}(\xi, \eta) \rfloor \, |\mathbf{J}(\xi, \eta)| \, \mathrm{d}\xi \, \mathrm{d}\eta \tag{2.41}$$

where the integration is now carried out over the *generic element*. Evaluation of the terms of $[\mathbf{k}]_e$ is a little more involved because of the spatial derivatives of pressure. The derivatives in expression (2.16) can be expressed in terms of local co-ordinates as follows:

$$\left\lfloor \dfrac{\partial}{\partial x} \dfrac{\partial}{\partial y} \right\rfloor^{T} = [\mathbf{J}]^{-1} \left\lfloor \dfrac{\partial}{\partial \xi} \dfrac{\partial}{\partial \eta} \right\rfloor^{T} \tag{2.42}$$

The integral on the generic element is carried out in a finite element computer program using Gauss-Legendre quadrature (e.g. Kreyszig 1999). For two dimensions

$$\int_{-1}^{+1} \int_{-1}^{+1} g(\xi, \eta) \, \mathrm{d}\xi \, \mathrm{d}\eta \approx \sum_{i=1}^{m} \sum_{j=1}^{n} H_i H_j \, g(\xi_i, \eta_j) \tag{2.43}$$

where ξ_i, η_j $(i = 1, 2, \ldots, m, \; j = 1, 2, \ldots, n)$ are the local co-ordinates of the points at which the integrand is evaluated, and H_i, H_j are the corresponding weights.

As before, once the global matrices $[\mathbf{K}_{\mathrm{a}}]$ and $[\mathbf{M}_{\mathrm{a}}]$ have been calculated, the eigenvalue problem is solved using standard numerical sub-programs. For the reverberation room example, the first four modal frequencies and modal pressure distributions are plotted in Figure 2.9. These have been calculated using a model comprising 108 elements, each having four nodes.

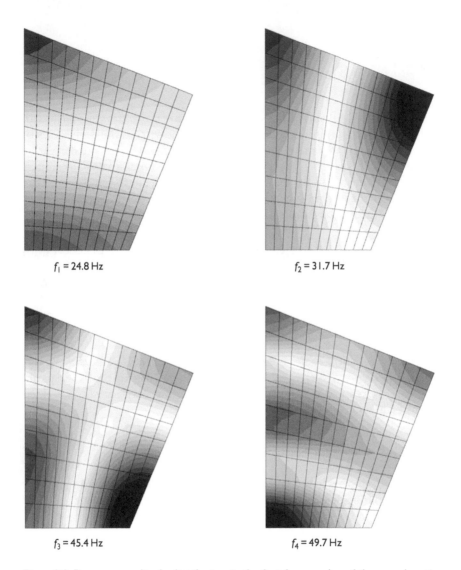

$f_1 = 24.8\,\text{Hz}$

$f_2 = 31.7\,\text{Hz}$

$f_3 = 45.4\,\text{Hz}$

$f_4 = 49.7\,\text{Hz}$

Figure 2.9 Pressure amplitude distribution in the first four modes of the reverberation room (omitting the uniform pressure solution).

2.3.3 A three-dimensional example: modes of a van interior

The method of constructing the global finite element matrices $[K_a]$ and $[M_a]$ for a three-dimensional cavity discretized using three-dimensional elements closely follows that of the two-dimensional example. The description of the process is, therefore, not repeated here.

By using a similar approach to that described above for two dimensions, Petyt *et al.* (1976) set out the theory for a three-dimensional isoparametric element which uses quadratic shape functions. The element, which therefore has 20 nodes, is shown in Figure 2.10.

Using the three-dimensional element, modal analysis of a scale model of a van's interior, 310 mm long, was performed (Petyt *et al.* 1976). The modes of the van have a plane of either symmetry or antisymmetry about the vertical longitudinal centre plane; therefore only half the van's interior was modelled. The finite element model used only eight elements, as shown in Figure 2.11. The nodal lines of the acoustic modes were identified experimentally using a small microphone probe in a model van in the laboratory; the position of these was compared with the mode shapes produced by the finite element analysis. Figure 2.12 shows this comparison for the first six modes. The frequencies of these modes from the finite element analysis and the measurements are compared in Table 2.1. These results show that, despite only using eight elements, differences of only a few per cent are obtained between the measured and the calculated modal frequencies of the cavity.

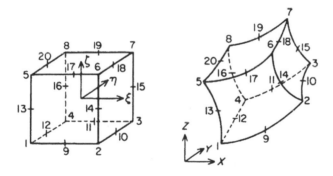

Figure 2.10 Three-dimensional isoparametric element with 20 nodes.

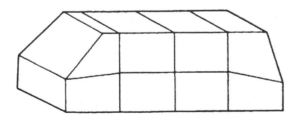

Figure 2.11 Model of half the van interior (reproduced from Petyt *et al.* 1976).

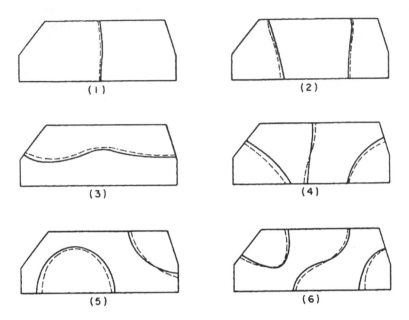

Figure 2.12 Nodal lines of the first six modes of the cavity (- - - - experimental, —— finite element) (reproduced from Petyt *et al.* 1976).

Table 2.1 Comparison of measured and calculated modal frequencies for the model van, symmetrical modes only (Petyt *et al.* 1976)

Mode	Experimental (Hz)	Finite element (Hz)	Percentage difference
1	606	593	−2.1
2	1174	1150	−2.1
3	1549	1556	+0.4
4	1613	1605	−0.5
5	1817	1829	+0.7
6	1992	2026	+1.7

2.3.4   An axisymmetric example: acoustic modes of a cylinder

In the three-dimensional example of Section 2.3.3, use was made of the fact that the cavity, and therefore the pressure field, was symmetrical in order to reduce the number of elements required to analyse the problem. Despite the increase in the memory capacity and processing speed of computers, the efficiency of a numerical model remains an important issue. The usefulness of two-dimensional models has already been noted (Section 2.3.2). In this section, the opportunity is taken of showing the efficiency that can be gained if the geometry of a cavity can be modelled as axisymmetric.

Figure 2.13 shows a closed volume, axisymmetric about the z-axis. Working from the formulation for a three-dimensional cavity expressed in equation (2.9), for the axisymmetric cavity, the integrals over the volume can be carried out for a differential element $dV = r\,dr\,d\theta\,dz$ where θ is the angle about the z-axis. In order to deal with generalised fields in an axisymmetric cavity, a harmonic variation of pressure with θ may be assumed, i.e. $p(r, \theta, z) = p(z, r)\cos n\theta, n = 0, 1, 2, \ldots$. Here, for simplicity we assume axisymmetry, i.e. $n = 0$. Then, for a rigid-walled cavity, equation (2.9) may be rewritten as

$$\int_A \left[\nabla_2 w \cdot \nabla_2 p - \left(\frac{\omega}{c_0}\right)^2 wp \right] r\,dA = 0 \tag{2.44}$$

where $\nabla_2 \equiv ((\partial/\partial z), (\partial/\partial r))$.

From this, it can be seen that a two-dimensional finite element analysis can be applied with only a small modification to the formulation.

To illustrate the use of the axisymmetric formulation, a simple case of the right circular cylindrical cavity shown in Figure 2.14 is analysed. Figure 2.14 shows the mesh of the radial generator area A of the cylinder using, for the sake of the example, a small number of two-dimensional, eight-noded quadrilateral elements (i.e. quadratic shape functions). Table 2.2 compares the calculated frequencies of a number of axisymmetric modes with those derived from an analytical solution for the modes of a cylindrical cavity. It can be seen that the first three predicted modal frequencies are in very close agreement with the analytical solution, whereas higher modal frequencies are predicted with increasing error.

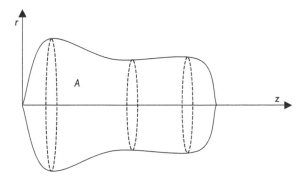

Figure 2.13 Axisymmetric cavity of area A in a radial section.

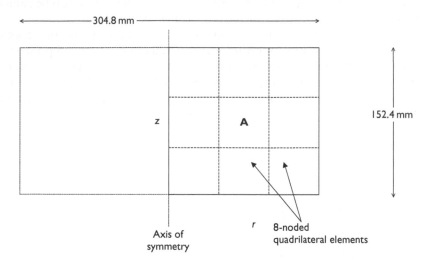

Figure 2.14 Dimensions of the example cylindrical cavity.

Table 2.2 Comparison of the modal frequencies of the cylindrical cavity of Figure 2.14 calculated using the finite element model and by analytical calculation

Mode identification			Frequency by FEM (Hz)	Frequency by analytical calculation (Hz)
m	*n*	*p*		
0	0	1	1116	1115
1	0	0	1361	1361
1	0	1	1760	1759
0	0	2	2255	2231
2	0	0	2513	2491

Notes
m Number of nodal circles.
n Number of nodal diameters.
p Number of longitudinal half-waves.

2.4 An example of the forced response of an acoustic cavity – analysis of a simple duct system

In this section, a simple duct system is analysed, where the normal velocity and impedance boundary conditions that are included in the formulation of equation (2.9) are utilised. This allows a solution for the frequency response to be calculated by solving the matrix equation (2.25).

An important application area for numerical modelling of an interior acoustic field is that of the calculation of the sound transmission loss of a duct. Such a duct may be part of an air conditioning system or an automotive silencer. These are often characterised by a series of chambers with a narrow inlet at one end and an outlet at the other. If the acoustic pressure and particle velocity are assumed to propagate as plane waves in the relatively narrow inlet and outlet tubes, their complex amplitudes may be denoted by p_i, v_i for the inlet and p_o, v_o for the outlet. The acoustic behaviour of the duct can then be determined by allowing the field at the outlet and that at the inlet to be connected using a four-pole parameter relationship in which the terms A_{11}, A_{12}, A_{21} and A_{22} are complex and frequency-dependent.

$$\begin{bmatrix} p_i \\ v_i \end{bmatrix} = \begin{bmatrix} A_{11} & A_{12} \\ A_{21} & A_{22} \end{bmatrix} \begin{bmatrix} p_o \\ v_o \end{bmatrix} \tag{2.45}$$

It can be shown that the transmission loss is calculated from the four pole parameters as

$$TL = 20 \log \left| \frac{1}{2} \left(A_{11} + \frac{A_{12}}{\rho_0 c_0} + A_{21} \rho_0 c_0 + A_{22} \right) \right| \tag{2.46}$$

Modelling the behaviour of such a duct therefore becomes the process of determining the parameters A_{11}, A_{12}, A_{21} and A_{22}. This may be achieved in two steps, first by setting the particle velocity at the outlet v_o to zero (as implied in the finite element formulation) and prescribing a unit velocity at the inlet. The parameters A_{11} and A_{21} are then calculated from the solution as p_i/p_o and $1/p_o$ respectively. The parameters A_{12} and A_{22} are calculated similarly by prescribing the pressure at the outlet p_o to be zero and solving for the velocity there, i.e. $A_{12} = p_i/v_o$ and $A_{22} = 1/v_o$ (with v_i still set to unity).

As an example, Figure 2.15 represents the finite element mesh of an axisymmetric silencer or duct with a single expansion chamber between an inlet and an outlet pipe (after Seybert and Cheng 1987). The elements have eight nodes with quadratic shape functions and the axisymmetric formulation is used. The expansion chamber therefore has the same dimensions and a similar discretization to the example of Section 2.3.4. In the example, all the walls are assumed to be rigid (i.e. $v_n = 0$) but the method can also be used with finite impedance boundary conditions.

Figure 2.16 presents the result of the calculation for the transmission loss of the duct both from the finite element model and from an analytical calculation that considers only plane longitudinal waves (Pierce 1989). The transmission-loss curves are close together at low frequency, and both show a minimum at the first longitudinal half-wavelength resonance of the expansion chamber. At higher frequencies, the plane wave solution displays a similar minimum at the second longitudinal resonance frequency whereas the finite element solution takes into account the full axisymmetric field of the cavity and exhibits minima at the other axisymmetric resonances (compare Table 2.2).

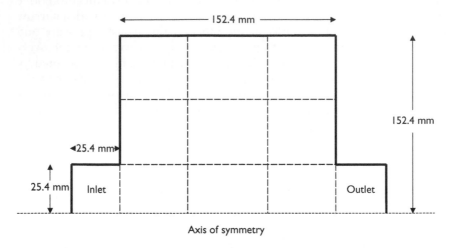

152.4 mm

152.4 mm

25.4 mm

25.4 mm   Inlet

Outlet

Axis of symmetry

Figure 2.15 Model of a simple axisymmetric duct.

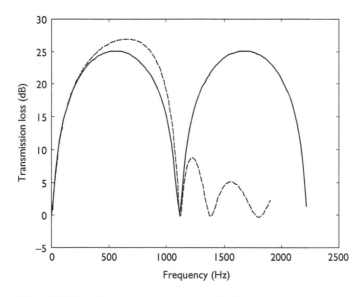

Figure 2.16 Predicted transmission loss for the axisymmetric duct compared with an analytical calculation that takes only longitudinal plane waves into account (- - - - - FEM, —— plane wave).

2.5 Finite element method for exterior problems

2.5.1 Statement of the problem

Consider a vibrating surface S_1 which is immersed in an infinite acoustic medium $V = V_1 + V_2$, as shown in Figure 2.17. Within V, the acoustic pressure p must satisfy the Helmholtz equation,

$$\nabla^2 p + \left(\frac{\omega}{c_0}\right)^2 p = 0 \tag{2.47}$$

and the Sommerfeld radiation condition (Nelson 1998).

$$\lim_{r \to \infty} \left[r \left(\frac{\partial p}{\partial r} + j \left(\frac{\omega}{c_0} \right) p \right) \right] = 0 \tag{2.48}$$

where r is the distance from a source of sound. Also, over the surface S_1, equality of normal velocity gives

$$\mathbf{n} \cdot \nabla p = -j\rho_0 \omega \mathbf{n} \cdot \mathbf{v} \tag{2.49}$$

These equations can easily be modified to represent the scattering of an incident pressure field p_i by a rigid obstacle S_1. If p now represents the scattered pressure, then equations (2.47) and (2.48) remain unchanged. The boundary condition expressed by equation (2.49) becomes

$$\mathbf{n} \cdot \nabla p = -\mathbf{n} \cdot \nabla p_i \tag{2.50}$$

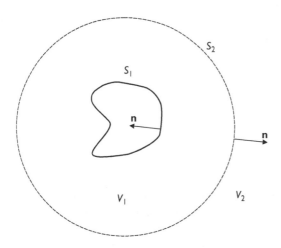

Figure 2.17 Radiation from a vibrating surface.

Therefore, the scattering problem can be treated as an equivalent radiation problem. In all that follows, the analysis is presented for the radiation problem.

2.5.2   Artificial boundary conditions

One way of obtaining an approximate solution to equations (2.47)–(2.49) using finite element methods is to enclose the vibrating surface S_1 completely by a spherical surface S_2 (see Figure 2.17). The radiation condition in equation (2.48) is replaced by an artificial boundary condition over S_2 which is designed to ensure that all acoustic waves propagate freely towards infinity and that no reflections occur at this boundary.

There are two types of boundary condition, local and non-local. A local boundary condition can be expressed as

$$n \cdot \nabla p = -j\rho_0 \omega \frac{p}{z_s} \quad \text{on } S_2 \tag{2.51}$$

where z_s is an assumed specific acoustic impedance. A finite element solution to equations (2.47), (2.49) and (2.51) is obtained by representing the volume V_1, bounded by the surfaces S_1 and S_2 (see Figure 2.17) by an assemblage of finite elements and proceeding, as in Section 2.2.3. The resulting equation is similar to equation (2.25).

If the radius of the sphere S_2 is large enough, the waves crossing it can be assumed to approximate to plane waves. In this case, $z_s = \rho_0 c_0$. Examples of the use of this approach can be found in Kagawa *et al.* (1977). The simplicity of this method of modelling acoustic radiation is offset by the fact that it results in a very large and computationally expensive finite element model. This can be overcome, to a certain extent, by using high-order local boundary conditions (Givoli 1992). Non-local artificial boundary conditions can also be applied to the surface S_2 but are more complicated. These have been surveyed by Givoli (1992). They are usually more accurate than local boundary conditions and guarantee good results with a small computational domain.

2.5.3   Infinite elements

An alternative approach to truncating the unbounded domain V is to model it in its entirety. This is accomplished by modelling V_1 using finite elements, as in Section 2.5.2, and V_2 by a single layer of infinite elements; that is, elements that extend from the surface S_2 to infinity. The pressure within an infinite element is expressed in terms of shape functions with a built-in amplitude decay and wave-like variation to model outgoing waves.

One of the earliest infinite elements was introduced by Bettess and Zienkiewicz (1977) who used it to study diffraction and refraction of surface waves in water. In one dimension, with co-ordinate r, the pressure and weight functions were taken to be of the form

$$p = \sum_{i=1}^{n} \phi_i(r) p_i, \quad w = \sum_{i=1}^{n} \psi_i(r) w_i \tag{2.52}$$

with $n = 3$ and

$$\phi_i(r) = \psi_i(r) = \exp(-jkr) \exp\left(-\frac{r}{L}\right) P_i(r) \tag{2.53}$$

where $k(= \omega/c_0)$ is the wave number, L is a length scale and $P_i(r)$ is a polynomial. The first term in expression (2.53) represents the basic wave shape and the second term the amplitude decay.

The above wave form of decay is inappropriate for acoustics. It should be $(1/r)$ in three dimensions and $(1/r^{1/2})$ in two dimensions. This was achieved by Göransson and Davidsson (1987) by using mapped infinite elements. An infinite element, co-ordinate r, is mapped into a finite element, co-ordinate ξ, using the transformation

$$\xi = 1 - 2\left(\frac{a}{r}\right) \tag{2.54}$$

where

$$a = x_2 - x_1 = x_1 - x_0, \qquad r = x - x_0 \tag{2.55}$$

(see Figure 2.18). This transformation maps a polynomial of the form

$$p = \sum_{i=0}^{n} \alpha_i \xi^i \tag{2.56}$$

into one of the form

$$\sum_{i=0}^{n} \frac{\beta_i}{r^i} \tag{2.57}$$

The form in equation (2.57) was used by Burnett (1994) and Burnett and Holford (1998a,b) when developing infinite elements in prolate and oblate spheroidal and ellipsoidal co-ordinates. In all cases, the weight functions were taken to be the same as the pressure functions. This results in the system matrices being symmetric. However, the integrands in these matrices involve complex exponentials and so special numerical integration techniques have to be used. One of the advantages of these elements is that as more terms in expression (2.57) are included, the infinite elements can

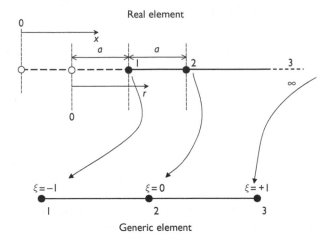

Figure 2.18 One-dimensional infinite element mapping.

intrude more deeply into the near field without loss of accuracy and the extent of the conventional finite element mesh can be reduced.

Another class of elements that have been proposed are referred to as conjugated elements. The first of these (Astley and Eversman 1983) were given the name wave envelope elements. These are elements which are large but finite in size. The pressure functions were taken to be of the form

$$\phi_i(r) = \exp(-jkr)\frac{1}{r}P_i(r) \qquad (2.58)$$

and the weight function is the complex conjugate of this, that is

$$\psi_i(r) = \phi_i^*(r) \qquad (2.59)$$

This choice removes the harmonic spatial variation from the integrals involved in the calculation of the system matrices. This means that Gauss-Legendre quadrature can be used. However, the system matrices are now asymmetric. Later, conjugated mapped infinite elements were proposed by Astley *et al.* (1994, 1998) and Astley (1998). In this case, the weight functions were taken to be

$$\psi_i(r) = \frac{1}{r^2}\phi_i^*(r) \qquad (2.60)$$

The factor $(1/r^2)$ was introduced to ensure that the elements of the system matrices were finite.

Surveys of infinite elements have been carried out by Bettess (1992) and Astley (2000).

2.6 Porous materials

In Section 2.2, sound-absorbing materials were assumed to be locally reacting. However, experience has shown that this assumption is not always valid, especially for thick materials. To overcome this problem, various finite element models have been developed to represent porous materials which are bulk-reacting.

One of the earliest finite element models (Craggs 1978, 1979) was based upon a generalised Rayleigh model of the material (Morse and Ingard 1968). This model assumes that the material is isotropic, the fibres are rigid and the resistivity is the same in all directions. Astley and Cummings (1987) also developed a similar finite element model for porous materials which is both homogeneous and isotropic. Empirical expressions (Delany and Bazley 1970) were used to represent the characteristics of the material.

Göransson (1995) developed a finite element model of a limp porous material by incorporating a simple correction for inertial effects. The derivation is based on the partial differential equations formulated by Ingard (1981). However, the element can only be used if the wavelength of the motion of the fibres is larger than the thickness of the absorbent.

A number of authors have developed finite element models of elastic porous materials based upon the equations developed by Biot (1956). Typical examples are Kang and Bolton (1995, 1996), Easwaran et al. (1996), Atalla et al. (1998) and Göransson (1998). These formulations differ in the choice of nodal degrees of freedom for the fluid, namely pressure, displacement or displacement potential.

2.7 Boundary element methods

The boundary element methods (BEM) can be used as an alternative numerical method to solve the wave equation for an acoustic cavity or for the radiation (or scattering) of an acoustic field by a surface. The formulation for the BEM is not derived from the differential wave equation; it is based on a surface integral equation (Kirkup 1998; Wu 2000). The BEM follows the finite element method in discretizing the domain of the integrals in the formulation and approximating the function of pressure. Initially, the solution for the acoustic pressure is required only on the bounding surface of the acoustic volume. Since the formulation contains terms for the geometry and field only on the boundary, the elements used in the descretization need only describe the geometry, pressure and boundary conditions on this surface. For radiation problems, therefore, the element mesh is finite thus avoiding the problem encountered in the conventional finite element method. Indeed, even for cavity problems, the number of elements required to model a surface is very much lower than that required by the finite element method to model a volume.

There are a number of alternative forms of the BEM, each of which has comparative advantages and disadvantages for the different systems being modelled. Here, the derivation of the simplest formulation, the 'direct method', is described for both the interior and the exterior domain cases in Sections 2.7.1 and 2.7.2. An alternative 'indirect formulation' is discussed in less detail in Section 2.7.3. For more detailed texts on alternative methods, the reader is directed to either Kirkup (1998) or Wu (2000).

A major issue in the application of any of the boundary element formulations for exterior problems is the method of overcoming a mathematical difficulty by which the solution is not uniquely defined at certain frequencies. The way in which the difficulty is overcome, different for each formulation, is a matter for control of the model by the user in addition to the element size considerations that are shared with the finite element method. It is therefore important that the user should understand both the BEM and the physics of the problem to which it is applied in order to construct valid and efficient models.

2.7.1   Direct methods for interior domains

Both the finite element method and the BEM are used to solve the acoustic wave equation. Working in the frequency domain, this reduces the wave equation to the familiar Helmholtz differential equation (2.1) that was the starting point for the derivation of the finite element method in Section 2.2.1. The starting point for the BEM is the integral representation of the wave equation. This is known as the Kirchhoff-Helmholtz integral equation (K-HIE) (Nelson 1998). For a volume V enclosed by the surface S (Figure 2.19)

$$\varepsilon p(\mathbf{x}) = \int_S \{g(\mathbf{x}, \mathbf{y})\nabla_y p(\mathbf{y}) - p(\mathbf{y})\nabla_y g(\mathbf{x}, \mathbf{y})\} \cdot \mathbf{n}\, \mathrm{d}S \qquad (2.61)$$

Note that in equation (2.58), and subsequently, vector notation is used to indicate the co-ordinate position of the points $\mathbf{x}$ and $\mathbf{y}$. The term $g(\mathbf{x}, \mathbf{y})$ is named as 'Green's function' after the nineteenth-century mathematician and physicist George Green. A Green's function is the solution at $\mathbf{x}$ for a unit source at point $\mathbf{y}$ and may also therefore be known as a unit solution or 'kernel function'. Green's function methods of solution are used in many areas of physics, i.e where the Green's function in each case is the unit solution for the problem in hand. The complete statement of the equation therefore includes the definition of the Green's functions.

For two-dimensional problems, the Green's function for the K-HIE is

$$g(\mathbf{x}, \mathbf{y}) = -\frac{j}{4}\, \mathrm{H}_0^{(1)}(kr) \qquad (2.62)$$

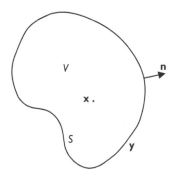

Figure 2.19 Definition of the domain for the direct boundary element formulation with the unit normal vector pointing out of the domain.

where $H_0^{(1)}$ is the Hankel function of the first kind and order 0. For three dimensions, the Green's function is

$$g(\mathbf{x}, \mathbf{y}) = \frac{e^{-jkr}}{4\pi r} \tag{2.63}$$

In both expressions, $r = |\mathbf{x} - \mathbf{y}|$. The coefficient ε of the left-hand side of equation (2.61) results from the integration over a spherical volume enclosing $\mathbf{x}$ in the limit as the volume becomes very small (Nelson 1998). $\varepsilon = 1$ for $\mathbf{x}$ in the volume V and 0 for $\mathbf{x}$ outside V. When $\mathbf{x}$ is on S, the coefficient ε can be shown to be 1/2 if S is smooth. More generally, for a non-smooth surface, it represents the solid angle subtended by the surface at the point $\mathbf{x}$ facing towards the domain.

2.7.1.1 The boundary integral equation

The first main step in deriving the boundary element formulation is to move the point $\mathbf{x}$ to the boundary S, to create a *boundary* integral equation. If the point $\mathbf{x}$ becomes the boundary point $\mathbf{y}'$, then the integral equation becomes

$$\frac{1}{2}p(\mathbf{y}') = \int_S^* \{g(\mathbf{y}', \mathbf{y})\nabla_y p(\mathbf{y}) - p(\mathbf{y})\nabla_y g(\mathbf{y}', \mathbf{y})\} \cdot \mathbf{n}\, dS \tag{2.64}$$

The $\int_S^*$ symbol indicates that special attention must be paid to the integration over the surface since both terms of the integrand are singular when $\mathbf{y}'$ and $\mathbf{y}$ coincide. (Here it is assumed, for simplicity, that the boundary is smooth at $\mathbf{y}'$ so that $\varepsilon = 1/2$.) The first term in the integrand can be expressed more simply as

$$\mathbf{n} \cdot \nabla_y p(\mathbf{y}) = \frac{\partial p}{\partial n} \tag{2.65}$$

Noting that

$$\frac{\partial p}{\partial n} = -j\omega\rho_0 v_n \tag{2.66}$$

where v_n is the normal surface velocity, it can be seen that in problems where the velocity everywhere on the surface is specified as the boundary condition, the first term in the integrand of equation (2.64) is specified and it therefore becomes an equation relating pressures at all points y on the surface to that at the surface point y'.

2.7.1.2 Boundary element discretization

The surface is now represented by an assemblage of elements similar to those used in the finite element method in Section 2.2.3. For the BEM, however, only one-dimensional elements are required for the analysis of a two-dimensional domain, i.e. they only have to represent a segment of the boundary represented by a line. For the analysis of a three-dimensional domain, two-dimensional elements are used to represent 'patches' of the surface.

Within each element, assume that

$$p = \lfloor N_a \rfloor_e \{p\}_e \tag{2.67}$$

and

$$\frac{\partial p}{\partial n} = \lfloor N_a \rfloor_e \left\{\frac{\partial p}{\partial n}\right\}_e \tag{2.68}$$

where, as before, $\lfloor N_a \rfloor_e$ is a row matrix of shape functions defined for a single element. The way in which this piece-wise approximation method may be used to evaluate the terms of the integrals in equation (2.64) has already been described for the finite element method in Section 2.3.2.

When these expressions are substituted into the surface integral equation, and it is evaluated for the point y' being at node i, the result is a row-matrix equation,

$$p_i = \sum_e \lfloor m \rfloor_e \left\{\frac{\partial p}{\partial n}\right\}_e - \sum_e \lfloor d \rfloor_e \{p\}_e \tag{2.69}$$

where

$$\lfloor m \rfloor_e = 2 \int_{S_e} g(y_i, y) \lfloor N_a \rfloor_e \, dS \tag{2.70}$$

and

$$\lfloor d \rfloor_e = 2 \int_{S_e} \frac{\partial g}{\partial n}(y_i, y) \lfloor N_a \rfloor_e \, dS \tag{2.71}$$

The two vectors $\{p\}_e$ and $\{\partial p/\partial n\}_e$ represent the unknown or prescribed pressure and, to within the factor indicated in expression (2.66), normal velocities at the nodes of the element. The integrals of expressions (2.70) and (2.71) are evaluated over S_e, i.e. the part of S represented by the element. For most of the elements, the terms of $[m]_e$ and $[d]_e$ are evaluated using the mapping technique and Gauss-Legendre quadrature that has already been described for the finite element method in Section 2.3.2. When, however, y_i is on the element S_e, the integrand has a singularity, i.e. it has an infinite value at the node i as $r \to 0$ in the Green's function. In this case, a different method for the numerical evaluation of the integral must be used. For the two-dimensional case, the Green's function varies as $\ln(r)$ as $r \to 0$ and the numerical integration is dealt with using a special Gauss quadrature rule that takes account of the logarithmic singularity. In the three-dimensional case, the integrands behave as $1/r$, and this singularity can be removed by taking the node point as the origin and changing to polar co-ordinates so that,

$$dS = r \, dr \, d\theta \tag{2.72}$$

The resulting integral is evaluated using Gauss-Legendre quadrature in r and θ.

The integral over the whole surface is implied in equation (2.69) by the summation over all the elements. In this process, the terms corresponding to the node numbers in the global system must be respected, as in the assembly of finite element matrices. Equation (2.69) might therefore be expressed as the row-matrix equation

$$p_i = \lfloor m \rfloor \left\{ \frac{\partial p}{\partial n} \right\} - \lfloor d \rfloor \{p\} \tag{2.73}$$

where the two vectors $\{p\}$ and $\{\partial p/\partial n\}$ represent the pressure and normal velocity amplitudes at the nodes of the global system, i.e. for the whole surface of, say, N nodes. $\lfloor m \rfloor$ and $\lfloor d \rfloor$ are now row matrices of length N.

Combining the set of equations in equation (2.73) produced by allowing the node i to become, in turn, each of the N nodes of the whole surface, produces a square set of equations

$$\{p\} = [M] \left\{ \frac{\partial p}{\partial n} \right\} - [D] \{p\} \tag{2.74}$$

where $[\mathbf{M}]$ and $[\mathbf{D}]$ are N by N matrices. As can be seen from their derivation, both $[\mathbf{M}]$ and $[\mathbf{D}]$ contain complex terms that are a function of wave number k. Unlike the global matrices of a finite element model, they must therefore be recalculated for every frequency for which the solution is required.

It is at this stage that the boundary conditions are applied. A boundary condition must be known for each part of the boundary represented by the nodal values in either $\{\mathbf{p}\}$ or $\{\partial\mathbf{p}/\partial n\}$. The pressure or the velocity may be prescribed (a rigid wall is set by $v_n = 0$). According to the boundary conditions set at each node of the surface, the equation (2.74) may be rearranged so that the unknowns are left in a single column vector $\{\hat{\mathbf{p}}\}$ and the matrix multiplication with the vector of known boundary condition values is carried out on the right-hand side. This results in a soluble set of global system equation:

$$[\mathbf{A}]\{\hat{\mathbf{p}}\} = \{\mathbf{B}\} \tag{2.75}$$

To illustrate, in the simplest case where the velocity is specified over the whole surface and the solution is required for the pressure, then

$$[\mathbf{A}] = [\mathbf{I}] + [\mathbf{D}] \tag{2.76}$$

and

$$\{\mathbf{B}\} = [\mathbf{M}]\left\{\frac{\partial\mathbf{p}}{\partial n}\right\} = -j\rho_0 c_0 k[\mathbf{M}]\{v_n\} \tag{2.77}$$

where $\{v_n\}$ is the vector of complex amplitudes of normal velocity specified at all the nodes at the wave number k.

2.7.1.3   Calculation of internal pressures

The solution of equation (2.75), when added to the prescribed boundary conditions, produces the complete set of normal velocities and pressures at the nodes of the surface. The K-HIE (2.61) may now be used to evaluate the acoustic pressure at any point in the domain – 'field points'. Simplifying the form of the equation from that of equation (2.61), for $\mathbf{x}$ in V

$$p(\mathbf{x}) = \int_S \left\{ g(\mathbf{x}, \mathbf{y})\frac{\partial p(\mathbf{y})}{\partial n} - p(\mathbf{y})\frac{\partial g(\mathbf{x}, \mathbf{y})}{\partial n} \right\} \, \mathrm{d}S \tag{2.78}$$

Since $p(\mathbf{y})$ and $\partial p(\mathbf{y})/\partial n$ are now known, $p(\mathbf{x})$ can be calculated using Gaussian quadrature routines which do not have to take account of any singularities as $\mathbf{x}$ does not lie on S.

In discretized form

$$p(\mathbf{x}_i) = \sum_e \lfloor \mathbf{m}_r \rfloor_e \left\{ \frac{\partial \mathbf{p}}{\partial n} \right\}_e - \sum_e \lfloor \mathbf{d}_r \rfloor_e \{ \mathbf{p} \}_e \tag{2.79}$$

where

$$\lfloor \mathbf{m} \rfloor_e = \int_{S_e} g(\mathbf{x}_i, \mathbf{y}) \lfloor \mathbf{N}_a \rfloor_e \, \mathrm{d}S \tag{2.80}$$

and

$$\lfloor \mathbf{d} \rfloor_e = \int_{S_e} \frac{\partial g}{\partial n} (\mathbf{x}_i, \mathbf{y}) \lfloor \mathbf{N}_a \rfloor_e \, \mathrm{d}S \tag{2.81}$$

The Gauss-Legendre quadrature method used for non-singular elements is used for evaluating expressions (2.80) and (2.81). However, if the field point lies very close to the discretized surface, the integrands in expressions (2.80) and (2.81) may become nearly singular and the results of the quadrature may become inaccurate. For this reason, where field point pressures are to be evaluated close to the boundary, the surface mesh should locally contain small elements. This is the responsibility of the software user and, as with other mesh-size considerations, may require some experimentation with the model of a particular analysis in order to be certain of the reliability of the results.

Combining the expressions for several field points gives

$$\{ \mathbf{p} \} = [\mathbf{M}_r] \left\{ \frac{\partial \mathbf{p}}{\partial n} \right\}_S - [\mathbf{D}_r] \{ \mathbf{p} \}_S \tag{2.82}$$

where $\{ \mathbf{p} \}_S$ and $\{ \partial \mathbf{p} / \partial n \}_S$ now represent the complete set of pressures and normal derivatives calculated from the particle velocities at each of the N nodes of the surface. For a number M of field points $[\mathbf{M}_r]$ and $[\mathbf{D}_r]$ will have dimensions M by N. Since there are only matrix multiplications and no matrix solution steps in the evaluation of field point pressures, a large number of field points may be evaluated at a computational cost which is small compared with that of the surface values solution.

2.7.2 Radiation and scattering

It is a simple matter to demonstrate that the boundary element formulation can be used for the solution of radiation problems. If the domain V is defined, as in Figure 2.20, such that it is bounded by two surfaces S_1 and S_2, then the outer surface may be enlarged and in the limit taken to infinity. Any physical acoustic field arising from a source of finite extent radiates a

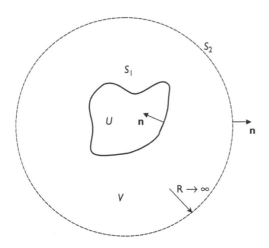

Figure 2.20 Radiation from a vibrating surface.

field which behaves like that of a point source at very large distances (R in Figure 2.20) so that

$$p \to \frac{e^{-jkR}}{R} \quad \text{as } R \to \infty \tag{2.83}$$

This is a statement of the Sommerfeld radiation condition. It can be shown that

$$\lim_{R \to \infty} \int_{S_2} \left\{ g(\mathbf{x}, \mathbf{y}) \frac{\partial p(\mathbf{y})}{\partial n} - p(\mathbf{y}) \frac{\partial g(\mathbf{x}, \mathbf{y})}{\partial n} \right\} \, dS_2 = 0 \tag{2.84}$$

That is, the outer part of the surface that has been taken to infinity contributes nothing to the integral and therefore the surface S_1 can be taken to be the surface for the purpose of equation (2.61), the boundary integral equation, with V external to it.

2.7.2.1 Non-uniqueness of the solution at certain frequencies

A problem arises with the boundary element formulation for the radiation problem because of the fact that the integral terms in the boundary integral equation (2.64) are common to both the interior and the exterior formulations, with only the direction of the normal being changed. The interior problem with $p = 0$ at all points on S is characterised by undamped modes, and at the corresponding wave numbers, equation (2.64) becomes singular for internal and external problems alike. When solving for radiation problems, where no modal behaviour is expected, the result is merely a breakdown of the solution at the problem wave numbers/frequencies. Since the global matrices in equation (2.74) become poorly conditioned close to these wave numbers, an inaccurate solution is inevitably obtained for a finite frequency range around the problem frequencies. For an arbitrarily shaped surface, the modes of the interior problem become more closely spaced at high frequencies and it becomes impossible to obtain accurate results at all. It should be emphasised that the difficulty that arises is purely mathematical and has no physical analogue.

A number of methods have been developed to overcome the non-uniqueness of the solution. One such method devised early on (Burton and Miller 1971) uses a second formulation produced by taking the derivative of equation (2.64) with respect to the normal vector **n**. The normal-derivative integral equation has the same solution for the external problem but has a set of non-unique frequencies which are different from those of the standard formulation, since they correspond to modes with $v_n = 0$. An improved formulation is produced by the linear combination of the standard and normal-derivative integral equations. This allows the solution of the external problem while mutually excluding both sets of non-unique frequencies. It has been shown that the use of the factor of j/k to combine the formulations results in a stable calculation at all frequencies; however, the normal-derivative integral equation contains strongly singular terms which require careful treatment before reduction of the formulation to a computable form.

A simpler technique, proposed by Schenck (1968) has been more popular amongst software writers. In this method, a number of points x_i are created in the interior domain where a zero pressure is required. That is, for a point x_i in the interior, U (Figure 2.20) defined as 'outside' V,

$$0 = \int_S \left\{ g(\mathbf{x}_i, \mathbf{y}) \frac{\partial p}{\partial n}(\mathbf{y}) - p(\mathbf{y}) \frac{\partial g(\mathbf{x}_i, \mathbf{y})}{\partial n} \right\} \, dS \tag{2.85}$$

which, when discretized, becomes

$$0 = \sum_e \lfloor \mathbf{m}_i \rfloor_e \left\{ \frac{\partial \mathbf{p}}{\partial n} \right\}_e - \sum_e \lfloor \mathbf{d}_i \rfloor_e \{ \mathbf{p} \}_e \tag{2.86}$$

where

$$\lfloor \mathbf{m}_i \rfloor_e = \int_{S_e} g(\mathbf{x}_i, \mathbf{y}) \lfloor \mathbf{N}_a \rfloor_e \, dS \tag{2.87}$$

$$\lfloor \mathbf{d}_i \rfloor_e = \int_{S_e} \frac{\partial g(\mathbf{x}_i, \mathbf{y})}{\partial n} \lfloor \mathbf{N}_a \rfloor_e \, dS \tag{2.88}$$

Combining equations for a number M of interior points gives

$$0 = [\mathbf{M}_i] \left\{ \frac{\partial \mathbf{p}}{\partial n} \right\} - [\mathbf{D}_i] \{\mathbf{p}\} \tag{2.89}$$

or

$$0 = -j\rho_0 c_0 k [\mathbf{M}_i] \{\mathbf{v}_n\} - [\mathbf{D}_i] \{\mathbf{p}\} \tag{2.90}$$

The combination of the global matrices representing the direct formulation, equation (2.74), and those representing the internal collocation points, equation (2.89), gives

$$\left[\overline{\mathbf{A}} \right] \{\mathbf{p}\} = \left[\overline{\mathbf{B}} \right] \tag{2.91}$$

where

$$\left[\overline{\mathbf{A}} \right] = \begin{bmatrix} \mathbf{I} + \mathbf{D} \\ \mathbf{D}_i \end{bmatrix}$$

$$\left[\overline{\mathbf{B}} \right] = \begin{bmatrix} \mathbf{M} \\ \mathbf{M}_i \end{bmatrix} \left\{ \frac{\partial \mathbf{p}}{\partial n} \right\} = -j\rho_0 c_0 k \begin{bmatrix} \mathbf{M} \\ \mathbf{M}_i \end{bmatrix} \{\mathbf{v}_n\} \tag{2.92}$$

Equations (2.91) and (2.92) illustrate the origin of an alternative name for the Schenck method, namely the 'CHIEF' method (Combined Helmholtz Integral Equation Formulation). For M interior points and N surface points the dimensions of the matrices $[\overline{\mathbf{A}}]$ and $[\overline{\mathbf{B}}]$ are $(M+N)$ by N; i.e. there are now more equations than unknowns. The system can be solved using a least squares method rather than a straightforward Gaussian solution.

There are no rules about the number or position of collocation points to use. In order to rule out high-order internal mode shapes, a number of collocation points should be used and these should not be regularly spaced as they will be ineffective if they coincide with a nodal pressure surface of a possible mode shape. For a known simple shape of cavity, it is possible to anticipate both the non-unique frequencies and the pressure distribution of the associated internal mode shape so that collocation points may be well placed. For an arbitrary shape of surface, however, the placement of the collocation points is again a matter for experimentation, testing and validation of a boundary element model to the satisfaction of the modeller. One consideration is that the number of internal collocation points used should be a small fraction of the number of nodes of the radiating surface. Unless this is so, the equations associated with the internal collocation points, represented by $[\mathbf{M}_i]$ and $[\mathbf{D}_i]$, are likely to distort the results of the least squares solution of the combined set of global equations away from that of the actual radiation problem, represented by $[\mathbf{M}]$ and $[\mathbf{D}]$.

A simple example of the use of the method is provided by an analysis of radiation from a pulsating cylinder. This is represented by a two-dimensional model consisting of 32 two-noded linear elements (Figure 2.21). At each of the nodes, a radial unit velocity is prescribed.

Rather than predicting the pressure at particular field points with this model, the radiation ratio σ has been calculated from the surface solution. σ is defined as

$$\sigma = \frac{W_{\mathrm{rad}}}{\left\langle \overline{v_n^2} \right\rangle S \rho_0 c_0} \tag{2.93}$$

where W_{rad} is the radiated sound power, $\left\langle \overline{v_n^2} \right\rangle$ is the space-averaged mean square normal vibration velocity of the surface and S is the surface area. The calculation of σ illustrates an effective way to use boundary element calculations in practice to inform other modelling methods. The radiation ratio has been produced for the model with and without a single internal collocation point on the axis of the cylinder shown in Figure 2.21. The results are shown in Figure 2.22. The general shape of the curve shows the expected convergence of the radiation ratio towards unity for high frequency and shows, below the critical frequency, a slope of 10 dB per decade that is characteristic of a line monopole source.

For the cylinder, the non-unique frequencies correspond to the zeros of the Bessel function $J_0(ka)$, where k is the wave number. These frequencies are indicated by the vertical lines on Figure 2.22. Before the internal collocation point is added, the solution shows a spurious peak and dip at each of the non-unique frequencies. There is no significance attached to the shape of these features, as the error in the solution of the

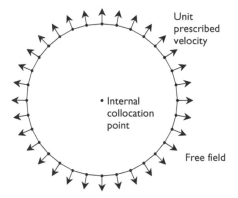

Figure 2.21 Two-dimensional boundary element model of a pulsating cylinder of radius 0.05 m comprising 32 two-noded line elements.

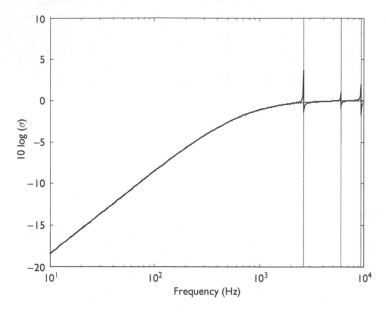

Figure 2.22 Radiation ratio for a pulsating cylinder of radius 0.05 m calculated with (······) and without (——) a single internal collocation point.

poorly conditioned matrices depends on the implementation of floating point arithmetic and the solution method used. After the internal collocation point is added, the solution can be seen to go smoothly through these frequencies. In this particular case, the single collocation point produces a good solution throughout the frequency range shown. The dotted line in Figure 2.22 would be indistinguishable on this scale of graph from the analytical solution (Beranek and Vér 1992), i.e.

$$\sigma = \frac{2}{\pi k a \left| H_1^{(1)}(ka) \right|^2} \tag{2.94}$$

where $H_1^{(1)}$ is the Hankel function of the first kind and of order one.

2.7.3 Indirect methods

Two fundamental approaches to the BEM have been developed. The direct BEM, studied already in Sections 2.7.1 and 2.7.2, is formulated from the K-HIE, itself derived from the wave equation by application of

Green's theorem (Nelson 1998). The other approach is called the indirect method. It has developed from the physical insight from electrostatics that the solution for a field can be expressed in terms of a source density function on the boundary described in terms of 'single' and 'double layer potentials'. In acoustics, the single-layer potential corresponds to a difference across the boundary in the normal gradient of pressure and the double-layer potential as the difference in acoustic pressure across the boundary. The jump in gradient of pressure or particle velocity across the surface can be regarded as due to a distribution of monopole sources and the difference in acoustic pressure as a distribution of dipole sources. The difference or 'jump' quantities then become the primary variables of the formulation which can then be rewritten in the form

$$\varepsilon p(\mathbf{x}) = \int_S \left\{ \sigma(\mathbf{y}) g(\mathbf{x}, \mathbf{y}) - \mu(\mathbf{y}) \frac{\partial g(\mathbf{x}, \mathbf{y})}{\partial n} \right\} dS \tag{2.95}$$

where $\sigma(\mathbf{y})$ represents the jump in the normal gradient of pressure and $\mu(\mathbf{y})$ the jump in acoustic pressure across the boundary (Filippi 1977; Saylii *et al.* 1981). ε has the same definition as for equation (2.61).

The equivalence of this statement (equation (2.95)) of the acoustic problem to the K-HIE (equation (2.61)) can be demonstrated as follows. Consider a thin structure such that the surface can be split into two parts $S = S^+ + S^-$, as shown in Figure 2.23. The K-HIE becomes

$$\varepsilon p(\mathbf{x}) = \int_{S^+} \left\{ g(\mathbf{x}, \mathbf{y}^+) \frac{\partial p(\mathbf{y}^+)}{\partial n^+} - p(\mathbf{y}^+) \frac{\partial g(\mathbf{x}, \mathbf{y}^+)}{\partial n^+} \right\} dS$$
$$+ \int_{S^-} \left\{ g(\mathbf{x}, \mathbf{y}^-) \frac{\partial p(\mathbf{y}^-)}{\partial n^-} - p(\mathbf{y}^-) \frac{\partial g(\mathbf{x}, \mathbf{y}^-)}{\partial n^-} \right\} dS \tag{2.96}$$

where, now, $\mathbf{y}^+$ represents one face of the thin body and $\mathbf{y}^-$ represents the other.

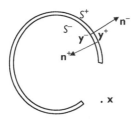

Figure 2.23 Representation of a thin radiating body with the boundary divided into two parts S^+ and S^-.

Neglecting a rigorous treatment of what happens at the non-smooth edges of the body, it can be seen that, in the limit as the thickness of the structure is reduced to zero,

$$\left.\begin{aligned} S^+ &= S^- = \bar{S} \\ \mathbf{y}^+ &= \mathbf{y}^- = \bar{\mathbf{y}} \\ \mathbf{n}(\mathbf{y}^+) &= -\mathbf{n}(\mathbf{y}^-) = \bar{\mathbf{n}} \\ g(\mathbf{x}, \mathbf{y}^+) &= g(\mathbf{x}, \mathbf{y}^-) \\ \frac{\partial g(\mathbf{x}, \mathbf{y}^+)}{\partial n^+} &= -\frac{\partial g(\mathbf{x}, \mathbf{y}^-)}{\partial n^-} \end{aligned}\right\} \tag{2.97}$$

and if we define the jump in the spatial derivative of pressure and the jump in pressure across the thin surface as

$$\sigma(\bar{\mathbf{y}}) = \frac{\partial p(\mathbf{y}^+)}{\partial \bar{n}} - \frac{\partial p(\mathbf{y}^-)}{\partial \bar{n}} \quad \text{and} \quad \mu(\bar{\mathbf{y}}) = p(\mathbf{y}^+) - p(\mathbf{y}^-) \tag{2.98}$$

then

$$p(\mathbf{x}) = \int_S \left\{ \sigma(\bar{\mathbf{y}}) g(\mathbf{x}, \bar{\mathbf{y}}) - \mu(\bar{\mathbf{y}}) \frac{\partial g(\mathbf{x}, \bar{\mathbf{y}})}{\partial \bar{n}} \right\} \, dS \tag{2.99}$$

which is the same as equation (2.95) for $\mathbf{x}$ in the domain.

Taking the point $\mathbf{x}$ to the boundary gives (dropping the bar notation)

$$\varepsilon p(\mathbf{y}') = \int_S \left\{ \sigma(\mathbf{y}) g(\mathbf{y}', \mathbf{y}) - \mu(\mathbf{y}) \frac{\partial g(\mathbf{y}', \mathbf{y})}{\partial n} \right\} \, dS \tag{2.100}$$

and taking the gradient of this equation and making use of the boundary condition

$$\mathbf{n} \cdot \nabla p(\mathbf{y}') = -j\rho_0 \omega v_n(\mathbf{y}') \tag{2.101}$$

gives a second equation

$$\varepsilon(-j\rho_0 \omega v_n(\mathbf{y}')) = \mathbf{n} \cdot \nabla_{\mathbf{y}'} \int_S \left\{ \sigma(\mathbf{y}) g(\mathbf{y}', \mathbf{y}) - \mu(\mathbf{y}) \frac{\partial g(\mathbf{y}', \mathbf{y})}{\partial n} \right\} \, dS \tag{2.102}$$

Equation (2.100) describes a formulation for parts of the surface on which the pressure is prescribed as a boundary condition. This part of the surface may be labelled S_1. The equation relates the variables $\sigma(\mathbf{y})$ and $\mu(\mathbf{y})$ inside the integral terms of the right-hand side of the equation to the known values of pressure on the surface. Equation (2.102) describes a formulation dealing with parts S_2 of the surface on which the velocity is prescribed. It relates the known values of velocity to the primary variables. Impedance boundary conditions can be treated by the development of further equations

(Vlahopoulos 2000). For simplicity, this further development is omitted here.

The equations (2.100) and (2.102) have to be solved to find $\sigma(y)$ and $\mu(y)$ on all parts of the surface. This is done by linking the two equations for S_1 and S_2 via the method of variational calculus in which the solution function is found by minimising a functional summed for the two parts of the surface (Coyette and Fyfe 1989; Vlahopoulos 2000).

The discretization of the resulting formulation using the finite element approximation technique

$$\sigma_e \approx \lfloor \mathbf{N}_\sigma \rfloor_e \{\boldsymbol{\sigma}\} \quad \text{and} \quad \mu_e \approx \lfloor \mathbf{N}_\mu \rfloor_e \{\boldsymbol{\mu}\} \tag{2.103}$$

leads to a global matrix equation of the form (Coyette and Fyfe 1989)

$$\begin{bmatrix} \mathbf{B} & \mathbf{C} \\ \mathbf{C}^{\mathrm{T}} & \mathbf{D} \end{bmatrix} \begin{Bmatrix} \boldsymbol{\sigma} \\ \boldsymbol{\mu} \end{Bmatrix} = \begin{Bmatrix} \mathbf{f}_\sigma \\ \mathbf{f}_\mu \end{Bmatrix} \tag{2.104}$$

from which the values of $\{\boldsymbol{\sigma}\}$ and $\{\boldsymbol{\mu}\}$ all over the surface are found. The matrices $\mathbf{B}$, $\mathbf{C}$ and $\mathbf{D}$ are functions of wavenumber, and $\mathbf{f}_\sigma$ and $\mathbf{f}_\mu$ arise from known boundary conditions. The pressure field may be solved by application of these into the discretized form of equation (2.95).

An advantage of the method over the direct formulation is that the solution is carried out simultaneously for the interior and the exterior domains and can be used, as implied by Figure 2.23, for thin structures where the interior and the exterior domains are connected. In this case, the surface S has an edge, and an extra constraint condition must be applied to ensure the jump of pressure and particle velocity is zero at the edge. With the application of further constraint conditions, the method can be made to apply to thin structures with multiple connections (Vlahopoulos 2000).

Like the direct formulation for exterior domains, the indirect formulation for the radiation problem also suffers from conditioning of the system matrices at certain frequencies because of the modal solution of the interior. In this case, the interior modes are damped out by adding an absorptive boundary condition to a surface on the interior. An extra-false interior surface consisting of a few elements is usually created for this purpose.

2.7.4 Example: prediction of the radiation ratio for a rail

It has already been shown that the BEM may be put to use to produce radiation ratios for components in vibroacoustic calculations. Here, a final illustrative example is presented. The radiation ratio is required for a vertically vibrating rail. It is known from calculations of the structural mechanics of the rail that, for frequencies above about 200 Hz, the wavelengths of vertical vibration in the rail are long compared to the wavelength in air and that there is a low rate of decay of amplitude of the vertical vibration of the rail along its length. The radiation of sound from the rail may therefore be treated as a two-dimensional problem. Figure 2.24 presents three plausible boundary element model surfaces for radiation from the rail. Figure 2.24a shows the rail in a free field. Figure 2.24b shows the rail attached to a surface, such as the sleeper, on which it rests via an elastomeric pad over at least some of its length. The hard reflecting surface of the sleepers is incorporated as a symmetry boundary condition in the boundary element formulation so that the rail therefore radiates into a half-space. A third possible model, Figure 2.24c, has the rail radiating immediately above a reflecting surface. This may be thought of as representing the condition between the sleeper supports.

The rail is modelled using 72 two-noded elements, and a unit velocity in the vertical direction is set as the boundary condition at all the nodes of the rail. This is resolved in the normal direction to the rail in order to prescribe v_n. The problem has been solved using the indirect BEM.

The results for the three models are presented in Figure 2.25. As for the cylindrical radiator of similar cross-sectional dimensions that was chosen for the example of Figure 2.22, above the critical frequency around 1000 Hz, the radiation ratio converges to a value around unity (0 dB). However, each of the rail models in this frequency range exhibiting significant peaks and

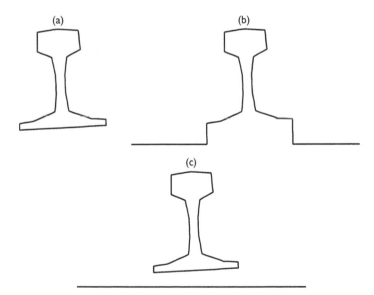

Figure 2.24 Three possible boundary element models for two-dimensional radiation from a rail (172 mm high rail section).

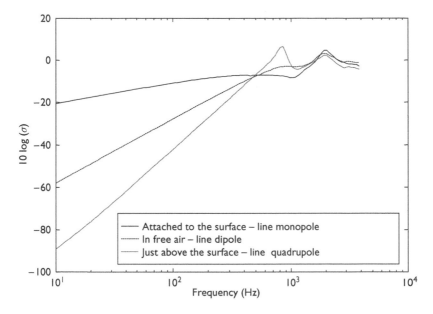

Figure 2.25 Radiation ratios calculated using three different two-dimensional models for radiation from a rail.

dips in the radiation ratio can be seen due to the effects of constructive and destructive interference between sound radiated from the 'rail head' and the 'rail foot'. At low frequency, the radiation ratio for the rail model of Figure 2.24b exhibits a slope with frequency of 10 dB per decade. This is characteristic of the radiation from a line monopole source that may also be seen in Figure 2.22. For the rail in the free field (Figure 2.24a) the slope is characteristic of a line dipole radiator, and for model (Figure 2.24c) it is close to that of a line quadrupole. The choice of model between three reasonably plausible options leads to a great difference in the estimation of radiated sound level from the rail. The example serves as a warning that a powerful technique to solve the acoustic field does not diminish the need for a good physical understanding, and perhaps a practical investigation, of the problem under consideration.

2.8 Structural–acoustic interaction

In previous sections, it has been assumed that the normal velocity distribution of the bounding surface of the region of interest is known *a priori*. This is rarely the case. If this bounding surface is the surface of a vibrating structure, then its normal velocity distribution can be predicted using finite element techniques for the structure. Following the methods described in

Chapter 10, assuming harmonic motion of frequency ω, the equations of motion of the structure take the form

$$\left[K - \omega^2 M + j\omega D\right]\{u\} = \{f\} \tag{2.105}$$

where $\{u\}$ is a column matrix of nodal displacements, $[K]$, $[M]$ and $[D]$ the stiffness, inertia and damping matrices, and $\{f\}$ a column matrix of equivalent nodal forces. Equation (2.105) can be solved for the nodal displacement $\{u\}$ at each frequency of interest. However, this column matrix represents the displacement at all the nodes in the model. It is therefore necessary to extract the components of displacement at the surface nodes, from which the normal component of displacement, and hence velocity (by multiplying by $j\omega$), can be determined. The acoustic response can then be obtained by substituting the normal velocity distribution into equations (2.25), (2.75), (2.91) or (2.104), depending on which method is being used.

In many situations, however, the acoustic pressures affect the structural response, which in turn affects the acoustic pressures. In this case, the simple uncoupled analysis, described above, cannot be used. Separating out the mechanical forces $\{f_m\}$ and the forces exerted by the acoustic medium on the structure $\{f_a\}$, equation (2.105) becomes

$$\left[K - \omega^2 M + j\omega D\right]\{u\} = \{f_m\} + \{f_a\} \tag{2.106}$$

Considering the virtual work of the acoustic pressures acting on a surface element of the structure shows that the equivalent nodal forces on an individual element are

$$\{f_a\}_e = \int_{S_e} \lfloor N_s(S) \rfloor_e^T \lfloor N_s(S) \rfloor_e \, dS \, \{p\}_e = \frac{1}{\rho_0} [S]_e^T \{p\}_e \tag{2.107}$$

Therefore, for the complete structure

$$\{f_a\} = \frac{1}{\rho_0} [S]^T \{p\} \tag{2.108}$$

Substituting equation (2.104) into (2.102) gives

$$\left[K - \omega^2 M + j\omega D\right]\{u\} - \frac{1}{\rho_0} [S]^T \{p\} = \{f_m\} \tag{2.109}$$

This equation can be solved in conjunction with either of the equations (2.25), (2.75), (2.91) or (2.104) (noting that $\{v\} = j\omega\{u\}$), for the structural response $\{u\}$ and the acoustic pressure $\{p\}$.

Further details and examples are given by Wilton (1978), Mathews (1986), Coyette and Fyfe (1989), Everstine and Henderson (1990), Jeans and Mathews (1990), Fyfe et al. (1991) and Roozen (1992).

2.9 Computer programs

An increasing number of developers of finite element computer programs for structural analysis are including finite and boundary element capabilities for acoustic analysis. There is also a trend for independent programs for acoustic analysis using the finite element and/or boundary element techniques, with an interface to a commercial structural finite element program. The techniques outlined in this chapter are therefore becoming increasingly useful in the industrial design process.

References

Astley, R.J. (1998) Mapped spheroidal wave-envelope elements for unbounded wave problems. *International Journal for Numerical Methods in Engineering*, **41**, 1235–1254.

Astley, R.J. (2000) Infinite elements for wave problems: a review of current formulations and an assessment of accuracy. *International Journal for Numerical Methods in Engineering*, **49**, 951–976.

Astley, R.J. and Cummings, A. (1987) A finite element scheme for attenuation in ducts lined with porous material: comparison with experiment. *Journal of Sound and Vibration*, **116**, 239–263.

Astley, R.J. and Eversman, W. (1983) Finite element formulations for acoustical radiation. *Journal of Sound and Vibration*, **88**, 47–64.

Astley, R.J., Macaulay, G.J. and Coyette, J.P. (1994) Mapped wave envelope elements for acoustical radiation and scattering. *Journal of Sound and Vibration*, **170**, 97–118.

Astley, R.J., Macaulay, G.J., Coyette, J.P. and Cremers, L. (1998) Three-dimensional wave-envelope elements of variable order for acoustic radiation and scattering. Part I: Formulation in the frequency domain. *Journal of the Acoustical Society of America*, **103**, 49–63.

Atalla, N., Panneton, R. and Debergue, P. (1998) A mixed displacement-pressure formulation for porelastic materials. *Journal of the Acoustical Society of America*, **104**, 1444–1452.

Beranek, L.L. and Vér, I.L. (eds) (1992) *Noise and vibration control engineering* (New York: John Wiley & Sons).

Bettess, P. (1992) *Infinite Elements* (Sunderland: Penshaw Press).

Bettess, P. and Zienkiewicz, O.C. (1977) Diffraction and reflection of surface waves using finite and infinite elements. *International Journal for Numerical Methods in Engineering*, **11**, 1271–1290.

Biot, M.A. (1956) Theory of propagation of elastic waves in a fluid-saturated porous solid. Part I. Low frequency range. *Journal of the Acoustical Society of America*, **28**, 168–178.

Burnett, D.S. (1994) A three-dimensional acoustic infinite element based on a prolate spheroidal multipole expansion. *Journal of the Acoustical Society of America*, **96**, 2798–2816.

Burnett, D.S. and Holford, R.L. (1998a) Prolate and oblate spheroidal acoustic infinite elements. *Computer Methods in Applied Mechanics and Engineering*, **158**, 117–141.

Burnett, D.S. and Holford, R.L. (1998b) An ellipsoidal acoustic infinite element. *Computer Methods in Applied Mechanics and Engineering*, **164**, 49–76.

Burton, A.J. and Miller, G.F. (1971) The application of integral equation methods to the numerical solution of some exterior boundary value problems. *Proceedings of the Royal Society*, London, **A323**, 201–210.

Coyette, J.P. and Fyfe, K.R. (1989) Solutions of elasto-acoustic problems using a variational finite element/boundary element technique. In R.J. Bernard and R.F. Keltie (eds), *Numerical Techniques in Acoustic Radiation* (ASME NCA – Volume 6), pp. 15–25.

Craggs, A. (1978) A finite element model for rigid porous absorbing materials. *Journal of Sound and Vibration*, **61**, 101–111.

Craggs, A. (1979) Coupling of finite element acoustic absorption models. *Journal of Sound and Vibration*, **66**, 605–613.

Craik, R.J.M. (1996) *Sound Transmission Through Buildings using Statistical Energy Analysis* (Aldershot, UK: Gower).

Delany, M.E. and Bazley, E.N. (1970) Acoustical properties of fibrous absorbent materials. *Applied Acoustics*, **3**, 105–116.

Easwaran, V., Lauriks, W. and Coyette, J.P. (1996) Displacement-based finite element method for guided wave propagation problems: application to porelastic media. *Journal of the Acoustical Society of America*, **100**, 2989–3002.

Everstine, G.C. and Henderson, F.M. (1990) Coupled finite element/boundary element approach for fluid–structure interaction. *Journal of the Acoustical Society of America*, **87**, 1938–1947.

Filippi, P.J.T. (1977) Layer potentials and acoustic diffraction. *Journal of Sound and Vibration*, **54**(4), 473–500.

Finlayson, B.A. (1972) *The Method of Weighted Residuals and Variational Principles* (London: Academic Press).

Fyfe, K.R., Coyette, J.P. and Van Vooren, P.A. (1991) Acoustic and elasto-acoustic analysis using finite element methods. *Sound and Vibration*, **25**(12), 16–22.

Givoli, D. (1992) *Numerical Methods for Problems in Infinite Domains* (Amsterdam: Elsevier Science).

Göransson, P. (1995) Acoustic finite element formulation of a flexible porous material – a correction for inertial effects. *Journal of Sound and Vibration*, **185**, 559–580.

Göransson, P. (1998) A three-dimensional, symmetric, finite element formulation of the Biot equations with application to acoustic wave propagation through an elastic porous medium. *International Journal for Numerical Methods in Engineering*, **41**, 167–192.

Göransson, P. and Davidsson, C.F. (1987) A three-dimensional infinite element for wave propagation. *Journal of Sound and Vibration*, **115**, 556–559.

Ingard, K.U. (1981) Locally and non-locally reacting flexible porous layers: a comparison of acoustical properties. Transactions ASME. *Journal of Engineering for Industry*, **103**, 302–313.

Jeans, R.A. and Mathews, I.C. (1990) Solution of fluid-structure interaction problems using a coupled finite element and variational boundary element technique. *Journal of the Acoustical Society of America*, **88**, 2459–2466.

Kagawa, Y., Yamabuchi, T. and Mori, A. (1977) Finite element simulation of an axisymmetric acoustic transmission system with a sound absorbing wall. *Journal of Sound and Vibration*, **53**, 357–374.

Kang, Y.J. and Bolton, J.S. (1995) Finite element modelling of isotropic elastic porous materials coupled with acoustical finite elements. *Journal of the Acoustical Society of America*, **98**, 635–643.

Kang, Y.J. and Bolton, J.S. (1996) A finite element model for sound transmission through foam lined double-panel structures. *Journal of the Acoustical Society of America*, **99**, 2755–2765.

Kirkup, S.M. (1998) *The Boundary Element Method in Acoustics* (Hebden Bridge, UK: Integrated Sound Software).

Kreyszig, E. (1999) *Advanced Engineering Mathematics* (8th edn) (New York: John Wiley & Sons).

Lyon, R.H. and DeJong, R.G. (1995) *Theory and Applications of SEA* (2nd edn) (Newton Massachusetts: Butterworth-Heinemann).

Mathews, I.C. (1986) Numerical techniques for three-dimensional steady-state fluid-structure interaction. *Journal of the Acoustical Society of America*, **79**, 1317–1325.

Morse, P.M. and Ingard, K.U. (1968) *Theoretical Acoustics* (New York: McGraw-Hill).

Nelson, P.A. (1998) An Introduction to acoustics. In F.J. Fahy and J.G. Walker (eds), *Fundamentals of Noise and Vibration* (London: E & FN Spon), pp. 1–59.

Pierce, A.D. (1989) *Acoustics* (2nd edn) (New York: Acoustical Society of America).

Petyt, M., Lea, J. and Koopman, G.H. (1976) A finite element method for determining the acoustic modes of irregular shaped cavities. *Journal of Sound and Vibration*, **45**, 495–502.

Roozen, N.B. (1992) Quiet by design: numerical acousto-elastic analysis of aircraft structures (PhD dissertation, Technical University of Eindhoven, The Netherlands).

Saylii, M.N., Ousset, Y. and Verchery, G. (1981) Solution of radiation problems by collocation of integral formulations in terms of single and double layer potentials. *Journal of Sound and Vibration*, **74**, 187–204.

Schenck, H.A. (1968) Improved integral formulation for acoustic radiation problems. *Journal of the Acoustical Society of America*, **44**, 41–58.

Seybert, A.F. and Cheng, C.Y.R. (1987) Application of the boundary element method to acoustic cavity response and muffler analysis. Trans. ASME. *Journal of Vibration and Acoustics*, **109**, 15–21.

van Nieuwland, J.M. and Weber, C. (1979) Eigenmodes in nonrectangular reverberation rooms. *Noise Control Engineering*, **13**(3), 112–121.

Vlahopoulos, N. (2000) Indirect variational boundary element method in acoustics. In T.W. Wu (ed.), *Boundary Element Acoustics: Fundamentals and Computer Codes* (Southampton: WITpress).

Wilton, D.T. (1978) Acoustic radiation and scattering from elastic structures. *International Journal for Numerical Methods in Engineering*, **13**, 123–138.

Wu, T.W. (ed.) (2000) *Boundary Element Acoustics: Fundamentals and Computer Codes* (Southampton: WITpress).

Chapter 3

Source identification and location

P.A. Nelson

3.1 Introduction

This chapter introduces and explains methods that can be used to locate and estimate the strength of acoustic sources. There are many approaches that can be taken to this problem, but the work presented here will principally address the techniques that make use of arrays of acoustic sensors and the methods used to process the signals output from the sensors. By far the most developed literature on this subject is that concerned with the location of underwater acoustic sources, where the sensors used are hydrophones and the signal processing methods are those associated with SONAR (SOund Navigation And Ranging). Here, however, whilst many of the basic results from this field will be referred to, the main emphasis will be on the use of microphone arrays for the location of sources of airborne sound.

The chapter begins with an introduction to beamforming methods and delineates the basic parameters that determine the ability of arrays to unambiguously resolve the angular location of far field sources. Simple methods for processing the signals from such arrays are also introduced and the influence of extraneous noise is dealt with. Least squares procedures are introduced together with matrix processing techniques when the problem posed is one of source strength estimation. Inverse methods are also dealt with in some detail and an attempt is made to clarify their relationship to Near-field Acoustic Holography (NAH). A description is also given of techniques used for the estimation of acoustic source strength distributions from far field measurements and a discussion is presented of the resolution limits of such procedures.

3.2 Spatial filtering

Many of the fundamental principles of source location methods can be explained in terms of spatial filtering. Consider, for example, the continuous line array lying on the x-axis depicted in Figure 3.1. Assume that a harmonic acoustic plane wave, having a wavelength λ, impinges with an angle θ upon

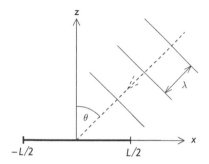

Figure 3.1 A continuous line array with an impinging plane wave.

this hypothetical line transducer, which produces an output in proportion to the acoustic pressure integrated along the line. The output of such an array, when assumed to have infinite length, can be written as

$$W(\theta) = \int_{-\infty}^{\infty} w(x)\,p(x)\,\mathrm{d}x \tag{3.1}$$

where $p(x)$ denotes the acoustic pressure on the x-axis and $w(x)$ denotes a weighting function which describes the extent to which different elements of the array contribute to the total output. The complex acoustic pressure due to the impinging plane wave can be written as

$$p(x) = e^{jkx\sin\theta} = e^{jk_x x} \tag{3.2}$$

where $k = \omega/c_0 = 2\pi/\lambda$ is the acoustic wavenumber associated with the plane wave having phase speed c_0 and a harmonic time dependence of the form $e^{j\omega t}$. The term $k\sin\theta$ describes the component k_x of the acoustic wavenumber in the x-direction (also known as the trace wavenumber). Substitution of equation (3.2) into equation (3.1) shows that the output of the array takes the form of a wavenumber transform given by

$$W(k_x) = \int_{-\infty}^{\infty} w(x)e^{jk_x x}\,\mathrm{d}x \tag{3.3}$$

where the output $W(\theta)$ has been written as $W(k_x)$ to emphasise the dependence on the wavenumber k_x. Note that this transform relationship is exactly analogous to the Fourier transform used to express the frequency content of a given time history $f(t)$ where, for example,

$$F(\omega) = \int_{-\infty}^{\infty} f(t)e^{-j\omega t}\,\mathrm{d}t \tag{3.4}$$

The simplest form of weighting function $w(x)$ associated with a continuous line array is the rectangular aperture function depicted in Figure 3.2a. This can be described by

$$w(x) = \begin{cases} (1/L) & -L/2 < x < L/2 \\ 0 & -L/2 > x > L/2 \end{cases} \qquad (3.5)$$

The output of such an array can therefore be written as

$$W(k_x) = \frac{1}{L} \int_{-L/2}^{L/2} e^{jk_x x} \, dx \qquad (3.6)$$

and the integral is readily evaluated to yield

$$W(k_x) = \frac{1}{jk_x L} \left(e^{jk_x L/2} - e^{-jk_x L/2} \right) \qquad (3.7)$$

Using the identity $e^{j\alpha} - e^{-j\alpha} = 2j \sin \alpha$ shows that the output of the array can be written as

$$W(k_x) = \frac{\sin\left[\left(\dfrac{kL}{2} \right) \sin \theta \right]}{\left(\dfrac{kL}{2} \right) \sin \theta} \qquad (3.8)$$

This is simply the spatial Fourier transform of the rectangular aperture function and is depicted in Figure 3.2b. When the angle of the impinging plane wave is θ_B, the first zero crossings of this 'sinc' function are given by

$$\left(\frac{kL}{2} \right) \sin \theta_B = \pm \pi \qquad (3.9)$$

and since $k = 2\pi/\lambda$ this shows that the range of angles between these zero crossings (for which the array produces an output greater than zero) is defined by

$$\theta_B = \pm \sin^{-1} \left(\frac{\lambda}{L} \right) \qquad (3.10)$$

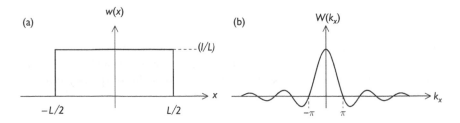

Figure 3.2 (a) The rectangular aperture function; (b) its spatial Fourier transform.

When the acoustic wavelength is much less than the length L of the array, then the range of angles defined by θ_B can be approximated by

$$\theta_B \approx \pm \frac{\lambda}{L} \quad (\lambda \ll L) \tag{3.11}$$

The range of angles θ_B lying between $-\lambda/L$ and $+\lambda/L$ defines the width of the *main lobe* (or *beam width*) of the array. Therefore, plane waves impinging from angles within this range produce a pressure distribution upon the array all the elements of which integrate together in order to result in a positive output $W(\theta)$. The array is most sensitive (produces greatest output) for waves impinging at an angle normal to the array (since all the contributions from the segments of the array are in phase). It is also clear from equation (3.11) that the width of the main lobe is in inverse proportion to the length of the array. Thus arrays which are long compared with the acoustic wavelength in question are necessary to produce a high degree of angular selectivity. This constitutes a fundamental physical limitation associated with beamformers intended to deduce the angular location of a far field source. A typical example of the directivity function of the continuous line array is depicted in Figure 3.3. This shows the output $W(k_x)$ plotted against the angle of incidence θ. The main lobe is clearly shown, together with 'side-lobes' which correspond to the ripples in the spatial Fourier transform depicted in Figure 3.2b. Note again that the line array is most sensitive to waves impinging normal to the array and therefore the maximum output is centred on angles given by $\theta = 0$ and $\theta = \pi$.

It is possible to suppress the side lobes appearing in the directivity pattern of the array response by 'shading' the array through the application of

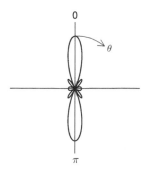

Figure 3.3 The directivity or pattern function of the continuous line array produced by plotting $W(k_x)$ against θ.

alternative weighting functions $w(x)$. For example, a simple function which results in side-lobe suppression is the triangular aperture function defined by

$$
w(x) = \begin{cases} 0 & |x| > L/2 \\ \dfrac{L}{2}\left(1 - \dfrac{|x|}{L/2}\right) & |x| < L/2 \end{cases} \tag{3.12}
$$

It can be shown (see, for example, Burdic 1984, Chapter 6, pp. 178–179) that this function results from the spatial convolution of the rectangular aperture function with itself. It therefore follows that the spatial Fourier transform describing the output is found by squaring that associated with the rectangular aperture function. The corresponding output can thus be written as

$$
W(k_x) = \frac{\sin^2\left[\left(\dfrac{kL}{2}\right)\sin\theta\right]}{\left[\left(\dfrac{kL}{2}\right)\sin\theta\right]^2} \tag{3.13}
$$

The aperture function and its resulting Fourier transform (or 'pattern function') are shown in Figure 3.4. There are many other more sophisticated 'spatial window' functions that give varying degrees of side-lobe suppression. These are described in more detail, for example, by Burdic (1984).

It should also be noted that the array can be 'steered' to be selective to sound impinging from a particular angular location. A simple means for achieving this is to define a weighting function given by

$$
w_s(x) = w(x)e^{jk_0 x} \tag{3.14}
$$

where θ_0 defines the selected angular location and $k_0 = k\sin\theta_0$. This weighting function ensures that the phase of the output of each segment of the array is shifted in order to match the phase of an impinging wave arriving from the angle θ_0. This, in turn, ensures that the net array output is maximised for this arrival angle, since the contributing outputs from

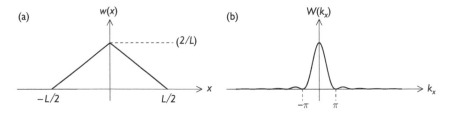

Figure 3.4 (a) The triangular aperture function; (b) the resulting pattern function.

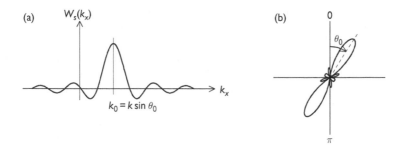

Figure 3.5 (a) The spatial Fourier transform associated with the continuous line array when weighted by e^{jk_0x}; (b) the resulting beam pattern function.

each element of the array will sum together in phase. It follows from the properties of Fourier transforms that

$$W_s(k_x) = W(k_x - k_0) \tag{3.15}$$

Thus, the output of the array is simply shifted as a function of k_x as depicted in Figure 3.5a. The corresponding directivity pattern is sketched in Figure 3.5b.

3.3 Discrete spatial sampling

Now consider the implications of deploying a number of discrete sensors in the array whose individual outputs are summed in order to produce the net output of the array. One may analyse this problem by analogy with the sampling of continuous time histories (see, for example, Nelson and Elliott 1992, Chapter 2, pp. 61–67). For example, the Fourier transform of a function of time $f(t)$ sampled at regular time intervals nT can be written as

$$F(e^{j\omega T}) = \int_{-\infty}^{\infty} \left[f(t) \sum_{n=-\infty}^{\infty} \delta(t - nT) \right] e^{-j\omega t} \, dt \tag{3.16}$$

where $\delta()$ denotes the Dirac delta function and n is an integer. It is well known that this Fourier transform can be written as

$$F(e^{j\omega T}) = \frac{1}{T} \sum_{m=-\infty}^{\infty} F(\omega - m\omega_s) \tag{3.17}$$

where $F(\omega)$ is the Fourier transform of the continuous signal $f(t)$ and $\omega_s = 2\pi/T$ denotes the sampling frequency. Thus, the Fourier spectrum of the sampled time history repeats periodically with frequency and consists of an infinite sum of the individual spectral contributions each centred on integer multiples m of the sampling frequency ω_s. Therefore, in order to prevent aliasing, it is important in the analysis of sampled time histories to ensure that contributions from the spectra for which m is non-zero do not appear in the frequency range between $-\omega_s/2$ and $+\omega_s/2$. Consequently, anti-aliasing filters are usually applied to continuous time signals prior to sampling in order to remove frequencies which are higher than half the sampling frequency.

Now consider the analogous situation that occurs when discrete samples of an acoustic field are taken at uniform intervals in space. The Fourier transform of the aperture function associated with an array of discrete transducers spaced at intervals $n\Delta x$ along the x-axis can be written in a form that is analogous to that given by equation (3.16):

$$W(e^{jk_x\Delta x}) = \int_{-\infty}^{\infty} \left[w(x) \sum_{n=-\infty}^{\infty} \delta(x - n\Delta x) \right] e^{-jk_x x}\, dx \tag{3.18}$$

where $w(x)$ again denotes the rectangular aperture function which is sampled as depicted in Figure 3.6a. This Fourier transform can be shown, by an analogous argument to that given for the sampling of time histories, to be given by

$$W(e^{jk_x\Delta x}) = \frac{1}{\Delta x} \sum_{m=-\infty}^{\infty} W(k_x - m(2\pi/\Delta x)) \tag{3.19}$$

where the spatial Fourier transform now repeats at integer multiples m of the spatial sampling frequency $2\pi/\Delta x$. This is sketched in Figure 3.6b. It is shown by Burdic (1984, Chapter 11, pp. 332–335) that in this case the summation of all these spectra can be written as

$$W(e^{jk_x\Delta x}) = \frac{1}{M} \frac{\sin\left[\left(kM\Delta\frac{x}{2}\right)\sin\theta\right]}{\sin\left[\left(k\Delta\frac{x}{2}\right)\sin\theta\right]} \tag{3.20}$$

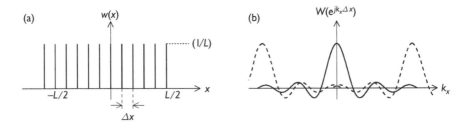

(a) $w(x)$ (b) $W(e^{jk_x\Delta x})$

Figure 3.6 (a) The discrete rectangular aperture function and (b) its spatial Fourier transform.

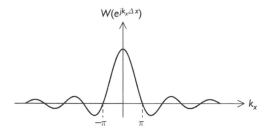

Figure 3.7 The beam pattern function of the discrete line array.

where M denotes the number of discrete sensors in the array. This function is illustrated in Figure 3.7. Polar plots showing this function explicitly as the angle of incidence θ is varied are also shown in Figure 3.8 for various separation distances Δx between the discrete sensors. Provided that Δx is less than one half of an acoustic wavelength, then the directional response of the array is dominated by the main lobe. However, once Δx exceeds $\lambda/2$, 'spatial aliasing' occurs and significant side-lobes are produced in the response of the

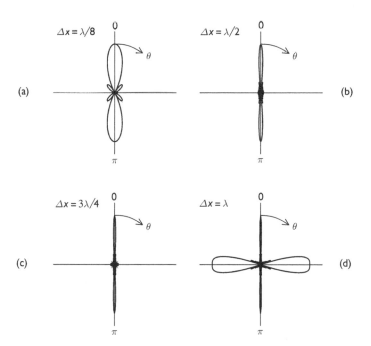

Figure 3.8 The pattern function of the discrete line array for a range of element spacings relative to the acoustic wavelength: (a) $\Delta x = \lambda/8$; (b) $\Delta x = \lambda/2$; (c) $\Delta x = 3\lambda/4$; (d) $\Delta x = \lambda$.

array. This is illustrated in Figure 3.8 which shows the pattern function of the array when Δx is set equal to $\lambda/8$, $\lambda/2$, $3\lambda/4$ and λ. This therefore shows a second important fundamental limitation associated with beamforming methods based on uniform spatial sampling; the ability to discriminate the angular location of a far field source at high frequencies (short wavelengths) is determined by the spacing between the discrete sensors. Recall that at low frequencies (long wavelengths), the ability to discriminate between arrival angles is determined by the total length of the array.

Finally, it should be noted that the discrete array can be steered in much the same way as the continuous line array described above. Thus, for example, provided the appropriate weighting function is applied to the elements of the array to ensure that their net outputs add coherently for a given arrival angle θ_0, it is again shown by Burdic (1984) that the output can be written as

$$W(e^{jk_x \Delta x}) = \frac{1}{M} \frac{\sin\left[\left(kM\Delta\frac{x}{2}\right)(\sin\theta - \sin\theta_0)\right]}{\sin\left[\left(k\Delta\frac{x}{2}\right)(\sin\theta - \sin\theta_0)\right]} \tag{3.21}$$

As in the case of the continuous line array, this simply results in a shift of the spatial Fourier transform with a resulting main lobe which is centred on the desired angle θ_0.

3.4 Discrete sampling in space and time

Practical beamformers are implemented by sampling the outputs of the sensors and operating on the resulting discrete time sequences. Before describing the details of the operations involved in various signal processing procedures, it is useful first to describe the sensor output sequences produced by a single acoustic source. The complex acoustic pressure produced in free space by a harmonic point monopole source having a volume acceleration v is given by

$$p(r) = \frac{\rho_0 v e^{-jkr}}{4\pi r} \tag{3.22}$$

where r represents the radial distance from the source to the receiver and ρ_0 is the mean density of medium. In the far field of the source, the approximation can be made that the acoustic pressure distribution along the x-axis (which is coincident with the axis of the sensor array) is given by

$$p(x) \approx \frac{\rho_0 v e^{-jk(r_0 - x\sin\theta)}}{4\pi r_0} \tag{3.23}$$

provided that $r_0 \gg L$, where r_0 is the distance from the source to the centre of the sensor array as shown in Figure 3.9. An alternative way of describing

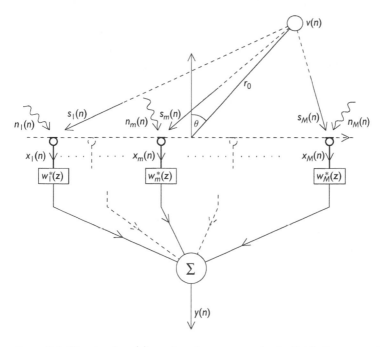

Figure 3.9 The signals $s_m(n)$ produced on an array in the field of an acoustic source having an output $v(n)$. The beamformer output is found by passing the sensor outputs $x_m(n)$ through digital filters having transfer functions $w_m(z)$.

this relation is to consider the Fourier transform of the time-dependent acoustic pressure to be given by $P(\omega) = G(\omega)V(\omega)$ where $V(\omega)$ represents the Fourier transform of the volume acceleration of the source and $G(\omega)$ describes an acoustic frequency response function, or Green function. This function can be written as a product of two terms, one relating to the distance of the source from the centre of the array and the other relating to the position on the array. Thus

$$G(\omega) = G_0(\omega)e^{j\omega x \sin\theta/c_0} \tag{3.24}$$

where $G_0(\omega) = (\rho_0 e^{-j\omega r_0/c_0})/4\pi r_0$, and the wavenumber $k = \omega/c_0$. It is useful to note that, by analogy with the fact that the inverse Fourier transform of $e^{j\omega\tau}$ is given by $\delta(t-\tau)$, the impulse response associated with the frequency response function component $G_0(j\omega)$ can be written as

$$g_0(t) = \frac{\rho_0}{4\pi}\delta\left(t - \frac{r_0}{c_0}\right) \tag{3.25}$$

Thus, this component of the frequency response function between the source volume acceleration and the acoustic pressure simply represents a pure delay equal to the travel time of acoustic disturbances propagating from the source to the centre of the sensor array. Similarly, the impulse response associated with the complete frequency response function $G(\omega)$ can be written as

$$g(t) = \frac{\rho_0}{4\pi} \delta\left(t - \frac{r_0}{c_0} + \frac{x\sin\theta}{c_0}\right) \tag{3.26}$$

Thus, in the far field approximation, sound will arrive at the different sensor positions with an additional time lag given by $x\sin\theta/c_0$ relative to the time delay r_0/c_0.

It will now be useful to consider the process of sound propagation from the source to the sensor array within the context of discrete time systems theory. Thus the assumption is made that the source volume acceleration $v(t)$ is sampled at regular intervals T such that it may be expressed as the sequence $v(nT)$. The acoustic pressure produced at the sensors can be equivalently expressed. The impulse response of the system relating the source volume acceleration sequence to the sequences generated at the sensors can be expressed as a series of digital filters having the impulse responses given by

$$g(nT) = \frac{\rho_0}{4\pi} \delta(nT - d_0 T + d_x T) \tag{3.27}$$

where $\delta(nT)$ now defines the digital impulse which has a value of unity for $nT = 0$ and a value of zero for all other values of n. The symbols d_0 and d_x denote the number of samples associated with the acoustic propagation delays such that

$$d_0 T = r_0/c_0, \qquad d_x T = x\sin\theta/c_0 \tag{3.28a,b}$$

It is important to note that one can only approximate d_0 and d_x as integers provided that $T \ll x\sin\theta/c_0$, in which case the appropriate number of samples delay is given by $d_0 = \text{int}[r_0/c_0 T]$, $d_x = \text{int}[x\sin\theta/c_0 T]$. Note that, formally speaking, the discrete time impulse response has been deduced here from the equivalent continuous time expression via an 'impulse invariant transformation'. It should also be noted that there are other methods for expressing a discrete time equivalent of a continuous time system (see, for example, Rabiner and Gold 1975, Chapter 4, pp. 211–226). The Fourier transform of the sequence $g(nT)$ can be written as

$$G(e^{j\omega T}) = \frac{\rho_0}{4\pi} e^{-j\omega(d-d_x)T} = G_0(e^{j\omega T}) e^{j\omega d_x T} \tag{3.29}$$

where $G_0(e^{j\omega T})$ is the discrete time equivalent of the continuous time transfer function $G_0(\omega)$.

Now consider the operation of the simple 'delay and sum beam-former' which takes the sampled sensor outputs and applies delays to these sequences in order to 'best match' the incoming wave field, thereby maximising the net beamformer output when all the channels are summed together. Assume that the mth sensor output is given by $x_m(nT)$ and that this has a Fourier transform $X_m(e^{j\omega T})$. In the absence of noise

$$X_m(e^{j\omega T}) = G_0(e^{j\omega T})V(e^{j\omega T})e^{j\omega d_m T} \tag{3.30}$$

where $d_m T = m\Delta \sin\theta_0/c_0$, such that d_m represents an integer number of samples delay associated with the mth sensor. Note that m varies from $-(M+1)/2$ to $(M+1)/2$ when there are M sensors in the array. The operation of the delay and sum beamformer is to apply a delay to the mth sensor output sequence given by d_m samples. This is equivalent to passing the mth sensor output sequence through a digital filter having the discrete time impulse response $\delta(nT - d_m T)$ and the Fourier transform $e^{-j\omega d_m T}$. Therefore the Fourier transform of the beamformer output can be written as

$$Y(e^{j\omega T}) = \sum_{m=-(M+1)/2}^{(M+1)/2} X_m(e^{j\omega T})e^{-j\omega d_m T} \tag{3.31}$$

Substitution of equation (3.30) describing the Fourier transform of each of the sensor output sequences then yields

$$Y(e^{j\omega T}) = MG_0(e^{j\omega T})V(e^{j\omega T}) \tag{3.32}$$

This demonstrates that the delay and sum beamformer produces an output whose Fourier transform is in direct proportion to the Fourier transform of the equivalent volume acceleration of a far field acoustic source.

The beamformer can be implemented by making use of the discrete Fourier transform (DFT) and its efficient implementation using the fast Fourier transform (FFT). First note that an estimate of the Fourier transform $Y(e^{j\omega T})$ can be made by computing the summation over N samples of the beamformer output sequence that is given by

$$Y(e^{j\omega T}) = \sum_{n=0}^{N-1} y(nT)e^{-j\omega_n T} \tag{3.33}$$

This amounts to estimating the DFT from a block of data consisting of N samples. The range of frequencies over which this estimate is made is chosen to be between zero and the sampling frequency given by $\omega_s = 2\pi/T$. This frequency range is then split into the same number, N, of discrete frequencies as there are time samples. Therefore N discrete frequencies are chosen at intervals $2\pi/NT$ and at the kth frequency (where k is used here as an index and does not refer to the acoustic wavenumber) the estimate of the Fourier transform is given by

$$Y(e^{jk(2\pi/NT)T}) = \sum_{n=0}^{N-1} y(nT)e^{-jk(2\pi/NT)nT} \tag{3.34}$$

This expression for the DFT is commonly abbreviated to

$$Y(k) = \sum_{n=0}^{N-1} y(n)e^{-j2\pi kn/N} \tag{3.35}$$

where the indices k and n denote discrete frequency and time, respectively. Since the sequence $y(n)$ is produced by delaying and summing the M sensor outputs, the DFT of the beamformer output can thus be written as

$$Y(k) = \sum_{m=-(M+1)/2}^{(m+1)/2} \sum_{n=0}^{N-1} x_m(n-d_m)e^{-j2\pi kn/N} \tag{3.36}$$

This expression can be rewritten in terms of the index $i = n - d_m$, or $n = i + d_m$ as

$$Y(k) = \sum_{m=-(M+1)/2}^{(M+1)/2} e^{-j2\pi kd_m/N} \sum_{i=-d_m}^{N-1-d_m} x_m(i)e^{-j2\pi ki/N} \tag{3.37}$$

This shows that, provided the number of samples delay associated with mth sensor output is much less than the number of points in the block of data used to compute the DFT (that is, $|d_m| \ll N$), then we may make the approximation

$$Y(k) = \sum_{m=-(M+1)/2}^{(M+1)/2} e^{-j2\pi kd_m/N} X_m(k) \tag{3.38}$$

Thus the Fourier transform of the beamformer output can be evaluated by using an effective phase shift (given by $e^{-j2\pi kd_m/N}$) to multiply each of the DFTs of the individual sensor outputs. More details of the process of implementing beamformers of this type in the frequency domain are given for example by Nielsen (1990), Chapter 2, pp. 51–94.

3.5 Source strength estimation in the presence of noise

The delay and sum beamformer described above is sometimes referred to as a 'data independent' array signal processing method since it takes no account of the particular form of the time histories detected at the sensors. It is now appropriate to consider the process of estimating the strength of an acoustic source in the presence of interfering noise. Again it will be assumed that all the data output from the sensors is sampled at a time interval T and that the output of the mth sensor $x_m(nT)$ is abbreviated to $x_m(n)$. Similarly, the Fourier transform of this sequence $X(e^{j\omega T})$ is abbreviated to $X(e^{j\omega})$. The mth sensor output sequence can be expressed as

$$x_m(n) = s_m(n) + n_m(n) \tag{3.39}$$

where $s_m(n)$ denotes the 'signal' sequence and $n_m(n)$ denotes the 'noise' sequence that interferes with the signals received at the sensor array. Working in discrete time it is assumed that the sensor output samples are passed through a digital filter having a frequency response function given by the Fourier transform $W^*(e^{j\omega})$, where the $*$ superscript denotes the complex conjugate. The use of the complex conjugate of the filters to be determined is conventional in the literature describing array signal processing techniques and helps to simplify the notation in what follows. First note that the beamformer output sequence $y(n)$ has the Fourier transform given by

$$Y(e^{j\omega}) = \mathbf{w}^H \mathbf{x} \tag{3.40}$$

where $\mathbf{w}$ and $\mathbf{x}$ are the complex vectors defined by

$$\mathbf{w}^H = [W_1^*(e^{j\omega}), W_2^*(e^{j\omega}) \dots W_M^*(e^{j\omega})] \tag{3.41}$$

$$\mathbf{x}^T = [X_1(e^{j\omega}), X_2(e^{j\omega}) \dots X_M(e^{j\omega})] \tag{3.42}$$

and where the superscript T denotes the transpose and the superscript H denotes the conjugate transpose. It will now be assumed that the signals $s_m(n)$ are generated by an isolated acoustic source having a source strength time history $v(n)$. Furthermore, as above, it will be assumed that the Green function relating the signal generated at the mth sensor to the source strength time history can be represented by a digital filter whose frequency response function can be written as $G_m(e^{j\omega})$. Therefore the vector of acoustic frequency response functions is defined by

$$\mathbf{g}^T = [G_1(e^{j\omega}), G_2(e^{j\omega}) \dots G_M(e^{j\omega})] \tag{3.43}$$

and the sensor output signal vector is given by

$$\mathbf{x} = \mathbf{s} + \mathbf{n} = V(e^{j\omega})\mathbf{g} + \mathbf{n} \tag{3.44}$$

A straightforward approach to the computation of the complex vector **w** is to find the processing scheme that minimises the cost function

$$J = E[|Y(e^{j\omega}) - V(e^{j\omega})|^2] \tag{3.45}$$

where $E[\ ]$ denotes the expectation operator. That is, we seek to produce an output from the beamformer whose Fourier transform is as close as possible to the Fourier transform of the acoustic source strength. Substitution of equation (3.40) then shows that

$$J = E\left[(\mathbf{w}^H\mathbf{x} - V(e^{j\omega}))(\mathbf{w}^H\mathbf{x} - V(e^{j\omega}))^*\right] \tag{3.46}$$

Noting that $(\mathbf{w}^H\mathbf{x})^* = (\mathbf{x}^T\mathbf{w}^*)^* = \mathbf{x}^H\mathbf{w}$ enables the cost function to be written as

$$J = \mathbf{w}^H E[\mathbf{x}\mathbf{x}^H]\mathbf{w} - \mathbf{w}^H E[\mathbf{x}V^*(e^{j\omega})] - E[\mathbf{x}^H V(e^{j\omega})]\mathbf{w} + E[|V(e^{j\omega})|^2] \tag{3.47}$$

The matrix of sensor output cross-spectra $\mathbf{S}_{xx}$, the vector of cross spectra between the sensor outputs and the acoustic source strength $\mathbf{s}_{xv}$, and the power spectrum of the source volume acceleration S_{vv} are, respectively, defined by

$$\mathbf{S}_{xx} = E[\mathbf{x}\mathbf{x}^H], \qquad \mathbf{s}_{xv} = E[\mathbf{x}V^*(e^{j\omega})], \qquad S_{vv} = E[|V(e^{j\omega})|^2] \tag{3.48}$$

The cost function can thus be written in the quadratic form

$$J = \mathbf{w}^H\mathbf{S}_{xx}\mathbf{w} - \mathbf{w}^H\mathbf{s}_{xv} - \mathbf{s}_{xv}^H\mathbf{w} + S_{vv} \tag{3.49}$$

The value of the vector **w** that minimises this function is given by

$$\mathbf{w}_{\text{opt}} = \mathbf{S}_{xx}^{-1}\mathbf{s}_{xv} \tag{3.50}$$

As described by Van Veen and Buckley (1998), this solution forms the basis of 'reference signal' beamformers where a reference or pilot signal is available that closely represents the desired signal $V(e^{j\omega})$ (see also Widrow *et al.* 1967; Widrow and Stearns 1985). Note that the vector of cross spectra between the sensor outputs and the acoustic source strength can be written as

$$\mathbf{s}_{xv} = E[\mathbf{x}V^*(e^{j\omega})] = E[(\mathbf{s} + \mathbf{n})V^*(e^{j\omega})] \tag{3.51}$$

In the particular case that the noise signal is uncorrelated with the signal vector **s**, and noting that $\mathbf{s} = V(e^{j\omega})\mathbf{g}$, shows that

$$\mathbf{s}_{xv} = \mathbf{g}S_{vv} \qquad \mathbf{S}_{xx} = S_{vv}\mathbf{g}\mathbf{g}^H + \mathbf{S}_{nn} \tag{3.52a,b}$$

where $S_{nn} = E[\mathbf{nn}^H]$ is the matrix of noise signal cross spectra. The solution for the optimal processor then becomes

$$\mathbf{w}_{opt} = [S_{vv}\mathbf{gg}^H + S_{nn}]^{-1}\mathbf{g}S_{vv} \tag{3.53}$$

Furthermore, if the noise signals at the sensors are all uncorrelated with each other, then the cross-power spectral matrix of noise signals can be approximated by $S_{nn} = S_{nn}\mathbf{I}$, where $\mathbf{I}$ is the identity matrix, and defining the signal to noise ratio $\gamma = S_{nn}/S_{vv}$ shows that

$$\mathbf{w}_{opt} = [\mathbf{gg}^H + \gamma\mathbf{I}]^{-1}\mathbf{g} \tag{3.54}$$

This expression can also be written in an alternative form by using the matrix inversion lemma

$$(\mathbf{A}+\mathbf{BCD})^{-1} = \mathbf{A}^{-1} - \mathbf{A}^{-1}\mathbf{B}(\mathbf{C}^{-1}+\mathbf{DA}^{-1}\mathbf{B})^{-1}\mathbf{DA}^{-1} \tag{3.55}$$

where in this case $\mathbf{A} = \gamma\mathbf{I}, \mathbf{B} = \mathbf{g}, \mathbf{C} = 1, \mathbf{D} = \mathbf{g}^H$. It therefore follows that

$$[\mathbf{g}^H\mathbf{g} + \gamma\mathbf{I}]^{-1} = (1/\gamma)\mathbf{I} - (1/\gamma)^2\mathbf{g}(1 + (1/\gamma)\mathbf{g}^H\mathbf{g})^{-1}\mathbf{g}^H \tag{3.56}$$

and, after some algebra, it then follows that

$$\mathbf{w}_{opt} = (\gamma \mid \mathbf{g}^H\mathbf{g})^{-1}\mathbf{g} \tag{3.57}$$

In the particular case when no noise is present, $\gamma = 0$, and the optimal filter vector becomes

$$\mathbf{w}_{opt} = (\mathbf{g}^H\mathbf{g})^{-1}\mathbf{g} \tag{3.58}$$

Since $Y(e^{j\omega}) = \mathbf{w}_{opt}^H\mathbf{g}V(e^{j\omega})$, this expression provides the solution that ensures that $Y(e^{j\omega}) = V(e^{j\omega})$ and therefore that the Fourier transform of the beamformer output is identical to the Fourier transform of the source strength output. Note that the mth element of the vector $\mathbf{g}$ is given by

$$G_m(e^{j\omega}) = G_0(e^{j\omega})e^{j\omega d_m} \tag{3.59}$$

and therefore the solution for the optimal filter vector in the noise free case reduces to the 'delay and sum' beamformer. It should also be noted that the solution for the optimal beamformer expressed by equation (3.58) is essentially that used in 'focused beamforming' where an attempt is made to map the distribution of acoustic source strength associated with a given source distribution by changing the assumed vector $\mathbf{g}$ of Green functions in accordance with the assumed source position. Some examples are given by Underbrink (2002), Bitzer and Simmer (2001), Billingsley and Kinns (1976), Dougherty (2002), Elias (1997), and Brooks *et al.* (1987).

The use of focused beamformers of this type has been extensive in recent years in attempting to locate the dominant sources of aerodynamically generated noise associated with aircraft. In particular, the airframe itself is now a major contributor to the noise generated by aircraft on landing (see Chapter 7), and a number of studies have been devoted to the use of beamforming techniques in this context. Some examples from the more recent aeroacoustics literature include those presented by Mosher (1996), Humphreys *et al.* (1998), Hayes *et al.* (1997), Piet and Elias (1997), Davy and Remy (1998), Sijtsma and Holthusen (1999), Michel and Qiao (1999), and Venkatesh *et al.* (2000). An example of some recent results of these endeavours is presented in Figure 3.10. This shows an array described by Stoker and Sen (2001) installed in a wind tunnel to investigate the sources of noise produced by a model of an aircraft wing. The array design used has microphones spaced so as to maximise the suppression of side-lobes in the directional response of the array and also uses different subsets of microphones in order to cover different ranges of frequencies. Low-frequency sound requires a larger aperture (widely spaced microphones) in order to enable a selective directional response whilst the high-frequency limit is determined by the inter-microphone spacing which dictates the onset of side-lobes due to spatial aliasing as described in Section 3.3. Figure 3.10 shows the output of the array at various frequencies as the focus of the array is scanned across the region of the model wing. Areas producing a high array output are clearly shown.

3.6 'Minimum variance' source strength estimation

Another approach that is prevalent in the beamforming literature is to choose the signal processing scheme in order to minimise the output of the array subject to the constraint of unit gain in the 'look-direction'. In the present context one wishes to ensure that the beamformer output remains equal to the volume acceleration of the acoustic source strength when the vector $\mathbf{w}$ is chosen such that $\mathbf{w}^H \mathbf{g} = 1$. Furthermore, the output power to be minimised can be written as $E[|Y_m(e^{j\omega})|^2]$ and, since $Y_m(e^{j\omega})$ is given by $\mathbf{w}^H \mathbf{x}$, the optimisation problem is described by

$$\text{Minimise} \quad \mathbf{w}^H \mathbf{S}_{xx} \mathbf{w} \quad \text{subject to} \quad \mathbf{w}_{\text{opt}}^H \mathbf{g} = 1 \tag{3.60}$$

This linearly constrained optimisation problem can be solved by using the method of Lagrange multipliers. Thus one seeks to minimise the quadratic function

$$J = \mathbf{w}^H \mathbf{S}_{xx} \mathbf{w} + \lambda \left(\mathbf{w}^H \mathbf{g} + (\mathbf{w}^H \mathbf{g})^* \right) \tag{3.61}$$

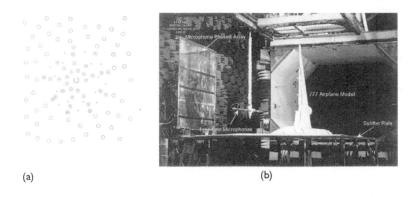

(a)   (b)

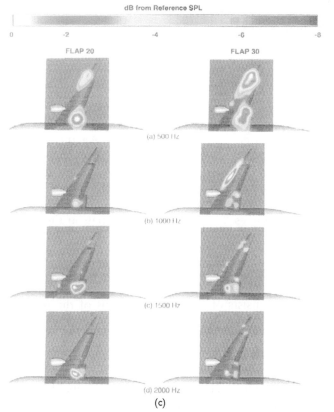

(c)

Figure 3.10 Results of a study of airframe noise by using a focused beamformer presented by Stoker and Sen (2001). (a) The microphone array used consisting of an outer low-frequency array and inner high-frequency array; (b) the experimental arrangement in a wind tunnel and (c) some results of the beamformer output as it is scanned across the area of the test object.

where λ denotes the Lagrange multiplier. The derivative of the function is given by

$$\frac{\partial J}{\partial \mathbf{w}} = 2\mathbf{S}_{xx}\mathbf{w} + 2\lambda\mathbf{g} \tag{3.62}$$

and setting $\partial J/\partial \mathbf{w} = 0$ shows that

$$\mathbf{w} = \lambda\mathbf{S}_{xx}^{-1}\mathbf{g} \tag{3.63}$$

Writing the constraint $\mathbf{w}_{opt}^H\mathbf{g} = 1$ as $\mathbf{g}^H\mathbf{w}_{opt} = 1$ then shows that

$$\lambda\mathbf{g}^H\mathbf{S}_{xx}^{-1}\mathbf{g} = 1 \tag{3.64}$$

and therefore that

$$\mathbf{w}_{opt} = \frac{\mathbf{S}_{xx}^{-1}\mathbf{g}}{\mathbf{g}^H\mathbf{S}_{xx}^{-1}\mathbf{g}} \tag{3.65}$$

The first obvious advantage of this solution over that given by equation (3.50) is that there is no requirement for prior knowledge of the source strength $V(e^{j\omega})$. Thus one can, in principle, deduce the solution for the optimal filter vector from the measured value of the cross-spectral matrix $\mathbf{S}_{xx}$ and an assumed value of the Green function vector $\mathbf{g}$. There are, of course, significant implications associated with the inversion of the cross-spectral matrix. First consider the form of the cross-spectral matrix $\mathbf{S}_{xx}$ when a number of sources S are present. The signals at the sensors can be written as

$$\mathbf{x} = \mathbf{Gv} + \mathbf{n} \tag{3.66}$$

where the vector $\mathbf{v}$ consists of Fourier transforms of the source strengths such that

$$\mathbf{v}^T = [V_1(e^{j\omega})V_2(e^{j\omega})\ldots V_s(e^{j\omega})] \tag{3.67}$$

and the matrix $\mathbf{G}$ has columns consisting of the S vectors of Green functions associated with each source such that

$$\mathbf{G} = \begin{bmatrix} G_{11}(e^{j\omega}) & G_{12}(e^{j\omega}) & \ldots & G_{1s}(e^{j\omega}) \\ G_{21}(e^{j\omega}) & G_{22}(e^{j\omega}) & \ldots & G_{2s}(e^{j\omega}) \\ \vdots & & & \\ G_{M1}(e^{j\omega}) & G_{M2}(e^{j\omega}) & \ldots & G_{Ms}(e^{j\omega}) \end{bmatrix} = [\mathbf{g}_1\ \mathbf{g}_2 \ldots \mathbf{g}_s] \tag{3.68}$$

Therefore the cross-spectral matrix of sensor signals can be written as

$$S_{xx} = E[(Gv + n)(Gv + n)^H] \tag{3.69}$$

Now, defining the cross-spectral matrices $S_{vv} = E[vv^H]$ and $S_{nn} = E[nn^H]$, and since v and n are uncorrelated, the expression for the cross-spectral matrix becomes

$$S_{xx} = GS_{vv}G^H + S_{nn} \tag{3.70}$$

Now note that the matrix S_{xx} is of order $M \times M$; since we have to invert this matrix it is important to determine its rank. The matrix G is of order $M \times S$ (rows $\times$ columns). The Singular Value Decomposition (SVD) can be used to expand G into the form

$$G = U\Sigma V^H \tag{3.71}$$

where the matrices U and V have columns consisting respectively of the right and left singular vectors such that

$$U = [u_1 u_2 \ldots u_M] \qquad V = [v_1 v_2 \ldots v_S] \tag{3.72a,b}$$

where u_i and v_i are respectively the ith left and right singular vectors. The matrix Σ is the diagonal matrix of singular values defined by

$$\Sigma = \text{diag}(\sigma_1, \sigma_2 \ldots \sigma_p) \quad p = \min\{M, S\} \tag{3.73}$$

where $\sigma_1 \geq \sigma_2 \ldots \geq \sigma_p \geq 0$ are the singular values of G. Note also that the matrices U and V are unitary and have the properties $U^H U = UU^H = I$ and $V^H V = VV^H = I$. The rank of a matrix is defined by the number n of non-zero singular values such that

$$\sigma_1 \geq \sigma_2 \ldots \geq \sigma_{r+1} = \ldots = \sigma_p = 0 \tag{3.74}$$

As an illustration of the SVD of the matrix G assume that there are $M = 3$ sensors and $S = 2$ sources. Then G can be expanded as

$$G = \begin{bmatrix} U_{11} & U_{12} & U_{13} \\ U_{21} & U_{22} & U_{23} \\ U_{31} & U_{32} & U_{33} \end{bmatrix} \begin{bmatrix} \sigma_1 & 0 \\ 0 & \sigma_2 \\ 0 & 0 \end{bmatrix} \begin{bmatrix} V_{11}^* & V_{21}^* \\ V_{12}^* & V_{22}^* \end{bmatrix} \tag{3.75}$$

By writing $\mathbf{U}$ and $\mathbf{V}$ in terms of the column vectors $\mathbf{u}_i$ and $\mathbf{v}_i$ as defined in equation (3.72a,b) the matrix $\mathbf{G}$ can be expressed in the form

$$\mathbf{G} = \sum_{i=1}^{r} \sigma_i \mathbf{u}_i \mathbf{v}_i^H \tag{3.76}$$

where the rank r denotes the number of non-zero singular values.

We now return to the matter of the rank of the matrix $\mathbf{S}_{xx}$. First note that the columns of the $M \times S$ matrix $\mathbf{G}$ are independent, and thus the matrix has rank S (assuming that the number of sources S is less than number of sensors M). The matrix $\mathbf{S}_{vv}$ is of order $S \times S$ and will be of rank S provided that the sources are either uncorrelated with each other or only partly correlated with each other. One therefore concludes that the $M \times M$ matrix $\mathbf{GS}_{vv}\mathbf{G}^H$ will have rank S.

For the purposes of further illustration, assume now that all the sources are uncorrelated and of the same strength such that $\mathbf{S}_{vv} = S_{vv}\mathbf{I}$ where S_{vv} is the power spectral density of the source volume acceleration. The application of the SVD to the matrix $\mathbf{G}$ then shows that

$$\mathbf{S}_{ss} = S_{vv}\mathbf{GG}^H = S_{vv}(\mathbf{U\Sigma V}^H)(\mathbf{U\Sigma V}^H)^H \tag{3.77}$$

Since $(\mathbf{U\Sigma V}^H)^H = \mathbf{V\Sigma U}^H$ and $\mathbf{V}^H\mathbf{V} = \mathbf{I}$, it follows that

$$\mathbf{S}_{ss} = S_{vv}\mathbf{U\Sigma\Sigma}^H\mathbf{U}^H \tag{3.78}$$

This is equivalent to an eigen decomposition of the matrix $\mathbf{S}_{vv}$ given by

$$\mathbf{S}_{ss} = \mathbf{E}_s\Lambda_s\mathbf{E}_s^H \tag{3.79}$$

where $\mathbf{E}_s$ is the matrix of normalised eigenvectors $\mathbf{e}_i$ which comprise the columns of $\mathbf{E}_s$, and Λ_s is the diagonal matrix of eigenvalues which in this case is given by

$$\Lambda_s = S_{vv}\Sigma^H\Sigma = S_{vv}\begin{bmatrix} \sigma_1^2 & 0 & \cdot & 0 & \cdot & \cdot & 0 \\ 0 & \sigma_2^2 & \cdot & 0 & \cdot & \cdot & \cdot \\ \cdot & \cdot & \cdot & \cdot & \cdot & \cdot & \cdot \\ 0 & \cdot & \cdot & \sigma_s^2 & \cdot & \cdot & \cdot \\ 0 & \cdot & \cdot & \cdot & 0 & \cdot & \cdot \\ \cdot & \cdot & \cdot & \cdot & \cdot & \cdot & \cdot \\ 0 & \cdot & \cdot & \cdot & \cdot & \cdot & 0 \end{bmatrix} \tag{3.80}$$

Thus the matrix $\mathbf{S}_{xx}$ will have S non-zero eigenvalues and $(M - S)$ zero eigenvalues.

Now we assume that all the noise signals are uncorrelated and have equal power spectra such that $S_{nn} = S_{nn}I$. The cross-spectral matrix of sensor signals can be written as

$$S_{xx} = S_{vv}U\Sigma\Sigma^H U^H + S_{nn}I \qquad (3.81)$$

Since $UU^H = I$ this expression can also be written as

$$S_{xx} = U(S_{vv}\Sigma\Sigma^H + S_{nn}I)U^H \qquad (3.82)$$

Thus the matrix Λ_x of eigenvalues of S_{xx} can be written as

$$\Lambda_x = S_{vv}\Sigma\Sigma^H + S_{nn}I \qquad (3.83)$$

The first S eigenvalues of S_{xx} are given by $(S_{vv}\sigma_1^2 + S_{nn})$ and the remaining $(M - S)$ eigenvalues are given by S_{nn}.

This argument can also be generalised to include the case where the sources are not assumed to be uncorrelated and of equal strength. Under these circumstances one can express the matrix S_{xx} as

$$S_{xx} = E_s\Lambda_s E_s^H + S_{nn}I \qquad (3.84)$$

where $E_s\Lambda_s E_s^H$ is again the eigen decomposition of S_{ss} and since $E_s E_s^H = I$, one may write

$$S_{xx} = E_s(\Lambda_s + S_{nn}I)E_s^H \qquad (3.85)$$

such that the eigenvectors of S_{xx} are the same as those of S_{ss}.

Since all the eigenvectors e_i are mutually orthogonal (i.e. $e_i^H e_j = 0$, $i \neq j$), one may view the eigen decomposition of S_{xx} as consisting of a 'signal subspace' and a 'noise subspace'. The signal subspace consists of the S non-zero eigenvalues of S_{ss} and the associated eigenvalues, whilst the noise subspace consists of the remaining $M - S$ eigenvectors which correspond to the zero eigenvalues of S_{ss}. Since the eigenvalues comprising the noise subspace are orthogonal to those in the signal subspace it follows that

$$GS_{vv}G^H e_j = \left(\sum_{i=1}^{M}\lambda_i e_i e_i^H\right)e_j = O \quad \text{for} \quad j = (S+1)\dots M \qquad (3.86)$$

Since S_{vv} and G are both of full rank, it therefore follows that

$$G^H e_j = 0 \qquad (3.87)$$

and the so-called 'steering vectors' g comprising the columns of G are orthogonal to the eigenvectors comprising the noise subspace.

These observations lead to a class of methods for dealing with the solution for the optimal vector $\mathbf{w}_{opt}$ given by equation (3.65). The *eigenvector method* evaluates the inverse matrix $\mathbf{S}_{xx}^{-1}$ by retaining only the eigenvalues associated with the noise subspace. Thus the solution given by equation (3.65) is written as

$$\mathbf{w}_{opt} = \frac{\mathbf{S}_{ev}^{-1}\mathbf{g}}{\mathbf{g}^H\mathbf{S}_{ev}^{-1}\mathbf{g}} \tag{3.88}$$

where

$$\mathbf{S}_{ev}^{-1} = \sum_{i=S+1}^{M} \frac{1}{\lambda_i}\mathbf{e}_i\mathbf{e}_i^H \tag{3.89}$$

Note that the power output of the beamformer may be written as

$$P = E[|Y(e^{j\omega})|^2] = \mathbf{w}^H\mathbf{S}_{xx}\mathbf{w} \tag{3.90}$$

which on substitution of equation (3.65) shows that

$$P = \frac{1}{\mathbf{g}^H\mathbf{S}_{xx}^{-1}\mathbf{g}} \tag{3.91}$$

Now, replacing $\mathbf{S}_{xx}^{-1}$ by $\mathbf{S}_{ev}^{-1}$ then shows that

$$P = \left[\mathbf{g}^H\left(\sum_{i=S+1}^{M} \frac{1}{\lambda_i}\mathbf{e}_i\mathbf{e}_i^H\right)\mathbf{g}\right]^{-1} \tag{3.92}$$

which can also be written as

$$P = \left[\sum_{i=S+1}^{M} \frac{1}{\lambda_i}|\mathbf{g}^H\mathbf{e}_i|^2\right]^{-1} \tag{3.93}$$

However, since by virtue of equation (3.87) all of the columns $\mathbf{g}$ comprising the matrix $\mathbf{G}$ are orthogonal to the noise subspace eigenvectors $\mathbf{e}_i$ (for $i = (S+1)\ldots M$), it follows that the term in square brackets of (3.93) is zero and an infinite value of P results. In other words, the power output is infinite in those 'look directions' defined by the vectors $\mathbf{g}$ in the signal subspace. The MUSIC (multiple signal classification) method due to Schmidt (1986) alternatively substitutes the inverse matrix $\mathbf{S}_M^{-1}$ into the solution where

$$\mathbf{S}_M^{-1} = \sum_{i=S+1}^{M} \mathbf{e}_i\mathbf{e}_i^H \tag{3.94}$$

and sets the small noise eigenvalues to a value of unity.

Whilst techniques based on methods such as those described above have been widely studied, it should be pointed out that this work has been conducted within the framework of the detection of far field sources rather than the accurate estimation of their strength *per se*. Thus the fact that the beamformer output tends to infinity in the look directions is of little help in the accurate estimation of the strength (volume acceleration) of the source.

3.7   Estimation of the strength of multiple sources

The solution given in Section 3.5 for the optimal estimation of acoustic source strength in the presence of noise is readily extended to the case of multiple sources. In this case the sensor output signals are defined to have Fourier transforms given by the elements of the vector

$$\mathbf{x} = \mathbf{s} + \mathbf{n} = \mathbf{Gv} + \mathbf{n} \tag{3.95}$$

where $\mathbf{s} = \mathbf{Gv}$ is the vector of signals due to the acoustic sources whose strengths (volume accelerations) are defined by the vector $\mathbf{v}$. The matrix $\mathbf{G}$ is again used to represent the frequency response functions (Green functions) relating the acoustic source strengths $\mathbf{v}$ to the signals $\mathbf{s}$ produced at the sensor array. The vector $\mathbf{n}$ denotes the Fourier transforms of the noise signals produced at the sensor array. It is now assumed that we wish to produce a vector of output signals whose Fourier transforms are the elements of the vector $\mathbf{y}$ which constitutes an optimal estimate of the source strength vector $\mathbf{v}$. For consistency with the above analysis, we therefore assume that the sensor output signals are processed via a matrix of filters defined by $\mathbf{W}^H$ such that

$$\mathbf{y} = \mathbf{W}^H\mathbf{x} \tag{3.96}$$

The least squares criterion may again be used and we seek to minimise the sum of squared error signals $\mathbf{e} = \mathbf{y} - \mathbf{v}$ and define a cost function given by

$$J = E[\mathbf{e}^H\mathbf{e}] = \text{Tr}\{E[\mathbf{ee}^H]\} \tag{3.97}$$

where Tr{} denotes the trace (sum of the diagonal terms) of a matrix. Thus minimisation of the sum of the diagonal terms of the matrix $E[\mathbf{ee}^H]$ is equivalent to the minimisation of the sum of the squared error signals. This cost function can therefore be written as

$$J = \text{Tr}\{E[(\mathbf{W}^H\mathbf{x} - \mathbf{v})(\mathbf{W}^H\mathbf{x} - \mathbf{v})^H]\} \tag{3.98}$$

which after expansion and use of the expectation operator results in

$$J = \text{Tr}\{\mathbf{W}^H\mathbf{S}_{xx}\mathbf{W} - \mathbf{W}^H\mathbf{S}_{vx}^H - \mathbf{S}_{vx}\mathbf{W} + \mathbf{S}_{vv}\} \tag{3.99}$$

where the spectral density matrices are defined by

$$S_{xx} = E[\mathbf{x}\mathbf{x}^H], \quad S_{vx} = E[\mathbf{v}\mathbf{x}^H], \quad S_{vv} = E[\mathbf{v}\mathbf{v}^H] \tag{3.100}$$

In order to find the minimum of J, we make use of two matrix identities involving the derivative of the trace of a matrix. Full details are given by Elliott (2001) who also shows that if

$$J = \text{Tr}\{\mathbf{ABA}^H + \mathbf{AC}^H + \mathbf{CA}^H\} \tag{3.101}$$

then the derivative of J with respect to the real and imaginary parts of $\mathbf{A}$ is given by

$$\frac{\partial J}{\partial \mathbf{A}_R} + j\frac{\partial J}{\partial \mathbf{A}_I} = 2\mathbf{AB} + 2\mathbf{C} \tag{3.102}$$

Thus, identifying $\mathbf{A}$ with $\mathbf{W}^H$ and setting the derivative of the cost function (equation (3.99)) to zero show that J is minimised when

$$2\mathbf{W}^H S_{xx} - 2S_{vx} = 0 \tag{3.103}$$

which shows that the solution for the optimal filter matrix can be written as

$$\mathbf{W}_{\text{opt}}^H = S_{vx}S_{xx}^{-1} \tag{3.104}$$

Note also that, since $S_{xx}^H = S_{xx}$ and that $S_{xx}^H = E[(\mathbf{v}\mathbf{x}^H)^H] = S_{xv}$, the solution can be written in the form

$$\mathbf{W}_{\text{opt}} = S_{xx}^{-1}S_{xv} \tag{3.105}$$

which shows that the optimal filter matrix consists of columns $\mathbf{w}_{\text{sopt}}$ given by

$$\mathbf{w}_{\text{sopt}} = S_{xx}^{-1}\mathbf{s}_{xs} \tag{3.106}$$

where the vector $\mathbf{s}_{xs}$ has elements consisting of the cross-spectra between the M sensor signals and the sth source signal (i.e. $S_{xs} = E[\mathbf{x}V_s^*(e^{j\omega T})]$). This demonstrates that the optimal solution for the estimation of the strength of multiple sources appears to consist of the superposition of all the solutions of the type given by equation (3.50) for the estimation of the strength of a single source. It should be recognised that the properties of the matrix S_{xx} will be determined by both the number and properties of the sources present in addition to the properties of the interfering noise.

Provided that the noise signals **n** are uncorrelated with the source signals **v**, then we may also write

$$S_{xx} = E[(Gv + n)(Gv + n)^H] = GS_{vv}G^H + S_{nn} \tag{3.107}$$

where the cross-spectral matrix of noise signals $S_{nn} = E[nn^H]$. Since, under these circumstances, we may also write

$$S_{vx} = E[v(Gv + n)^H] = S_{vv}G^H \tag{3.108}$$

then the optimal solution may be written as

$$W_{opt}^H = S_{vv}G^H[GS_{vv}G^H + S_{nn}]^{-1} \tag{3.109}$$

Thus the solution for the optimal filter matrix is clearly seen to depend on the form of the matrix S_{vv} of cross-spectra associated with the multiple sources. The properties of this solution have been explored by Stoughton and Strait (1993) and by Nelson (2001) who points out the procedures necessary to use this solution in order to reconstruct the time histories of the acoustic source strengths. It is also worth noting that this result may be written in an alternative form through application of the matrix inversion Lemma given by equation (3.55). This shows that we may write

$$W_{opt}^H = S_{vv}G^H \left[S_{nn}^{-1} - S_{nn}^{-1}G \left(S_{vv}^{-1} + G^H S_{nn}^{-1}G \right)^{-1} G^H S_{nn}^{-1} \right] \tag{3.110}$$

As described by Kailath *et al.* (2000) the terms in the square brackets can then be manipulated to show that

$$\begin{aligned}
S_{nn}^{-1} - &S_{nn}^{-1}G \left(S_{vv}^{-1} + G^H S_{nn}^{-1}G \right)^{-1} G^H S_{nn}^{-1} \\
&= \left[S_{nn}^{-1} \left[\left(S_{vv}^{-1} + G^H S_{nn}^{-1}G \right)^{-1} G^H S_{nn}^{-1} \right]^{-1} - S_{nn}^{-1}G \right] \\
&\quad \times \left(S_{vv}^{-1} + G^H S_{nn}^{-1}G \right)^{-1} G^H S_{nn}^{-1} \\
&= \left[G^{H-1} \left(S_{vv}^{-1} + G^H S_{nn}^{-1}G \right) - S_{nn}^{-1}G \right] \\
&\quad \times \left(S_{vv}^{-1} + G^H S_{nn}^{-1}G \right)^{-1} G^H S_{nn}^{-1}
\end{aligned} \tag{3.111}$$

Substitution of this result into equation (3.110) then shows that

$$\begin{aligned}
W_{opt}^H = &[S_{vv} \left(S_{vv}^{-1} + G^H S_{nn}^{-1}G \right) - S_{vv}G^H S_{nn}^{-1}G] \\
&\left(S_{vv}^{-1} + G^H S_{nn}^{-1}G \right)^{-1} G^H S_{nn}^{-1}
\end{aligned} \tag{3.112}$$

and, since the term in the square brackets reduces to the identity matrix, it follows that the optimal solution may be written as

$$\mathbf{W}_{\mathrm{opt}}^{\mathrm{H}} = \left[\mathbf{S}_{vv}^{-1} + \mathbf{G}^{\mathrm{H}}\mathbf{S}_{nn}^{-1}\mathbf{G}\right]^{-1}\mathbf{G}^{\mathrm{H}}\mathbf{S}_{nn}^{-1} \tag{3.113}$$

This form of the solution is sometimes referred to as being expressed in 'information form', since it is written in terms of the inverses of the relevant cross-spectral matrices (see Kailath *et al.* 2000 for a further discussion).

A particularly important special case of the solution follows from this result. If we have very little prior knowledge of the source strength cross-spectral matrix $\mathbf{S}_{vv}$, it is argued by Kailath *et al.* (2000) that $\mathbf{S}_{vv} = \alpha\mathbf{I}$ say where $\alpha \gg 1$ and thus in the limit $\alpha \to \infty$, the solution reduces to

$$\mathbf{W}_{\mathrm{opt}}^{\mathrm{H}} = \left[\mathbf{G}^{\mathrm{H}}\mathbf{S}_{nn}^{-1}\mathbf{G}\right]^{-1}\mathbf{G}^{\mathrm{H}}\mathbf{S}_{nn}^{-1} \tag{3.114}$$

This solution is precisely that which results from the Gauss–Markov Theorem describing the optimal unbiased linear least-mean-squares estimator of the source strength vector $\mathbf{v}$ when this is assumed to be a deterministic (rather than a random) variable (Kailath *et al.* 2000). The practical significance of the use of equation (3.114) rather than equation (3.113) can be examined by assuming respectively that $\mathbf{S}_{w} = S_{vv}\mathbf{I}$ and $\mathbf{S}_{nn} = S_{nn}\mathbf{I}$ and again defining the signal to noise ratio $\gamma = S_{nn}/S_{vv}$ such that equation (3.113) becomes

$$\mathbf{W}_{\mathrm{opt}}^{\mathrm{H}} = \left[\gamma\mathbf{I} + \mathbf{G}^{\mathrm{H}}\mathbf{G}\right]^{-1}\mathbf{G}^{\mathrm{H}} \tag{3.115}$$

and thus when $\gamma \to 0$ and the noise signal is vanishingly small compared with the source strength signal, then

$$\mathbf{W}_{\mathrm{opt}}^{\mathrm{H}} \to \left[\mathbf{G}^{\mathrm{H}}\mathbf{G}\right]^{-1}\mathbf{G}^{\mathrm{H}} \tag{3.116}$$

Therefore, under these circumstances, the estimation of the acoustic source strength depends only on the inversion of the matrix of frequency response functions relating the source strength fluctuations to the sensor outputs.

3.8   Source strength estimation with inverse methods

We will now consider further the estimation of acoustic source strength using the result given by equation (3.116). Early applications of this approach include those by Fillipi *et al.* (1988), Kim and Lee (1990), Veronesi and Maynard (1989) and Tester and Fisher (1991). More recently there have been applications to problems in aeroacoustics such as those presented by Grace *et al.* (1996a,b) and Fisher and Holland (1997). First note that equation (3.116) is precisely the result that follows from a deterministic least squares approach to the estimation of v. Thus we seek the solution for the source strength vector v that minimises the sum of squared errors between the signals x received at the sensors and the signals s = Gv produced by the assumed model of (or indeed measurements of) the sound transmission process. We therefore seek to minimise

$$J = (x - s)^H(x - s) = (x - Gv)^H(x - Gv) \tag{3.117}$$

which results in the quadratic form

$$J = v^H G^H G v - v^H G^H x - x^H G v + x^H x \tag{3.118}$$

that is minimised by

$$v_{opt} = [G^H G]^{-1} G^H x \tag{3.119}$$

If v_{opt} is now interpreted as the output of the filter matrix W_{opt}^H such that $v_{opt} = y = W_{opt}^H x$, it becomes clear that the solution for W_{opt} is given by equation (3.116). Note that $W_{opt}^H = [G^H G]^{-1} G^H = G^+$ where G^+ is known as the 'pseudo inverse' of G. Of central importance to the estimation of source strength through the use of equation (3.119) is the 'condition number' of the matrix G. The singular value decomposition of the matrix G given by equation (3.71) shows that the expression for the optimal filter matrix may be written as

$$W_{opt}^H = \left[(U\Sigma V^H)^H (U\Sigma V^H)\right]^{-1} (U\Sigma V^H)^H \tag{3.120}$$

and upon use of the properties of the unitary matrices U and V given by $U^H U = U U^H = I$ and $V^H V = V V^H = I$, this expression becomes

$$W_{opt}^H = [V\Sigma^H \Sigma V^H]^{-1} V\Sigma^H U^H \tag{3.121}$$

which further reduces to

$$W_{opt}^H = V\Sigma^+ U^H \tag{3.122}$$

where $\Sigma^+ = [\Sigma^H \Sigma]^{-1} \Sigma^H$. Note that using the example given in equation (3.75) it can be shown that this matrix has the form

$$\Sigma^+ = \begin{bmatrix} \dfrac{1}{\sigma_1} & 0 & 0 \\ 0 & \dfrac{1}{\sigma_2} & 0 \end{bmatrix} \tag{3.123}$$

and therefore small singular values give rise to large terms in the solution for the inverse matrix. Specifically it can be shown (see, for example, Golub and Van Loan 1989) that if the signal s is perturbed by an amount δs, the perturbation δv that results in the solution for the optimal source strength vector, where $\mathbf{G}(\mathbf{v} + \delta \mathbf{v}) = \mathbf{s} + \delta \mathbf{s}$, is given by

$$\frac{||\delta \mathbf{v}||}{||\mathbf{v}||} \leq \kappa(\mathbf{G}) \frac{||\delta \mathbf{s}||}{||\mathbf{s}||} \tag{3.124}$$

where $\kappa(\mathbf{G})$ is the condition number of the matrix $\mathbf{G}$ defined by

$$\kappa(\mathbf{G}) = ||\mathbf{G}|| \, ||\mathbf{G}^{-1}|| = \sigma_{\max} / \sigma_{\min} \tag{3.125}$$

and $||\cdot||$ denotes the two-norm of the matrix or vector (see, for example, Golub and Van Loan 1989). The condition number is thus defined by the ratio of the maximum to the minimum singular value of the matrix $\mathbf{G}$. A well-conditioned matrix $\mathbf{G}$ with a low value of $\kappa(\mathbf{G})$ will therefore produce relatively small errors in the estimated solution, whereas if $\kappa(\mathbf{G})$ is large, the errors in the solution resulting from errors in detecting the signals will be amplified in proportion to the condition number.

A simple example that illustrates the dependence of the condition number of the Green function matrix associated with acoustic radiation is that associated with a linear array of point sources. Figure 3.11 shows a two-dimensional arrangement of harmonic point sources and point sensors where it is assumed that the sources are in a free field and radiate in accordance with the Green function

$$G(kr) = \rho_0 e^{-jkr} / 4\pi r \tag{3.126}$$

Figure 3.12 shows the dependence of the singular values of the matrix $\mathbf{G}$ as the sensor array is moved away from the source array. It is clear that, for

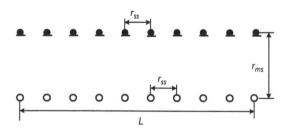

Figure 3.11 A two-dimensional arrangement of point sources and sensors used to illustrate the estimation of acoustic source strength by inverse methods.

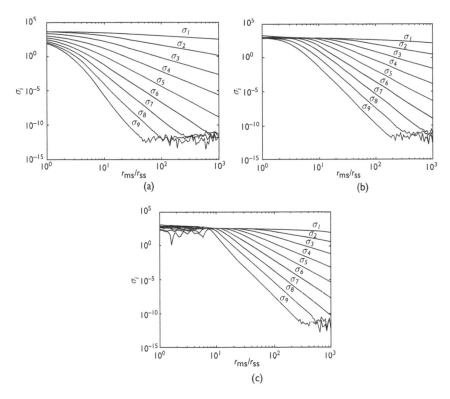

Figure 3.12 Decay of the singular values of the Green function matrix **G** associated with the source/sensor arrangement shown in Figure 3.10 when (a) $r_{ss} = \lambda/10$; (b) $r_{ss} = \lambda/2$; (c) $r_{ss} = 2\lambda$.

the case shown, the smallest singular values of **G** decay extremely rapidly. In fact the decay appears to be well approximated by a straight line over much of the logarithmic scale shown, implying an exponential decay of the singular values with increasing distance of the sensor array from the source array. It is therefore apparent from this simple example that there will be instances where the matrix of Green functions is poorly conditioned and therefore extremely prone to errors in the solution for the reconstructed source strength.

One approach to this difficulty is to simply ignore the small singular values in the reconstruction process and omit them from the matrix Σ^+ in equation (3.122). This 'singular value discarding' approach can prove effective in arriving at a useful solution for the source strength but the question remains as to how to choose which singular values to discard, unless of course there is an obvious cut-off where the singular values suddenly become relatively very small. Another approach is to regularise the problem

(Tikhonov and Arsenin 1977). This amounts to adding a term to the cost function J such that equation (3.117) is rewritten in the form

$$J = (\mathbf{x} - \mathbf{s})^{\mathrm{H}}(\mathbf{x} - \mathbf{s}) + \beta \mathbf{v}^{\mathrm{H}}\mathbf{v} \tag{3.127}$$

such that the solution for the optimal filter matrix becomes

$$\mathbf{W}_{\mathrm{opt}}^{\mathrm{H}} = [\mathbf{G}^{\mathrm{H}}\mathbf{G} + \beta \mathbf{I}]^{-1}\mathbf{G}^{\mathrm{H}} \tag{3.128}$$

The action of the regularisation parameter β can be explained by examining this solution within the context of the SVD. Repetition of the operations leading to equation (3.122) shows that the solution can be written in the form

$$\mathbf{W}_{\mathrm{opt}}^{\mathrm{H}} = \mathbf{V}\Sigma_R^+\mathbf{U}^{\mathrm{H}} \tag{3.129}$$

where $\Sigma_R^+ = [\Sigma^{\mathrm{H}}\Sigma + \beta \mathbf{I}]^{-1}\Sigma^{\mathrm{H}}$, the subscript R denoting the regularised solution. Again using the example given in equation (3.75), it can be shown that

$$\Sigma_R^+ = \begin{bmatrix} \dfrac{\sigma_1}{\sigma_1^2 + \beta} & 0 & 0 \\[2mm] 0 & \dfrac{\sigma_2}{\sigma_2^2 + \beta} & 0 \end{bmatrix} \tag{3.130}$$

and therefore if σ_2, say, is very small, the term $\sigma_2/(\sigma_2 + \beta) \to 1/\beta$ as $\sigma_2 \to 0$. This therefore prevents the term from 'blowing up' and disproportionately amplifying any noise present.

The effectiveness of regularisation in retrieving useful solutions to ill-conditioned inversion problems can again be illustrated with reference to the two-dimensional radiation example described above. Figure 3.13 shows the result of using equation (3.12) to deduce the estimates of the strengths of the sources when one of the sources has unit strength and the other sources have zero strength. The results for the reconstructed source strength are shown as a function of the regularisation parameter β at a particular frequency and particular value of separation distance between the source array and the receiver array. It is evident that a very small value of β produces significantly large errors in the solution. As β is increased, the solution converges to a more accurate (but still imperfect) estimate of the source strength distribution. Further increases in β result in a further deterioration in the accuracy of the solution.

It also appears from these results that the introduction of regularisation reduces the effective spatial resolution of the technique. That is to say, the reconstructed source strength distribution is a 'smeared' image of the actual source distribution; only one source has a finite strength and yet

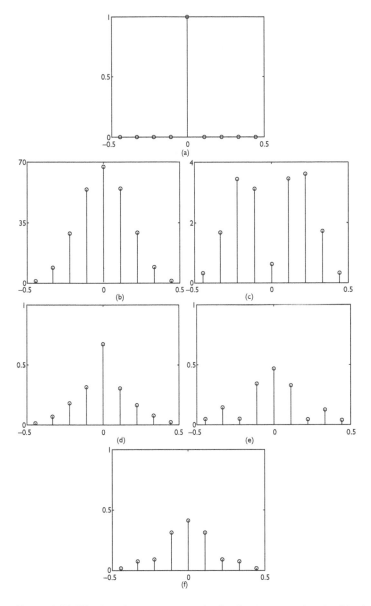

Figure 3.13 The 'true' source strength distribution associated with the geometry shown in Figure 3.10 is shown in (a). Estimated values of the strengths of the sources are shown when the regularised inverse solution is used with (b) $\beta = 0$; (c) $\beta = 0.1$; (d) $\beta = 1$; (e) $\beta = 100$; (f) $\beta = 10^4$, when the source and sensor array are separated by $r_{ms} = 2L$ and $r_{ss} = \lambda/2$. The measured data have been computed assuming that a 10 per cent noise signal has been added to the pressures produced.

the neighbouring sources are also attributed with a finite strength (see, for example, the results shown in Figure 3.13d). The matter of spatial resolution will be returned to later, but it is worth emphasising here that the SVD can be used to describe this phenomenon in terms of spatial frequency. Thus, in the noise free case, the signals $\mathbf{x}$ produced at the sensor array are given by $\mathbf{x} = \mathbf{Gv}$ and use of the SVD together with the property $\mathbf{U}^H\mathbf{U} = \mathbf{I}$ shows that

$$\mathbf{U}^H\mathbf{x} = \Sigma\mathbf{V}^H\mathbf{v} \tag{3.131}$$

This can also be written in the form $\tilde{\mathbf{x}} = \Sigma\tilde{\mathbf{v}}$ where $\tilde{\mathbf{x}} = \mathbf{U}^H\mathbf{x}$ and $\tilde{\mathbf{v}} = \mathbf{V}^H\mathbf{v}$ can be regarded as spatially transformed variables. Thus the ith element of $\tilde{\mathbf{x}}$ (given by $\tilde{x}_i$, say) is related to the ith element of $\tilde{\mathbf{v}}$ by $\tilde{x}_i = \sigma_i\tilde{v}_i$ and the singular value σ_i relates the amplitude of the ith 'spatial mode' of the source distribution to the ith 'spatial mode' of the sensor distribution. The form of these spatial modes, as basis functions, is described by the left and right singular vectors $\mathbf{u}_i$ and $\mathbf{v}_i$ comprising the columns of the matrices $\mathbf{U}$ and $\mathbf{V}$. As noted by Nelson (2001) there is a tendency for the higher spatial frequencies to be associated with the smaller singular values. Thus the suppression of the contribution of these as a result of regularisation will act to limit spatial resolution.

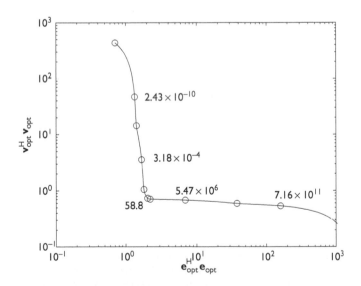

Figure 3.14 The L-curve plotted as a function of the regularisation parameter β when the source and sensor array depicted in Figure 3.10 are separated by $r_{ms} = 2L$ and $r_{ss} = \lambda/2$. It has again been assumed that the pressure signals measured include 10 per cent noise.

3.9   Regularisation techniques

As implied by the simulations presented above, there is often an optimal choice of regularisation parameter β. In essence, the balance that has to be struck is to choose β so that the solution is stabilised, but without undue deterioration in the accuracy of the solution (as manifested by, for example, undue suppression of the high spatial frequency information as described above). A very simple method for examining this balance is to plot the 'L-curve' that results from plotting (on a logarithmic scale) the magnitude of the regularised solution (given by $v_{opt}^H v_{opt}$ where $v_{opt} = W_{opt}^H x$) against the magnitude of the minimum mean squared error (given by $e_{opt}^H e_{opt}$ where $e_{opt} = x - G v_{opt}$). The L-curve associated with the solution for the same estimation problem as that depicted in Figure 3.13 is plotted as a function of β in Figure 3.14.

It is evident that as β is increased, there is a steady fall in the magnitude of the solution ($v_{opt}^H v_{opt}$) which is represented by the vertical part of the L in the curve that results. The optimal value of β results in a knee of the L-curve and increase of β beyond this optimal value simply increases the magnitude of the error ($e_{opt}^H e_{opt}$) without decreasing the magnitude of the solution. The horizontal part of the L-curve then results. More details of the use and limitations of the L-curve method are presented by Hansen (1992, 1998) and Hansen and O'Leary (1993).

A further range of techniques for the determination of optimal regularisation parameters is provided by cross-validation methods. These have their origins in the 'leaving one out' method (Allen 1974) that was originally applied to statistical problems such as fitting curves to uncertain data (Wahba and Wold 1975; Craven and Wahba 1979; Wahba 1990). The basis of the approach is to find the optimal vector of complex source strengths $v_R(\beta, k)$ that minimises

$$J(\beta, k) = \sum_{\substack{m=1 \\ m \neq k}}^{M} [x_m - s_m]^2 + \beta v^H v \tag{3.132}$$

which is the regularised cost function of equation (3.127) with the kth values of measured and modelled sensor signals omitted. The optimal value of $v_R(\beta, k)$ that minimises this function is then used to predict the value of the sensor signal x_k that was left out of the calculation. Denoting this predicted value as $x_k(\beta, k)$ we then define the Ordinary Cross Validation (OCV) function by

$$V_{ord}(\beta) = \frac{1}{M} \sum_{k=1}^{M} [x_k - x_k(\beta, k)]^2 \tag{3.133}$$

which is a summation over all the sensors of the errors between the predicted signals (when left out) and the actually measured signals. This calculation is therefore repeated for each sensor in turn in order to define the OCV function. It can be shown (see Golub *et al.* 1979 or Yoon and Nelson 2000) that this function can be expressed in terms of the solution $x_{opt} = Gv_{opt}$ that results from the minimisation of the full cost function defined in equation (3.127). Denoting the kth element of x_{opt} as $x_{opt,k}$, it can be shown that

$$V_{ord}(\beta) = \frac{1}{M} \sum_{k=1}^{M} \left[\frac{x_k - x_{opt,k}}{1 - b_{kk}} \right]^2 \tag{3.134}$$

where the terms b_{kk} are the diagonal elements of the matrix

$$B(\beta) = G(G^H G + \beta I)^{-1} G^H \tag{3.135}$$

which is known as the influence matrix. The essence of the OCV method is thus to find β that minimizes $V_{ord}(\beta)$. That is, the value of β is sought that effectively minimises the sum of squared deviations of the predictions of the sensor signals that result from the regularised solution.

Golub *et al.* (1979) extended the OCV technique with the introduction of Generalised Cross Validation (GCV). They suggested that the choice of β should be invariant to an orthogonal transformation of the sensor signals. They therefore use the SVD to write the relationship between the sensor signals and the source strengths as

$$U^H x = \Sigma V^H v + U^H e \tag{3.136}$$

and then apply the transformation of pre-multiplication by the Fourier matrix F such that

$$FU^H x = F\Sigma V^H v + FU^H e \tag{3.137}$$

where the (s, m)th element of F is given by $(1/\sqrt{M})e^{2\pi j sm/M}$. Thus the transformed sensor signals $x_{tr} = FU^H x$; similarly we can define a transformed Green function matrix G_{tr}, source strength vector v_{tr} and error vector e_{tr} such that

$$x_{tr} = G_{tr} v_{tr} + e_{tr} \tag{3.138}$$

Application of the OCV method to this transformed model then results in the GCV function given by

$$V_{gen}(\beta) = \frac{\dfrac{1}{M}||(I - B(\beta))x||_e^2}{\left[\dfrac{1}{M} \text{Tr}(I - B(\beta)) \right]^2} \tag{3.139}$$

where $\| \ \|_e^2$ denotes the Euclidean norm and $\text{Tr}()$ denotes the trace. The relationship of the GCV function to the OCV function can be demonstrated by writing the GCV function as

$$V_{\text{gen}}(\beta) = \frac{1}{M} \sum_{k=1}^{M} \left[\frac{x_k - x_{\text{opt},k}}{1 - b_{kk}} \right]^2 w_{kk} \qquad (3.140)$$

where the weights w_{kk} are given by

$$w_{kk} = \left(\frac{2 - b_{kk}}{1 - \dfrac{1}{M} \text{Tr}(\mathbf{B}(\beta))} \right)^2 \qquad (3.141)$$

The GCV function again attempts to strike the balance between the sum of squared errors (which can be shown to be proportional to the numerator of equation (3.139)) and the perturbation of the matrix to be inverted by the action of regularisation, as quantified by the denominator of the expression. Simple search techniques can be used to find the maximum of the GCV function although, as described by Hansen (1998), the function can have relatively broad minima which makes the optimal value of β less easy to define. The L-curve on the other hand, very often gives a clear definition of the optimal regularisation parameter, but numerical methods for establishing this value are less straightforward. An illustration of the GCV functions computed for the example given above is shown in Figure 3.15 and this tends to confirm Hansen's observation. Practical examples of the performance of the cross-validation technique are presented by Yoon and Nelson (2000) (see also, Nelson and Yoon 2000; Nelson 2001).

A good example of the application of the inverse method to a problem of considerable practical significance is that presented by Schumacher *et al.* (2003). This work describes the reconstruction of the source strength distribution associated with a vehicle tyre in contact with a laboratory-based rolling road. The boundary element method was used to determine the matrix of Green functions relating the tyre surface source strength distribution to the radiated field, and the inversion of this matrix was optimally regularised using the L-curve method. Some results of this process are shown in Figure 3.16.

3.10 The Fourier transform relationship between source and field

A further approach to the source strength estimation problem is to make use of the classical Fourier transform relationships between acoustic source distributions and their far field. This can be illustrated with reference to the

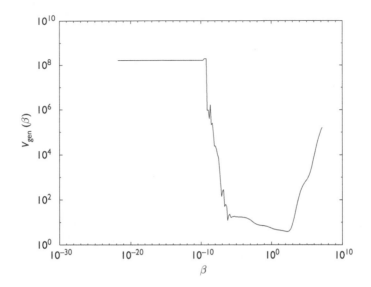

Figure 3.15 The GCV function when the source and sensor array depicted in Figure 3.10 are separated by $r_{ms} = 2L$ and $r_{ss} = \lambda/2$. A 10 per cent noise signal is again assumed to contaminate the measured pressures.

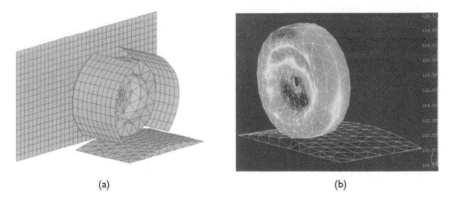

(a)   (b)

Figure 3.16 Results of the application of inverse methods to the reconstruction of the source strength distribution associated with a rolling tyre presented by Schumacher *et al.* (2003). (a) The boundary element model used in computing the forward problem; (b) results for the reconstruction of the surface velocity. Results are shown for a tyre velocity of 80 km/h at a frequency of 60 Hz.

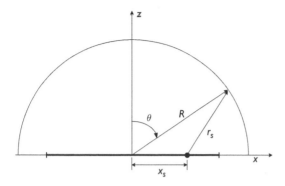

Figure 3.17 A line monopole source distribution on the x-axis.

monopolar line source distribution and the geometry shown in Figure 3.17. The complex acoustic pressure $p(R, \theta)$ generated by a line source distribution having a volume acceleration per unit length given by $v(x_s)$ is given by

$$p(R, \theta) = \int_{-\infty}^{\infty} \frac{\rho_0 v(x_s) e^{-jkr_s}}{4\pi r_s} \, dx_s \qquad (3.142)$$

where x_s represents the position of an elemental source on the x-axis and r_s is the distance from this source to the field point. The coordinates (R, θ) are sufficient since the radiated field is cylindrically symmetric about the x-axis. It follows from simple geometry that

$$r_s = \left[R^2 \cos^2 \theta + (R \sin \theta - x_s)^2 \right]^{\frac{1}{2}} \qquad (3.143)$$

which can also be written as

$$r_s = \left[R^2 + x_s^2 - 2Rx_s \sin \theta \right]^{\frac{1}{2}} = R \left[1 + \frac{x_s^2}{R^2} - \frac{2x_s}{R} \sin \theta \right]^{\frac{1}{2}} \qquad (3.144)$$

Use of the binomial expansion $(1 + \alpha)^{\frac{1}{2}} = 1 + (\alpha/2) - (\alpha^2/8) \dots$ then shows that as $R \to \infty$, we can make the approximation $r_s = R - x_s \sin \theta$ in the exponential term in equation (3.142). Also approximating $(1/r_s)$ by $(1/R)$ then shows that in the far field of the source distribution

$$p(R, \theta) = \frac{\rho_0 e^{-jkR}}{4\pi R} \int_{-\infty}^{\infty} v(x_s) e^{jkx_s \sin \theta} \, dx_s \qquad (3.145)$$

Now, defining $\hat{k} = k \sin \theta$ makes it clear that the far field pressure is a spatial Fourier transform of the acoustic source strength distribution. That is, we may write

$$p(R, \theta) = \frac{\rho_0 e^{-jkR}}{4\pi R} V(\hat{k}) \qquad (3.146)$$

where $V(\hat{k})$ is the Fourier transform

$$V(\hat{k}) = \int_{-\infty}^{\infty} v(x_s)e^{j\hat{k}x_s} \, dx_s \tag{3.147}$$

It also follows that $v(x_s)$ can in turn be deduced by the inverse transform relationship

$$v(x_s) = \frac{1}{2\pi} \int_{-\infty}^{\infty} V(\hat{k})e^{-j\hat{k}x_s} \, d\hat{k} \tag{3.148}$$

It is therefore possible, in principle, to deduce $v(x_s)$ from a knowledge of the far field pressure $p(R, \theta)$ since

$$v(x_s) = \frac{1}{2\pi} \left[\frac{4\pi R e^{jkR}}{\rho_0} \right] \int_{-\infty}^{\infty} p(R, \theta)e^{-j\hat{k}x_s} \, d\hat{k} \tag{3.149}$$

A crucial observation can be made at this juncture by evaluating the expression for $V(x_s)$ that is produced if it is assumed that the pressure field $p(R, \theta)$ is simply that due to a point monopole source of unit strength at the origin. In this case, $p(R, \theta)$ is a uniform function of θ and is given simply by $\rho_0 e^{-jkR}/4\pi R$. Substitution into equation (3.149) then shows that

$$v(x_s) = \frac{1}{2\pi} \int_{-\infty}^{\infty} e^{-j\hat{k}x_s} \, d\hat{k} \tag{3.150}$$

In the far field however, the wavenumber $\hat{k} = k \sin \theta$ varies only over the range from $-k$ (i.e. when $\theta = -\pi/2$) to $+k$ (i.e. when $\theta = +\pi/2$) and therefore

$$v(x_s) = \frac{1}{2\pi} \int_{-k}^{k} e^{-j\hat{k}x_s} \, d\hat{k} \tag{3.151}$$

Evaluation of this integral (see Section 3.2) then shows that

$$v(x_s) = \frac{-1}{2\pi j x_s} \left(e^{-jkx_s} - e^{jkx_s} \right) \tag{3.152}$$

and therefore that

$$v(x_s) = \frac{\sin kx_s}{\pi x_s} \tag{3.153}$$

Thus the 'image' of the source distribution deduced by inverse wavenumber transformation of the far-field-radiated pressure field is actually a 'smeared' representation of the true (point) source distribution (the form of the function is represented in Figure 3.2b). This in turn implies a limit to which source distributions can be resolved by making use of far field data. The first zero of the function given by equation (3.153) occurs when $kx_s = \pi$ and thus when $x_s = \lambda/2$. It is well known in optics that the Rayleigh resolution limit precludes the clear identification of sources separated by less than a distance of one-half wavelength when one makes use of the exactly analogous Fourier transform relationship between source and field that governs the propagation of light. In acoustics however, it is also well known that making measurements closer to the source (in the near-field rather than the far field) can produce a resolution of the source distribution that is better than this classical Rayleigh limit. This will be returned to later.

In order to describe this procedure formally, we first define the far field directivity function (or normalised far field pressure) by $p(\hat{k}) = 4\pi R e^{jkR} p(R, \theta)/\rho_0$ such that equation (10.8) can be written as

$$v(x_s) = \frac{1}{2\pi} \int_{-\infty}^{\infty} p(\hat{k}) e^{-j\hat{k}x_s} \, d\hat{k} \tag{3.154}$$

Now assume that the microphones are used to sample the data at an integer number m of equispaced intervals $\Delta\hat{k}$ in the far field. This results in the evaluation of the Fourier transform of the sampled data defined by

$$v(e^{j2\pi x_s/\Delta\hat{k}}) = \frac{1}{2\pi} \int_{-\infty}^{\infty} \left[p(\hat{k}) \sum_{m=-\infty}^{\infty} \delta(\hat{k} - m\Delta\hat{k}) \right] e^{-j\hat{k}x_s} \, d\hat{k} \tag{3.155}$$

and exactly as described in Section 3.2, this Fourier transform repeats periodically such that

$$v(e^{j2\pi x_s/\Delta\hat{k}}) = \frac{1}{\Delta\hat{k}} \sum_{q=-\infty}^{\infty} v\left(x_s - q\frac{2\pi}{\Delta\hat{k}} \right) \tag{3.156}$$

where q is an integer and $2\pi/\Delta\hat{k}$ is the effective sampling frequency. Since the field is sampled only at a finite number of, say M, microphone positions, the estimate of this Fourier transform that can actually be retrieved when the integral is evaluated in equation (3.155) is given by

$$\hat{v}(e^{j2\pi x_s/\Delta\hat{k}}) = \frac{1}{2\pi} \sum_{m=0}^{M-1} p(m\Delta\hat{k}) e^{-jm\Delta\hat{k}x_s} \tag{3.157}$$

It is possible to evaluate this summation for different chosen values of x_s and thus build up an image of the source distribution along the x-axis. As emphasised above, however, this image will be limited in spatial resolution by 'wavenumber windowing'. Note also that we may choose a number of, say S, source positions along the x-axis spaced at intervals $s\Delta x$ such that at the sth position we evaluate

$$\hat{v}(e^{j2\pi_s\Delta x/\Delta\hat{k}}) = \frac{1}{2\pi} \sum_{m=0}^{M-1} p(m\Delta\hat{k})e^{-jm\Delta\hat{k}s\Delta x} \qquad (3.158)$$

This can be written in matrix notation as

$$\hat{v} = \mathbf{H}\mathbf{p} \qquad (3.159)$$

where the (s, m)th element of $\mathbf{H}$ is given by $(1/2\pi)e^{-jm\Delta\hat{k}s\Delta s}$. It is also worth noting that this relationship reduces to a DFT for a particular choice of the number of source positions S along the x-axis. It is necessary, however, for the number S to be equal to the number M of microphone positions. This follows since the sampling interval $\Delta\hat{k}$ results from splitting the range $-k$ to k into $2k/M$ equispaced samples. Thus, in the exponential term on the right hand side of equation (3.158), we can write

$$e^{-jm\Delta\hat{k}s\Delta x} = e^{-jm2k_s\Delta x/M} \qquad (3.160)$$

This term can be made equal to $e^{-j2\pi ms/M}$ provided that $2k\Delta x = 2\pi$; since $k = 2\pi/\lambda$, this will be satisfied provided that $\Delta x = \lambda/2$. That is, when the source positions along the x-axis are chosen to be separated by one-half of the acoustic wavelength at the frequency of interest, the matrix $\mathbf{H}$ becomes equal to the Fourier matrix $\mathbf{F}$.

Now consider the use of the inverse Fourier transform relationship given by equation (3.149) in order to deduce the distribution of source strength $v(x_s)$ from measurement of the acoustic pressure made at, for example, discrete microphone locations in the far field. The mere fact that θ varies only between $-\pi/2$ and $\pi/2$ automatically ensures that a wavenumber window is effectively applied to the continuous far field pressure data and this results in the spatial resolution limit described above. The act of *sampling* the data (by using discrete microphone positions) also introduces the possibility of aliasing. Thus, if the far field data fluctuates spatially as a function of $\hat{k} = k\sin\theta$, steps must be taken to ensure that the spacing between microphones in the far field is sufficient to capture the highest spatial frequency fluctuations in the data. It also makes sense to sample the data as a function of θ in order to ensure that $\hat{k}$ is sampled at equal intervals. Thus one might choose to place microphones in the far field at equal increments of $\hat{k}$ (and thus equal increments of $\sin\theta$) as illustrated in Figure 3.18. Then one has to be sure that these sampling points are chosen to be sufficiently close together to guarantee that the spatial sampling frequency is higher than twice that of the highest spatial frequency variation

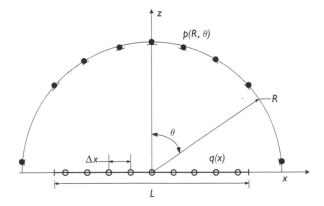

Figure 3.18 A discrete monopole source distribution and far field sensors placed at equal increments $\hat{k} = k \sin \theta$ in the far field.

present (two sample points are required per spatial wavelength). Compliance with this criterion therefore depends crucially on the far field directivity pattern of the source in question: the task of source strength reconstruction will require fewer microphones for sources whose directivity is a smoothly varying function of θ.

A further important observation that can be made within this context is the extent to which source activity is transmitted to the far field. Assume that the source distribution $v(x_s)$ is described as shown in Figure 3.19a by a travelling wave having a speed of propagation c_p and wavenumber $k_p = \omega/c_p$, so that $v(x_s) = Ae^{-jk_p x_s}$, where A describes the amplitude of the source strength fluctuation. Equation (3.146) for the far field pressure produced by such a source distribution confined to the x-axis between $x_s = -L/2$ and $x_s = L/2$ is given by

$$p(R, \theta) = \frac{\rho_0 A e^{-jkR}}{4\pi R} \int_{-L/2}^{L/2} e^{j(\hat{k} - k_p)x_s} \, \mathrm{d}x_s \tag{3.161}$$

which, upon evaluation of the integral, reduces to

$$p(R, \theta) = \frac{\rho_0 A e^{-jkR}}{4\pi R} \frac{\sin(\hat{k} - k_p)L}{(\hat{k} - k_p)} \tag{3.162}$$

The function $\sin(\hat{k} - k_p)L/(\hat{k} - k_p)$ is plotted in Figure 3.19b. The part of the function of $\hat{k}(= k \sin \theta)$ that lies between $-k$ and $+k$ defines the far field directivity function: if $k_p > k$, the main lobe of the directivity function falls outside this range. This will occur if the speed of propagation

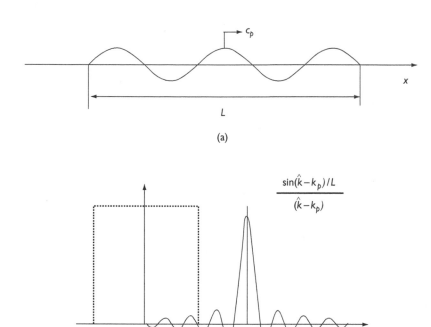

Figure 3.19 (a) An illustration of a travelling wave line source distribution; (b) the far field directivity function plotted as a function of $\hat{k} = k \sin \theta$.

c_p of the waves comprising the source distribution is less than the sound speed c_0. As the length of the source distribution L increases, the function $\sin(\hat{k} - k_p)L/(\hat{k} - k_p)$ becomes taller and narrower, the first zero crossings on either side of the main lobe being given by $(\hat{k} - k_p) = \pm \pi/L$. In fact, in the limit $L \to \infty$, equation (3.161) reduces to

$$p(R, \theta) = \frac{\rho_0 A e^{-jkR}}{2R} \delta(\hat{k} - k_p) \tag{3.163}$$

since it follows from the theory of Fourier transforms that

$$\int_{-\infty}^{\infty} e^{j(\hat{k} - k_p)x_s} \, dx_s = 2\pi \delta(\hat{k} - k_p) \tag{3.164}$$

and the directivity function is localised to a specific value of $\hat{k}$. Therefore, in this hypothetical limiting case of an infinite line source distribution supporting waves having a propagation speed c_p, it is only when $c_p > c_0$

that $\delta(\hat{k} - k_p)$ lies in the range of $\hat{k}$ between $-k$ and $+k$ that defines the far field. Thus, only supersonically travelling waves on the source distribution radiate sound to the far field. The subsonically travelling waves (for which $c_p < c_0$) radiate no sound. Of course when the source has a finite length, there will be some wavenumber components in the range $-k$ to $+k$ even when $c_p < c_0$ because the ripples of the function $\sin(\hat{k} - k_p)L/(\hat{k} - k_p)$ spill over into this range, as illustrated in Figure 3.19b.

It therefore becomes apparent that only limited information regarding the source distribution is actually available in the far-field-radiated sound and that one might expect only to be able to reconstruct the radiating wavenumber components in the source distribution from a measurement of the far field. There is however an argument that originates from the optics literature (see, for example, Harris 1964) that suggests that for bounded sources, the far field wavenumber spectrum is unique to that source and that in principle a knowledge of the far field over the range of $\hat{k}$ from $-k$ to k should be sufficient to reconstruct the Fourier transform over the entire range of $\hat{k}$. In practice, however, the presence of uncertainty in the measurement (for example, produced by contaminating noise) prevents accurate extrapolation of the function and precludes the accurate reconstruction of the source distribution for all wavenumbers. See Castleman (1996) or Andrews and Hunt (1977) for a further discussion of this point.

3.11   The polar correlation technique

A particular method of source strength identification which makes use of the relationships developed in the last section is the polar correlation technique first introduced by Fisher *et al.* (1977) and subsequently used extensively in the study of jet noise and the noise radiated from jet engines. The method involves measurement of the cross-spectral density (or strictly the cross-correlation function) between a number of microphones placed on a polar arc in the far field. Using the geometry illustrated in Figure 3.17, it follows from equation (3.145) that a monopole source strength distribution along the x-axis will result in a pressure at a microphone placed at $\theta = 0°$ that is given by

$$p(R, 0) = \frac{\rho_0 e^{-jkR}}{4\pi R} \int_{-\infty}^{\infty} v(x_s)\, dx_s \tag{3.165}$$

and the pressure at a microphone at the angle θ is given by

$$p(R, \theta) = \frac{\rho_0 e^{-jkR}}{4\pi R} \int_{-\infty}^{\infty} v(x_s) e^{jkx_s \sin\theta}\, dx_s \tag{3.166}$$

The cross-spectral density between these two microphones can be written as

$$S_{pp}(0, \theta) = E[p(R, 0)p^*(R, \theta)] \tag{3.167}$$

and therefore

$$S_{pp}(0, \theta) = \left(\frac{\rho_0}{4\pi R}\right)^2 E\left[\int_{-\infty}^{\infty} v(x_s)\,dx_s \int_{-\infty}^{\infty} v^*(x_s')e^{-jkx_s'\sin\theta}\,dx_s\right] \qquad (3.168)$$

Use of the expectation operator then results in

$$S_{pp}(0, \theta) = \left(\frac{\rho_0}{4\pi R}\right)^2 \int_{-\infty}^{\infty}\int_{-\infty}^{\infty} S_{vv}(x_s, x_s')e^{-jkx_s'\sin\theta}\,dx_s'\,dx_s \qquad (3.169)$$

where $S_{vv}(x_s, x_s') = E[v(x_s)v^*(x_s')]$ is the source strength cross-spectral density function. In the particular case that the source strength distribution is uncorrelated (having zero correlation length) such that $S_{vv}(x_s, x_s') = S_{vv}(x_s)\delta(x_s - x_s')$ then performing the integration with respect to x_s' yields

$$S_{pp}(0, \theta) = \left(\frac{\rho_0}{4\pi R}\right)^2 \int_{-\infty}^{\infty} S_{vv}(x_s)e^{-jkx_s\sin\theta}\,dx_s \qquad (3.170)$$

Under these circumstances there is a simple Fourier transform relationship between the power spectral density of the assumed uncorrelated source strength distribution and the cross power spectrum between a reference microphone at $\theta = 0°$ and another moveable microphone position in the far field. The source strength power spectral density can therefore be deduced by using the inverse transform relationship in an exactly analogous way, and with the same limitations, to that described in the previous section. Fisher *et al.* (1977) are careful to point out the intrinsic assumptions of the method (a monopole source strength distribution having zero correlation length) and also describe in detail alternative interpretations of the results of employing the method to produce an image of the equivalent source strength distribution. The technique has proved to have good practical utility in the examination of the noise output from gas turbine engines and some typical results are presented in Figure 3.20. Full practical details of the implementation of the technique are described by Fisher *et al.* (1977).

3.12 Near-field acoustic holography

There is another transform relationship between an acoustic source distribution and its radiated field that has also proved useful in quantifying acoustic source strength from sound field measurements. The relationship is most easily described within the context of planar near-field acoustic holography which, for example, enables the reconstruction of the distribution of normal surface acceleration field of a vibrating plane surface. (See the basic papers authored by Maynard *et al.* (1985) and Veronesi and Maynard (1987).)

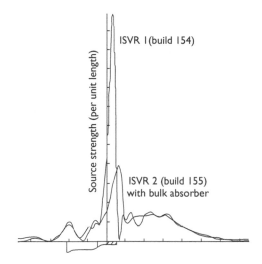

Figure 3.20 Some typical results of the application of the polar correlation technique to the estimation of the source strength distribution associated with a Jet engine presented by Fisher *et al.* (1977). A sketch of the engine shape is shown below the horizontal axis. The results show the source strength per unit length associated with an RB211 engine with and without a bulk absorber fitted with a 5 ft extension (shown hatched) to the unlined core nozzle of the engine.

The technique relies on the application of the two-dimensional wavenumber transform pair described by

$$V(k_x, k_y) = \int_{-\infty}^{\infty} \int_{-\infty}^{\infty} v(x, y) e^{j(k_x x + k_y y)} \, dx \, dy \qquad (3.171)$$

$$v(x, y) = \frac{1}{(4\pi^2)} \int_{-\infty}^{\infty} \int_{-\infty}^{\infty} V(k_x, k_y) e^{-j(k_x x + k_y y)} \, dk_x \, dk_y \qquad (3.172)$$

where $v(x, y)$ describes the source acceleration in the x, y plane and $V(k_x, k_y)$ is its two-dimensional spatial Fourier transform. The wavenumber variables k_x and k_y represent spatial frequency in the x and y directions that define the plane of the source.

It can be shown that there is a simple relationship between the transformed source acceleration $V(k_x, k_y)$ and the equivalently transformed acoustic pressure $P(k_x, k_y, z)$ measured over a planar surface parallel to, and a distance z from, the source plane. In order to demonstrate this, the transform operation described by equation (3.171) is first applied to the Helmholtz equation governing the complex acoustic pressure which, in rectangular cartesian coordinates, is

$$(\nabla^2 + k^2) p(x, y, z) = 0 \qquad (3.173)$$

The two-dimensional spatial Fourier transform of this equation can be written as

$$\int_{-\infty}^{\infty}\int_{-\infty}^{\infty} \left(\frac{\partial^2}{\partial x^2} + \frac{\partial^2}{\partial y^2} + \frac{\partial^2}{\partial z^2} + k^2 \right) p(x,y,z) e^{j(k_x x + k_y y)} \, dx \, dy = 0 \qquad (3.174)$$

Noting that the derivatives with respect to the z-coordinate can be taken outside the integral and that the derivatives with respect to x and y integrate to $-k_x^2 P(k_x, k_y, z)$ and $-k_y^2 P(k_x, k_y, z)$ yields (Junger and Feit 1986)

$$\left(k^2 - k_x^2 - k_y^2 + \frac{\partial^2}{\partial z^2} \right) P(k_x, k_y, z) = 0 \qquad (3.175)$$

Solutions of this equation take the form

$$P(k_x, k_y, z) = A e^{jk_z z} \qquad (3.176)$$

where A is an arbitrary constant and the wavenumber k_z is given by

$$k_z = \sqrt{k^2 - k_x^2 - k_y^2} \qquad (3.177)$$

It is easy to establish the relationship between the transformed pressure $P(k_x, k_y, z)$ and the transformed source acceleration $V(k_x, k_y)$ by using the linearised equation of momentum conservation given by $\rho_0 v(x,y) = -\partial p(x,y,z)/\partial z$ which, when transformed, reduces to

$$\rho_0 V(k_x, k_y) = -\frac{\partial}{\partial z} P(k_x, k_y, z) \qquad (3.178)$$

Substitution of the solution given by equation (3.176) into this equation shows that the constant $A = \rho_0 V(k_x, k_y)/jk_z$ and therefore that

$$P(k_x, k_y, z) = \frac{\rho_0 V(k_x, k_y)}{jk_z} e^{-jk_z z} \qquad (3.179)$$

This relationship therefore demonstrates that the transformed pressure field at a distance z above the surface of the source has a simple relationship to the transformed distribution of source acceleration. The most significant aspect of this solution is that the wavenumber k_z may under some circumstances become imaginary. This occurs when $(k_x^2 + k_y^2) > k^2$ in which case the solution given by equation (3.179) represents an exponential decay of the field in the z-direction.

At this juncture, the discussion is assisted by assuming that the acceleration field varies with only one spatial coordinate. If it is assumed that the surface acceleration is independent of the y-coordinate, the transformed acceleration field given by equation (3.171) reduces to

$$V(k_x, k_y) = 2\pi\delta(k_y) \int_{-\infty}^{\infty} v(x)e^{jk_x x}\mathrm{d}x = 2\pi\delta(k_y)V(k_x) \qquad (3.180)$$

where the delta function of k_y results from the integration with respect to k_y. The expression for the acceleration distribution when applying the inverse transform given by equation (3.172) is

$$v(x, y) = \frac{1}{(2\pi)^2} \int_{-\infty}^{\infty}\int_{-\infty}^{\infty} 2\pi\delta(k_y)V(k_x)e^{-j(k_x x + k_y y)} \, \mathrm{d}k_x \, \mathrm{d}k_y \qquad (3.181)$$

which, upon evaluating the integral with respect to k_y, reduces to

$$v(x) = \frac{1}{2\pi} \int_{-\infty}^{\infty} V(k_x)e^{jk_x x} \, \mathrm{d}k_x \qquad (3.182)$$

Thus, these relationships imply that assuming the velocity is independent of the y-coordinate direction amounts to evaluating the spatial Fourier transforms at zero wavenumber k_y (which in turn implies infinite spatial wavelength; see Williams 1999).

Therefore the relationship between transformed pressure and transformed acceleration given by equation (3.179) can be written as

$$P(k_x, 0, z) = \frac{\rho_0 V(k_x, 0)}{j\sqrt{k^2 - k_x^2}} e^{-j\sqrt{k^2 - k_x^2}z} \qquad (3.183)$$

This makes clear that an exponential decay of the transformed pressure field occurs when $k_x > k$; that is, when the spatial frequency in the source distribution is greater than the wavenumber k associated with the propagation of acoustic plane waves. For example, assume that at a frequency ω, the wavenumber $k_x = \omega/c_p$ where c_p is the speed of propagation of waves on the source in the x-direction. Then the condition $k_x > k$ implies $c_p < c_0$ which in turn implies that subsonically travelling source waves generate an exponentially decaying field. On the other hand, exactly as discussed in the previous section, supersonically travelling waves generate radiation that propagates to the far field. The observation was also made in the previous section that since the 'subsonic' wavenumber components do not radiate into the far field, there is little hope of capturing these details of the source mechanism from far field measurements. However, by making measurements close to the source, in the exponentially decaying 'near-field', there is the possibility of reconstructing the source distribution in finer detail.

The process of source strength reconstruction from near-field measurements can be described by using equation (3.183) to show that

$$V(k_x, 0) = \left(j\sqrt{k^2 - \frac{k_x^2}{\rho_0}} \right) P(k_x, 0, z) e^{j\sqrt{k^2 - k_x^2} z} \qquad (3.184)$$

Thus one first has to measure the complex acoustic pressure at a distance z from the surface of the source to evaluate the wavenumber transform $P(k_x, 0, z)$. The process of back propagation is then applied by multiplying each wavenumber component k_x in turn by the factor $e^{j\sqrt{k^2 - k_x^2} z}$. The acceleration distribution $v(k)$ is then deduced by inverse wavenumber transformation of $V(k_x, 0)$. Whilst this procedure appears straightforward enough in principle, an enormous amount of care is required in its practical application.

A comprehensive discussion of the issues involved is presented by Williams (1999) and only a brief account will be given here. It is important first to note that as k_x increases progressively beyond the value of k, the greater will be the rate of decay in the associated exponentially decaying sound field. Consequently, in the process of back propagation, the 'amplification' of the measured wavenumber component is proportionately larger. Thus, rapidly decaying high wavenumber components are more prone to contamination by measurement noise and more likely to give completely erroneous results when exponentially increased during the process of back propagation. This makes the process of 'wavenumber filtering' to remove wavenumbers above a certain limit essential if good results are to be achieved. Furthermore, the fact that sampling the pressure field at discrete microphone positions can lead to aliasing places a further limitation on the technique; the spatial sample rate must be chosen to ensure that the highest wavenumber components dealt with are sampled adequately. Finally, the field is necessarily sampled over a restricted aperture. This convolves the wavenumber spectrum with the wavenumber transform of the spatial window of the aperture (see Figure 3.2b).

Within the context of the two-dimensional example presented above, the reconstruction process at a series of discrete wavenumbers $n\Delta k_x$ is described by equation (3.184) which can be written as

$$V(n\Delta k_x) = G(n\Delta k_x)P(n\Delta k_x) \qquad (3.185)$$

where it is assumed that $P(n\Delta k_x)$ is evaluated at the distance z above the surface of the source and that the inverse propagator is given by

$$G(n\Delta k_x) = \left(j\sqrt{k^2 - \frac{(n\Delta k_x)^2}{\rho_0}} \right) e^{j\sqrt{k^2 - (n\Delta k_x)^2} z} \qquad (3.186)$$

Now consider the numerical procedure for the evaluation of $P(n\Delta k_x)$. First, the sampling of the complex pressure that is measured as a function of x at a distance z above the source plane results in the spatial Fourier transform

$$P(e^{jk_x\Delta x}) = \int_{-\infty}^{\infty} \left[p(x) \sum_{m=-\infty}^{\infty} \delta(x - m\Delta x) \right] e^{jk_xx} \, dx \qquad (3.187)$$

which as described in Section 3.2 is a periodically repeating replica of the Fourier transform $P(k_x)$ of the original data. Application of a window function to the original data $p(x)$ also results in a modification to this spectrum. Thus if the rectangular aperture function is used over a distance L in the x-direction then the spectrum that results is a convolution of $P(e^{jk_x\Delta x})$ with the spatial Fourier transform of the rectangular aperture function as described in Section 3.2. Denoting this windowed result by $\hat{P}(e^{jk_xx})$ and evaluating the result at N discrete values of k_x, where N is chosen to be the same number as the discrete number of positions at which the sound field is sampled, results in

$$\hat{P}(e^{jn\Delta k_x\Delta x}) = \sum_{m=0}^{N-1} p(m\Delta x)e^{-jn\Delta k_x m\Delta x} \qquad (3.188)$$

where $\Delta k_x - 2\pi/N\Delta x$. This increment is chosen so that the spatial frequency range between zero and the spatial sampling frequency $2\pi/\Delta x$ is split into N equispaced values. This then results in the discrete Fourier transform which can be written compactly as

$$\hat{P}(n) = \sum_{m=0}^{N-1} p(m)e^{-j2\pi nm/N} \qquad (3.189)$$

Note that this operation can be written in matrix form as the product $\mathbf{Fp}$ say, where $\mathbf{F}$ is the Fourier matrix and $\mathbf{p}$ is the vector whose elements are given by the sampled pressure $p(m)$. It also follows that if a wavenumber filter is to be applied to this data, this can be viewed as multiplying the vector $\mathbf{Fp}$ by a diagonal matrix $\mathbf{D}_F$ to avoid the undue amplification of the high wavenumber components. The process of back propagation can further be written as multiplication by a further diagonal matrix $\mathbf{D}_G$ whose elements are given by $G(n\Delta k_x)$ as defined in equation (3.186). The entire process can therefore be written as

$$\mathbf{V} = \mathbf{D}_G\mathbf{D}_F\mathbf{Fp} \qquad (3.190)$$

where $\mathbf{V}$ is the vector whose components are given by $V(n\Delta k_x)$. The sampled acceleration distribution is then given by operating on this vector with the inverse discrete Fourier transform such that

$$\mathbf{v} = \mathbf{F}^{-1}\mathbf{V} = \mathbf{F}^{-1}\mathbf{D}_G\mathbf{D}_F\mathbf{Fp} \qquad (3.191)$$

where $\mathbf{v}$ is the vector of discrete values of velocity $v(n\Delta x)$ reconstructed along the source plane. Since the Fourier matrix $\mathbf{F}$ has the property $\mathbf{F}^H\mathbf{F} = \mathbf{I}$, then this relationship can also be written in the form

$$\mathbf{v} = \mathbf{F}^H\mathbf{D}_G\mathbf{D}_F\mathbf{F}\mathbf{p} \tag{3.192}$$

Note the similarity between this expression and that resulting from the application of inverse methods described in Section 3.8 where if it is assumed that $\mathbf{v} = \mathbf{G}\mathbf{p}$ (where $\mathbf{G}$ is square) and the SVD is applied, then it follows that

$$\mathbf{v} = \mathbf{V}\Sigma_R^+\mathbf{U}^H\mathbf{p} \tag{3.193}$$

Therefore the process of back propagation with associated wavenumber filtering (through the choice of $\mathbf{D}_F$) is seen to be directly analogous to the process of singular value discarding or regularisation in treating the inverse problem. The relationship between these procedures is a topic of current research (see, for example, Williams 2000; Ih 2000). There are also many practical examples of the application of NAH that have been presented in the literature. Good introductions are given by Hald (1989, 1995) and Ginn and Hald (1989). More recent work exploring the relationship between NAH and inverse methods has been presented by Schumacher (1999) and Schumacher and Hansen (2001).

3.13 Conclusions

Least squares estimation methods have been described and explained in order to provide a common framework for understanding the range of array signal processing techniques available for the estimation of acoustic source strength. Direct methods based on beamforming or focused beamforming are now being applied extensively to produce qualitative maps of the source strength associated with real acoustic source distributions. However, inverse methods offer the possibility of producing more accurate quantitative information, albeit within resolution limits associated with the conditioning of the matrix to be inverted and the noise environment in which the measurements are made. Near-field acoustic holography can also be viewed in a similar way and further developments in this field are expected as a consequence of the application of inverse techniques and methods for the regularisation of ill-posed inverse problems.

References

Allen, M. (1974) The relationship between variable selection and data augmentation and a method for prediction. *Technometrics*, **16**, 125–127.

Andrews, H.C. and Hunt, B.R. (1977) *Digital Image Restoration*, Englewood Cliffs, NJ: Prentice-Hall.

Billingsley, J. and Kinns, R. (1976) The acoustic telescope. *Journal of Sound and Vibration*, 48, 485–510.

Bitzer, J. and Simmer, K.U. (2001) Superdirective microphone arrays. In: Brandstein, M. and Ward, D. (eds), *Microphone Arrays*, Berlin: Springer-Verlag.

Brooks, T.F., Marcolini, M.A. and Pope, D.S. (1987) A directional array approach for the measurement of rotor noise source distributions with controlled spatial resolution. *Journal of Sound and Vibration*, 112(1), 192–197.

Burdic, W.S. (1984) *Underwater Acoustic System Analysis*, Englewood Cliffs, NJ: Prentice-Hall.

Castleman, K.R. (1996) *Digital Image Processing*, Upper Saddle River, NJ: Prentice-Hall.

Craven, P. and Wahba, G. (1979) Smoothing noisy data with spline functions – estimating the correct degree of smoothing by the method of generalised cross-validation. *Numerische Mathematik*, 31, 377–403.

Davy, R. and Remy, H. (1998) Airframe noise characteristics of a 1/11 scale airbus model. *4th AIAA/CEAS Aeroacoustics Conference*, Toulouse, France, AIAA Paper 98-2335.

Dougherty, R.P. (2002) Beamforming in acoustic testing. In: Mueller, T.J. (ed.), *Aeroacoustic Measurements*, Berlin: Springer-Verlag.

Elias, G. (1997) Experimental techniques for source location. In: *Aeroacoustics and Active Noise Control, Lecture Series 1997–07*, von Karman Institute for Fluid Dynamics.

Elliott, S.J. (2001) *Signal Processing for Active Control*, London: Academic Press.

Fillipi, P.J.T., Habault, D. and Piraux, J. (1988) Noise source modelling and identification procedure. *Journal of Sound and Vibration*, 14, 285–296.

Fisher, M.J. and Holland, K.R. (1997) Measuring the relative strengths of a set of partially coherent acoustic sources. *Journal of Sound and Vibration*, 201, 103–125.

Fisher, M.J., Harper-Bourne, M. and Glegg, S.A.L. (1977) Jet engine noise source location: the polar correlation technique. *Journal of Sound and Vibration*, 51, 23–54.

Ginn, K.B. and Hald, J. (1989) STSF-practical instrumentation and application. *Brüel and Kjær Technical Review No. 2*.

Golub, G.H. and Van Loan, C.F. (1989) *Matrix Computations*, 2nd edn, Bathmore: John Hopkins University Press.

Golub, G.H., Heath, M. and Wahba, G. (1979) Generalised cross-validation as a method for choosing a good ridge parameter. *Technometrics*, 21, 215–223.

Grace, S.P., Atassi, H.M. and Blake, W.K. (1996a) Inverse aeroacoustic problem for a streamlined body, Part 1: Basic formulation. *AIAA Journal*, 34, 2233–2240.

Grace, S.P., Atassi, H.M. and Blake, W.K. (1996b) Inverse aeroacoustic problem for a streamlined body, Part 2: Accuracy of solutions. *AIAA Journal*, 34, 2241–2246.

Hald, J. (1989) STSF – a unique technique for scan-based near-field acoustic holography without restrictions on coherence. *Brüel and Kjær Technical Review No. 1*.

Hald, J. (1995) Spatial transformation of sound fields (STSF) techniques in the automotive industry. *Brüel and Kjær Technical Review No. 1*.

Hansen, P.C. (1992) Analysis of discrete ill-posed problems by means of the L-curve. *SIAM Review*, 34, 561–580.

Hansen, P.C. (1998) Rank deficient and discrete ill-posed problems. *Society for Industrial and Applied Mathematics*, Philadelphia.

Hansen, P.C. and O'Leary, D.P. (1993) The use of the L-curve in the regularization of discrete ill-pose problem. *SIAM J. Sci. Comput*, **14**, 1487–1503.

Harris, J.L. (1964) Diffraction and resolving power. *Journal of the Optical Society of America*, **54**, 931–936.

Hayes, J.A., Horne, W.C., Soderman, P.T. and Best, P.H. (1997) Airframe noise characteristics of a 4.7% scale DC-10 model. *3rd AIAA/CEAS Aeroacoustics Conference*, Atlanta, GA, USA, AIAA Paper 97-1594.

Humphreys, W.H., Brooks, T.F., Hunter, W.W. and Meadows, K.R. (1998) Design and use of microphone directional arrays for aeroacoustic measurements. *36th Aerospace Sciences Meeting and Exhibition*, Reno, NV, USA, AIAA Paper 98-0471.

Ih, J.G. (2000) Meanings of SVD and wave-vector filtering in the near-field acoustical holography using the inverse BEM (abstract). *Journal of the Acoustical Society of America*, **108**, 2528.

Junger, M.C. and Feit, D. (1986) *Sound Structures and their Interaction*, 2nd edn, Cambridge, MA: MIT Press.

Kailath, T., Sayed, A.H. and Hassibi, B. (2000) *Linear Estimation*, Upper Saddle River, NJ: Prentice-Hall.

Kim, G.T. and Lee, B.H. (1990) 3-D sound reconstruction and field projection using the Helmholtz integral equation. *Journal of Sound and Vibration*, **136**, 245–261.

Maynard, J.D., Williams, E.G. and Lee, Y. (1985) Nearfield acoustic holography: I. Theory of generalised holography and the development of NAH. *Journal of the Acoustical Society of America*, **78**, 1395–1413.

Michel, U. and Qiao, W. (1999) Directivity of landing-gear noise based on flyover measurements. *5th AIAA/CEAS Aeroacoustics Conference*, Bellevue, WA, USA, AIAA Paper 99-1956.

Mosher, M. (1996) Phased arrays for aeroacoustic testing: Theoretical development. *2nd AIAA/CEAS Aeroacoustics Conference*, State College, PA, USA, AIAA Paper 96-1713.

Nelson, P.A. (2001) A review of some inverse problems in acoustics. *International Journal of Acoustics and Vibrations*, **6**(3), 118–134.

Nelson, P.A. and Elliott, S.J. (1992) *Active Control of Sound*, London: Academic Press, 416–420.

Nelson, P.A. and Yoon, S.H. (2000) Estimation of acoustic source strength by inverse methods: Part I: Conditioning of the inverse problem. *Journal of Sound and Vibration*, **233**, 643–668.

Nielsen, R.O. (1990) *Sonar Signal Processing*, London: Artech House.

Piet, J.F. and Elias, G. (1997) Airframe noise source localization using a microphone array. *3rd AIAA/CEAS Aeroacoustics Conference*, Atlanta, GA, USA, AIAA Paper 97-1643.

Rabiner, L.R. and Gold, B. (1975) *Theory and Application of Digital Signal Processing*, Englewood Cliffs, NJ: Prentice-Hall.

Schmidt, R.O. (1986) Multiple emitter location and signal parameter estimation. *IEEE Transactions on Antennas and Propagation*, **AP-34**(3), 276–280.

Schumacher, A. (1999) Practical application of inverse boundary element method to sound field studies of tyres. *Proceedings of Internoise 99*.

Schumacher, A. and Hansen, P.C. (2001) Sound source reconstruction using inverse BEM. *Proceedings of Internoise 2001*.

Schumacher, A., Hald, J., Rasmussen, K.B. and Hansen, P.C. (2003) Sound source reconstruction using inverse boundary element calculations. *Journal of the Acoustical Society of America*, **113**(1), 114.

Sijtsma, P. and Holthusen, H. (1999) Source location by phased array measurements in closed wind tunnel test sections. *5th AIAA/CEAS Aeroacoustics Conference*, Bellevue, WA, USA, AIAA Paper 99-1814.

Stoker, R.W. and Sen, R. (2001) An experimental investigation of airframe noise using a model-scale Boeing 777. *39th AIAA Aerospace Sciences Meeting and Exhibition*, Reno, NV, AIAA Paper 2001-0987.

Stoughton, R. and Strait, S. (1993) Source imaging with minimum mean-squared error. *Journal of the Acoustical Society of America*, **94**, 827–834.

Tester, B.J. and Fisher, M.J. (1991) Engine noise source breakdown: Theory, simulation and results. *AIAA 7th Aeroacoustics Conference*, Palo Atto, California, USA, AIAA-81-2040.

Tikhonov, A. and Arsenin, V. (1977) *Solution of Ill-Posed Problems*, Washington, DC: Winston.

Underbrink, J.R. (2002) Aeroacoustic phased array testing in low speed wind tunnels. In: Mueller, T.J. (ed.), *Aeroacoustic Measurements*, Berlin: Springer-Verlag.

Van Veen, B. and Buckley, K.M. (1998) Beamforming techniques for spatial filtering. In: Madisetti, V.K. and Williams, D.B. (eds), *The Digital Signal Processing Handbook*. Boca Raton, FL: CRC Press.

Venkatesh, S.R., Polak, D.R. and Narayanan, S. (2000) Phased array design, validation, and application to jet noise source localization. *6th AIAA/CEAS Aeroacoustics Conference*, Lahaina, HI, USA, AIAA Paper 2000-1934.

Veronesi, W.A. and Maynard, J.D. (1987) Nearfield acoustic holography (NAH) II. Holographic reconstruction algorithms and computer implementation. *Journal of the Acoustical Society of America*, **81**, 1307–1321.

Veronesi, W.A. and Maynard, J.D. (1989) Digital holographic reconstruction of sources with arbitrarily shaped surfaces. *Journal of the Acoustical Society of America*, **85**, 588–598.

Wahba, G. (1990) Society for Industrial and Applied Mathematics. Spline models for observation data. *CBMS-NSF Regional Conference Series in Applied Mathematics*, **59**.

Wahba, G. and Wold, S. (1975) A completely automatic French curve: Fitting splines by cross validation. *Comm. Statist.* **4**, 1–17.

Widrow, B. and Stearns, S.D. (1985) *Adaptive Signal Processing*, Englewood Cliffs, NJ: Prentice-Hall.

Widrow, B., Mantey, P.E., Griffiths, L.J. and Goode, B.B. (1967) Adaptive antenna systems. *Proceedings of the IEEE*, **55**, 2143.

Williams, E.G. (1999) *Fourier Acoustics: Sound Radiation and Nearfield Acoustical Holography*, London: Academic Press.

Williams, E.G. (2000) Regularisation of inverse problems with evanescent waves (abstract). *Journal of the Acoustical Society of America*, **108**, 2503.

Yoon, S.H. and Nelson, P.A. (2000) Estimation of acoustic source strength by inverse methods: Part II: Experimental investigation of methods for choosing regularisation parameters. *Journal of Sound and Vibration*, **233i**, 669–705.

Chapter 4

Modelling of sound propagation in the ocean

P.R. White

4.1 Introduction

Sound propagation in the ocean is strongly influenced by the fact that sound speed is inhomogeneous, leading to significant refraction effects (Tolstoy and Clay 1966; Urick 1983; Leighton 1998). The variations in sound speed may be relatively small, of the order of 3 per cent, but in most applications these variations have a major effect. There are several reasons why one may seek methods to compute the sound field resulting from an acoustic source in the ocean. The most obvious reason is to predict the perform-ance of sonar systems, a task that depends critically on one's ability to estimate the transmission loss (TL) between a source and a receiver. This probably represents the greatest source of error in performance calculations and is strongly dependent on the spatial variability of the sound speed. A second use for propagation models is in inversion schemes, where one makes measurements in the ocean and then uses the propagation model to construct estimates of the environmental parameters. This approach is attractive to oceanographers since it offers the potential to survey large tracts of ocean at one time, a goal which is unrealisable using conventional surveying techniques.

4.2 Background

To calculate the sound field in an environment remote from any acoustic sources we seek solutions of the linearised wave equation for sound pressure

$$\nabla^2 p - \frac{1}{c^2}\frac{\partial^2 p}{\partial t^2} = 0 \tag{4.1}$$

where the acoustic pressure $p = p(x, y, z, t)$ is a function of position and time and, in general, the speed of sound $c = c(x, y, z)$ and is assumed to be time-independent. Note that here we shall adopt the convention that z denotes depth, so that positive values of z refer to negative vertical displacements. The Laplacian operator ∇^2 is defined as

$$\nabla^2 = \frac{\partial^2}{\partial x^2} + \frac{\partial^2}{\partial y^2} + \frac{\partial^2}{\partial z^2} \quad \text{in Cartesian coordinates}$$

$$\nabla^2 = \frac{1}{r}\frac{\partial}{\partial r}\left(r\frac{\partial}{\partial r}\right) + \frac{1}{r^2}\frac{\partial^2}{\partial\theta^2} + \frac{\partial^2}{\partial z^2} \quad \text{in cylindrical polar coordinates}$$

If a harmonic wave field is assumed, such that

$$p(x, y, z, t) = p(x, y, z)\, e^{j\omega t} \tag{4.2}$$

$p(x, y, z)$ is the complex pressure amplitude at the position (x, y, z). Substitution of equation (4.2) into equation (4.1) yields the Helmholtz equation

$$\nabla^2 p + \frac{\omega^2}{c^2} p = \nabla^2 p + k^2 p = 0 \tag{4.3}$$

To solve this equation, one has also to specify the boundary conditions. Except for the very simplest boundary conditions and uniform media, it is not possible to obtain a complete analytic solution for equation (4.3). Thus, one is forced to make simplifying assumptions and/or use numerical methods. Models used to solve either equation (4.1) or (4.3) are called propagation models and form the subject of the following sections.

4.2.1 Dimensions of propagation models

There are many types of propagation models. Most common models only solve the wave equation in two dimensions (normally in terms of range r and depth z) and are referred to as 2-D or 'range dependent' models. This implies that the speed of sound c is at most a function of the spatial coordinates r and z and we write $c = c(r, z)$. For a truly 2-D model, the boundary conditions can only be functions of range so these models cannot predict sound propagation around 3-D structures such as seamounts. The resulting model simulates sound propagation through a vertical slice in the ocean. An underlying assumption of such a model is that a ray leaving a source will remain within the vertical plane.

Greater simplification results if one is willing to assume horizontal stratification (sound speed is uniform at a given depth) so that c only depends on z; in this case we write $c = c(z)$. Such models are sometimes called 'range independent'. These models only require information pertaining to a single sound speed profile, that is the function $c(z)$. One consequence of assuming horizontal stratification is that one immediately implies that the ocean floor is plane and horizontal (so that the boundary condition there is also range-independent).

One of the major differences between range-dependent and range-independent models is the amount of data required to run the model. Ideally, a range-dependent model requires knowledge of the sound speed at all depths for all ranges. In practice, one can, at best, obtain a selection of depth profiles at various ranges. In addition to the information about the sound speed variations, one also has to provide such models with data concerning the depth profile (bathymetry); for some types of model one also needs to specify information about the sub-bottom profile. For long-range propagation problems, a major oceanographic survey maybe required before modelling can take place.

There are models which go even further and accommodate sound speed variations in all three spatial parameters and allow for sea-bottom variations in two directions. These are termed 3-D models, but are not widely used since they require large amounts of data to specify the sound speed variations and the bathymetry. This type of model requires very much more computational power than the 2-D model.

4.2.2   Models used for propagation modelling

There are many techniques used for propagation modelling, most of which can be thought of as variations on the following four methods:

4.2.2.1   Ray tracing (Tolstoy and Clay 1966; Kinsler et al. 1982; Brekhovskikh and Lysanov 1991)

This involves following the paths of a set of rays as they leave the source and tracking them as they propagate through the medium. Ray tracing models can be used for range-dependent and range-independent problems, but are most commonly used for range-independent problems, since in this case, very efficient algorithms can be constructed. They are most useful for short-range, high-frequency modelling.

4.2.2.2   Normal mode modelling (Tolstoy and Clay 1966; Kinsler et al. 1982; Brekhovskikh and Lysanov 1991)

In this method the sound field is expressed as a sum of a series of normal modes. The sound field at a distant point can be calculated by summing all the modal contributions at that point, although often only a few of these modes are significant. This method requires the assumption of a stratified ocean (some modifications can be made to allow slowly varying range dependence). There is also a class of methods called fast field programs (or Green's function models) (Buckingham 1992; Etter 1995) which are closely related to the method of normal modes.

4.2.2.3 Finite element method (Pack 1986)

In this method, the derivatives in the wave equation are approximated by differences of the field variables at discrete spatial points on a mesh. These are probably the most flexible and accurate propagation models. The meshes used by these models require about 10 grid points per wavelength; thus, even over moderate distances, there are a large number of grid points, and so these methods impose a heavy computational load.

4.2.2.4 Parabolic equation model (Buckingham 1992; Etter 1995)

This method is based on separating the wave equation into two differential equations, one of which has a standard solution (Bessel functions); the second equation can be approximated by a parabolic equation, which is comparatively easy to solve. This method is implicitly approximate and is generally valid over a narrow range of angles, although it is computationally efficient.

In the following sections, we shall consider ray tracing and the method of normal modes in more detail.

4.3 Ray tracing

In the following, the mathematical justification for using ray theory is presented in which the limitations of the method are also illuminated (Tolstoy and Clay 1966; Brekhovskikh and Lysanov 1991). We first express the complex sound pressure amplitude of a harmonic sound field in polar form

$$p(x, y, z) = A(x, y, z) e^{jk_0 \varphi(x,y,z)} \tag{4.4}$$

where $A(x, y, z)$ is the modulus (amplitude) of $p(x, y, z)$, $\varphi(x, y, z)$ is the argument (phase) and k_0 is a reference wavenumber. Using equation (4.2), the pressure at a point (x, y, z) is given by

$$p(x, y, z, t) = A(x, y, z) e^{j\omega\left(t + \frac{\varphi(x,y,z)}{c_0}\right)} \tag{4.5}$$

with $c_0 = \omega/k_0$ representing reference sound speed. Substitution of equation (4.4) into the Helmholtz equation (4.3) yields

$$\nabla^2 A e^{jk_0\varphi} + 2jk_0 \nabla A \cdot \nabla\varphi e^{jk_0\varphi} + jAk_0 \nabla^2 \varphi e^{jk_0\varphi}$$
$$- Ak_0^2 \nabla\varphi \cdot \nabla\varphi e^{jk_0\varphi} + k^2 A e^{jk_0\varphi} = 0$$

Cancelling common terms and equating real and imaginary parts yields the two equations

$$\nabla^2 A - A k_0^2 \nabla\varphi \cdot \nabla\varphi + k^2 A = 0 \tag{4.6}$$

and

$$2\nabla A \cdot \nabla\varphi + A\nabla^2\varphi = 0 \tag{4.7}$$

If we simplify by making the assumption

$$\frac{\nabla^2 A}{A} \ll k^2 \tag{4.8}$$

the first term in equation (4.6) may be neglected to yield

$$\nabla\varphi \cdot \nabla\varphi = \left(\frac{k}{k_0}\right)^2 = n^2 \tag{4.9}$$

where n is referred to as the 'refractive index' and is defined as

$$n = \frac{k}{k_0} = \frac{c_0}{c}$$

Equation (4.9) is termed the 'Eikonal equation' and its companion, equation (4.7), is called the 'transport equation'. The Eikonal equation implies that, at least locally, the wavefront propagates as a plane wave. To see this, we consider the equation defining a plane wave in three dimensions, namely

$$P(x, y, z, t) = Ae^{j\left(\omega t - k_x x - k_y y - k_z z\right)}$$

Comparing this with equation (4.5) we see that

$$\varphi(x, y, z) = -\frac{k_x x + k_y y + k_z z}{k_0} \tag{4.10}$$

Substituting this into the Eikonal equation yields

$$\frac{k_x^2 + k_y^2 + k_z^2}{k_0^2} = n^2 \quad \Rightarrow \quad k_x^2 + k_y^2 + k_z^2 = k^2$$

which we know to be true for a plane wave. The surfaces $\varphi(x, y, z) = $ constant define the wavefronts. These wavefronts move along the normal directions defined by $\nabla\varphi$.

The preceding results all rely upon the applicability of the assumption expressed by equation (4.8). This assumption imposes several restrictions on the physics, which in turn limit the applicability of ray theory. These conditions are:

(i) *The amplitude must not vary significantly over a wavelength.* For example, this theory does not directly allow for diffraction about a solid object. The theory is invalid because in the shadow of the object the pressure field has regions which exhibit large spatial variations.

(ii) *The speed of sound must not vary significantly over a wavelength.* This implies that ray theory may not be valid when there are abrupt changes in the sound speed. There is nothing in the above theory which accounts for reflection.

We wish to be able to plot the ray paths. In order to achieve this, an expression for the variation of the vector $\nabla\varphi$ with position along the ray is useful. This is obtained in derivative form by deriving an expression for

$$\frac{\mathrm{d}}{\mathrm{d}s}(\nabla\varphi) \tag{4.11}$$

where s is the measure of distance along the ray, that is, the path length.

From equation (4.9) we see that the length of the vector $\nabla\varphi$ is n while the direction of $\nabla\varphi$ is normal to the surface of constant phase (wavefront). If we define a unit vector e which is perpendicular to the wavefront (see Figure 4.1), then, by construction

$$\nabla\varphi = ne \tag{4.12}$$

We shall proceed by differentiating this with respect to s. It is noted that since s advances in a direction perpendicular to the wavefront then

$$\frac{\mathrm{d}}{\mathrm{d}s} = e\cdot\nabla \tag{4.13}$$

which means that differentiation with respect to s is equivalent to applying the grad operator and then projecting the answer onto the e direction.

Applying this relation to equations (4.13) and, then, (4.12) yields

$$\frac{\mathrm{d}}{\mathrm{d}s}(\nabla\varphi) = (e\cdot\nabla)\nabla\varphi = \frac{(\nabla\varphi\cdot\nabla)\nabla\varphi}{n} \tag{4.14}$$

However, it can be shown that, for an arbitrary scalar field F,

$$(\nabla F\cdot\nabla)\nabla F = \frac{\nabla(\nabla F\cdot\nabla F)}{2} = \frac{\nabla(\nabla F)^2}{2}$$

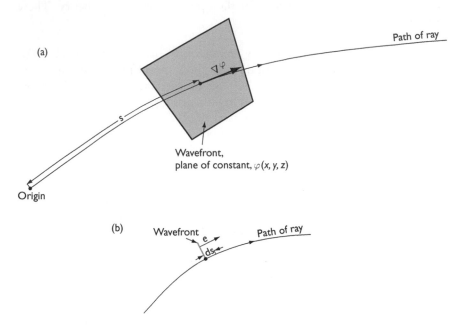

Figure 4.1 Wavefront geometry: (a) 3-D view; (b) section through.

Using equation (4.14) becomes

$$\frac{d}{ds}(\nabla\varphi) = \frac{\nabla(\nabla\varphi)^2}{2n} = \frac{\nabla n^2}{2n} = \frac{2n\nabla n}{2n} = \nabla n \qquad (4.15)$$

where once again we have used equation (4.9). This describes how variations in the sound speed (appearing here as the refractive index n) translate into changes in the direction of propagation of a wavefront.

4.3.1 The 2-D stratified model

If we are only interested in a 2-D model of the ocean and we are also willing to accept the assumption that the ocean is stratified (a range-independent model) then the preceding results simplify considerably. The phase function (locally) is

$$\varphi(x, z) = -\frac{k_x x + k_z z}{k_0} = -\frac{k\cos\theta x + k\sin\theta z}{k_0}$$

so that

$$\nabla\varphi = -n\cos\theta\, i - n\sin\theta\, k$$

Because the model is stratified then $n = n(z)$, in which case

$$\nabla n = 0i + \frac{dn}{dz}k$$

Using equation (4.15)

$$\frac{d}{ds}(n\cos\theta)\, i + \frac{d}{ds}(n\sin\theta)\, k = 0i - \frac{dn}{dz}k$$

Equating the i components gives

$$\frac{d}{ds}(n\cos\theta) = 0$$

Integrating both sides and remembering the definition of n gives

$$\frac{c_0\cos\theta}{c} - \text{constant} \quad \Rightarrow \quad \frac{\cos\theta}{c} = \text{constant}$$

This is simply Snell's law. Indeed we can regard equation (4.15) as a generalisation of Snell's law into three dimensions; it describes the rate at which the normal to the wavefront departs from the ray path (see Figure 4.1a).

4.3.2 Radius of curvature for a ray

The radius of curvature R_c is defined as, see Figure 4.2a,

$$R_c = \left| \frac{ds}{d\theta} \right|$$

Figure 4.2 The radius of curvature: (a) arbitrary curve; (b) circle.

For a ray in a 2-D range-independent environment, Snell's law tells us that

$$\frac{\cos\theta}{c} = \frac{\cos\theta_0}{c_0}$$

where c_0 and θ_0 are defined at some fixed reference points on the ray (possibly the source). Differentiating this expression with respect to z yields

$$-\sin\theta\frac{d\theta}{dz} = \frac{\cos\theta_0}{c_0}\frac{dc}{dz}$$

As $ds\,\sin\theta = dz$, then

$$\frac{1}{R_c} = \left|\frac{d\theta}{ds}\right| = \frac{\cos\theta_0}{c_0}\left|\frac{dc}{dz}\right| \tag{4.16}$$

This states that the inverse of the radius of curvature of a ray is proportional to the rate of change of wave velocity with depth, and that the constant of proportionality is the same constant as that found on the right-hand side of Snell's law.

If we assume a linear sound speed profile so that

$$c(z) = bz + c_1$$

then the radius of curvature is a constant and the ray path is circular.

4.3.3   Range and travel time along a ray

If we consider the ray shown in Figure 4.3a, leaving a source at depth of z_0 at angle θ_0, then we first aim to obtain an expression for the horizontal distance travelled by that ray (also see Brekhovskikh and Lysanov 1991). Consideration of a ray element, such as that shown in Figure 4.3b, shows that

$$\frac{dz}{dr} = \tan\theta \quad\text{or}\quad dr = \frac{dz}{\tan\theta}$$

To obtain the horizontal distance travelled as the ray traverses a finite depth, from z_1 to z_2 then

$$r = \int_{z_1}^{z_2} \frac{dz}{\tan\theta} \tag{4.17}$$

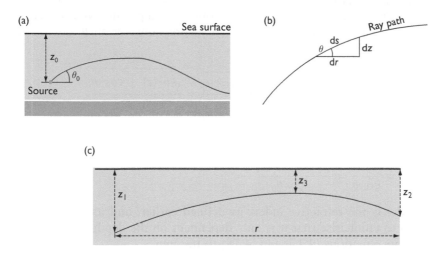

Figure 4.3 Ray geometry.

Equation (4.17) is only valid if the ray does not turn. If the ray does have a turning point in the region of interest we have to divide the integral into two parts (see Figure 4.3c), so the integral becomes

$$r = \int_{z_1}^{z_3} \frac{dz}{\tan \theta} - \int_{z_3}^{z_2} \frac{dz}{\tan \theta}$$

The negative sign is to take account of the fact that the first integral is in the direction of increasing z and the second in decreasing z. In general, if a ray has m turning points one would have to divide the integral into $m+1$ separate components (Medwin and Clay 1998).

In a similar fashion, one can obtain the time taken to travel along the ray. The time taken to traverse an infinitesimal element is

$$dt = \frac{ds}{c}$$

or, with reference to Figure 4.2b,

$$dt = \frac{dz}{c \sin \theta}$$

so the total time taken is

$$t = \int_{z_1}^{z_2} \frac{dz}{c \sin \theta} \tag{4.18}$$

In general, equations (4.17) and (4.18) are not in terms of quantities to which one has direct access, such as the sound speed profile. We can rectify this by the use of Snell's law, to obtain expressions for $\sin \theta$ and $\tan \theta$, namely

$$\sin \theta = \sqrt{1 - \cos^2 \theta} = \sqrt{1 - \frac{c(z)^2 \cos^2 \theta_0}{c_0^2}} = \frac{\sqrt{n^2(z) - \cos^2 \theta_0}}{n(z)}$$

and

$$\tan \theta = \frac{\sin \theta}{\cos \theta} = \frac{\sqrt{n(z)^2 - \cos^2 \theta_0}}{\cos \theta_0}$$

where $n(z)$ is the refractive index, as defined earlier, and its argument is included to emphasise that it is only dependent on the depth z. Hence we can rewrite equations (4.17) and (4.18) to become

$$r = \int_{z_1}^{z_2} \frac{\cos \theta_0 \, dz}{\sqrt{n(z)^2 - \cos^2 \theta_0}} \tag{4.19}$$

and

$$t = \frac{1}{c_0} \int_{z_1}^{z_2} \frac{n(z)^2 \, dz}{\sqrt{n(z)^2 - \cos^2 \theta_0}} \tag{4.20}$$

These two equations allow us to calculate the travel time and horizontal distance travelled for any ray, given only knowledge of the velocity profile and the ray's launch angle.

4.3.4   Focusing factors and caustics

We begin by deriving a result relating the intensity at a point to the ray. If we consider two close rays leaving the source at launch angles θ_0 and $\theta_0 + d\theta_0$ (see Figure 4.4) then the acoustic power flowing between these two rays is assumed to remain between them. The incremental change in range dr, due to the change in launch angle is given by

$$dr = \frac{\partial r}{\partial \theta_0} d\theta_0$$

From Figure 4.4 we see that the perpendicular height, h, of the ray tube at its end is

$$h = dr \sin \theta = \frac{\partial r}{\partial \theta_0} d\theta_0 \sin \theta$$

(a)   (b)

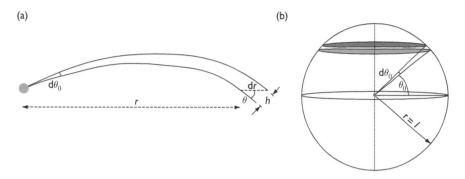

Figure 4.4 Rays from source in 3D.

We must bear in mind that the source is radiating in three dimensions. The total power between the rays of interest is the total power radiating through the shaded area in Figure 4.4b. The surface area of the shaded region is $2\pi \cos\theta_0 d\theta_0$ and the surface area of the whole sphere is 4π. Hence the power passing through the shaded region is $W\cos\theta_0 d\theta_0/2$ where W is the total output power of the source and it has been assumed that the source radiates uniformly. By the time the rays reach a range of r this power is distributed over an area of $2\pi rh$, and thus the intensity is

$$I = \frac{W\cos\theta_0}{4\pi r \dfrac{\partial r}{\partial\theta_0}\sin\theta}$$

The focusing factor f is defined as this intensity divided by the intensity one would observe in a homogeneous medium, that is, the intensity for a spherical spreading model. Since the intensity assuming spherical spreading would be $I_{sph} = W/(4\pi r^2)$, the focusing factor can be expressed as

$$f = \frac{I}{I_{sph}} = \frac{r\cos\theta_0}{\dfrac{\partial r}{\partial\theta_0}\sin\theta} \tag{4.21}$$

Focusing factors greater than unity correspond to regions in the sound field where rays tend to converge. Conversely, focusing factors less than unity correspond to regions where the rays diverge. In regions where f takes extreme values then ray theory is often not applicable. From equation (4.21) we note that the focusing factor becomes infinite if $\partial r/\partial\theta_0 = 0$. The condition on the derivative is true if two adjacent rays converge (see Figure 4.5a).

This type of condition occurs in many situations in the ocean. One example occurs when a family of rays turn and cause a line along which

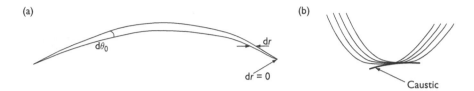

Figure 4.5 Converging rays and caustics.

the intensity, as predicted by this theory, is infinite. Such a line is called a *caustic* (see Figure 4.5b). Conditions of infinite intensity never occur in practice. Ray theory, which predicts these events, begins to break down at the points where the amplitude increases rapidly (see the conditions used to derive the Eikonal equation). It is possible to add corrections to ray theory so that it can cope with caustics. Real caustics in the ocean manifest themselves as extended regions of high intensity rather than points, or lines, of infinite intensity.

4.3.5   Convergence zones and shadow zones

Consider a sound fixing and ranging (SOFAR) channel (Leighton 1998) of sufficient depth that the sound speed below the channel axis reaches that at the surface. Then two commonly observed features are convergence zones and shadow zones. Consider the configuration shown in Figure 4.6 with a source located near the surface of the ocean. Rays grazing the ocean surface are bent deep into the SOFAR channel to eventually be returned at such an angle that once again they graze the surface. Such a ray is labelled as Ray 1 in Figure 4.6. Now, consider a downward pointing ray which grazes the bottom of the SOFAR channel; this ray will also graze the sea surface when it returns and an example is Ray 2 shown in Figure 4.6.

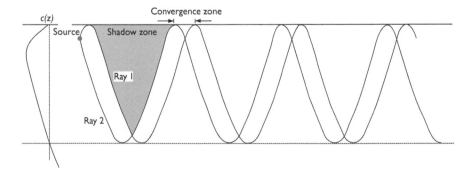

Figure 4.6 Shadow zones and convergence zones.

All the rays leaving the source at angles between the launch angles for Ray 1 and Ray 2 will be contained within the envelope defined by those two rays. No such ray will enter the shaded region shown in the figure, and this region is called the *shadow zone*. The only acoustic energy entering this region is due to scattering of rays which lie outside the envelope defined by Ray 1 and Ray 2. Within a shadow zone, the focusing factor is small.

In contrast to the shadow zone, convergence zones exist near the surface, where the focusing factor is large. The rays contained in the envelope all turn within a short distance of each other producing a caustic close to the surface. In this region the acoustic intensity is large. Typically, these convergence zones occur at 30–35 mile intervals in the real ocean (Leighton 1998) and are commonly observed (Urick 1983). Such convergence zones have widths of 5–10 per cent of the range to the target, the zones nearest to the target being narrower than those further away. The differences in signal intensity between the convergence and the shadow zones can range from 10 to 30 dB.

4.3.6   Practical ray tracing models

In principle, we already possess the tools for calculating the path along which a ray leaving a source from a given launch angle θ_0 will travel. Specifically, equation (4.19) gives us range as a function of depth for any ray. To perform the integration in equation (4.19), analytically, one needs to know the sound speed profile exactly. In practice, the sound speed profile $c(z)$ is not completely known. Most commonly, one will only have access to a few measurements of the sound speed at different depths. Even if the complete sound speed profile were available, then, in all but a few examples, the integration in equation (4.19) is intractable, and numerical integration methods have to be used. Such techniques are not too computationally demanding if only a few rays are to be traced, but often many rays are required to produce an accurate picture of the sound field, and then the use of numerical integration routines becomes too demanding. For these reasons, ray-tracing programmes which are used in practice are not based on equation (4.19) but are built around a different principle. These methods conceptually divide the ocean into distinct layers (Medwin and Clay 1998). (Remember that, at present, we are only interested in range-independent models.) The sound speed profile within these layers is assumed to take a simple form, so the ray paths also take a simple form. A ray is then traced by calculating where it will enter one layer and where it will leave that layer to enter the next.

The simplest example of this strategy is to assume that the ocean consists of homogeneous layers (that is, layers in which the sound speed profile is constant); one such model is shown in Figure 4.7. Within one of these layers a ray will travel in a straight line. Upon meeting a boundary between

True sound
speed profile

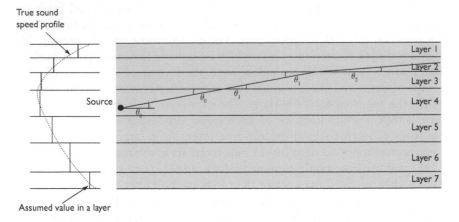

Assumed value in a layer

Figure 4.7 Example of a model of a stratified ocean using layers with constant sound speeds.

layers the ray is refracted (according to Snell's law). To calculate the angle of refraction θ_{i+1} from the angle of incidence θ_i, one simply uses

$$\cos\theta_{i+1} = \frac{c_{i+1}\cos\theta_i}{c_i}$$

If the right-hand side of this expression has a modulus greater than unity then the wave is not refracted but is reflected.

One controls the accuracy of these models by adjusting the number of layers. As the number of layers tends to infinity the model sound speed profile approaches the true sound speed profile everywhere, and the results should converge to the theoretical ray paths given by equation (4.19). However, many layers are often required to get a good approximation. A more efficient method for modelling the layers is not to assume a uniform sound speed profile within a layer but to assume a linear sound speed profile, so that the velocity in the ith layer is modelled as

$$c_i(z) = a_i + b_i z$$

This model tends to model sound speed profiles more accurately and requires the use of fewer layers. As an example, consider a region of the deep ocean, where there is a well-defined deep isothermal layer (Urick 1983; Burdic 1991). Within the deep isothermal layer the ocean temperature is constant and the variation in sound speed is almost entirely due to variations in hydrostatic pressure. Sound speed increases, very nearly, linearly with pressure (and hence with depth) so that a single linear sound speed layer might accurately be used to model the whole deep isothermal layer. Whereas, if uniform layers are to be used, then one might have to employ many such layers to get an accurate representation of the sound field.

We have already seen that within one of these linear sound speed segments the rays travel along arcs of circles. Equation (4.16) indicates that the radii of these circles are given by

$$R_i = \frac{c_0}{b_i \cos \theta_0}$$

where c_0 and $\cos \theta_0$ are measured at any point in the layer. If we consider a ray entering the medium, then at that point we know its grazing angle and the sound speed at the boundary. So we can calculate the radius of the circle along which the ray will travel. By repeating this across many layers, a ray can be traced as it travels through the medium.

Models with layers whose velocity profiles are linear for the basis of the most common type of ray tracing program are computationally efficient, since they only require one to work out simple geometric relations, rather than to perform numerical integrations. Even so, in long-range propagation problems, the acoustic field at a point results from the sum of contributions from many rays. For these problems the computational burden associated with ray tracing becomes prohibitive because of the many rays that need to be traced.

One further problem which can limit the usefulness of ray tracing is the creation of spurious caustics. There is a tendency for the models using layers with linear sound speed profiles to create their own caustics. These caustics arise, not because of the underlying physics, but are purely a consequence of the approximations used in the model. The specific reason is that, at the boundaries between layers, the gradient of the modelled sound speed profile is discontinuous and it is this discontinuity which gives rise to the spurious caustics.

4.4 The method of images (Kinsler *et al.* 1982)

Ray tracing is an adequate method for modelling sound transmission in deep-water environments. The phrase 'deep water' is a relative term; at high frequencies deep water may be only a few metres deep, whilst at low frequencies deep water only occurs off the continental shelf. Ray theory tends to be inappropriate for situations where rays undergo many reflections. Traditionally, normal mode methods have been used to model these 'shallow water' problems; these methods form the subject of the next section. In this section we present the method of images which can be viewed as forming a connection between the fundamentally different approaches of ray tracing and normal mode methods.

Consider a portion of shallow ocean of depth H in which the sound speed is constant. Further assume that this ocean is bounded at the surface by a perfect pressure release boundary and that the sea floor consists of a semi-infinite homogeneous medium. Such a section of ocean is depicted in Figure 4.8. The acoustic pressure measured at the receiver is the sum of the contributions of all the rays arriving at that point. For a receiver distant from the source there will be a large number of rays making significant contributions. The reflections are replaced by virtual images, as shown in Figure 4.9. The images are numbered according to the number of bottom reflections that the rays undergo. With the exception of the case of the two rays, which do not intersect the bottom, there are four different rays which interact with the bottom a given number of times. The reflection coefficient for the bottom generally depends on the angle of incidence and is denoted by $\rho(\theta)$. The phase within each of the rays can be appropriately modelled

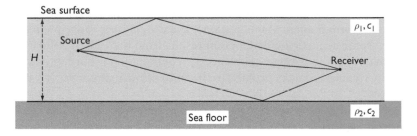

Figure 4.8 A section of isospeed shallow water ocean.

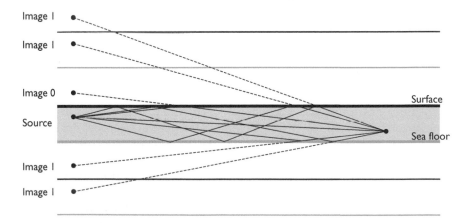

Figure 4.9 Images for source in a shallow water environment.

as random. This is a consequence of several factors including: small-scale inhomogeneities in the ocean and the complex nature of acoustic interactions with the sea surface and the sea floor. As a result, the contributions from the rays are summed incoherently; that is to say, we sum the intensities.

If r denotes the total distance travelled by a ray (as opposed to R which denotes the horizontal distance travelled), then $I(r)$ is the intensity at that range, and is given by

$$I(r) = \frac{I(1)}{r^2}$$

where $I(1)$ is the intensity at the reference range of $1\,\mathrm{m}$. The intensity of an image source which has been reflected by the sea floor m times can be approximated by

$$I_m(R) = \frac{I(1)\rho(\theta_m)^{2m}}{R^2 + (2mH)^2}$$

where θ_m is the angle of incidence for the mth ray. Thus the total intensity is

$$I_{\text{tot}}(R) = 2I_0(R) + 4 \sum_{m=1}^{\infty} I_m(R)$$

Hence we can write

$$\frac{I_{\text{tot}}(R)}{I(1)} = \frac{2}{R^2} \left(1 + 2 \sum_{m=1}^{\infty} \frac{\rho(\theta_m)^{2m}}{1 + \left(\dfrac{2mH}{R} \right)^2} \right) \tag{4.22}$$

The summation can be expressed in terms of the angle of incidence θ_m using

$$\cos^2 \theta_m = \frac{1}{1 + m^2 \left(\dfrac{2H}{R} \right)^2} \tag{4.23}$$

leading to

$$\frac{I_{\text{tot}}(R)}{I(1)} = \frac{2}{R^2} \left(1 + 2 \sum_{m=1}^{\infty} \cos^2 \theta_m \, \rho(\theta_m)^{2m} \right)$$

The summation in the above can be evaluated relatively simply on a computer, assuming that the reflection coefficient is known. To obtain analytic results it is easier to convert this summation into an integral. This requires that the other variables in the expression, which depend on m, in this case θ_m, vary slowly as m is incremented. This translates to requiring that $R \gg H$, so that the range is assumed to be much greater than the depth of the ocean. Under this assumption, the following approximation can be justified

$$\sum_{k=1}^{\infty} \cos^2 \theta_m \, \rho(\theta_m)^{2m} \approx \int_{\nu=1}^{\infty} \cos^2 \theta \, \rho(\theta)^{2\nu} \, d\nu$$

where the variable ν plays the role of m and we note that θ is now a continuous variable. One should bear in mind that θ and ν are still interdependent. It can be shown from equation (4.23) that

$$\nu = \frac{R}{2H} \tan \theta$$

If we now rewrite the integral in terms of the angle θ we obtain

$$\frac{R}{2H} \int_{\nu=\frac{2H}{R}}^{\frac{\pi}{2}} \rho(\theta)^{\frac{R \tan(\theta)}{H}} \, d\theta \qquad (4.24)$$

where we have assumed that $R \gg H$ so that $\tan^{-1}(2H/R) \approx 2H/R$. To evaluate this integral we need to be able to express the reflection coefficient as a function of the angle of incidence. If we assume Rayleigh reflection, meaning that we assume that the bottom can be treated as a fluid and is unable to support stress, then

$$\rho(\theta) = \frac{z_2 \sin \theta_t - z_1 \sin \theta_i}{z_2 \sin \theta_t + z_1 \sin \theta_i} \qquad (4.25)$$

where z_1 is the characteristic acoustic impedance of the ocean and z_2 is the characteristic acoustic impedance of the sea floor. The angles θ_t and θ_i are the angles of transmission and incidence respectively. Snell's law can be employed to eliminate the angle of transmission since

$$\sin \theta_t = \sqrt{1 - \frac{c_2^2 \cos^2 \theta_i}{c_1^2}}$$

but making this substitution does not make the final result any more tractable. To proceed we consider three cases:

(i) *Rigid bottoms*
   No acoustic energy enters a rigid bottom, so

$$\rho(\theta) = 1 \qquad \forall \theta$$

In this case one can evaluate equation (4.22) using the approximation described by equation (4.24) to give

$$\frac{I(R)}{I(1)} = \frac{2}{R^2} \left(\frac{\pi}{2} \frac{R}{H} - 1 \right)$$

which means that the TL is (Urick 1983)

$$TL = 10 \log_{10} \left(\frac{I(1)}{I(R)} \right) \approx 10 \log_{10} R + 10 \log_{10} \frac{H}{\pi}$$

Hence in this situation the TL corresponds to a cylindrical spreading model.

(ii) *Fast bottoms*

For the case $c_2 \gg c_1$, this might be a sea floor consisting of solid rock. Since the density of the sea floor will be greater than that of the sea (otherwise it would tend to float), the acoustic impedance of the floor will also be large in comparison with that of the sea, so that $z_2 \gg z_1$. Under these circumstances we model the reflection coefficient as

$$\rho(\theta) = 1 \qquad |\theta| < \theta_c$$

where θ_c represents a critical angle. Equation (4.24) can then be approximated by

$$\frac{R\theta_c}{2H}$$

so that one obtains

$$TL \approx 10 \log_{10} (R) + 10 \log_{10} \left(\frac{H}{2\theta_c} \right)$$

(iii) *Slow bottoms*

Sea floors which consist of sand are often said to constitute a slow bottom, for which $c_2 \ll c_1$. In this case the reflection coefficient, for small angle of incidence, is approximately

$$\rho(\theta) = e^{-\gamma\theta}$$

where γ is a parameter determined by the bottom characteristics. If this model is substituted into equation (4.24) then one obtains

$$\frac{R}{2H} \int_{\frac{2H}{R}}^{\frac{\pi}{2}} e^{-\frac{\gamma\theta R \tan \theta}{H}} \, d\theta$$

We assume that the primary contributions to this integral occur when θ is small, so that $\tan\theta \approx \theta$ and that the upper limit of integration can be replaced by ∞. If further we assume that the range is sufficiently large so that H/R can be approximated by zero, then this integral simplifies to

$$\frac{R}{2H} \int_0^\infty e^{\frac{-\gamma R\theta^2}{H}} \, \mathrm{d}\theta$$

Upon making the substitution $\nu = \sqrt{(\gamma R/H)}\,\theta$ this integral becomes

$$\frac{1}{2} \sqrt{\frac{R}{\gamma H}} \int_0^\infty e^{-\nu^2} \, \mathrm{d}\nu$$

It is a standard result that $\int_0^\infty e^{-\nu^2} \, \mathrm{d}\nu = \sqrt{\pi/2}$. Thus from equation (4.22) we obtain

$$\frac{I(R)}{I(1)} = \frac{2}{R^2} \left(1 + \frac{1}{2} \sqrt{\frac{\pi R}{\gamma H}} \right)$$

and as R becomes large we obtain

$$TL \approx 15 \log_{10} R + 5 \log_{10} \frac{\gamma H}{\pi}$$

This corresponds to an intermediate spreading model, which can be regarded as a compromise between the excessive attenuation predicted by spherical spreading and the underestimated losses predicted by a cylindrical spreading.

4.5 Normal mode methods

The technique of normal modes (Tolstoy and Clay 1966; Kinsler *et al.* 1982; Medwin and Clay 1998) is closely related to that of the method of images. Indeed, for some simple problems, it is possible to obtain an expression in terms of modes by rearranging the solution obtained using the method of images. One real advantage the normal mode approach possesses over the method of images is its flexibility. Using normal modes one does not need to assume a uniform sound speed within the channel. Further, there is greater scope for specifying boundary conditions, which allows for better modelling of the sea floor reflections. These facilities are available with any reasonable, computer-based, normal mode package. However, the analytical treatment of such problems becomes difficult, if not impossible.

So to keep the analysis tractable, and to retain the essence of the approach, we make several assumptions, two of which are:

(i) that the sea surface is a perfect pressure release boundary, with reflection coefficient of -1;
(ii) that the speed of sound within the channel is uniform.

To generate a modal solution we begin with the Helmholtz equation, equation (4.3):

$$\nabla^2 p + \frac{\omega^2}{c^2} p = \nabla^2 p + k^2 p = 0$$

where p is a function of cylindrical polar coordinates, so that $p = p(r, z)$.

To proceed, we assume that the solution is separable, so that it can be expressed as the product $p(r, z) = U(r) V(z)$, where $U(r)$ is solely a function of r and $V(z)$ is only a function of z. In this case the Helmholtz equation can be written as

$$\frac{U''(r)}{U(r)} + \frac{1}{r} \frac{U'(r)}{U(r)} = -\frac{V''(z)}{V(z)} - k^2$$

The right-hand side of this equation is independent of z and similarly the left-hand side is independent of r. This allows us to conclude that both sides must be equal to a constant. This constant is called 'the constant of separation' and in this case we denote it as $-\gamma^2$. Hence our partial differential equation reduces to the pair of ordinary differential equations:

$$U''(r) + \frac{1}{r} U'(r) + \gamma^2 U(r) = 0$$

$$V''(z) + \left(k^2 - \gamma^2\right) V(z) = 0$$

Both of these are common forms of differential equations; the first is a zeroth order Bessel's equation and the second is a harmonic equation, the general solutions of which are

$$U(r) = A J_0(\gamma r) + B Y_0(\gamma r)$$

$$V(z) = C \sin \kappa z + D \cos \kappa z, \quad \kappa^2 = k^2 - \gamma^2 \tag{4.26}$$

where A, B, C and D are all arbitrary constants and $J_0(.)$ and $Y_0(.)$ are order Bessel functions of zero order of the first and the second kinds, respectively. Since $Y_0(\kappa r) \to \infty$ as $r \to 0$, and because we do not expect to see infinite pressures in our field, then B must be zero. There are also the boundary conditions at the sea surface and the floor to consider. In this simple case, we shall assume that at the surface $(z = 0)$

$$p(r, 0) = 0 \quad \Rightarrow \quad V(0) = 0 \quad \Rightarrow \quad D = 0$$

which is consistent with a pressure release boundary, and at sea floor $(z = H)$

$$\frac{\partial p(r, H)}{\partial z} = 0 \quad \Rightarrow \quad V'(H) = 0 \quad \Rightarrow \quad C\kappa \cos(\kappa H) = 0$$

which is consistent with a perfectly rigid bottom, and implies that

$$\kappa = \frac{\pi}{2H}, \frac{3\pi}{2H}, \frac{5\pi}{2H}, \ldots$$

Thus there are infinite number of discrete values that κ can take; these are termed the eigenvalues of the system. Each of these discrete values corresponds to a mode of propagation. We more succinctly write the set of permissible κ as

$$\kappa_m = \left(m - \frac{1}{2} \right) \frac{\pi}{H}, \qquad m = 1, 2, 3, \ldots$$

from equation (4.25) so that these values can be related back to the constant γ via

$$\kappa_m^2 = k^2 - \gamma_m^2$$

where k is the wavenumber and is independent of m. From this we see that as m gets larger, γ_m eventually becomes imaginary, the significance of which we shall discuss later.

There is thus an infinite number of solutions (modes) of the differential equation. To obtain the most general solution we form the sum of all the modes as

$$p(r, z) = \sum_{m=1}^{\infty} A_m \sin(\kappa_m z) J_0(\gamma_m r)$$

where

$$\kappa_m = \left(m - \frac{1}{2} \right) \frac{\pi}{2H}, \qquad \gamma_m^2 = \frac{\omega^2}{c^2} - \kappa_m^2$$

and the A_m is a series of arbitrary constants. It is common practice to use the asymptotic form of the Bessel function; specifically, for large r,

$$J_0(\gamma_m r) \approx \sqrt{\frac{2}{\pi \gamma_m r}} \cos\left(\gamma_m r - \frac{\pi}{4} \right)$$

The only problem left to solve is how to evaluate the constant A_m. To do this one uses the remaining physical feature which has not been employed in the solution so far, namely, the source characteristics. Using the fact that the modes are normal one can show that

$$A_m = \frac{QV_m(z_0)}{\nu_m}, \qquad \nu_m = \int_{z=0}^{H} V_m(z)^2 \, dz$$

where z_0 is the source depth, $V_m(z)$ is the depth-dependent part of the solution for the mth mode and Q is the volume velocity of the harmonic source. To obtain this expression, we have implicitly assumed that the sound speed is constant within the water column. It is relatively simple to generalise the result for the case where the sound speed varies with depth.

The final expression for the pressure field at long range is

$$p(r, z) = \sum_{m=1}^{\infty} \sqrt{\frac{2}{\pi \gamma_m r}} \frac{\sin(\kappa_m z_0) \sin(\kappa_m z) \cos\left(\gamma_m r - \frac{\pi}{4}\right)}{\nu_m} \qquad (4.27)$$

Let us now try and place a physical interpretation on this approach. The two constants κ_m and γ_m represent the components of the wavenumber vector when resolved into vertical and horizontal directions, respectively. Since $k^2 = \kappa_m^2 + \gamma_m^2$, one can write

$$\kappa_m = k \sin \theta_m \quad \text{and} \quad \gamma_m = k \cos \theta_m$$

where θ_m represents the direction of propagation of the component waves of the mth mode, measured relative to the horizontal. As m increases, so does κ_m, implying that the component plane wave forming the higher modes propagates in directions closer to the vertical. For sufficiently large m, γ_m becomes imaginary, as we have already mentioned. This implies that these modes have pressure fields that reduce exponentially with range; these are termed evanescent modes. At long ranges the effects of such modes are negligible. So, at sufficiently long ranges, one need only consider the modes which satisfy

$$\kappa_m = \left(m - \frac{1}{2}\right)\frac{\pi}{H} \le k \quad \Rightarrow \quad m \le \frac{Hk}{\pi} + \frac{1}{2} = \frac{2H}{\lambda} + \frac{1}{2} \qquad (4.28)$$

This means that the summation in equation (4.27) need not extend to infinity but need only include non-evanescent modes. This reveals another major advantage of the normal mode methods over the method of images; namely that at long ranges the normal mode method is computationally less demanding. This is because one need only include a few modes in the sound field calculation, whereas one may need to include many more images.

We now briefly consider equation (4.28) as a function of frequency. Consider a specific mode, for example, the first mode, so $m = 1$. This mode will be evanescent if $\omega/c \geq \kappa_1 = \pi/2H$. Thus, below the frequency $\omega_{c,1}$ defined by

$$\omega_{c,1} = \frac{\pi c}{2H}$$

the first mode will not contribute significantly to the pressure field a large distance from the source. This frequency is called the cut-off frequency for the first mode. In general, the cut-off frequency for the mth mode is given by

$$\omega_{c,m} = c\kappa_m = \frac{c\pi}{H}\left(m - \frac{1}{2}\right)$$

The group velocity for an individual mode is given by $c_g = c\cos\theta_m$, which corresponds to the horizontal component of velocity for that mode. One can show that

$$\cos\theta_m = \sqrt{1 - \left(\frac{\omega_{c,m}}{\omega}\right)}$$

This implies that, within a specific mode, the lower frequencies travel down the channel more slowly than the higher frequencies. A system in which waves of different frequencies travel at different speeds is called a dispersive system. The propagation speed c is the same for all frequencies; it is only the speed at which the disturbance progresses down the channel which varies with frequency. This is termed *geometric dispersion*, to distinguish it from *intrinsic dispersion*, such as that observed in bending waves in a metal bar, where it is the case that speed of propagation is frequency-dependent. In the ocean, this produces noticeable effects. A loud impulse, such as an explosion, when heard at great distances, becomes distorted. Specifically, rather than being a sharp bang, it initially sounds like a rather muffled 'thud' followed by low-frequency rumblings.

Example

Consider a section of shallow ocean, 10 m in depth, which is bounded below by solid rock and above by a perfect pressure release boundary and assume that the speed of sound is constant. In the following we show the acoustic field generated by a point source, sited at a depth of 5 m in the water column, over a range of 200–300 m. The source is emitting a tonal signal whose angular frequency is equal to the speed of sound, so that,

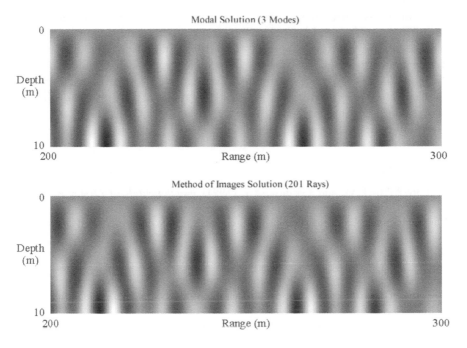

Figure 4.10 Acoustic pressure fields computed using normal modes and method of images (dark shades represent high pressure amplitudes).

numerically, $\omega = c$. From this we see that $m = 1$. Using equation (4.28) the highest order of non-evanescent mode is

$$\frac{10}{\pi} + \frac{1}{2} \approx 3.7$$

so that using normal modes one need only consider the first three modes.

Figure 4.10 shows the pressure amplitude as a function of range and depth, as computed using the method of images and normal mode methods; note that the ranges considered are relatively large. The normal mode approximation is computed using only three modes, whilst the method of images employs 201 rays. This illustrates one advantage that modal solutions have over ray-based methods at long ranges; namely that one can compute the acoustic field more efficiently.

4.6   Conclusion

This chapter has described the background to two of the most basic oceanic sound propagation models. Ray tracing, which provides an efficient method for high-frequency, relatively short-range problems, and normal mode methods that can be used for long-range, low-frequency problems. Both methods suffer from significant limitations, even when applied to problems over favourable ranges. In the ray tracing approach it is difficult to incorporate realistic boundary conditions, and normal mode methods are strictly only applicable in range-independent problems. Further, in practice, one often encounters problems in which the range is too long for the efficient use of ray tracing methods and the range is too short for efficient use of modal techniques; in which case, one may seek to employ one of the alternative methods mentioned in Section 4.2.2.

References

Brekhovskikh, L.M. and Lysanov, Yu. (1991) *Fundamentals of Ocean Acoustics* (Berlin: Springer-Verlag).

Buckingham, M.J. (1992) Ocean-acoustic propagation models. *Journal D'Acoustique*, 5, 223–287.

Burdic, W.S. (1991) *Underwater Acoustic System Analysis* (New York: Prentice-Hall).

Etter, P.C. (1995) *Underwater Acoustic Modelling: Principles, Techniques and Applications* (London, New York: E & FN Spon).

Kinsler, L.E., Frey, A.R., Coppens, A.B. and Sanders, J.V. (1982) *Fundamentals of Acoustics* (3rd edn) (Boston: John Wiley & Sons).

Leighton, T.G. (1998) Fundaments of signal processing. In Fahy, F. and Walker J. (eds), *Fundamentals of Noise and Vibration* (London, New York: E & FN Spon).

Medwin, H. and Clay, C.S. (1998) *Fundamentals of Acoustical Oceanography* (Boston: John Wiley & Sons).

Pack, P.M.W. (1986) *The Finite Element Method in Underwater Acoustics*, Ph.D. Thesis, University of Southampton.

Tolstoy, I. and Clay, C.S. (1966) *Ocean Acoustics: Theory and Experiment in Underwater Sound* (New York: McGraw-Hill).

Urick, R.J. (1983) *Principles of Underwater Sound* (New York: McGraw-Hill).

Environmental and industrial acoustics

Chapter 5

Environmental noise management

I.H. Flindell and J.G. Walker

5.1 Introduction

5.1.1 The need for environmental noise management

Environmental noise is a significant and increasing problem for large numbers of people in many parts of the world. Widespread mechanisation and industrialisation, together with growing populations, particularly in cities, tend to increase the total amount of noise generated and the number of people exposed to it. Depending on noise levels, the effects range from mild annoyance and occasional sleep disturbance in the lower range of possible noise levels, through increasing degrees of interference with quality of life with increasing noise levels, up to the possibility of noise-induced hearing loss and other non-auditory health effects in the most noise-susceptible individuals at the highest noise levels, particularly in some of the more recently industrialised nations with less-developed environmental noise management policies.

People vary in their sensitivity to different levels of environmental noise. Many people seem to be able to tolerate a limited amount of noise without suffering any obvious ill-effects. People who appear to be insensitive to environmental noise may be genuinely unaffected, or they may have changed or adapted their lifestyle (possibly at some cost) so that whatever interference or disturbance was caused by noise is no longer a problem for them. There are many other people who give every appearance of having been very seriously affected. Quite often, only a minority of residents living around major sources of environmental noise such as main roads, industrial sites or airports, ever complain, but this does not necessarily mean that the rest of the exposed population is unaffected. For example, noise can become a very significant issue when assessing the likely consequences of major infrastructure development projects such as new roads and airports.

5.1.2 Balancing costs and benefits

Noise management is about balancing the costs and inconvenience of noise control against the likely or assumed benefits. As with most fields of human endeavour, the costs incurred usually increase in proportion to the benefits gained, although not necessarily according to any linear relationship. In this situation, the distinction between 'acceptable' and 'unacceptable' amounts of noise can depend as much on the overall costs as on the benefits, but any equation here can be complicated by the fact that the costs and benefits usually accrue to different people and are measured in different currencies. The costs of alternative noise control engineering options can usually be added up in monetary terms, but the benefits of noise control, or the costs of *not* doing any noise control in terms of the effects on people, are much harder to quantify in monetary terms.

Decision making sometimes depends as much on political and social imperatives as on technical arguments about noise levels and noise effects. One of the main problems here is that noise management is mostly about predicting what will happen in the future – which of two or more alternative noise management plans is likely to be most cost-effective over the longer term. Usually, it is much easier to predict the engineering or physical consequences of any proposed action than to predict the actual effects on people's attitudes and opinions. Noise management decisions can only normally be based on anticipated effects, rather than on actual effects, which can only be measured retrospectively, after the decision has been put into effect.

5.1.3   *Approaches to environmental noise management*

There are, in general, two broad approaches to this problem. The simpler of the two approaches is to set fixed noise limits or benchmarks above which environmental noise, either now or in the future, is deemed unacceptable and below which environmental noise is correspondingly deemed acceptable. Any approach based on fixed noise level benchmarks has the considerable advantage of simplicity and straightforwardness. Noise management decisions can be based purely on technical assessments and comparisons. For the public, fixed noise level benchmarks are easy to understand and relatively easy to test for compliance. Based on either measurement or prediction, it is easy to determine precisely how much additional noise attenuation may be required to provide an adequate degree of protection for noise-exposed properties, or over which specific time period additional restrictions should be imposed. Existing standards and regulations can be read and applied exactly as they are written, and unnecessary paperwork can (usually) be kept to an absolute minimum. On the other hand, merely maximising administrative convenience does not automatically imply that the best compromise decisions will be reached for the individuals and organisations involved.

The second approach recognises that there are often no very good justifications for setting fixed noise level benchmarks at any particular, and usually arbitrarily selected, noise level; that noise level alone can rarely explain more than a small proportion of the total variance in community response; and that overall acceptability depends as much on the differences between the likely noise effects associated with the available noise management alternatives as it does on the absolute noise levels associated with any one noise management option. This approach requires that decisions might vary in different situations depending on 'informed flexibility'. Informed flexibility describes a general attitude to environmental noise management whereby noise control engineers and decision makers are prepared to weigh up all available information before reaching the best compromise decision in any particular case, rather than just deciding on the basis of fixed noise level benchmarks. Informed flexibility can lead to apparent inconsistency when tested against fixed noise level benchmarks because management action focuses on achieving the best compromise solution in any particular case, rather than just minimising the numbers of people exposed above the noise level benchmarks.

The 'correct' application of informed flexibility can require considerable expertise over a range of different technical and scientific disciplines on the part of decision makers. Informed flexibility almost invariably requires extensive consultation and it can be particularly vulnerable to legal challenge and appeal. The more complex and apparently arcane the decision making process, the more likely that objectors to any decision may not fully understand the decision making process. Different situations can arise under different circumstances giving at least an appearance of inconsistency. In any particular circumstance, where the available regulations and standards permit a choice, which of the two approaches (fixed benchmarks or informed flexibility) will help decision makers to reach the best compromise solutions is usually a matter of opinion rather than a foregone conclusion.

Either approach will usually benefit from a clear appraisal of the relative advantages and disadvantages of alternative noise management options, which can range from doing nothing at all to completely shutting down the offending noise source. This is where the current interest within the environmental noise management industry on 'noise mapping' may be useful. Noise mapping is the process of either measuring or calculating physical noise exposure levels and/or some indicator of the effects of noise on a distributed geographical basis. Noise maps can be configured to show noise exposure level contours overlaid on conventional maps; but the overall process of noise mapping can also include counting the number of houses or the number of people situated within particular noise exposure contour levels. Simply counting the number of people exposed to levels of environmental noise that are considered likely to cause significant disturbance or annoyance, will show the overall extent of noise problems on a strategic basis, although noise mapping is potentially much more useful when it is used to show the relative effects of alternative environmental noise management options.

5.2 UK strategic noise policy

5.2.1 Existing standards and regulations

In the United Kingdom, the Government Planning Policy Guidance Note PPG24, 'Planning and Noise' (DoE 1994)[1] sets out general guidance on how local planning authorities can minimise the adverse impact of noise without placing unreasonable restrictions on development or adding unduly to the costs and administrative burdens of business. Annex 3 of PPG24 summarises the existing and separate approaches for noise from road traffic, railways, aircraft, industrial and commercial developments, construction sites, recreational and sporting activities and landfill waste disposal sites, based on a number of existing standards and regulations, and using a range of noise indicators, not all of which are entirely consistent with current initiatives by the EC and the WHO. Further details of these specific approaches are provided in Section 5.6 as representative examples of practical applications.

The most recent National Society for Clean Air survey of local authorities throughout the UK (NSCA 2002) concluded that noise problems persist and, for many sources, complaints continue to increase, affecting the perceived or reported quality of life of a large proportion of the population. There is a general perception among local authority officers in many areas that environmental noise is increasing, that people have higher expectations of quiet, and that many people's aspirations are not being met. However, the extent to which noise problems may be increasing because of new noisy development, increased activity of existing noise sources, or new residential development in existing noisy areas has not been established.

5.2.2 Current initiatives by DEFRA

Governments and regulatory bodies are responding to these continuing environmental noise problems. The UK Department for Environment, Food and Rural Affairs (DEFRA) recently published a consultation document, 'Towards a National Ambient Noise Strategy' (DEFRA 2001) seeking views on the Government's proposed three-phase approach, which is to:

- assess the number of people adversely affected by different levels of noise;
- evaluate and identify options for prioritising alternative action plans;
- proceed with the most appropriate action.

As background work leading up the UK National Ambient Noise Strategy, the UK National Noise Incidence Study 2000–2001 (Skinner and Grimwood 2002) which was carried out by the Building Research Establishment (BRE) for DEFRA found that while changes in average noise level and noise exposure between 1990 and 2000 have been mostly quite small, by looking at a range of different indicators it has been possible to show that a number of rather more subtle changes have taken place. Measured outside a large sample of representative dwellings (1160 measurement sites), average daytime ambient noise levels ($L_{Aeq,16h}$ and $L_{A10,16h}$) were found to have decreased by fractions of a decibel while average night-time background noise levels ($L_{A90,8h}$) were found to have increased by similar fractional amounts. The overall pattern was consistent with an assumption that, on average, noise levels from individual events had decreased, but the frequency with which such events occur had increased, particularly at night. The mean sound levels outside homes throughout the UK in 2000 were 57.1 dB $L_{Aeq,12h}$ (0700 h–1900 h daytime), 53.1 dB $L_{Aeq,4h}$ (1900 h–2300 h evening) and 48.2 dB $L_{Aeq,8h}$ (2300 h–0700 h night-time).

A separate analysis for Greater London showed that the London area was consistently noisier (on average) than the rest of the country and that average night-time background noise levels had similarly increased, albeit with a generally shorter night-time period than outside London.

The UK National Noise Attitude Survey 1999/2000 (Grimwood *et al.* 2002), also carried out by the BRE for DEFRA, found that while the majority of respondents were not particularly disturbed by noise, a sizeable minority were, and that the small minority who reported that they were seriously affected added up to several million people over the country as a whole. In detail, 69 per cent of respondents reported general satisfaction with their noise environment, 57 per cent reported that noise did not at all spoil their home life, and noise was on average only rated 9th in a list of 12 environmental problems. On the other hand, 21 per cent reported that noise spoiled their home life at least 'to some extent' and 8 per cent reported that noise spoiled their home life 'quite a lot' or even more. A smaller but unstated percentage reported that noise 'totally' spoiled their home life. Nearly three times as many respondents reported that road traffic noise at their homes had got worse in the last 5 years (28 per cent) compared with the percentage who reported that it had got better (10 per cent). Thirty-seven per cent reported that they were bothered, annoyed or disturbed by noise from neighbours to some extent. Overall, the proportion who reported being adversely affected by noise from neighbours had increased over the last 10 years and only a small proportion of these respondents had complained to the local authority. Based on these findings, the BRE report went on to suggest that noise complaint statistics collected by local authorities could be significantly underestimating the true extent of community dissatisfaction with noise.

5.3   European directive on environmental noise

5.3.1   Historical context

There has been a long history of European Directives concerned with the management of noise emissions by specific sources such as road transport vehicles, outdoor machinery and aircraft, mostly from the point of view of defining harmonised-type approval noise limits or otherwise regulating noise at source. While these various directives have arguably contributed to a general reduction in noise levels, they have paid no attention to the overall effect on noise levels received by European residents, or to any beneficial effects that may have ensued. As noted in the UK National Noise Incidence Study (Skinner and Grimwood 2002), described above, there is good evidence that the reductions in noise emissions generated at source have, in many cases, been offset by significant increases in overall traffic, leading to negligible reductions in noise exposure, as measured outside houses, and even some increases in some areas, particularly at night.

5.3.2   Control of noise exposure vs noise emission

Having recognised this as a problem, the European Commission started working in 1996 towards a new directive to address perceptions of, or exposure to, noise; that is, on achieving measurable reductions in noise level exposure as measured at the receivers or as perceived by European residents, and/or measurable reductions in adverse effects such as annoyance, sleep disturbance or other effects on general health and well-being. Noise exposure (that is, as measured at the receiver, or 'received' noise) is described as noise 'immission' in some European documents related to the new directive. The preparatory work has recently culminated in a new European Directive on the Assessment and Management of Environmental Noise (EU 2002). The new directive is based on the conclusions of a European Green Paper dating from 1996 (EC 1996) which was in turn based on an inventory of European environmental noise problems, first published in 1994 (Lambert and Vallet 1994). After receiving comments from many parties from all member states, the Commission organised two conferences about European Future Noise Policy, first in The Hague and then in Munich in 1997, following which the Commission established a number of working groups to contribute technical advice regarding important details of the new directive. The terms of reference of the working groups were agreed at a further conference on European Future Noise Policy held in Copenhagen in 1998. The Commission published brief overviews for the benefit of the general public (EC 2000a,b). Most of the more recent progress since those dates has been largely administrative.

5.3.3   Environmental noise directive

The stated aims of the new environmental noise directive are as follows:

> to define a common approach intended to avoid, prevent, or reduce on a prioritised basis the harmful effects, including annoyance, due to exposure to environmental noise. To that end the following actions shall be implemented progressively:

(a) the determination of exposure to environmental noise, through noise mapping, by methods of assessment common to the Member States;

(b) ensuring that information on environmental noise and its effects is made available to the public;

(c) adoption of plans by the Member States, based upon noise mapping results, with a view to preventing and reducing environmental noise where necessary and particularly where exposure levels can induce harmful effects on human health and to preserving noise quality where it is good.

The directive also aims 'at providing a basis for developing community measures to reduce noise emitted by the major sources, in particular road and rail vehicles and infrastructure, aircraft, outdoor and industrial equipment and mobile machinery'. The directive does not apply to noise caused by the exposed person, noise from domestic activities, noise created by neighbours, noise at workplaces or noise inside means of transport or due to military activities in military areas.

The directive imposes deadlines on member states for the production of strategic noise maps for all large centres of population and alongside major roads, railways and airports. 'Competent authorities' are to be engaged to draw up action plans designed to manage, within their territories, noise issues and effects, including noise reduction, where necessary.

The directive also requires the European Commission to report progress after 7 years, with particular emphasis on:

(a) setting long-term goals for the protection of the public against the harmful effects of environmental noise, taking into account different climates and different cultures;

(b) additional noise reduction measures for specific sources such as outdoor equipment, means and infrastructures of transport and certain categories of industrial activity;

(c) the protection of quiet areas in open country.

The directive lays down that 'the reduction of harmful effects and the cost-effectiveness ratio shall be the main criteria for the selection of the strategies and measures proposed'.

Annex I defines standardised noise indicators (L_{den} and L_{night}). L_{den} is simply the 24 h L_{Aeq} with the evening period (this is assumed to be 1900 h–2300 h unless otherwise defined) weighted by 5 dB and the night period (this is assumed to be 2300 h–0700 h unless otherwise defined) weighted by 10 dB. L_{night} is the 8 h L_{Aeq} for the night period (2300 h–0700 h, the same as for the night part of L_{den}) unweighted, the precise conditions under which they should be determined, and the special conditions under which additional or supplementary noise indicators may be required; Annex II defines harmonised methods for computation and measurement; Annex III defines harmonised methods for assessing harmful effects based on agreed dose–effect relationships; Annex IV defines minimum requirements for strategic noise mapping; Annex V defines minimum requirements for action plans; Annex VI defines data to be sent to the European Commission. The data required by the Commission include estimates of the number of people living within defined noise level contour bands with additional information on the number of people benefiting from increased noise insulation or from quiet facades to their dwellings.

5.4   World Health Organisation guidelines

5.4.1   Historical context

The World Health Organisation (WHO) has taken an interest in environmental noise over many years. Following various drafts produced by a number of international working groups which met several times throughout the 1970s, the WHO published a comprehensive review of environmental noise problems as an Environmental Health Criteria Document No. 12 (WHO 1980). This document included an authoritative review of the available literature, and then recommended a number of specific criteria for the avoidance of hearing damage risk (75 dB $L_{Aeq,8h}$), for good speech intelligibility indoors (45 dB L_{Aeq}), for the avoidance of sleep disturbance at night (35 dB L_{Aeq} measured in the bedroom), for the avoidance of significant community annoyance (55 dB L_{Aeq} daytime measured outdoors) and to meet sleep criteria at night (45 dB L_{Aeq} night-time measured outdoors).

Following a period of further widespread international consultation, during which a number of intermediate drafts were published, the original 1980 guideline values were extended and updated by a new 'Guidelines for Community Noise' document published in 2000 (Berglund *et al.* 2000). Users of the new guidelines document should note that the new guidelines do not have legal force except where they have been separately approximated into national regulations and standards. The inside cover sheet of the printed document states that 'This document is not a formal publication of the World Health Organisation . . . ' and 'The authors alone are responsible for the views expressed in this document'. On the other hand, the foreword states that; 'This document is the outcome of consensus deliberations of the WHO Expert Task Force'. This task force was convened in April 1999 in London following an earlier task force meeting held in 1992 in Dusseldorf, and following widespread comments received on an earlier draft published by Stockholm University and the Karolinska Institute in 1995 (Berglund and Lindvall 1995).

5.4.2   Current guideline values

The current WHO guideline values are set out in Table 5.1 (reproduced from Table 4.1 of the WHO guidelines document published in 2000).

Note that there are no WHO guideline values (at the time of writing – (Spring 2003) for protecting the public against possible non-auditory health effects such as hypertension and ischaemic heart disease, immune system deficiencies, mental health disorders or other possible stress-related health problems. This is because there is no scientific consensus on dose–effect relationships for these possible effects. At the time of writing, the WHO regional office for Europe was working towards 'operational guidelines' in these areas (WHO Regional Office for Europe July 2002).

5.4.3   Commentary on current WHO guidelines

The current guideline values are based on the previous WHO recommendations (WHO 1980), but have been developed and extended to represent the most recent research findings, as reported to the various task forces and ultimately to the named authors responsible for publication. As a general principle, they represent the lowest thresholds at which each noise effect has been observed in either laboratory or field studies. Depending on the effect concerned, the prevalence cannot be determined just from the observation threshold alone; it is also perfectly possible that many of these observed effects might not become particularly significant in terms of overall quality of life until much higher sound levels are reached. Recent estimates of community noise exposure across the European Union suggest that around half the total population live in areas where the current WHO guideline values for annoyance and sleep disturbance are exceeded under today's conditions. Only a minority of these people actually report significant disturbance. This implies that the majority of these people have acclimatised to the noise, although this does not necessarily mean that acclimatisation has occurred at no cost to the individuals concerned. It may also be important to note that individual sensitivity to noise varies over a wide range.

According to the recent UK National Noise Incidence Study (see Section 5.2.2), slightly more than half of the UK population are exposed to noise levels which exceed those set out in the current WHO guidelines; 54 per cent are exposed to noise levels which exceed the current WHO guideline value for the avoidance of 'serious annoyance' of 55 dB L_{Aeq} daytime (that is, only 46 per cent are exposed below the guideline value); 67 per cent are exposed to noise levels which exceed the current WHO guideline value for the avoidance of sleep disturbance of 45 dB L_{Aeq} night-time (that is, only 33 per cent are exposed below) (the error bands are approximately plus or minus 3 per cent on all percentages reported). While reported annoyance can be assumed to increase for noise levels which exceed 50–55 dB L_{den}, less than 10 per cent of respondents actually reported

Table 5.1 WHO guideline values for community noise (Berglund et al. 2000) (reproduced with the permission of the World Health Organization)

Environment	Critical health effect	dB L_{Aeq}	dB L_{Amax}
Outdoor living area	Serious annoyance, daytime and evening	55	
	Moderate annoyance, daytime and evening	50	
Dwelling, indoors	Speech intelligibility and moderate annoyance, daytime and evening	35	
Inside bedrooms	Sleep disturbance, night-time	30	45
Outside bedrooms	Sleep disturbance, window open	45	60
School classrooms	Speech intelligibility	35	
Pre-school bedrooms, indoors	Sleep disturbance	30	45
School playground	Annoyance (external source)	55	
Hospital wards	Sleep disturbance	30	40
Hospital treatment rooms	Interference with rest and recovery	As low as possible	
Industrial, commercial, shopping, etc.	Hearing impairment	70	110
Ceremonies, entertainments	Hearing impairment for infrequent patrons	100	110
Public address systems	Hearing impairment	85	110
Headphones	Hearing impairment (free-field values)	85	110
Toys, fireworks	Hearing impairment – adults		140
	Hearing impairment – children		120
Parklands and conservation areas	Disruption of tranquillity	Preserve existing quiet areas and maintain low intrusion	

themselves as 'highly annoyed' at these noise levels. In addition, it is also interesting that the contemporaneous UK National Noise Attitude Study (also see Section 5.2.2) found that the majority of the population do not report significant noise disturbance.

These findings suggest that, for the UK population considered as a whole, the current WHO guidelines would appear to be somewhat conservative or even stringent. On the other hand, there is some uncertainty over the true nature of the effects being considered. In respect of environmental matters, the precautionary principle suggests action wherever there is any doubt, rather than waiting for definitive scientific evidence to become available. However, it is often forgotten that the precautionary principle should apply both ways; if there is no strong evidence of the likely benefits of noise control in any particular case, then decision makers should also be careful of the costs or other inconveniences or adverse effects of inappropriate noise control. For example, the damage to overall quality of life that might result from closing down a noisy factory, which might thereby put a lot of people out of work, could outweigh the benefits achieved by improving the noise environment.

5.4.4 Overview of UK, EU and WHO initiatives

National governments, the European Union and international bodies such as the WHO are paying increasing attention to the problem of environmental and community noise. There is some evidence that current actions addressed towards limiting noise levels generated at source have not been completely successful in reducing noise levels at noise-sensitive receivers. Whilst the National Ambient Noise Strategy in the UK, the European Directive on the Assessment and Management of Environmental Noise and the most recent WHO Guidelines for Community Noise have focused on establishing the extent of exposure against stated observation thresholds for a wide range of different possible noise effects, it is not yet clear by which means environmental noise might actually be reduced in the future, or perhaps more importantly (given the competition for scarce resources which would otherwise be spent on other things), with what priority.

5.5 Noise effects

5.5.1 Environmental noise

So what is environmental noise? Any sound existing outdoors and defined as unwanted for any reason can generally be classified as environmental noise. The WHO uses the term 'community noise' to mean much the same thing. Unwanted sound generated as part of the normal activities of a household can be classified as 'domestic noise', while domestic noise generated next-door and passing through the walls or windows into someone else's house can be classified as 'neighbour noise'. Noise generated in the work-place and affecting workers is generally defined as 'occupational noise' and does not concern us here, except that where it breaks out from the work-place and then affects nearby residents it is known as 'industrial noise'. In existing standards and regulations, industrial noise is often dealt with as a separate category of environmental noise. Construction noise, road traffic noise, railway noise and aircraft noise can all be considered as occupa-tional noise in respect of workers employed on construction sites or around transport infrastructure and they can all be considered as different types of environmental noise in respect of nearby residents. Amplified speech or music or the sound of people talking or shouting at each other in a restaur-ant, theatre or discotheque is usually defined as 'entertainment noise' with varying implications, depending on whether we are considering customers, employees or nearby residents.

When operating, all machinery generates noise in differing degrees, which can then adversely affect any person residing or working in the vicinity. In the middle range of sound levels, which can occur in typical residential envir-onments, the relative significance or importance of the noise can depend as much on the context in which the noise is heard as on the actual sound levels. In the lower range of environmental sound levels, it is the extent to which any specific sound can actually be heard or not that is probably the most important factor. As will be seen later on, the relative audibility of any specific sound, as compared to any masking background noise present, can be estimated objectively. At increasing specific sound levels, the character of the sound and the context in which it is heard become increasingly import-ant as different individual attitudes and opinions come into play. Then, at some upper but unknown end of the range of environmental noise levels typically encountered, it is reasonable to assume that noise has become so intolerable on account of its sound level alone that the character and context have become irrelevant again. In practice, the middle range in which character and context are important can extend over a very wide range of actual noise levels.

In summary, for real-life environmental noise assessment, guideline values based on the assumed magnitude or importance of any resulting effects may be required for any or all of the following possibilities:

- 'just audible': mere audibility is a necessary but not a sufficient condi-tion for genuine noise nuisance to occur.
- 'middle range': the prevalence and severity of noise nuisance and other effects are assumed to increase as noise levels increase above audi-bility thresholds. In addition, individual differences, context and situ-ation become increasingly important. Most environmental assessments have to balance overall costs and overall benefits in this wide middle range where a limited amount of noise nuisance or other effects might

be considered tolerable when balanced against the overall costs and possible inconvenience of noise control.

• 'completely intolerable': noise nuisance and other effects are so severe that they must be avoided irrespective of cost.

The assessment of environmental noise can often be made more difficult because people sometimes use the same words to describe a number of different things depending on the context. All noises are sounds but not all sounds are noises. The word 'sound' can refer to either 'physical sound' or 'subjective sound'. Physical sound is a small fluctuating disturbance in atmospheric pressure that propagates as an acoustic wave (which can exist irrespective of whether there is any listener present or not); subjective sound is the sensation or perception of a sound. Noise is just sound with different meanings attached, but the word 'noise' can similarly mean either 'subjective noise' or 'physical noise'. Subjective noise is the subjective sensation or perception arising when the noise is present; physical noise is the pressure disturbance that has been defined as noise, presumably because of some real or assumed unwanted or undesired effect. The main distinction between sound and noise is subjective; the only physical factors which can be used to distinguish between sound and noise are the place and context in which they exist. Neither the place nor the context can be identified from the physically measurable characteristics of the sound alone.

In addition, confusion can also arise when using words such as 'unacceptable' or 'intolerable' because what is unacceptable to one individual may be perfectly acceptable to another. To maximise administrative convenience, objective tests of noise acceptability or tolerability could be extremely helpful, but the difficulty here is that different individuals might not agree.

5.5.2 Physical energy levels

The amount of physical energy involved in even quite high levels of environmental noise is quite small (although this is not always true for certain kinds of industrial or military noise exposure). An A-weighted continuous sound level of 90 dB (this sound level exceeds WHO guideline values for hearing damage risk if continued for extended periods of time) is equivalent to an A-weighted acoustic intensity of only 1 milliwatt per square metre. The A-weighted acoustic intensity at the mean sound level measured outside homes throughout the UK of 57.1 dB $L_{Aeq,12h}$ (0700 h–1900 h daytime) is less than half a microwatt per square metre. At an A-weighted continuous sound level of 90 dB, the amount of sound power entering the small cross-sectional area of the ear canal is very small, and at an A-weighted continuous sound level of only 57.1 dB the A-weighted sound power entering the cross-sectional area of the ear canal is nanoscopic.

These amounts of physical sound energy in typical environmental noise are much too small to cause any measurable physical disruption to the tissues or even to cause any significant heating effects. There is no physical mechanism by which these small amounts of acoustic energy can have any cumulative effect in the environment. It is clear that environmental noise is not the same as chemical pollution where waste products can accumulate in the environment to cause significant long-term damage. The physical magnitude can be measured at the time that the sound occurs by using an appropriate instrument such as a sound level meter but no instrument has yet been invented that can measure sound levels retrospectively (after the sound has ceased).

Because unwanted sound has no cumulative effect in the physical environment, and because most sound is either neutral or even beneficial (for example, environmental sounds caused by leaves rustling in the wind or a river flowing past or wanted sounds such as speech and music are not usually classed as harmful), sounds can only be defined as noise because of some undesired effect on people. It is debatable to what extent people should actually be present for any sound to be classified as noise, although there are some effects of noise on animals. There are no known effects of general environmental noise on plants, although it should be noted that there is always a theoretical possibility of hitherto undiscovered effects. In laboratory experiments on the possible effects of noise on plants, it could be difficult to control for any possible effects of noise on the human experimenters in order to be able to separate out any real effects on the plants alone.

So, in order to understand noise pollution, we must understand the possible unwanted effects of excessive sound on people. These effects are often thought of as being independent of the source and the context in which the noise is heard, but as has been explained above, this is not really true. The main reasons why different types of sound might be defined as unwanted and thereby classified as noise, or not, vary in different situations and contexts, as further explained.

5.5.3   The effects of environmental noise on people

In the UK, the first serious attempt at understanding the then growing problem of environmental noise was carried out by the Wilson Committee on the problem of noise in the early 1960s (HMSO 1963). While some of the research that was carried out to support the committee's deliberations at the time may seem a little dated by present day standards, it is quite difficult to see where any really significant progress has been made since that time. There has been a considerable amount of research carried out in the intervening 40 years or so, but no hitherto unsuspected effects of noise have been discovered since the Wilson Report was published, and there is still considerable uncertainty regarding the scale or prevalence of any effects (Berry et al. 1998).

The main effects of environmental noise vary across the different ranges of noise level which can occur in different types of environment. At the lowest levels of environmental noise, the main effects are determined simply by whether or not the particular noise is audible or not above other background noise sources present. Any noise which cannot be heard should not be considered as producing a genuine noise problem. It may perhaps be surprising to some readers that inaudibility does not preclude the possibility that knowledge of the presence of the noise source gained by some means other than audition could be a cause for complaint, rather than the noise levels generated by that source. If the noise is clearly noticeable (that is, the sound level for the specific noise source exceeds the masked threshold level determined by other background noise sources present by a sufficient margin to ensure clear audibility) then there is at least a possibility that a listener might at first notice the noise and then, once the noise has been noticed, there is a further possibility that the listener might object to the noise or become bothered or annoyed by it. The threshold for 'noticeability' is generally 5–10 dB higher than the lowest auditory masked threshold level because the lowest auditory threshold represents the level at which the specific sound can only just be detected at a slightly better than chance probability.

As the sound level increases above the masked threshold level for that situation, the available evidence suggests that listeners are increasingly likely to report mild, moderate or even significant annoyance attributable to the presence of the specific noise. Reported annoyance could arise from actual interference with, or disruption to, a whole range of wanted or desired activities; or listeners may simply be registering their disapproval at the mere presence or existence of the noise source. The types of activities that can be disturbed by unwanted noise include listening to wanted sounds such as speech and music which could be partially or completely masked by the unwanted noise, disturbance or interruption to rest and relaxation or disturbance to sleep.

Depending on an individual's psychological orientation to the presence of the noise, there is at least a theoretical possibility that a perceived inability to influence the operation of the noise source (this is described by psychologists as a 'lack of control') could contribute to some form of long-term psychological stress. Experts disagree over both the definition and the long-term health implications of psychological stress, but it could in theory contribute to a number of possible long-term 'stress-related illnesses' such as hypertension and other cardiovascular diseases, immune system deficiencies and mental health problems. There is also a separate possibility of direct or autonomic physiological responses to noise having some bearing on long-term health. Unexpected sudden loud noises are known to cause acoustic startle effects characterised by increased heart rate and blood pressure and a general orientation of the endocrine and autonomic nervous systems towards 'fight' or 'flight'. While these effects are not necessarily harmful in themselves, the theoretical possibility that transient increases in blood pressure, resulting from acoustic startle or similar effects could become permanent if repeated often enough, has at least an appearance of biological plausibility. On the other hand, the short-term physiological after-effects of acoustic startle are not very different from the short-term physiological after-effects of exercise, which is generally considered to be beneficial. What seems to be much more certain in this general area is that both habituation and expectation can be important.

5.5.4 Masking effects: just-audible thresholds

The extent to which any specific sound which has been the subject of complaints can be heard within or above a fixed background noise level attributable to other noise sources present, determines whether or not that specific sound can be considered to be the cause of a genuine noise problem. In cases where people have complained about sounds which cannot actually be heard, the most likely alternative explanation would be some kind of tinnitus, although there is also a possibility that low-frequency vibration (when present) could be perceived as low-frequency sound. It is unlikely that engineering noise control applied to any noise source, about which complaints are received but which is inaudible, would be effective. Neither is it unknown for complainants to become highly sensitised to particular sounds which other people can hear but do not particularly notice, and which might not therefore be considered as a justifiable nuisance by others. This type of situation can be particularly distressing for the complainant because they are then unable to obtain any kind of redress. On the other hand, it would perhaps be unfair for potentially costly engineering noise control to be imposed or required by authorities in situations where most people do not consider the noise to be a problem.

When a specific sound cannot be heard in the presence of a steady background noise level attributable to other noise sources present, the sound level attributable to that specific sound (when measured separately or in the absence of any other sounds present) is said to be below the auditory masked threshold level for that specific sound under those conditions. The amount in decibels by which the sound level of a specific sound exceeds the auditory masked threshold level is defined as the 'sensation level' of that specific sound. The relative prominence or intrusiveness of the specific sound can then be related to its sensation level.

Auditory masking occurs when the frequency components of a specific sound and a simultaneous masking sound overlap along the vibration-sensitive basilar membrane in the inner ear. There might then be no differences in the excitation patterns of the sensitive hair cells along the basilar membrane between the two sounds, and consequently no information available to the brain, which would otherwise allow it to discriminate between the two sounds. Auditory masking is most effective when the specific sound

and the masking sound have the same or overlapping frequency components, and least effective when the frequency components are completely
different, or where there is no overlap in time. The relative audibility of
any simple pure-tone specific sounds can be estimated from objective physical measurements to within relatively narrow limits by taking into account
the known frequency selectivity of the normal human ear, but only when
the specific and masking sounds can be measured separately. This can be
difficult or impossible under real-life field conditions. Estimating auditory
masked thresholds where either the specific or background sounds (or both)
have time-varying multiple pure-tone and broadband frequency components
present can be more difficult still. Existing theoretical models of auditory
masking are reasonably accurate for simple situations but many of the more
complex situations which can arise under real-life field conditions have not
been adequately dealt with by current research. However, because auditory
masking depends mainly on the physiological and bio-mechanical properties of the human auditory system, it should be much more amenable to
estimation based on objective measurements than attitudinal effects such as
reported annoyance.

On the other hand, even if adequate theoretical modelling capabilities
existed, any frequency analyser systems used will generally require much
narrower frequency resolution than the 1/3rd octave band frequency resolution commonly provided with hand-held sound level meters, and the
measurement limitations imposed by physical time–frequency uncertainty
would also have to be taken into account.

As a practical alternative, and provided only that the specific sound can
be turned on and off, it may be possible to determine auditory masked
thresholds by direct listening. A panel of listeners with normal hearing
would be asked to judge whether or not they can hear or detect the specific
sound ideally over a range of presented sound levels, but if necessary only
when the specific sound is either on or off. Variability between different
listeners and over time can be reduced by averaging over multiple observations, although care should be taken to avoid averaging out across genuine
systematic differences. The main weakness of direct listening methods is that
separate adjustment of either the specific target sound or the masking background sound might not be possible in practical situations. Direct listening
methods are sometimes more easily applied to recordings or simulations of
the actual sounds but this can introduce unknown differences between the
real-life and recorded or simulated situations.

For practical applications, auditory masked thresholds can be particularly relevant to certain kinds of neighbour noise and fixed industrial and
commercial plant noise problems. Masked threshold levels are not generally at issue when assessing transportation noise sources, which are often
clearly audible above any steady background noise present. Transportation noise, which is only just audible, would not normally command a

very high priority for noise management effort. On the other hand, for specific sounds which are only just audible, complaints are more likely to be provoked by particular acoustic features such as tones or impulses than by the overall sound level alone. The sensation level of any acoustic feature present provides a direct indication of by how many decibels that feature might have to be reduced to achieve inaudibility (Porter *et al.* 1993). This principle is reflected in the relative approach to noise assessment adopted in the British Standard method for rating industrial noise affecting mixed residential and industrial areas (BS 4142 1997). Even if reported annoyance or actual interference with perceived quality of life is not strongly related to sensation level in all cases, it is difficult to see how any specific sound which is below the masked threshold level, set by the existing background noise attributable to other noise sources, and therefore deemed to be inaudible, could be classified or assessed as a genuine noise problem.

5.5.5   Masking effects: speech masking

The masking of wanted speech (or music) sounds by unwanted background noise intrusion has considerable similarities to the masking of unwanted specific sounds by general background noise. However, in this case, it is the speech or music signal that is wanted, and the background noise which is not; that is, the general situation is reversed. As for the just-audible thresholds of specific sounds, it is possible to estimate the amount of 'speech masking' present from objective physical measurements of the time-varying frequency spectra of the speech and background noise. However, the overall importance or significance of any particular degree of short-term or transient speech masking to general human activities also depends on the context or situation in which the speech masking occurs. For example, the overall significance of a continuous background noise at a level which is sufficient to cause partial masking of wanted speech, could be greater than an intermittent background noise which causes complete speech masking at those times when the background noise is present, but causes no masking at all for the rest of the time when it is not present. This depends on the extent to which the wanted speech signals contain redundant information. Sometimes it is vitally important that people should be able to hear a single spoken message clearly and effectively, but in everyday life it is much more common that a limited amount of speech masking can be tolerated without significantly affecting overall message intelligibility. People can adapt their method of communication to accommodate intermittent interruption to speech communication simply by waiting for the interruption to pass by. Any annoyance caused by interference with speech communication can have as much to do with the context in which the interference occurs as on the quantitative loss of information content. In addition, attitudes and opinions towards the source of the interference can be very important determining factors.

The 'articulation index' and the 'speech transmission index' are internationally standardised methods of measuring the speech signal-to-masking noise ratio in separate 1/3rd octave frequency bands used for the purpose of calculating overall weighted speech signal-to-masking noise ratios, based on the relative importance of each frequency band present in normal speech to the overall 'speech audibility' (ANSI 1997; BS 6840-16 1998; IEC 1998). The signal-to-noise ratio measured in decibels in each frequency band is linearly scaled from $+1$ at $+18\,dB$ to 0 at $-12\,dB$ (this represents a 30 dB dynamic range) and then the relative contributions of each band are weighted and added appropriately to give an overall index ranging from zero to one to cover the range from zero to perfect *audibility*. The intelligibility of real speech, which is defined as the percentage or proportion of words or sentences correctly understood, also depends on familiarity with the language used in addition to the audibility (and freedom from non-linear distortion) of that speech.

Because of speech masking that can occur in natural environments, human speech and hearing have evolved in such a way that speech intelligibility can be maintained even where significant proportions of the overall speech frequency bandwidth have been masked by unwanted noise. Human speech is said to be highly redundant in terms of information content. In addition, intermittent masking noise can completely mask speech when it is present, but it may still be possible to understand speech messages perfectly well from the time segments of speech that are left. For longer-duration intermittent noise such as aircraft noise or railway noise, it is usually possible to just stop talking while the noise is present and then resume talking when the noise has ceased without any loss of message intelligibility at all. This may cause annoyance or fatigue even where there is no actual speech disturbance because speech has been avoided for the duration of the noise.

Engineers and administrators often wish to measure the speech masking effect of a particular noise. This can vary under different conditions because the speech masking effect depends on the type of speech and the purpose for which the speech communication is taking place. *Behavioural speech intelligibility* can be measured by counting the proportion of separate messages or speech phonemes (a phoneme is the smallest segment of speech that conveys separate meaning such as a vowel or a consonant sound) which have been correctly understood. Behavioural speech intelligibility can be quite high for meaningful sentences about a common theme or it might be quite low for separate unconnected nonsense syllables where the preceding and the following speech gives no contextual clues to the precise meaning. Listeners can also be asked to provide a subjective judgement or rating of *perceived speech intelligibility* which could be a subjective impression of the proportion of speech correctly understood, or it could be a subjective rating of the amount of conscious effort required in an attempt to understand the speech signals. Finally, objective measurement methods such as the articulation index are sometimes mistakenly described as measures of speech intelligibility whereas in fact they are not measures of either behavioural or perceived speech intelligibility at all, even if they might be correlated with them to some extent.

In summary, speech masking can be considered as an objective phenomenon, which can be related to frequency-band-limited speech signal-to-masking noise ratios at the time that it occurs. Any resulting loss of speech or message intelligibility can be more difficult to measure because this also depends on the context in which the speech is heard. Intermittent aircraft noise could potentially cause significant speech masking when present while the overall loss of message intelligibility might be very small. Steady road traffic noise at the same overall long-term average sound level (L_{Aeq} or L_{den}) might cause lower instantaneous speech masking effects when compared to the instantaneous speech masking effects occurring at the maximum noise level during an aircraft flyover, but the overall loss of message intelligibility or any resulting annoyance could be higher.

The WHO guideline value of 35 dB L_{Aeq}, which is proposed as the observation threshold value for measurable speech intelligibility effects in school classrooms, seems highly conservative when compared with a typical A-weighted background noise level for a desktop PC cooling fan of around 40 dB, and with typical raised voice levels from a teacher of 60–70 dB, even at the back of the classroom. On the other hand, even relatively low level sounds entering a classroom from outside could cause significant distraction or inattention, but this kind of disruption is highly sensitive to context and to individual differences in application and motivation.

5.5.6 Activity interference

Direct interference with particular activities such as rest, relaxation, work or concentration is often claimed as a particular consequence of excessive noise, but there is considerable variability regarding the actual noise levels at which these effects are likely to occur. There are no consistent data to support reliable guideline values.

The extent to which noise can interfere with various activities depends on the type and level of the noise, on the particular activity being interfered with, and on the individual concerned. Unwanted noise can certainly interfere with *rest and relaxation* and can cause considerable distraction from any task requiring significant *concentration*, particularly where there might be some particular meaning or information content present in the noise. On the other hand, a certain amount *of steady background noise* can be used in open-plan offices to mask other sounds which might otherwise cause considerable distraction. In these cases, background noise can improve task performance or workplace productivity. Highly motivated individuals can be resistant to any kind of task or activity interference caused by unwanted noise, while less-motivated or tired individuals might be quite sensitive.

In theory, different types of background noise could contribute to heightened or reduced levels of *psycho-physiological arousal*, which in turn might contribute to heightened or reduced task performance, depending on the optimum level of arousal required for that particular task.

It is well known that different types and styles of music can have particular effects on mood and motivation, and this knowledge is often exploited by the designers of background music for railway stations, fashion shops, and supermarkets at different times of the day. Where designers are mainly interested in the effects on customer satisfaction and sales figures, it is better to measure these variables directly under different types of background music condition rather than expend any effort in measuring reported attitudes and preferences for different types of music. For commercial reasons, designers do not usually publish the results of any tests carried out. In addition, the particular style or genre of any background music which has been deployed for the specific purpose of contributing to increased sales figures should be seen as an essential component of the overall commercial strategy of any particular kind of store. What works in one store in 1 year may not work in another store the next year.

In summary, it seems clear that background noise of various types can either interfere with, or even enhance, a wide range of different types of activity, and that it is very difficult or impossible to come up with any general rules. Background noise is not necessarily always bad in this context. Background music in shopping centres (known in the UK as muzak), however much it might be disliked by certain sections of the population, is only there because commercial data have shown that it has a positive effect on sales. The management appears to be convinced that it makes a positive contribution to sales. WHO has not developed guidelines specifically to protect people against activity interference caused by noise, and it seems unlikely that any simple and consistent guideline could ever be developed anyway.

5.5.7 Sleep disturbance

Almost everyone has personal experience of sleep disturbance caused by unwanted or unfamiliar noise. On the other hand, there is also considerable evidence that sleep disturbance can be subject to significant adaptation and habituation. Relatively modest acoustic signals in the unfamiliar surroundings of a sleep laboratory can cause transient disturbances in electroencephalogram (EEG) signals whilst asleep, whilst significant but familiar events at home sometimes cause no response at all (Robertson *et al.* 2000). Measured sleep disturbances can range from minor transient disturbances, premature changes in sleep stage, behavioural awakening or even objectively measurable next-day sleepiness or fatigue. Transient disturbances are not necessarily of any significance in a whole night context, while objectively measurable next-day sleepiness has not yet been demonstrated in the context of typical levels of environmental or community noise. There is also a possibility of perceived or reported sleep disturbance, which may or may not be correlated with any objectively measurable disturbance. Given these uncertainties, it is difficult to arrive at any consensus as to which specific sleep disturbance effect or outcome measure should be the focus of future standards and guideline values (Porter *et al.* 2000).

The design of alarm clocks and fire alarm systems takes into account the fact that at least the *peripheral auditory system* remains fully operational even while the listener is asleep. It is more useful for survival of the species that individuals will be awakened by sudden or unexpected noise rather than not waking up and then possibly being eaten by an undetected predator. Of course, undetected man-eating predators are not a feature of modern western civilisation, but these changes have been too recent for human evolution to have had enough time to catch up. On the other hand, people can get used to familiar noises at night without waking up all the time. The long-term survival advantages of continuing to wake up every time a regular and known-to-be harmless noise occurs must be rather doubtful, and so in this context, *habituation* is probably a helpful biological adaptation. For habituation to familiar night-time noises to occur, there must be some kind of auditory processing going on at a subconscious level. If all forms of perceptual processing were completely shutdown while a person is asleep, there would be no possibility of recognising familiar noises and then classifying them as no-threat stimuli, not requiring conscious (awake) attention for more considered appraisal.

One of the major scientific problems here is that research has not provided a fully convincing explanation of why sleep appears to be necessary for health and well-being. It has not been scientifically demonstrated that sleep is not merely a *biological adaptation* to the day–night cycle to encourage humans (and many other types of animals) to curl up and hide away from potential sources of danger at night. We know that people wake up briefly at regular intervals throughout the night, presumably just to make a quick check that all is well, even though they do not usually remember these awakenings in the morning. We also know that some people can manage perfectly well on much less sleep than others without apparently suffering any ill-effects. Nor are people who sleep more than average noticeably any healthier.

Unfamiliar noise or other unexpected events such as a light being turned on and off can disrupt the regularly repeated sleep cycle which usually proceeds from the initial stages of falling asleep into deep sleep and back into light sleep again with intervening short periods of awakening at approximately 90-minute periods throughout the night. By convention, each sleep cycle can be divided into separate periods of sleep stages one to four roughly corresponding to increasingly deep sleep, with the separate shorter periods which are normally associated with dreaming known as rapid eye movement

(REM) sleep. The different sleep stages have been defined in accordance with clearly distinguishable differences in brain wave patterns and frequencies as recorded by EEG measurements. EEG systems record small voltages generated across electrodes, which are temporarily attached to the skin at various positions around the head and face. The most dominant frequencies observed in EEG recordings collected at different stages throughout the repeated sleep cycle are used to discriminate between the different sleep stage classifications. EEG records can be used to demonstrate significant transient disturbances in brain wave patterns and frequencies associated with unexpected noise events (presented while the subject is asleep) which do not necessarily cause any overt change in the sleep cycle or lead to behavioural awakening.

Unfamiliar noise or other unexpected events can cause changes in the regular sleep cycle to occur before they might have been expected in an undisturbed regular sleep cycle. On the other hand, the time spent in different stages of sleep varies naturally, and normal sleep includes repeated short awakenings as normal occurrences. For each separate apparent sleep disturbance event detected, there are no objective methods for discriminating between events, which are definitely caused by extraneous events, and events which would have occurred anyway. Discriminating between the two types of event can only be done by comparing event probabilities for short time periods (known as epochs in some noise and sleep research) with extraneous events present and not present. This can require large amounts of data for statistical reliability.

Sleep experiments have shown that the probability of actually being awakened by noise increases towards the end of the sleep period and also varies in different parts of the sleep cycle. Some of the research carried out in specialist sleep laboratories using unfamiliar noise suggests much greater sensitivity to sleep disturbance caused by noise at night than research carried out in people's own homes where any noises present are more likely to be familiar to the research study participants.

The current WHO guideline values for community noise at night are based mainly on *laboratory research* using unfamiliar noises. The recommended guideline values of 30 dB L_{Aeq} or 45 dB L_{Amax} measured inside the bedroom are inconsistent with the results of many *field studies* carried out in people's own homes. These generally suggest that most people can tolerate much higher noise levels without suffering from any overall loss of sleep or any significant next-day effects. Transient sleep disturbance is not necessarily harmful unless it is a precursor of some more significant overall loss of sleep. Also, depending on the context, unfamiliar noise at much lower levels than the WHO guideline values can wake people up.

There is an additional problem in that reported or perceived (subjective) sleep disturbance is not very well correlated with actual or objectively observed sleep disturbance. People may believe that they have been woken up or prevented from going to sleep by some external noise even if objective evidence based on EEG measurements shows this not to have been the case. In western democracies, subjective opinions can be more important than objectively determined scientific 'facts'. It is widely accepted that people cannot form reliable subjective judgements of external events that occur while they are asleep, although there is limited evidence that people can retain some memory of events which took place while they were asleep during the previous night.

Finally, it should be noted that just as there is no scientific consensus about the precise biological or evolutionary function of normal sleep, there is also no scientific consensus about which specific aspects of sleep disturbance are the most important in terms of effects on health or quality of life. Unwanted noise can cause transient disturbances to the normal EEG patterns observed throughout the night which do not necessarily progress to cause any disruption to overall sleep patterns or any increase in the number of transient awakenings that would normally occur even in the complete absence of any intruding noise. Objectors to night-time noise sources are likely to argue that any form of disruption to normal EEG patterns, even if these are only transient in nature, are potentially damaging to long-term health. Apologists for, or proponents of, night-time noise sources are equally likely to argue that transient disturbances to EEG records are merely indicators of normal biological function and that there is no cause for concern unless night-time disturbance leads to some kind of significant next-day effect.

Significant next-day effects caused by unwanted noise such as objectively determined next-day sleepiness or any kind of behavioural or performance decrement have not been demonstrated in the context of typical levels of environmental noise. Next-day sleepiness can be measured subjectively by questionnaire, or objectively by using sleep latency tests. Sleep latency tests measure the time taken to fall asleep at specified times throughout the next day following night-time disturbance. Subjects must be immediately woken as soon as they have fallen asleep to prevent recovery from any loss of sleep incurred during the previous night. One of the problems here is the possibility that, if there is any significant loss of sleep caused by night-time noise, various compensatory or adaptive mechanisms might start to operate through subsequent nights leading to reduced sensitivity to night-time disturbance through those following nights. When averaged over several nights, there might then be no statistical differences between next-day effects measured after nights with noise events and nights without noise events. Compensation or adaptation, if it exists, could in theory incur additional or hidden costs to long-term health, but there is no scientific evidence to support this suggestion in respect of night-time noise.

However, it is known that significant next-day effects such as next-day sleepiness (which can be dangerous for drivers or airline pilots, for example) or lowered productivity at work can definitely arise following loss of sleep from other causes. This means that night-time environmental noise is taken seriously, even in the absence of definitive scientific evidence, demonstrating significant next-day effects from this cause alone.

5.5.8   Health effects

To the general public, the generic concept that excessive levels of community or environmental noise may have adverse effects on health seems plausible enough. However, the underlying hypothesis that excessive noise contributes to psychological stress which in turn contributes to various maladaptive physiological responses, and on which the concept is based, has been hard to demonstrate on a statistical basis. This hypothesis has never been satisfactorily explained in terms of plausible causative mechanisms. Partly because there is insufficient definitive experimental data available, and partly because of general uncertainty about possible theoretical explanations for these hypothesised relationships, there is no consensus on which any specific standards or guideline values might be based (Berry *et al.* 1998).

Because there is no consensus on these issues, the current WHO guidelines do not include any recommendations for the avoidance of adverse health effects, other than the avoidance of noise-induced hearing loss, for which a very conservative guideline value of 70 dB L_{Aeq} is recommended (see Section 5.4.2). It should be noted that the new Environmental Noise Directive (EC 2002) specifically mentions the possibility of 'harmful effects on human health' (see Section 5.3.3), although it should also be noted that this does not necessarily mean there is any scientific consensus that there are any such effects. National and European occupational noise standards for the avoidance of noise-induced hearing loss are set at much higher levels, namely 85 and 90 dB L_{Aeq} extrapolated to an 8-hour working day (Noise at Work Regulations UK and EC directive). Residential exposures to community and environmental noise do not normally reach such high noise levels even when measured outdoors, although the UK National Noise Incidence Study 2001 (Skinner and Grimwood 2002) indicates that around 2 per cent of the UK population are resident in areas where the 70 dB L_{Aeq} WHO guideline value is exceeded. However, even if outdoor noise levels are high enough to constitute a risk of noise-induced hearing loss, it is unlikely that residents would be present outdoors long enough for there to be any identifiable risk in practice.

Anecdotal evidence suggests that the public generally understand the concept that unwanted noise might contribute to physiological or psychological stress which might in turn contribute to the development of persistent changes in the neuro-physiological, endocrine, sensory, cardiovascular or immune systems of the body. While small changes need not necessarily be considered as being of any clinical significance, any of these so-called stress effects could accumulate to become clinically significant outcomes over the longer term. Unfortunately, the stress hypothesis does not differentiate between beneficial and harmful stress. A certain amount of stress seems to be necessary or desirable to maintain arousal in the short term and basic fitness over the longer term. There is no clear mechanism by which the observable short-term physiological responses to noise, such as acoustic startle, might be assumed to translate into potentially damaging chronic changes in the level of basic physiological functions. On the other hand, the fact that no satisfactory theoretical mechanism has yet been proposed which might explain the possible development of long-term adverse health effects of noise is not a proof that such effects do not exist.

Laboratory studies using animal subjects have clearly demonstrated the possibility of adverse health effects 'caused' by prolonged exposure to very high noise levels by comparing noise-exposed and non-noise-exposed animals. Epidemiological studies comparing people resident in noisy and quiet areas have been less convincing, but there are a number of significant methodological problems which would need to be overcome before any further progress can be expected in this area, for the following reasons:

- the expected relationships are probably quite weak and may therefore be hard to detect;
- individual susceptibility is likely to vary over a wide range. Because there is no method available for identifying the most susceptible individuals in advance by some independent means, any real effects in the most susceptible individuals might be concealed by averaging across the population as a whole;
- there are numerous 'confounding factors' such as diet, socio-economic status and lifestyle which might well have much more significant effects on health outcomes;
- individual exposure to noise can be very difficult to measure;
- even if cross-sectional studies were to show statistical associations between the level of noise and the prevalence of effects, this kind of study cannot be used to demonstrate which is the cause and which is the effect.

5.5.9 Reported annoyance

Reported noise annoyance is measured by asking people how disturbed, bothered or annoyed they are by the noise, using direct questionnaires administered by post, telephone or in direct face-to-face interviews. Reported noise annoyance has obvious face validity and at least an appearance of being directly meaningful, but it can be highly context-dependent and it is not always clear how it can be related to perceived quality of life or overall amenity issues. The concept of 'percentage highly annoyed' has been promoted as a relatively stable indicator of average community annoyance (see for example, Meidema 2001), but even this concept can be hard to pin down. There are large individual differences which are clearly dependent on both individual attitudes and opinions and on the context in which the noise is heard (Flindell and Stallen 1999).

Given the various difficulties in estimating some of the other potential effects of noise, the main outcome measure used for noise management purposes is usually the reported or assumed noise annoyance. Despite many attempts to define precisely what is meant by noise annoyance over the last 40–50 years, it means different things to different people, and depends both on personal characteristics and on the context in which the noise is heard. For example, noise annoyance can be understood as some perceived or imagined property of the noise source, or of the sound as heard by the listener, or it can be understood as an indicator of the attitudinal response of an individual to the presence or intrusion of that noise (in other words, it is not always clear whether noise annoyance should be related to the source or to the effect of the noise).

Many researchers have attempted to pin down the precise meaning by using precise wordings in questionnaires when measuring annoyance. Different wordings used in different questionnaires often lead to different results. For those researchers who assume that noise annoyance is a universal concept which can be translated between different cultures and languages, there are additional difficulties when translating words like 'annoyance', 'bother' and 'disturbance' into other languages. More recently, international committees have been established to recommend specific wordings for noise annoyance questionnaires for use in different languages (Fields et al. 2001). Of course, even if appropriate descriptors can be found in different languages with the same precise meaning, none of these initiatives can solve the basic problem of *context dependency*. Reported annoyance (researchers rarely if ever measure actual annoyance) varies depending on the context in which it is asked about. For example, the two alternative questionnaire wordings set out below would normally be expected to return significantly different patterns of results:

Q1 Do the aircraft (around here) EVER make you feel annoyed or bothered, and if so, please record your degree of bother or annoyance by using this scale?

Q2 Considered overall, how annoying is the average amount of aircraft noise around here compared to any other things about living around here and which might be annoying to some extent or other?

Q1 focuses on aircraft noise annoyance independent of any other sources of annoyance or disamenity which might be present in the same area.

Q2 attempts to place reported aircraft noise annoyance within the general context of whatever other sources of annoyance or disamenity might be present. Usually, Q1-type questionnaire items find higher reported annoyance than Q2-type questionnaire items.

Most research studies, comparing reported annoyance against noise levels, have been able to show some form of level dependency whereby average annoyance increases with noise level. However, it should be noted that *large statistical uncertainties* are normally present, arising both from individual differences and from separate uncertainties in measuring the dependent and the independent variables involved. Further meta-analyses of combined data-sets have demonstrated a general level of consistency between different studies, although combining the data on to single dose–effect relationship curves always requires numerous assumptions about the comparability of the input data and measuring instruments between different studies. It is difficult or impossible to test the validity of any normalising assumptions made, except by any resulting convergence of the separate results obtained from the separate studies included in the analyses. This test of convergence implicitly assumes that there is an underlying consistency of noise dose–annoyance effect relationships between different studies, which can only be revealed by making the appropriate normalising assumptions, whereas it is also possible that there is no underlying consistency, and that any convergence obtained is essentially illusory. For this reason, the international scientific community appears to have divided into two schools of thought. The first accepts the existence of underlying universal noise dose–annoyance effect relationships and the second has not (yet) been convinced.

The current WHO guidelines for community noise are based on a consensus view of particular noise level threshold values, below which various annoyance and other effects have not been observed to be particularly significant. This consensus view is not entirely satisfactory because it cannot be used to estimate noise level threshold values, at which the prevalence of particular levels of annoyance or other effects becomes significant to society as a whole. For the WHO guidelines, 50 dB L_{Aeq} measured outdoors is proposed as the threshold below which *moderate annoyance* is infrequent, and 55 dB L_{Aeq} measured outdoors is proposed as the equivalent threshold below which *serious annoyance* is infrequent.

As an alternative, many regulatory or administrative agencies rely on pragmatic evidence of community effects at different noise levels based on *noise complaints* or other forms of community action. Actual noise complaints can form a much more effective stimulus to provoke noise management action than some theoretical prediction of the percentage likely to express 'high annoyance'. On the other hand, it is well known that, despite many claims which might be made to the contrary, noise complainants are not always particularly representative of the *silent majority* who are similarly exposed to the noise but who do not take any particular action. Without asking them directly, it is impossible to determine whether the silent majority are in fact seriously disturbed by the noise but are too busy or lack the necessary self-confidence to complain, or that the silent majority remain silent because they are not at all concerned. The approach adopted by the EC environmental noise directive (that the aggregate effects of noise across a community as a whole can be estimated by applying assumed universal noise dose–annoyance response relationships to calculated noise level contours overlaid on geographically referenced residential population maps) will only give a correct indication of the true scale of any noise problem if the assumed noise dose–annoyance response relationships are correct. Regulatory or administrative agencies will have to decide for themselves on which basis to first consider and then assess noise management action; this is where the two alternatives of (i) decisions based on informed flexibility and (ii) decisions based on fixed criteria or noise limits, may have to be critically compared.

5.5.10 Justification for specific guideline values

The key features that can be discerned from all the available information about the effects of noise on man are that:

- noise effects tend to increase with increasing noise levels;
- most noise effects vary between and within different individuals;
- noise effects can be influenced by a large number of non-acoustic context and situation variables;
- even where reliable dose–effect relationships can be found, there are no discrete steps which could be used to distinguish between acceptable and unacceptable noise exposure.

On the other hand, there is a strong demand for specific guideline values to assist decision makers. As stated in Section 5.4.2, and for these reasons, the specific guideline values recommended by the WHO are based on effects thresholds; that is, noise levels below which specific effects are assumed to be either negligible or relatively unimportant. As such, modest exceedances of the WHO guideline values need not necessarily be considered as particularly significant within the overall context of the many possible effects of some development. This clearly leaves a gap where national and/or international authorities may perceive a need for noise level benchmarks set at some higher noise level, where the consequences of exceedance are sufficiently severe for the development to be considered unacceptable. Wherever any such higher noise level benchmarks have been defined (for example, to define entitlement to additional outdoor-to-indoor noise insulation, or for some other form of compensation), the precise sound levels at which any such benchmarks have been set are essentially arbitrary. If it is necessary to define a cut-off point in terms of noise exposure for some purpose or other, then a cut-off point will be defined, but this does not mean that people just above the cut-off point are significantly worse off than people just below it.

5.6   Practical applications

5.6.1   Legal framework

In the UK, there is a complex statutory framework applicable to different types of noise problem. The existing law developed on a piecemeal basis from the original common law principles based on a court decision of what is 'reasonable' and what is not. For noise nuisance, different individuals have different ideas about what is reasonable and what is not. The resulting uncertainty has been resolved to some extent by statutory provision, but not everyone will agree with particular decisions made by the courts.

Current laws empower authorities to take action to control statutory noise nuisances as they arise while also granting immunities to noise-makers, provided they comply with agreed procedures. Quite often, authorities are responsible for promoting new development and managing existing development, at the same time as being responsible for setting environmental limits. For example, at the time of writing (Spring 2003) the Department for Transport (DfT) is responsible both for developing and managing transport infrastructure and for managing the resulting environmental impacts, while the DEFRA is responsible for environmental limitation (where appropriate), although in many cases without any clear jurisdictional boundary between the two areas of responsibility. For example, while the current Planning Policy Guidance Note, Planning and Noise, PPG 24 (DoE 1994; see Section 5.2.1) remains within the purview of the current DEFRA, much of the content was either contributed by or agreed between the then DoE and DoT. Various other organisations such as the UK Civil Aviation Authority, the Strategic Rail Authority, the Highways Agency, County Transportation Departments and others, all have similarly divided responsibilities between catering for increased traffic demand and meeting environmental limitations.

According to the traditional common law principles based on reasonableness which have now been incorporated into various Acts of Parliament, Courts of Law are empowered to consider 'best practicable means', and BATNEEC – 'best available techniques not entailing excessive cost' – when deciding what is fair and reasonable in any particular case. For example, while the current planning system is designed to ensure that the environmental consequences of all proposed new developments are taken fully into account, the system cannot protect all residents against all possible noise impacts, and does not automatically ensure that developers have adopted 'best practicable means' as part of their noise management policies. Full compliance with existing planning permission is not a defence against individual actions for statutory nuisance, but specific immunities may have been granted under the various relevant acts.

5.6.2 Amplified music

The various noise problems caused by amplified music provide good examples of the different types of complicated situations that can arise in practice. Many entertainment venues, such as discotheques and night-clubs, use sound systems capable of generating sound power outputs in the range from 1 to 100 watts and more. While these acoustic outputs might be considered relatively modest in terms of physical energy, they can also represent very high A-weighted sound pressure levels, which can often considerably exceed 120 dB inside the venue. While this might seem excessive in terms of possible noise nuisance or hearing damage risk, the installation of high-power audio systems can usually be justified by strong commercial reasons. High-power audio systems are generally part of the overall business strategy to attract and retain certain types of customers who may be discouraged from visiting the premises again if the audio systems are under-powered. So, in this case (as in many other business decisions), commercial profit must be offset against possible noise nuisance or long-term health risks.

Except for unamplified dramatic performances or those types of musical concert where the audience is either used to sitting quietly or can be persuaded to listen quietly, high levels of amplification are often required so that speech and music can still be heard above the self-noise generated by the crowd. Crowd noise can reach surprisingly high sound levels when several hundred people are shouting at each other, all at the same time. A non-amplified scream by only a single human voice can exceed an A-weighted sound level of 120 dB at 1 m distance for a few seconds at a time, and sustained shouting can approach these levels. A further problem is that, in discotheques and live music venues, sound levels tend to increase during the course of an evening as customers gradually increase their voice levels in an effort to be heard, and the sound system volume is adjusted to compensate. If customers are also experiencing an increasing degree of temporary threshold shift as the evening progresses, this can only make the problem worse. A few hours spent in a noisy discotheque can be fatiguing for the voice as well as being potentially risky for the ears. It is possible to install sound level limiting devices which prevent sound levels from increasing indefinitely, but this does not really solve the problem unless the customers can also be persuaded or induced to keep their own voice levels down.

In addition, when it is time to leave the venue, some customers continue to use a raised voice level outside, partly because they could be suffering from a degree of temporary threshold shift, and partly because they might not have fully adapted to the quieter conditions outside. Noise generated by crowds leaving a venue very late at night can be just as, or even more, disturbing for nearby residents than music noise escaping through the walls and windows of the premises.

When entertainment noise can be heard inside neighbouring residential premises, the sounds that were wanted inside the venue can cause considerable annoyance to unwilling listeners outside, even if the sounds are only just audible after frequency-dependent attenuation through the intervening structures. Sometimes it is the absolute level of the noise which is most annoying, but more often, it is the repetitive low-frequency bass content which is reported as being the biggest problem, notwithstanding the relative insensitivity of the human ear to low-frequency sound. There are three factors here which exacerbate this problem:

(1) The original amplified music is likely to be bass-heavy not just for reasons of individual preference or aesthetic sensitivity, but also because the crowd has the same relative insensitivity to low- and very low-frequency sound as the annoyed neighbours.

(2) The intervening structures between the venue and the neighbouring premises will almost always be much less effective at attenuating low-frequency sound than mid- and high-frequency sounds, leading to a particularly bass-heavy sound character audible inside the neighbouring premises.

(3) As the evening progresses, general background noise levels are usually decreasing at the same time as the sound levels inside the entertainment venue are increasing. This means that the relative audibility of the entertainment noise inside neighbouring premises is likely to increase from the time when it first becomes audible until the venue closes down, possibly many hours later. Any resulting annoyance may be increased not only in direct proportion to the steadily increasing audibility of the noise, but also by the listener's anticipation that the problem is likely to get worse before it gets better.

Practical experience suggests that some individuals can become increasingly sensitised to this type of auditory stimulus, so that even where they might at first have been able to ignore noise intruding at several decibels above the masked threshold level, they may, in time, find that they are becoming increasingly disturbed by any identifiable music noise which is only just audible. If this happens, the only practical solution would be either to render the intruding noise completely inaudible above the prevailing background noise levels due to other non-annoying noise sources present at the same time, or for the annoyed person to move away. It should be noted that, providing the steady background noise is not a significant source of annoyance, then increasing it to the extent that the annoying sound becomes completely masked could be a viable technical solution, even though no noise is in fact removed.

When approving planning applications for new development, including entertainment venues likely to require speech and music amplification, local planning authorities may be able to impose specific conditions which are intended to prevent amplified music noise from ever becoming a problem in the first place. However, there is a large body of existing legislation and case law in the UK which may constrain the type of planning conditions which may be legally enforceable, and developers may find themselves able to avoid the more onerous or restrictive type of conditions if they go to appeal. In Newbury District Council vs Secretary of State for the Environment (1981) AC 578 the House of Lords decided that planning conditions must comply with the following tests:

(a) they must be imposed for a planning purpose and not for an ulterior one;
(b) they must fairly and reasonably relate to the development permitted;
(c) they must not be so unreasonable that no reasonable authority could have imposed them.

Note that the words 'fair' and 'reasonable' appear often. If a developer appeals against a condition restricting levels of noise permitted to break-out from an otherwise approved development, it may be possible to argue that the noise levels are so low, as compared with other noise sources affecting that environment, that the condition is unreasonable. There can also be a problem with enforcement. Unfortunately, any listener who has become particularly sensitised to the noise may find themselves as being considerably disturbed by noise break-out from the development which is only just audible. As discussed in Section 5.5.4, and mainly because of a combination of the frequency selectivity of the ear and the typical variation in background noise levels from one minute to the next and over longer periods, it may be very difficult to confirm the exact sound levels of any just-audible noise, simply in terms of A-weighted or 1/3rd octave band sound levels. Defining audibility objectively in terms of much narrower frequency band measurements requires particular expertise and equipment which is not usually available to most local authorities, except if they are able to employ external acoustic consultants. Audibility could be tested subjectively by a qualified local authority officer with normal hearing, but legal advisers who might not have fully understood the technical difficulties associated with objective measurement at threshold are likely to be wary of subjective judgement in this context. This means that it may be difficult to devise enforceable conditions which restrict noise break-out to below the limits of audibility in neighbouring residential premises.

The alternative of setting conditions which can be tested objectively using a simple measurement of A-weighted sound levels may allow a significant amount of noise break-out (that is, sufficient break-out above the existing steady background noise level to support reliable objective measurement) to occur. In addition, the separate contributions to the steady background noise levels made by successive developments, which are each permitted to contribute small but measurable increments to the overall ambient noise environment, could lead to significant increases overall. Another approach might be to set absolute noise limits which are so low that any noise break-out complying with these limits would be effectively inaudible at the nearest receiver sites, but the same problems of measurement for enforcement also apply. Given these difficulties, and wherever possible, it is generally better to impose conditions defining structural details which if complied with, will effectively prevent excessive noise break-out than to attempt to specify target noise limits which might not be measurable in practice. The main difficulty here might be if the developer disagrees about the extent of any structural enhancements required. Many of these difficulties can be overcome by establishing good working relationships between developers and planning authorities, so that both parties fully understand the other's problems, aspirations and constraints.

5.6.3   Protecting noise-sensitive development against existing environmental noise

In the UK, transportation noise is dealt with very differently from amplified music emanating from entertainment venues. PPG24 (DoE 1994), described in Section 5.2.1, sets out guidance for local planning authorities when determining applications for residential and other potentially noise-sensitive developments in areas affected by existing transport and so-called 'mixed noise' sources, but does not offer any suggestions regarding restricting the noise sources themselves. One reason for this is that there is no provision under the existing planning system for regulating the activities of existing noise sources that are not themselves within the scope of the development being applied for. The law courts might be expected to take a dim view of any attempts to apply the planning system retrospectively to restrict or curtail noisy activities which were previously permitted, because of the serious economic and social implications that this might have for established commercial or residential uses of property including land.

As an example, consider the case of new residential development adjacent to a busy (and therefore noisy) existing airport. Is it reasonable for new residential occupiers to be able to restrict the future development of the airport for the purpose of protecting the amenity of their own properties? What is reasonable and what is not, does not just depend on the amount of noise. At the lower end of the scale, where airport noise levels might not be particularly significant when considered against all other noise sources present at those sites, any specific noise restrictions which apply just to airport activity alone might be very hard to justify. At the upper end of the scale, in a possible hypothetical situation where airport noise levels are so high that they might be considered as a significant health risk in their own right, the need for specific noise restrictions might be considered as over-riding any established rights to continue operating and developing the airport. Then, the economic costs of new restrictions would appear as a direct consequence of an earlier wrong decision to allow residential development within an existing excessively noisy environment.

Unfortunately, according to current scientific knowledge, it is difficult to define any specific noise level at which airport noise levels are high enough to impose significant risks on long-term health. This means that there is a very wide range of noise levels over which noise effects can be assumed as increasing roughly in proportion to physical noise levels but only with a relatively weak correlation. There is no specific cut-off point at which noise effects increase more steeply with noise levels, meaning that judging the overall significance of calculated or measured noise levels and any associated effects is often a matter of opinion. An airport facing the possibility of severe restrictions imposed for the protection of new residential development, which is nothing to do with the airport, might have a different opinion on these matters than airport noise objectors.

The specific guidance in PPG24: 1994 (DoE 1994) for permitting residential and other noise-sensitive development is based around four defined noise exposure categories (NECs), which are shown in Table 5.2.

Recommended noise levels, in L_{Aeq}, for these categories are given in Table 5.3.

The divisions between daytime and night-time are set at 2300 and 0700 h. This is partially consistent with the new European Directive which further divides the 0700–2300 h daytime into 12-h day (0700–1900 h) and 4-h evening (1900–2300 h) periods. For night-time (2300–0700 h) a footnote to the noise exposure category table states that sites where individual noise events regularly exceed 82 dB L_{Amax} (S time weighting) several times in any hour should be treated as being in NEC C regardless of the overall night-time L_{Aeq} (except where the $L_{Aeq,8h}$ already puts the site in NEC D). Note that local authorities are encouraged to apply

Table 5.2 The noise exposure categories defined in PPG24: 1994 (DoE 1994)

A	Noise need not be considered as a determining factor
B	Noise should increasingly be taken into account
C	A strong presumption against planning permission – special conditions required
D	Planning permission should normally be refused

Table 5.3 Recommended noise levels in L_{Aeq} for each Noise Exposure Category defined in PPG24: 1994 (DoE 1994)

		A	B	C	D
Road traffic	Day	<55	55–63	63–72	>72
	Night	<45	45–57	57–66	>66
Rail traffic	Day	<55	55–66	66–74	>74
	Night	<45	45–59	59–66	>66
Air traffic	Day	<57	57–66	66–72	>72
	Night	<48	48–57	57–66	>66
Mixed sources	Day	<55	55–63	63–72	>72
	Night	<45	45–57	57–66	>66

the recommended guidelines with some flexibility to account for differences in local circumstances which might affect housing demand and/or the practicality of alternative noise management options.

In PPG24, no specific guidance levels are given for industrial noise, construction noise or neighbour noise; neither is any specific guidance offered in respect of new noise sources in noise-sensitive areas. The document provides references to relevant noise insulation regulations and other documents which already exist.

The overall approach set out in PPG24 has been criticised both because it is too rigid and because it is not rigid enough. The main problem is that not all noise problems have 'win–win' solutions where both noise makers and noise sufferers are completely happy with the outcome of any decisions made. Many residents may be quite happy to live in a nice house which is convenient for local services and places of employment regardless of the overall amount of environmental noise present, while others would not be able to accept even the slightest amount of noise. The recommended flexibility can be interpreted either to the benefit or disbenefit of parties involved in any dispute. If the element of flexibility currently included within PPG24 were to be taken out, then this might work in favour of some parties and against others, who might then be expected to react accordingly.

5.6.4   Road traffic noise

5.6.4.1   Operation of new road schemes

The UK Highways Agency publishes comprehensive guidance setting out recommended procedures for assessing road traffic noise in the relevant chapters of a 'Design manual for roads and bridges' (Highways Agency 1994). Volume 11, Section 3, Part 7 (1994) defines appropriate noise level indicators for a range of different conditions and sets out noise assessment procedures for the environmental appraisal of new road schemes in terms of their effects on nearby residents. The three recommended noise level indicators are L_{Aeq}, L_{A10} and L_{A90}. Based on a combination of previous research and accepted custom and practice, and where road traffic is the most dominant source of noise, steady state road traffic noise 'nuisance' is assumed to be most closely related to the arithmetic average of the separate $L_{A10,1h}$ values from 0600 h to 2400 h local time. Quite apart from any technical debate over the relative merits of L_{A10}- and L_{Aeq}-type noise level indicators, there is an obvious difficulty here, in that new road schemes almost always involve some change in road traffic noise levels which are no longer 'steady state' as such. This difficulty is addressed in the Highways Agency recommendations by proposing that any change in road traffic noise nuisance can equally be related to the change in L_{A10}. This recommendation was not based on any substantive body of research, but was merely a common-sense interpretation of the available information at that time. It should be noted that there is some research evidence that while reported annoyance can be expected to increase if noise levels increase, reported annoyance does not necessarily decrease in the same way if noise levels go down. The problem here is that reported annoyance can depend as much on respondents' overall attitudes to the new road scheme and the ways in which it has been represented to them by the responsible authorities as it does on noise levels alone. The Highways Agency recommendations require that all road traffic noise calculations or measurements are carried out according to the UK standard methods set out in the Government publication, 'Calculation of Road Traffic Noise' (DoT 1988).

The Highways Agency recommendations cover two additional situations where road traffic noise is not the most dominant source of noise. In situations where the overall or ambient noise climate is dominated by undefined distant noise sources, then the L_{A90} background noise level is recommended as the best noise level indicator to use. In situations where the overall or ambient noise climate is dominated by aircraft or railway noise sources, then it is recommended that the effects of these noise sources should first be excluded from any measurement and then either the L_{A10} or the L_{A90} of the residual noise used, depending on whether the residual noise climate is dominated by road traffic noise or by undefined distant noise sources. These particular sections of the Highways Agency recommendations, while intended to be helpful in terms of dealing with situations where road traffic noise is not the most dominant noise source, are not based on any substantive research and have been found to have contributed to professional disagreements at planning inquiries which cannot easily be resolved. In addition, there is no standard method for calculating road traffic noise levels in terms of the L_{A90} background noise level indicator.

One of the key features of the Highways Agency recommendations is the comparison of the relative effects of 'do-minimum' and 'do-something' options, taking into account any further increases in road traffic predicted over the next 15 years from the bringing into use of the new scheme. The do-minimum option implies that no new scheme is brought into use, although where a limited amount of minor road traffic management schemes might be envisaged even where the new scheme does not take place, then these minor developments should be taken into account as part of the do-minimum option. The alternative 'do-nothing' option is not considered in the assessment because it is not considered to be a realistic option for comparison, although it should be noted that comparison of do-something against do-minimum is likely to show a smaller incremental effect of the proposed development than comparison of do-something against do-nothing. Wherever road traffic noise levels are predicted to increase, developers are encouraged to consider all possible noise mitigation options such as re-siting the route further away from noise-sensitive areas, erecting noise barriers and installing noise absorbent road surfaces.

The Highways Agency recommendations set out a detailed three-stage assessment procedure which is carried out to increasing levels of detail as the initial outline scheme progresses through a number of more detailed design stages. For the initial Stage 1 assessment, it is considered sufficient to identify noise-sensitive locations to either side of the proposed route and then to count the numbers of houses within 300 m to either side of any route likely to experience a 25 per cent or bigger change in road traffic flows. This does not require any noise measurements or calculations, because there is an implied assumption here that any change in road traffic flows of less than 25 per cent would have only an insignificant impact on road traffic noise levels. Measurements or calculations are required for the next and more detailed Stage 2 assessment, where the numbers of residential properties likely to be exposed to changes in road traffic noise level (up or down) exceeding 1 dB are to be counted. For the final Stage 3 assessment, the developer is expected to have discussed and agreed all mitigation measures, assessed noise levels at all identified noise-sensitive properties and prepared appropriate summary tables in a framework format showing all information considered relevant to the assessment.

In an ideal world, noise management decisions regarding alternative mitigation options would be decided based on a comprehensive cost-benefit or cost-effectiveness analysis of all available options, where the increasing numbers of residents likely to be exposed to the lower ranges of road traffic noise levels would be taken into account. In practice, most noise management decisions are based on the number of residential properties likely to be entitled to noise insulation grants under the Noise Insulation Regulations 1975 (HMSO 1975, 1988) and the likely costs of providing the specified noise insulation. These regulations were written according to the Land Compensation Act 1973 which allows for compensation for loss in value consequent upon the bringing into use of some new noisy development such as a new road or airport. The cost of noise insulation provided, according to the Noise Insulation Regulations, is offset against any other compensation paid. Highways developers often expect engineers to balance the likely costs of noise barriers and other mitigation options such as low-noise road surfaces against the likely costs saved through reducing the number of residential properties entitled to noise insulation grants.

5.6.4.2 Construction of new road schemes

There is a separate section in the Highways Agency's Design Manual for Roads and Bridges (Highways Agency 1994) dealing with the potential disruption due to the construction of any new road schemes, before the scheme is actually brought into use. Developers are encouraged to consider noise insulation of houses as mitigation against construction noise at an early stage in the development, particularly for those houses which would eventually have qualified for statutory noise insulation when the new road scheme is brought into use, and for those houses where noise insulation is considered to be justified purely on the basis of the construction noise alone.

5.6.4.3 Statutory method of measuring road traffic noise

The DoT published the UK statutory method for calculating road traffic noise in 1988 (DoT 1988). This method is used for calculating entitlement for statutory noise insulation under the Noise Insulation Regulations 1975, as amended in 1988. Road traffic noise levels from both new and existing roads must be measured (or calculated) at identified noise-sensitive receiver façades by adding 2.5 dB to sound levels measured or calculated for free-field (incident sound) conditions to account for the effects of an acoustic reflection from the façade.

Noise-sensitive properties then qualify for statutory noise insulation if the calculated or measured road traffic noise level exceeds 68 dB $L_{A10,18h}$, if there has been an increase in calculated or measured sound levels of at least 1 dB and if the contribution to the increased road traffic noise levels by the new or altered highway also exceeds 1 dB. The effect of these additional qualifications is to severely restrict the overall numbers of noise-sensitive properties that actually receive any compensation. Existing noisy roads, however noisy, do not qualify. Increases in road traffic noise caused by increased traffic associated with a new road being developed elsewhere do not qualify either, even if the increases in traffic flow are a direct result of that development. Most residential properties that are currently exposed at road traffic noise levels above 68 dB $L_{A10,18h}$ have never qualified for statutory noise insulation because one or more of the three qualification clauses have not been met.

The standard road traffic noise insulation package, as provided under the Noise Insulation Regulations 1975, as amended in 1988, includes wide-spaced (up to 200 mm if possible) internal secondary glazing with absorbent-lined reveals, powered acoustic ventilators to allow the windows to be kept closed and additional loft insulation where site surveys suggest that noise break-in through the roof could be a problem. The existing (external) windows need to be in good condition with fully functioning seals around all opening lights and between the window frame and the surrounding walls for the standard noise insulation package to be effective. Outdoor-to-indoor sound insulation measurements carried out *in situ* have shown that modern high-performance replacement window systems fitted with narrow-spaced sealed double-glazing units can be as effective as, or more effective than, internal secondary glazing fitted inside existing windows, which are in less than perfect condition, and will often be preferred by residents. However, high-performance replacement windows are more expensive than internal secondary glazing. It is not always possible to provide effective noise insulating windows in buildings of historical or architectural importance where current planning regulations prevent any alterations to the existing window structures.

The road traffic noise calculation method is based on dividing the road scheme up into separate discrete segments and then aggregating overall noise levels from all the separate segment noise levels. The calculation method is based on experimental data and is usually assumed to be accurate to within 3 or 4 dB, with the expected level of accuracy decreasing at increasing distances from the road. The method is based on simple formulae which are also implemented as a series of charts which were originally provided so that the method could be used by the mathematically literate general public. The 'basic noise level' at a reference distance of 10 m from the source line is calculated from the traffic flow and composition, and then attenuated according to distance, screening, reflections and angle of view. The method has now been implemented in a number of commercially available noise mapping packages.

There are separate recommendations for a measurement method, which is only permitted in unusual circumstances. In many situations, unless the measurements are continued for extended periods of time, usually at least 2 weeks or more, sample measurements of environmental noise can be less representative of long-term average conditions than calculated noise levels. The method requires that measurements are taken at a height of 1.2 m above the road surface and at a convenient distance of between 4 m and 15 m from the edge of the carriageway, with the attenuation from the roadside to the receiver position calculated based on the distance and any screening, reflection and angle of view corrections. Measurements are only permitted at the noise-sensitive receiver position where there are definitely no other noise sources present which might otherwise contaminate the measurement. In practice, measured and calculated road traffic noise levels are always likely to differ by at least 1 or 2 dB, and this might then create problems if one result was above the qualifying criterion and the other below. On the other hand, an aggrieved householder who did not qualify for statutory noise insulation because calculated noise levels were below 68 dB $L_{A10,18h}$, and who then subsequently commissions measurements which suggest actual noise levels above 68 dB $L_{A10,18h}$, might feel somewhat cheated.

5.6.5 Railway noise

In the UK, statutory entitlement to noise insulation in respect of new railway schemes is dealt with under The Noise Insulation Regulations (railways and other guided transport systems) 1995 (HMSO 1995). There is a separate statutory method for the calculation of railway noise (DoT 1995) which follows the same general model as was previously adopted for road traffic noise in the UK. To be entitled to statutory insulation, noise levels from both new and existing tracks must exceed 68 dB $L_{Aeq,18h}$ (day) or 63 dB $L_{Aeq,16h}$ (night); the increase in noise levels attributable to the new scheme must be at least 1 dB, and the contribution to the increase from the new or altered tracks must exceed 1 dB.

As for existing noisy roads, existing noisy railway tracks or increases in traffic on existing but unaltered railway tracks do not qualify, however noisy they are. Again, as for existing noisy roads, only a very small proportion of railway noise-exposed properties have ever qualified for statutory noise insulation. The method of calculation follows the road traffic noise calculation method as far as possible, except where the use of L_{A10}-type noise indicators used for road traffic is completely inappropriate. Specific source reference noise levels are given in terms of the A-weighted Sound Exposure Level (SEL) at a standard reference distance of 25 m. Owing to the large number of different types of railway vehicle which might be encountered, specific-source noise level measurements are encouraged.

5.6.6 Aircraft noise

The overall number of people affected by aircraft noise is much less than the number of people affected by road traffic noise, but the aircraft noise problem has a much higher profile in the national media. Many people living around airports seem to habituate to aircraft noise and often appear to hardly notice it all, unless or until it is brought to their specific attention, whereas a significant and often vocal minority tend to get very annoyed by aircraft noise, particularly at night. In questionnaire surveys, people tend to respond with higher annoyance ratings compared to road traffic noise at the same long-term average noise level (L_{Aeq}), possibly because individual aircraft noise events are likely to be much more intrusive than individual road traffic noise events, or because of other differences, such as fear of crashes, *but* there are always gaps between aircraft events when there is no aircraft noise at all, and the extent to which these 'respite' periods contribute to apparent habituation is not known.

Civil aviation has an additional international dimension compared with road traffic or railway noise and has therefore had to be governed by bilateral agreements between different countries. Separate national regulations, which permitted a noisy aircraft type to take-off from one country but not to land in another, might create significant international difficulties, particularly where national flag carriers are involved. If legal action on the grounds of nuisance were permitted by individual residents in the UK, then this could make bilateral international agreements unworkable. In the UK, the Civil Aviation Act 1982 (and more recent legislation such as the Transport Act 2000) imposes duties on the government to ensure that the environmental effects of civil aviation are taken fully into account in exchange for granting immunity against civil actions for nuisance. Individual residents do not always feel that their particular interests have been taken fully into account when decisions are made by government, which might give at least an appearance of favouring airline profits over environmental concerns.

In the UK, the relevant government department (currently the DfT) defines the noise management regime at 'designated airports'. These are the main UK airports such as Heathrow, Gatwick and Stansted. None of the smaller regional airports are presently 'designated' although the government has powers to designate any airport if it so chooses. Day-to-day noise management has increasingly been delegated locally, even at the designated airports, which now deal with noise complaints, noise levels and flight track monitoring and local consultative committees. Overall policy is defined by government, in terms of noise limits, specific restrictions for noisier aircraft types and similar constraints. Airports cannot impose additional restrictions or fines on noisier aircraft without being instructed or permitted to do so by the relevant government department. However, airports can and do negotiate with operators to discourage noisier operations when it is in both the airport and the operators' interest to do so. In addition, and at many airports including many of the non-designated airports, there is often a relatively complex system of specific planning conditions or other local agreements, which effectively regulate the use of different areas on the airports concerned for different purposes, although the detailed application of particular schemes of mitigation can vary to some considerable degree.

A number of aviation matters are dealt with under international agreements made under the auspices of the International Civil Aviation Organisation (ICAO). This organisation deals with the regulation of airworthiness, flight safety and navigation and similar matters, and includes aircraft-type variant noise certificates within its overall remit. Aircraft-type variants are allocated Chapter 2, 3 or 4 noise certificates based on field measurements carried out under tightly controlled reference conditions, and older noisier aircraft types have now been phased out from operation in many parts of the world based on international (multiple bilateral) agreement. Additional restrictions have been applied in Europe and other geographically defined parts of the world based on ICAO-type variant noise certificates. ICAO working parties have also developed so-called low-noise approach and departure procedures and other contributions to aircraft noise management and mitigation. One difficulty often quoted by environmental activists is that ICAO agreed type variant noise limits and/or approved approach and departure procedures do not always result in the lowest achievable noise levels because of an emphasis within ICAO on avoiding 'excessive' economic cost.

5.6.7   Surface mineral workings

Surface mineral workings require the use of heavy plant and machinery to extract and process the minerals. The UK government published guidance notes (MPG11) in 1993 (DoE 1993) to assist site operators and local planning authorities to reach balanced agreements regarding permissible noise limits and mitigation measures. The guidance recommends surveys of existing background noise levels to inform environmental appraisal which is then based on calculated noise levels using established procedures for heavy plant and machinery (see BS 5228 discussed in Section 5.6.12). In addition to those factors mentioned in BS 5228, the minerals planning guidance note suggests that acoustic attenuation over soft ground and meteorological variation can be particularly relevant at minerals sites. The guidance recommends that site operators and local authorities should reach agreements on noise monitoring, noise control measures and, where appropriate, on-site boundary noise limits based on defined exceedances above measured background noise levels and/or on somewhat arbitrarily defined noise level benchmark criteria such as 55 dB $L_{Aeq,1h}$ daytime and 42 dB $L_{Aeq,1h}$ nighttime. Users of the guidance note are sensibly cautioned over the possibility of additional annoyance attributable to tonal features (heavy plant reversing bleepers can cause particular noise nuisance problems) and over the need to adopt a certain amount of local flexibility, where justified by local condition.

5.6.8   Renewable energy: wind farms, etc.

Heavy or large-scale fixed plant and machinery which have been installed to exploit renewable energy resources such as wind and wave power can be noisy, depending on the installed power and the distances to the nearest noise-sensitive receivers. The UK government has published guidance notes (Planning policy guidance PPG22: renewable energy) to assist operators

and local planning authorities to agree about noise assessments and mitigation measures (DoE 1993). The guidance suggests that predicted noise
levels should be compared with existing background noise levels, generally
following the principles set out in BS 4142 (see Section 5.6.10) but also,
and rather unhelpfully, cautions that this standard is not always the most
appropriate method to use. Operators are commended to use 'best practicable means' to reduce or avoid noise nuisance, which implies that a limited
amount of residual noise nuisance might be acceptable where the overall
environmental benefits are judged to be more significant. The guidance
recognises that special consideration may need to be given to the particular
circumstances extant in the predominately rural areas where this type of
plant is most likely to be installed.

5.6.9 *Neighbourhood and domestic noise*

In the UK there is a relatively complex series of specific legislation covering
different aspects of neighbourhood and domestic noise. These are outlined
below:

Control of Pollution Act 1974, Part III (England Wales and Scotland)
Pollution Control and Local Government (Northern Ireland) Order 1978
For general construction noise, Section 60 empowers local authorities to
serve notices on developers or contractors to reduce noise, with penalties
for non-compliance. Section 61 allows for developers/contractors to obtain
immunity from prosecution for noise nuisance by the local authority under
Section 60 provided that they comply with noise management procedures
agreed in advance with the local authority. Section 71 authorises approved
codes of practice. Existing codes of practice cover noise and vibration
control on construction and open sites (BS 5228); noise from ice cream
chimes; noise from audible intruder alarms; noise from model aircraft.

Civic Government (Scotland) Act 1982
Section 54 empowers uniformed police officers to request sound levels to
be reduced or to confiscate loudspeakers if polite requests are ignored.

Environmental Protection Act 1990, Part III (England and Wales)
(HMSO 1990)
Sections 79–81 of the Act (HMSO 1990) empower local authorities to
take action to control statutory noise nuisance arising from fixed premises,
including both buildings and land. Section 82 empowers individuals to take
action against statutory noise nuisance through the courts. 'Statutory noise
nuisance' can be any noise nuisance which has been defined as such by
a court of law under the provisions of this Act. A noise nuisance is not
a statutory noise nuisance unless or until it has been defined as such by a
court.

Noise and Statutory Nuisance Act 1993 (England and Wales)
This Act amended Sections 79–82 of the Environmental Protection Act
1990 to empower local authorities to control statutory noise nuisance
arising in the street. It empowers local authorities to control loudspeakers
used in the street and noisy burglar alarms and to recover expenses incurred
(provided that the courts agree).

Environment Act 1995 (England Wales and Scotland)
This act extended the provisions for statutory notices in the Environmental
Protection Act 1990 to Scotland.

Noise Act 1996 (England Wales and Northern Ireland)
This Act introduced a new Night Noise Offence with fixed penalties and
new procedures for the seizure and forfeiture of noise making equipment
such as audio and disco equipment. Local authorities are formally required
to adopt the procedures set out in the Act before they are permitted to use
any of the powers provided by the Act. Few local authorities have done so,
probably because of the resource implications of the requirement that they
must appoint night duty officers who are then available to visit premises
and resolve noise complaints using the Act. There is an objective noise
level benchmark above which an offence is deemed to occur.

Crime and Disorder Act 1998 (England Wales and Scotland)
This Act empowers police and local authorities to tackle anti-social
behaviour including noise nuisance. There is provision for criminal penalties
of up to 5 years imprisonment.

5.6.10 Industrial noise

The British Standard method for rating industrial noise affecting mixed
residential and industrial premises (BS 4142 1997) really has two inter-
linked parts. The first part sets out a method for determining industrial
noise levels at noise-sensitive receiver positions, and the second part sets
out a method for assessing the 'likelihood of complaints'. Interestingly, the
foreword to the standard explicitly states that the 'quantitative assessment of
annoyance is beyond the scope of this standard'. The standard is not explicit
regarding which level of 'likelihood of complaints' should be considered as
having exceeded the bounds of acceptability. BS 4142 is apparently one of
the most widely used of all British Standards and has often been applied
outside of its stated scope.

The first part of the standard defines the 'specific noise level' as the
noise level attributable to the specific industrial noise source, in this case
measured using $L_{Aeq,1h}$ daytime and/or $L_{Aeq,5min}$ at night. The specific noise
level then becomes the 'rating level' when adjusted (or not) for the presence
of distinguishing acoustic features such as tones, impulses, irregularity, etc.
by adding a $+5\,dB$ correction factor. The 'ambient noise' is defined as all

sound from all sources present and the 'residual noise' is defined as the ambient noise with the specific noise excluded. The 'background noise level' is defined as the L_{A90} of the residual noise, although if the specific noise is intermittent the background noise level can be measured while the specific source is still operating.

The assessment method requires that the rating level (of the specific noise) is compared with the background noise level. The standard then assumes that the greater the margin by which the rating level exceeds the background noise level, the greater the likelihood of complaints. An exceedance of +10 dB suggests that complaints are likely, an exceedance of +5 dB is of 'marginal significance', and an exceedance of −10 dB (a rating level 10 dB below the background noise level) is deemed to suggest that complaints are unlikely.

The available research evidence suggests that local authorities and acoustic consultants tend to use BS 4142 only where it supports their subjective assessment of the relative severity of any noise problem investigated, and avoid using it where it seems to give contrary advice. As a consequence, technical debates over the 'proper' application and interpretation of the standard can take up a great deal of time in courts and planning inquiries. The assessment method is not based on any substantive research. The detailed wording of the standard allows for alternative interpretation in some situations where the responsible standards committee was unable to reach a consensus view.

5.6.11   Sound insulation and noise reduction in buildings

The British Standard Code of Practice for sound insulation and noise reduction in buildings (BS 8233 1999) cross-references many other standards with more specific applications. The Code of Practice sets out recommended sound levels for inside educational buildings, for inside industrial buildings, for the avoidance of interference with speech communications in offices and for avoiding interference with acoustical privacy in consulting rooms used for health and welfare purposes (Table 5.4 and 5.5).

Minimum values of sound insulation for educational buildings expressed in sound level difference (D) are given below.

25 dB between rooms assigned for noisy use
35 dB between classrooms, etc.
45 dB between rooms needing quiet
>45 dB to protect a quiet use against a noisy workshop, etc.

Table 5.4 Recommended sound levels for educational buildings (indoors) (BS 8233: 1999)

Circulation areas and workshops	50 L_{Aeq}
Tutorial rooms and libraries	45 L_{Aeq}
Classrooms	40 L_{Aeq}
Lecture rooms and language labs	35 L_{Aeq}
Music and drama	30 L_{Aeq}

Extracts from BS 8233: 1999 are reproduced with the permission of BSI under licence number 2003DH0300. British Standards can be obtained from BSI Customer Services, London.

Table 5.5 Recommendations for offices. Maximum noise levels for effective speech communication (BS 8233: 1999)

Distance (m)	Normal voice (L_{Aeq})	Raised voice (L_{Aeq})
1	57	62
2	51	56
4	45	50
8	39	44

Extracts from BS 8233: 1999 are reproduced with the permission of BSI under licence number 2003DH0300. British Standards can be obtained from BSI Customer Services, London.

5.6.12   Construction noise

The British Standard Code of Practice for noise and vibration control on construction and open sites (BS 5228 1997) has four parts dealing with basic information and procedures (Part 1), relevant legislation (Part 2), surface coal extraction by open-cast methods (Part 3) and noise and vibration control applicable to piling operations (Part 4). There are approximately 150 pages of general advice. There is a relatively simple method for calculating construction plant and machinery noise levels included in the code of practice. The method of calculation has generally been found to be particularly useful and has accordingly been implemented in a number of commercial noise mapping packages. The Code of Practice does not specify site boundary noise limits, as this is a matter which should be agreed locally (if site boundary noise limits are appropriate at all), but does include a significant amount of general advice on noise management and monitoring procedures.

It should be noted that site boundary noise limits are often proposed and agreed without any clear understanding of precisely what can and cannot be achieved by their adoption. Site boundary noise limits are only of any practical use if they (a) constrain site noise levels to within acceptable limits and (b) can be separately monitored or otherwise determined to ensure compliance. Setting noise limits low enough to avoid possible disturbance often means that they are too low in comparison to other noise sources present for effective monitoring. If noise limits are set high enough that they can easily be monitored in comparison with other noise sources present, then they might not provide enough protection for nearby residents to ensure acceptability. Given these difficulties, it is often better to base noise mitigation schemes on a comprehensive appraisal carried out at the project planning stage with subsequent monitoring limited to ensuring that all proposed mitigation options are properly followed, or applied specifically where unexpected problems might have occurred.

5.6.13   General environmental noise

British Standard BS 7445: Parts 1–3: 1991 (BS 7445 1991) implements the corresponding ISO 1996 Parts 1–3: from 1982 and 1987. Part 1 defines required quantities (noise level indicators) and procedures. Part 2 provides guidance regarding the acquisition of noise level data pertinent to land use management. Part 3 offers guidance on the selection and application of noise limits. All three parts of ISO 1996 are undergoing a protracted process of updating and revision at the time of writing, and may eventually be incorporated into a revised version of BS 7445 in due course. The ISO definitions of specific quantities set out in BS 7445 Part 1 have often been found to be useful in the UK, although the methods set out in BS 7445 Parts 2 and 3 have not, in general, been found to be particularly useful in a UK context.

5.7   General methods of assessment of environmental noise

5.7.1   Defining the problem

So what are the alternatives? In practice, unless there has been significant complaint *and* there is at least a possibility of noise control action being considered there is little or no point in carrying out any arbitrary assessments of an existing static situation. Environmental noise management is about choosing between realistic or feasible alternative courses of action which, while they may include taking no action at all, should at least include some options which potentially make some difference to the amount of noise. In these cases, the most useful guideline values and/or noise assessment procedures are those which inform the selection of the most appropriate or the best compromise noise management solutions.

It is well known that different kinds of health risk are perceived differently by the public at large. Neighbour noise has been directly implicated as a contributory factor in numerous disputes leading to actual violence, but most of these cases are relatively isolated and mainly concern individuals rather than the general public as a whole. For aircraft noise, the available evidence suggests that many people, while they do not believe that their own health or quality of life is affected, report that they believe that other people in their neighbourhood have their health or quality of life affected (Diamond *et al.* 2000). The extent to which noise management is becoming a general priority for the public could be taken as an indication of the overall level of noise acceptance by the public as a whole.

While, in theory at least, noise guideline values should be based on the severity of adverse effects caused by the noise, they must, in practice, be based on physical measurements of the noise on the assumption that meaningful relationships between noise exposure and noise effects exist. When the amount of noise has been described in numeric terms, three generic types of numeric comparison may be applicable (Flindell *et al.* 1997). It should be noted that there is no theoretical reason why all three comparisons should necessarily agree:

- *Change* – Is the situation better or worse after development compared with before (in other words, have noise levels gone up or down)?
- *Context* – Specific noise levels can be compared with ambient, residual or background noise levels.
- *Benchmarks* – Specific noise levels can be compared against fixed noise limits laid down in standards and regulations.

Each of these numeric comparisons could use a whole range of different sound level indicators to reflect the relative importance in each particular case of different factors such as the time of day or night, any special characteristics of either the noise itself or of the residual, ambient or background noise sources present, or the number of specific events compared against the average sound levels of those events. Current standardisation on noise indicators based on A-weighted maximum sound levels and long-term averages such as L_{Aeq} tends to be more a matter of convenience than any reflection on the strength of any assumed underlying dose–effect relationships.

It is also important to consider the precise reasons why any particular noise assessment is being done. In practice, decisions can only take one of the three generic forms as follows:

- *refuse* application for development;
- *permit with conditions* attached;
- *permit* with no conditions.

To be of any use at all, any noise assessment method has to be able to inform decisions made between these three alternatives. It is relatively easy to set fixed, but essentially arbitrary, benchmark levels above which all planning applications should be refused because of intolerable risks to health, and below which all applications for development should be allowed without conditions because it is unlikely that anyone would even be able to hear the noise. However, there is a very wide range of noise levels in between where there might be no reason to refuse an application, providing that appropriate conditions are imposed in mitigation. It is in this middle range that the greatest problems arise, where it is important to balance noise effects against other aspects of the development, both positive and negative. For example, some developments might increase both noise and employment. How far should noise effects be balanced against economic benefits?

5.7.2   Practical solutions

For clear and consistent decision making, all of the relative advantages and disadvantages of any proposal must be identified, measured and then weighed in a proper balance. In many cases the 'best compromise solution' may be a 'matter of opinion', which can in turn be very much dependent on individual weightings of the relative importance of all contributory information. Where there are complex issues present, a fully informed opinion may require considerable background knowledge and experience, and should not be based on any simplified overview. This requires expert knowledge and experience. But on the other hand, such matters as these should not be dealt with only by experts working behind closed doors. There seem to be a number of arguments in favour of dealing with issues like these in public, such as in public inquiries (in the UK), but at the same time, improving the procedures to overcome the current deficiencies of the system. Improvements may be required to overcome the problems caused by applying adversarial legal systems to differences of opinion and by the generally disjointed and contradictory nature of much of the available standards and guidelines in this area.

On the assumption that everyone would be assisted by 'better' standards and guidelines, British Standards Institution (BSI) sub-committee EH/1/3, 'Residential and Industrial Noise' is working towards a new standard entitled 'British Standard Guidelines for Environmental Noise Management'. This new standard will follow on from a range of initiatives prepared by others (for example, see Turner et al. 2002) and may substantially replace the existing BS 4142 1997 'Method for rating industrial noise affecting mixed residential and industrial areas' (BSI 1997). BS 4142 lies within the purview of sub-committee EH/1/3, as do a number of other standards and guidelines documents.

The consultant's brief now under consideration by the BSI administration sets out the following requirements:

- the guidelines should provide clear guidance as to which existing standards and regulations should apply (if any) in any particular situation;

- where existing standards conflict, the guidelines shall provide clear guidance regarding ways of presenting information (and *which* information) to allow decision makers to reach the best compromise solutions;
- the guidelines should assist in defending UK best practice based on *informed flexibility* against possibly unwelcome innovations arriving via ISO and CEN standards routes.

It remains to be seen to what extent it is in fact possible to write authoritative guidelines for an environmental noise document which can define effective 'rules' for noise assessment without unduly restricting informed flexibility.

Note

1 It should be noted that the Department of the Environment (DoE) was replaced in 2002 by the Department for the Environment, Food and Rural Affairs (DEFRA).

References

ANSI S3.5-(1997) (R2002) *American National Standard Methods for Calculation of the Speech Intelligibilty Index*. American National Standards Institute, New York.

Berglund, B. and Lindvall, T. (1995) *Community Noise*, Archives of the Centre for Sensory Research, Volume 2, Issue 1, Stockholm University and Karolinska Institute.

Berglund, B., Lindvall, T., Schwela, D.H. and Goh, K.-T. (2000) *Guidelines for Community Noise*, World Health Organisation, Geneva and Ministry of the Environment, Singapore.

Berry, B.F., Porter, N.D. and Flindell, I.H. (1998) Health effects-based noise assessment methods; a review and feasibility study, NPL Report CMAM 16, July.

BS 7445 (1991) *Description and measurement of environmental noise, Part 1. Guide to quantities and procedures, Part 2. Guide to the aquisition of data pertinent to land use, Part 3. Guide to application to noise limits*, British Standards Institution, London.

BS 4142 (1997) *Method for rating industrial noise affecting mixed residential and industrial areas*, British Standards Institution, London.

BS 5228 (1997) *Noise and vibration control on construction and open sites, Part 1. Code of practice for basic information and procedures for noise and vibration control, Part 2. Guide to noise and vibration control legislation for construction and demolition including road construction and maintenance, Part 3. Code of practice applicable to surface coal extraction by opencast methods*, British Standards Institution, London.

BS 6840-16 (1998) *Sound system equipment. Objective rating of speech intelligibility by speech transmission index*. British Standards Institution, London.

BS 8233 (1999) *Code of practice for sound insulation and noise reduction in buildings*. British Standards Institution, London.

DEFRA (2001) *Towards a national ambient noise strategy – A consultation paper from the Air and Environmental Quality Division*, Department for Environment, Food and Rural Affairs, London, November.

Department of the Environment and The Welsh Office (1993) *MPG11 Minerals planning guidance: The control of noise at surface mineral workings*, HMSO, London.

Department of the Environment and The Welsh Office (1993) *PPG22 Planning policy guidance: Renewable energy*, HMSO, London.

Department of the Environment and The Welsh Office (1994) *Planning Policy Guidance, Planning and Noise*, PPG24.

Department of Transport (1995) *Calculation of Railway Noise*, HMSO, London, ISBN 0-11-551754-5.

Department of Transport and The Welsh Office (1988) *Calculation of Road Traffic Noise*, HMSO, London, ISBN 0-11-550847-3.

Diamond, I., Stephenson, R., Sheppard, Z., Smith, A., Hayward, S., Heatherley, S., Raw, G. and Stansfield, S. (2000) *Perceptions of aircraft noise, sleep and health*, Report to UK Civil Aviation Authority, December.

European Commission (1996) *Future noise policy – European Commission Green Paper*, COM(96) 540 final, ISBN 92-78-10730-1.

European Commission (2000a) *The noise policy of the European Union – towards improving the urban environment and contributing to global sustainability*, Office for official publications of the European Communities, Luxembourg.

European Commission (2000b) *The noise policy of the European Union – Year 2 (1999–2000)*, Office for Official Publications of the European Communities, Luxembourg, 2000, ISBN 92-828-9304-9.

European Union (2002) *Directive of the European Parliament and of the Council relating to the assessment and management of environmental noise*, European Commission, 2000/0194 (COD), Brussels, April.

Fields, J.M., De Jong, R.J., Gjestland, T., Flindell, I.H., Job, R.F.S., Kurra, S., Lercher, P., Vallet, M., Yano, T., Guski, R., Fleshcher-Suhr, U. and Schumer, R. (2001) Standardised general-purpose noise reaction questions for community noise surveys: research and a recommendation, *Journal of Sound and Vibration*, 242(4), 641–679.

Flindell, I.H. and Stallen, P.J. (1999) Non-acoustical factors in environmental noise, *Noise and Health*, Vol 3, Apr–Jun, 11–16.

Flindell, I.H., Porter, N.D. and Berry, B.F. (1997) The assessment of environmental noise – the revision of ISO 1996, Acoustics Bulletin, November/December, pp. 5–10.

Grimwood, C.J., Skinner, C.J. and Raw, G.J. (2002) The UK national noise attitude survey 1999/2000, Noise Forum Conference 20 May 2002, The Chartered Institute of Environmental Health, London.

Highways Agency (1994) Highways Agency, *Design manual for roads and bridges, Volume 11, section 3, part 7, Traffic noise and vibration*, HMSO, London.

HMSO (1963) *Committee on the problem of noise – Final Report*, HMSO, London.

HMSO (1990) *Environmental Protection Act 1990*, HMSO, London.

IEC 60268-16 (1998) *Sound system equipment. Objective rating of speech intelligibility by speech transmission index.* International Electrotechnical Commission, Geneva.

Lambert, J. and Vallet, M. (1994) *Study related to the preparation of a communication on a future EC noise policy*, LEN Report No 9420, Institut National De Recherche Sur Les Transports et Leur Securite, France.

Meidema, H.M.E. (2001) Noise and Health: How does noise affect us? Proceedings of Inter-Noise 2001, The Hague, The Netherlands, August.

Newbury District Council *vs* Secretary of State for the Environment (1981) AC 578.

NSCA (2002) *Trends in noise management across the UK*, National Society for Clean Air, Brighton.

Porter, N.D., Flindell, I.H. and Berry, B.F. (1993) *An acoustic feature model for the assessment of environmental noise*, Acoustics Bulletin, November/December, pp. 40–48.

Porter, N.D., Kershaw, A.D. and Ollerhead, J.B. (2000) *Adverse effects of night-time aircraft noise*, Department of Operational Research and Analysis, National Air Traffic Services Ltd, London, R&D Report 9964.

Robertson, K., Flindell, I.H., Wright, N., Turner, C. and Jiggins, M. (2000) Aircraft noise and sleep, 1999 UK trial methodology study, Proceedings of Inter-Noise 2000, Nice, France, August.

Skinner, C.J. and Grimwood, C.J. (2002) *The UK national noise incidence study 2000/2001*, Noise Forum Conference 20 May 2002, The Chartered Institute of Environmental Health, London.

Statutory Instruments, 1975 No. 1763, *Building and buildings: the noise insulation regulations*, HMSO.

Statutory Instruments, 1988 No. 2000, *Building and buildings: the noise insulation (amendment) regulations*, HMSO.

Statutory Instruments, 1995, *The Noise Insulation (Railways and other guided transport systems) Regulations*, HMSO.

Turner, S., Collins, K., Baxter, M., Fuller, K. and Parry, G. (2002) *Guidelines for noise impact assessment – consultation draft*, Institute of Environmental Management and Assessment and Institute of Acoustics, April.

WHO (1980) *Environmental Health Criteria 12, Noise*, World Health Organisation, Geneva.

Chapter 6

Vehicle noise

D.J. Thompson and J. Dixon

6.1 Introduction

For most people, the noise which has the greatest impact on their environment comes from transport (European Commission 1996). The huge increases in mobility achieved by technological development in the last century have meant that high background noise levels, particularly due to road traffic, have become a feature of our society. It is estimated that, in the European Union, 80 million people (20 per cent) are exposed to noise levels greater than 65 dB L_{Aeq}, and a further 170 million are exposed to more than 55 dB. Road traffic noise is the dominant source of noise for nine-tenths of those exposed to more than 65 dB, whilst about 1.7 per cent of the population are exposed to railway noise over this limit. These large numbers of people affected by transport noise have been the driving force behind legislation, first introduced in the 1970s, which has attempted to limit the noise emission of individual road vehicles. Similar legislation for railway vehicles seems likely for the future and is already being introduced in some European countries.

Noise is also important for people who are users of transport, although here it is not legal requirements but customer demand that provides the main thrust for reduced noise. The noise inside vehicles crucially affects the level of comfort, or fatigue, experienced by the occupants. For road vehicles, in particular, therefore, the requirements in terms of interior noise are an integral part of the specification, with manufacturers competing for sales of their product on the basis of acoustic environment as well as more traditional features such as performance.

This chapter focuses on the noise from, and within, railway and road vehicles. It introduces the various noise sources that are present in each case and discusses means to control these noise sources. Attention is concentrated on the vehicles themselves, so that the reduction of environmental noise by, for example, conventional noise barriers is not discussed. The reduction of noise at source is, in general, a complex topic, requiring considerable engineering expertise; the results that are presented represent the outcome of many years of research effort. In the context of this book, the control of railway and automotive noise sources should be of general interest to the reader as an extended case study in noise control at source, as well as being of specific interest to those working in the particular fields.

Section 6.2 deals with railway noise sources, concentrating mainly on exterior noise. This is followed in Section 6.3 by a discussion of automotive noise sources, where the main thrust of the chapter, as of the industry, is on the control of interior noise.

6.2   Railway noise

6.2.1   Sources of railway noise

A variety of noise sources is present in railway operations. The dominant noise from mainline railways is usually the rolling noise generated by the wheel and rail. The sound pressure level increases with train speed V at a rate of roughly $30 \log_{10} V$, as indicated in Figure 6.1, which draws on data from

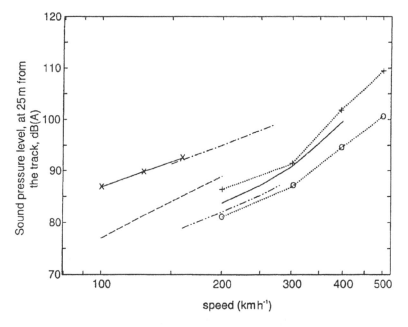

Figure 6.1   A-weighted sound pressure level at 25 m from running trains as a function of train speed. x———x BR MkII (tread-braked) vehicles, - - - BR MkIII (disc-braked) coaches, — ·· — ·· Talgo (articulated, drum-braked), — · —· TGV first generation (articulated, tread- plus disc-braked), + ··· + TGV Atlantique power cars (articulated, tread- plus disc-braked), o ··· o TGV Atlantique trailer cars (articulated, disc-braked), — ICE (disc-braked).

a number of sources (Hemsworth 1979; King III 1990; Mauclaire 1990; Wettschurek and Hauck 1994). In this figure, results for several types of train are presented in terms of the A-weighted sound pressure level at 25 m from the track, as a function of train speed. At high speeds it can be seen that the sound pressure level increases more rapidly with speed. This may be attributed to aerodynamic sources which exhibit a speed dependence of typically $60 \log_{10} V$ or more. The speed above which aerodynamic sources dominate depends on the design of the train, but as seen here, for modern high-speed trains it is around $300 \, \text{km h}^{-1}$.

On jointed track, or at points and crossings, the noise from the wheel and rail is augmented by impact excitation. This produces considerably higher instantaneous sound pressure levels compared with rolling noise on continuously welded rail and, as a result, also an increase in the average level.

Curve squeal noise is a tonal phenomenon associated with sharp curves, and brake squeal is a similar type of noise that occurs during stopping. These sources of noise can be particularly annoying, far more than a measure of their sound level would suggest, due to their tonal and intermittent nature, although they are generally confined to particular localities.

Noise from the power unit is generally not a function of train speed; for example, a diesel electric locomotive operates at a throttle setting that depends on the power required rather than on the train speed. Generally, power unit noise only contributes significantly to the overall sound emitted by a train at low speeds. Similarly, the noise from warning horns is independent of speed and usually of lesser importance. An exception to this is found on North American freight railways where it is obligatory to sound the horn in an extended sequence on the approach to road crossings. There are many such crossings, especially in populated areas and so it is a major source of annoyance, particularly from operations at night (Meister and Saurenman 2000). Other noise, such as shunting noise from freight yards or noise from station operations, are more localised (Wettschurek and Hauck 1994). Although these may be a problem for local inhabitants, they affect far fewer people than the noise from running trains.

For passengers inside the train, rolling noise is also the dominant noise in most situations (Hardy and Jones 1989). Power unit noise is important for multiple unit stock with under-floor-mounted diesel engines, whilst aerodynamic noise is significant in high-speed trains. Air-conditioning noise can also be significant in modern rolling stock due to the limited space in which to package the air-conditioning unit and ducts.

Although there are many sources of noise due to railway operations, the major sources are seen to originate from the wheel–rail interaction. Rolling noise has been the subject of extensive research since the 1970s and will be discussed first, and in greatest depth, before turning attention briefly to other sources of noise.

6.2.2 Rolling noise

6.2.2.1 Roughness excitation

Since the widespread introduction of continuously welded rail, dating from the late 1960s and the 1970s, the noise from railway wheels running on the track is no longer dominated by the 'clickety-clack' of rail joints. Nevertheless, the noise that remains, which has a broad-band random character, is the dominant noise from railways. It has been seen in Figure 6.1 that rolling noise increases with train speed approximately as $30\log_{10} V$, which corresponds to a 9 dB increase per doubling of speed.

This figure also shows that there are significant differences between the sound level at a given speed for various types of train. For example, modern high-speed trains are no noisier at $300\,\mathrm{km\,h^{-1}}$ than older trains at $150\,\mathrm{km\,h^{-1}}$. The main reason for such differences turns out to be their braking system; wheels fitted with cast-iron brake blocks are found to develop a quasi-periodic roughness on their running surface due to a complex process involving the formation of hot-spots during braking. This has a wavelength of around 50–100 mm and an amplitude (peak-to-trough) of typically 10–30 μm and, as described below, this induces higher noise levels than occur for wheels that are fitted with disc brakes.

The roughness of the rail is also an important parameter in determining the noise level. In some circumstances a severe periodic roughness may develop on the rail head, known as rail corrugation. There are many forms of corrugation, but the most common on modern high-speed passenger lines has a wavelength of about 50 mm and an amplitude of 50–100 μm (Grassie and Kalousek 1993). This leads to an increase in noise levels of about 10 dB for cast-iron tread braked rolling stock and up to 20 dB for disc-braked rolling stock compared with smooth rails, causing the noise levels of these two types of rolling stock to become similar.

Although such roughness amplitudes might seem small, when the wheel runs on the rail at operational speeds, this roughness forms a significant dynamic excitation of the wheel and rail by inducing them to vibrate relative to one another. For a wavelength λ (m), traversed at a speed v ($\mathrm{ms^{-1}}$), the frequency of excitation is given by $f = v/\lambda$ (Hz). Thus, a wavelength of 50 mm at a speed of $44\,\mathrm{ms^{-1}}(160\,\mathrm{km\,h^{-1}})$ corresponds to an excitation frequency of about 900 Hz. Moreover, a displacement amplitude of 10 μm at 900 Hz corresponds to a vibration velocity amplitude of $0.06\,\mathrm{ms^{-1}}$ or an acceleration amplitude of $300\,\mathrm{ms^{-2}}$. Such amplitudes, when induced in the wheel or the track, are sufficient to produce the sound pressure levels shown in Figure 6.1.

This mechanism of relative displacement excitation is at the heart of theoretical models for rolling noise generation (Thompson 1993a; Thompson *et al.* 1996). Figure 6.2 shows a schematic noise path diagram for such a model. The dynamic forces generated at the wheel–rail contact by roughness excitation cause the wheel and/or rail to vibrate. For simplicity,

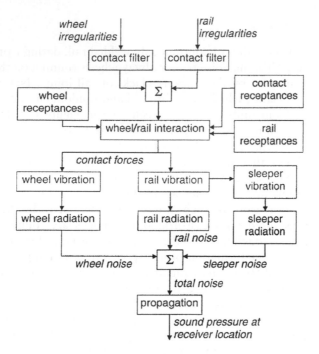

Figure 6.2 Flowchart of theoretical model for railway rolling noise.

a model involving coupling in the vertical direction only is presented here. A mobility coupling approach is used (see Chapter 9), based on the equilibrium of forces and continuity of displacement at the contact. Representing the wheel and rail by their point mobilities (that is, the velocity at the driving point per unit force as a function of frequency), the vibration velocity of the rail v_R and of the wheel v_W at a given frequency ω are given by

$$v_R = \frac{-j\omega r Y_R}{Y_R + Y_W + Y_C} \tag{6.1}$$

$$v_W = \frac{j\omega r Y_W}{Y_R + Y_W + Y_C} \tag{6.2}$$

where r is the roughness amplitude at this frequency, Y_R is the point mobility of the rail, Y_W is the point mobility of the wheel and Y_C is the mobility of the contact spring. The latter represents local elasticity at the contact, which, although it is non-linear, can be approximated by a linear model for a given nominal preload provided that the displacement is small compared with the overall deflection of the contact (Thompson 1993a).

From equations (6.1) and (6.2) it may be observed that, at frequencies where the rail has the highest mobility of the three, $v_R \approx -j\omega r$. This means that the rail vibration velocity is approximately equal to the roughness excitation velocity. Similarly, at a wheel resonance where the wheel mobility is highest, $v_W \approx j\omega r$. Typical (predicted) mobilities of a wheel, a track and the contact spring are shown in Figure 6.3. The wheel represents a standard freight wheel of diameter 920 mm, while the track comprises concrete sleepers and a rail pad stiffness 350 MN m^{-1}. From this it can be seen that the track has the highest mobility over most of the range 100–1000 Hz; so in this range the rail vibration at the contact point will be equal to the roughness excitation. In much of the range above 1000 Hz the contact spring has the highest mobility. At frequencies for which this is the case, the contact spring will tend to 'absorb' the roughness, attenuating its effect on the wheel and rail. At wheel resonances, however, the wheel mobility is much greater than that of the contact spring or the rail. This leads to considerable wheel vibration around these resonance frequencies, especially in the frequency range 1500–5000 Hz (Thompson 1993b). The

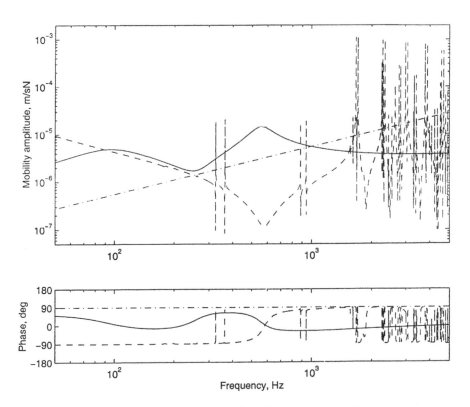

Figure 6.3 Vertical point mobilities of a standard 920 mm freight wheel (- - -), a track with concrete sleepers and rail pads of stiffness 350 MN m^{-1} (—) and a contact spring of stiffness 1.14 × 10^9 N m^{-1} (— · —).

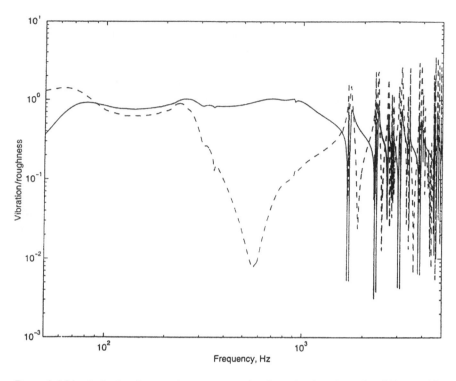

Figure 6.4 Vertical vibration at the contact point for wheel and track of Figure 6.3.
- - - wheel, — rail.

resulting wheel and rail vibrations predicted by the model are presented in
Figure 6.4.

In practice, the wheel and rail are also coupled in other directions than
vertical, and equations (6.1) and (6.2) may be replaced by matrix equations (Thompson 1993a). Nevertheless, the above analysis remains generally
valid. In fact, Figure 6.4 was predicted with coupling between the wheel
and rail in the lateral as well as vertical direction. The result is that the rail
is the dominant source of noise up to around 1500 Hz, whereas the wheel
contributes the main noise radiation above this frequency. The rail noise is
broad-band in nature; the wheel noise spectrum contains peaks associated
with modes of vibration of the wheel. The noise spectra predicted for the
above example are presented in Figure 6.5 in one-third octave form.

6.2.2.2   Track noise

Equations (6.1) and (6.2), and their matrix equivalent, determine the vibration of the rail at the contact point. Since the excitation acts in the vertical

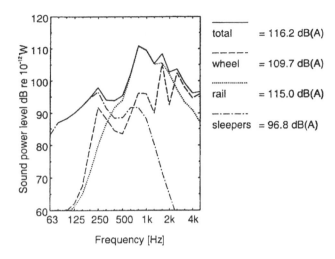

total	= 116.2 dB(A)
wheel	= 109.7 dB(A)
rail	= 115.0 dB(A)
sleepers	= 96.8 dB(A)

Figure 6.5 Predicted one-third octave band sound power components for wheel and track of Figure 6.3 for a train speed of 120 km h^{-1}.

direction, the lateral vibration at the contact is usually at least 10 dB less than the vertical vibration. This vibration propagates along the rail in the form of structural waves – vertical and lateral bending, torsion and, at higher frequencies, other more complex waveforms involving deformation of the rail cross-section (Thompson 1997).

The sound power radiated by a vibrating structure, W_{rad} can be expressed as

$$W_{rad} = \rho_0 c_0 \sigma S \left\langle \overline{v^2} \right\rangle \tag{6.3}$$

in which $\rho_0 c_0$ is the characteristic-specific acoustic impedance of air, $\left\langle \overline{v^2} \right\rangle$ is the spatially averaged mean-square normal velocity of the vibrating surface, S is its surface area and σ is the radiation ratio (also known as radiation efficiency). Since, in the case of the rail, vibration is transmitted along its length, the average squared vibration is required over a sufficiently large length; for example, the length over which the amplitude decays by 10 dB.

The extent to which the various waves are damped as they propagate is crucial in the determination of the radiated noise level. Since the decay of vibration amplitude with distance is roughly exponential, this damping can be expressed as a decay rate in dB m^{-1}. Some typical results are shown in Figure 6.6 (Thompson *et al.* 1999).

The decay rate of both vertical and lateral bending waves is large at low frequencies (typically 10 dB m^{-1}). At these frequencies, the rail vibration is localised around the excitation point in the form of non-propagating waves. Above a certain frequency, the rail becomes decoupled from the foundation

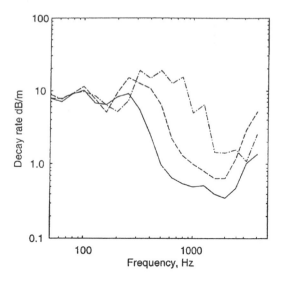

Figure 6.6 Measured rates of decay with distance of vertical waves in the track. — track with soft pad (120 MN m^{-1}), - - - track with medium stiffness pad (270 MN m^{-1}), — · —· track with stiff pad (1000 MN m^{-1}).

due to the flexibility of the rail pads that are inserted between the rail and the sleepers. Waves then start to propagate and the decay rate falls to around 1 dB m^{-1} or less, rising again only slightly at higher frequencies. It can be seen from Figure 6.6 that the frequency at which propagation begins depends on the rail pad stiffness, as does the level to which the decay rate falls. Softer rail pads result in a lower decoupling frequency, which can be understood as the resonance frequency of the mass per unit length of the rail on the stiffness of the rail pads (per unit length of track). This can also be seen as a peak in the rail mobility (500 Hz in Figure 6.3). This frequency is given by

$$f = \frac{1}{2\pi}\sqrt{\frac{k_p}{\mu_r d}} \tag{6.4}$$

where k_p is the pad stiffness, μ_r is the rail mass per unit length and d is the distance between support points. For the cases given in Figure 6.6, $d = 0.6$ m, $\mu_r = 60$ kg m^{-1}, so for the pad stiffnesses listed this gives frequencies 290, 440 and 840 Hz, respectively.

Softer rail pads also lead to lower damping of the rail. For a stiffness k_p and a loss factor η_p the damping force is proportional to $k_p \eta_p$. Most pads have loss factors around 0.1–0.2, whereas stiffnesses can vary by more than a factor of 10. Therefore, soft pads lead to lower decay rates in the rail in the frequency region in which propagating waves occur. For a track with soft rail pads, the noise from the rail is caused mainly by its vertical motion. With stiff pads, however, the vertical decay rate has a large, broad peak of around 500–1000 Hz, where the sleeper mass acts as a dynamic absorber as it vibrates on the rail pad stiffness. The pad stiffness is lower in the lateral direction than in the vertical direction, so lateral waves have a lower decay rate. Consequently, although the amplitude of lateral motion is smaller than that of the vertical motion at the contact, when averaged over the whole length of the rail it is the lateral motion that is more important with such pads.

Another consequence of rail pad stiffness is the degree of coupling to the sleeper. At low frequencies the rail and sleeper move with approximately the same amplitude. Since the sleeper is broader than the rail, it radiates more noise in this region (see Figure 6.5). This is due to both a larger area S and a higher radiation ratio σ (see equation (6.3)). At the decoupling frequency defined above, the sleeper can vibrate with a greater amplitude than the rail. Above this frequency the amplitude of the sleeper becomes relatively small as it becomes dynamically isolated from the rail. Since the rail pad stiffness determines the decoupling frequency, with a soft pad the sleeper will radiate less noise than with a stiff pad. These trends are summarised in Figure 6.7

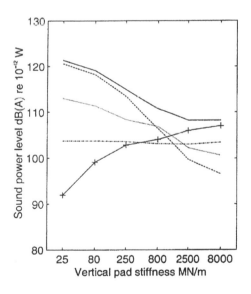

Figure 6.7 Dependence of sound power components of rolling noise on rail pad stiffness. — total track noise, - - - rail vertical component, ··· rail lateral component, + — + sleeper component, ···· wheel noise. Based on wheel roughness of tread-braked wheels, standard freight wheel design, 100 km h⁻¹.

(see Vincent *et al.* 1996; Thompson *et al.* 1999). Note that changes to the track usually have only a very small influence on the component of noise from the wheel and, similarly, changes to the wheel have only a small effect on the track noise.

6.2.2.3   Wheel noise

Wheel vibration is characterised by its natural modes (Thompson 1993b), most of which involve predominantly axial (out-of-plane) motion. They can be identified by the number of nodal diameters and nodal circles. In addition, a set of modes exist which are predominantly radial (in-plane), again with various numbers of nodal diameters. Below 5 kHz a typical railway wheel has about 25–30 modes of vibration. Since the excitation is primarily in the vertical direction, it is the modes with a significant vertical component at the contact point that are excited. Moreover, the modes with less than two nodal diameters generally have higher damping due to coupling with the axle, and are therefore less important. Consequently, the modes of most importance to rolling noise are the radial and 1-nodal-circle axial modes with 2, 3, 4, 5, . . . nodal diameters. These modes occur in the frequency region above about 1500 Hz, and can be identified as a series of sharp peaks in the wheel mobility curve in Figure 6.3. The noise from the wheel is dominated by the noise in this part of the frequency spectrum, as seen in Figure 6.5. In fact, the mobility of Figure 6.3 includes the effect of wheel rotation. This leads to a splitting of modal peaks into two, one above and one below the resonance frequency. However, this has only minor consequences for the radiated noise (Thompson 1993c).

From Figure 6.4, it is clear that the peaks in the vibration response are not as sharp as those in the wheel mobility. This is a consequence of coupling between the wheel and the rail, the rail acting effectively as a damper connected to the wheel; its mobility has a phase close to zero over much of the frequency range (see Figure 6.3). This has the effect of increasing the apparent damping of the wheel. Any damping treatments added to the wheel must increase its damping by more than this amount if they are to reduce rolling noise, even though they may appear effective when the wheel is tested in isolation. In contrast, much smaller amounts of damping may be added to a wheel to suppress squeal noise.

6.2.3   Noise reduction measures for rolling noise

In order to reduce rolling noise at source, the theoretical model (Figure 6.2) can be used to identify potential solutions. These fall into a number of categories, many of which have been investigated in large-scale research projects sponsored by the EU (Hemsworth *et al.* 2000).

6.2.3.1   Reduction of the excitation by reduction of the roughness

Wheel roughness can be reduced by replacing cast-iron brake blocks by disc brakes or composition blocks, giving reductions of up to around 10 dB. Rail roughness is usually less than the wheel roughness on tread-braked wheels with cast-iron blocks unless corrugation has formed on the rail surface. If this is the case, the usual remedy is to grind the rail head using a special grinding train. It may be noted from Figure 6.2 that, while the noise in a particular situation may be dominated by vibrations of, for example, the track, it is not necessarily the rail roughness that is responsible. Similarly, noise due to wheel vibrations may be caused by roughness of either the rail or the wheel. This makes it difficult to separate the influence of the vehicle and the track on the noise.

6.2.3.2   Reduction of track noise by structural modifications

As seen above, the rail pad stiffness is one parameter of the track affecting the noise that can be adjusted. As the pad stiffness increases, the rail component of noise decreases and the sleeper component increases (see Figure 6.7). Where these two components are equal, an 'optimum' pad stiffness can be identified (Vincent *et al.* 1996). However, such an optimum corresponds to a rather stiff pad (greater than $1000\,MN\,m^{-1}$) whereas the trend in track design is to use softer rail pads in order to reduce the dynamic loads acting on the track which cause damage to the track structure. Increasing the rate of decay of rail vibration with distance by use of a tuned damper is a promising technique that has been shown to be capable of reducing track noise by about 6 dB in field tests (Thompson *et al.* 2000). This is particularly suited to tracks with a low stiffness pad and allows the advantages of such a pad to be realised without an increase in noise.

6.2.3.3   Reduction of wheel noise by shape optimisation and/or added damping

In order to minimise coupling between the vertical excitation and the axial response of the wheel that radiates most of the sound, the wheel should be made as symmetrical as possible in cross-section. Traditionally, to accommodate thermal expansion during braking, railway wheels have been constructed with a curved web profile (the region between the tread and the hub). Making the wheel symmetrical implies that this should be made straight, although this is only possible for disc-braked wheels, where the requirements for thermal expansion do not apply. Furthermore, a smaller diameter and thicker web are beneficial since these raise the natural frequencies of the modes that are excited by a vertical force, particularly the 1-nodal-circle modes. Although there are practical limitations on the change in wheel diameter for retrofit solutions, this can be used effectively for new designs.

Added damping, if greater than the 'rolling damping', leads to the reduction of the response in the high-frequency region where the resonances of the wheel are dominant (see Figure 6.2). Damping treatments used successfully on wheels include constrained layers (typically 1 mm thick visco-elastic material plus a 1 mm thick steel or aluminium constraining plate) and tuned absorbers (damped mass-spring systems added to the wheel). The latter produce an appreciable increase in the mass of the wheel. In Germany, wheel absorbers have been used successfully for many years and have been found to produce reductions in overall rolling noise of 6 dB on Intercity trains (Hölzl 1994). A wheel with a combination of an optimised shape, slightly smaller diameter and added damping has been shown to reduce the wheel component of noise by about 7 dB in tests (Hemsworth *et al.* 2000). Some examples of wheel treatments are shown in Figure 6.8.

6.2.3.4 Reduction of sound radiation from the track

For a given vibration level, the radiated sound power can be reduced by the use of a smaller surface area (see equation (6.3)). Practical changes in the size of the rail, however, produce only very limited changes in surface area. Nevertheless, a reduced size also leads to a reduction in the radiation ratio in the low-frequency region and such changes can be more appreciable. A narrower rail foot leads to reduced radiation from the vertical vibration

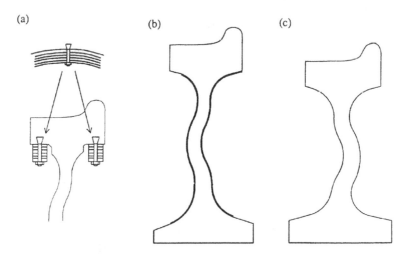

Figure 6.8 Examples of noise reduction measures for railway wheels: (a) tuned absorbers; (b) constrained layer damping applied to the wheel web; (c) reduced diameter and thicker web.

whilst a low-height rail reduces the noise from lateral motion. Moderate reductions are possible from such a technique; for example a 2–3 dB reduction in rail noise was achieved by narrowing the rail foot from 150 to 100 mm in an experimental implementation (Hemsworth *et al.* 2000).

6.2.3.5   Reduction of sound radiation from the wheel

Apart from a reduction in radiating area, changes in the wheel size do not have an appreciable effect on the sound radiation, as the modes of interest lie in the high-frequency region where the radiation ratio is close to 1. Another possibility for reducing the radiation ratio is to introduce holes in the wheel web, which allows the sound from the front and back surfaces to interfere destructively. Unfortunately, due to the thickness of a wheel, this is only effective below about 1 kHz, so it is not significant for the overall level of wheel noise.

Another technique for reducing the wheel noise that has been found to be quite effective is to shield the web with a resiliently mounted panel. Using a thin panel for this shield minimises the sound radiation from its own vibration. With an effective shield over the wheel web, the sound radiation from the wheel is reduced to that from the tyre region, and this has been found to give reductions in wheel noise of around 7 dB (Hemsworth *et al.* 2000).

6.2.3.6   Local shielding measures

A combination of low, close trackside barriers and bogie-mounted shrouds (Figure 6.9) can be effective in shielding the noise from the wheel–rail region. Due to the need to allow for differential movement between the vehicle body, bogie and infrastructure, substantial clearances have to be maintained. The success of this technique relies on minimising the inevitable

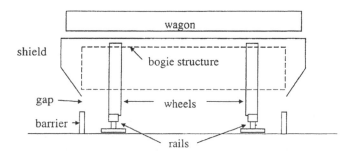

Figure 6.9 Local shielding in the form of a bogie shroud added to the vehicle and a low barrier close to the rail.

gap between the two parts of the shield and adding absorption within the bogie shroud. It is more effective for tracks with a high decay rate of rail vibration as a low decay rate leads to significant leakage of rail noise out of the bogie region. Field tests, for example in the UK, have demonstrated that reductions of 8–10 dB are possible (Jones 1994).

6.2.3.7   Summary

Starting from the development of theoretical models, a number of techniques have been developed in recent research projects that should allow rolling noise levels to be reduced. Their effectiveness relies on the application of a suitable *combination* of measures, chosen according to the relative importance of the wheel and track as sources of noise in the situation considered. However, there are many practical obstacles to be overcome before this type of solution becomes commonplace.

6.2.4   Impact noise

Discontinuities in the rail surface, for example rail joints, dipped welds, points and crossings, lead to impulsive noise. Similarly, wheel 'flats' induced by sliding during braking produce a repeated impulsive noise. Published information on this phenomenon is very limited (Vér *et al.* 1976; Remington 1987).

In extreme cases, momentary loss of contact occurs between the wheel and the rail and when it is re-established a high impact load occurs. Even without loss of contact, a discontinuity produces a partial unloading followed by a severe increase in wheel load over a short time period. This dynamic load produces wheel and rail vibration in the same way as the force induced by the random roughness discussed above; the excitation is in the vertical direction and the wheel–track system responds in a similar manner. Since the impact load can vary between zero and four times the nominal wheel load, the assumption of a linear contact stiffness cannot be used. The inclusion of a non-linear element in the system means that the frequency-domain approach outlined above for roughness-induced rolling noise can no longer be used; instead a time-domain approach is required (Newton and Clark 1979; Wu and Thompson 2001).

Prevention of the occurrence of discontinuities is clearly the most effective measure to minimise impact noise. Detection systems to identify wheel flats are coming into use so that the vehicles can be taken out of service and the wheels reprofiled. Jointed track has given way to continuously welded rail on most main lines in Europe, although it should be realised that significant impact noise can still be generated by welded rail if the welds are not straightened. Points and crossings with a movable crossing nose can help overcome much of the impact loading. In addition, the techniques discussed in Section 6.2.3, that are designed to reduce rolling noise by structural modification or local shielding, can be expected to be equally effective for impact noise.

6.2.5 Curve squeal

6.2.5.1 Mechanism of curve squeal

When a railway vehicle traverses a sharp curve, an intense tonal noise known as curve squeal may be produced. The frequency of this noise, which may be as low as 250 Hz or as high as 8 kHz, corresponds to a resonance of modal vibration of the wheel. Usually the mode is one whose mode shape contains a large axial (lateral) displacement at the tread, notably the 0-nodal-circle modes with various numbers of nodal diameters. This points to an excitation mechanism acting in the lateral direction. Indeed, the cause of curve squeal is usually lateral slip of the wheel on the rail head. On a straight track, the coned profile of railway wheels (along with the properties of the bogie frame) is responsible for their stable running. When a wheelset moves to one side, one wheel becomes effectively larger than the other ('rolling radius difference') and this steers the wheelset back towards the track centre.

In a curve, a similar mechanism causes the wheelsets to attempt to adopt a radial attitude, the rolling radius difference allowing the outer wheel to travel the larger distance required. In a sharp curve, however, the wheelsets in a bogie cannot adopt a purely radial attitude owing to the stiffness of the bogie frame and limitations on the rolling radius difference. The flange of the front outer wheel comes into contact with the rail and this wheelset has a considerable angle of attack relative to the rail: it tries to run out of the curve (see Figure 6.10). The difference between the velocity vector of the

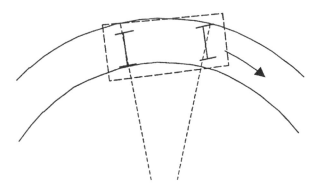

Figure 6.10 Typical attitude of a bogie in a sharp curve showing the high angle of attack of the front wheelset.

train (around the curve) and the direction of rolling (straight on) can be seen as a relative lateral motion. On the outer wheel a restoring force is provided by flange contact whereas on the inner wheel lateral slip occurs between the wheel and the rail. Owing to the characteristics of the friction behaviour, a stick-slip type of excitation is generated, much the same as that occurring when a violin is bowed. A feedback loop between this stick-slip mechanism and the vibration of the wheel results in a 'self-excited' oscillation of a lightly damped mode of the wheel. Which mode is excited depends upon its modeshape at the excitation point and upon damping (Heckl 2000; de Beer et al. 2000). Since the force acting on the wheel also acts on the rail (in the opposite direction), the rail will also vibrate at the tone of the squeal noise. However, since the mobility of the wheel at resonance is much larger than that of the rail, the corresponding vibration of the rail is small compared with that of the wheel and its noise radiation will also be negligible.

6.2.5.2   Reducing squeal

Because squeal noise is caused by an instability of a mode of vibration of the wheel, mitigation measures usually eliminate, rather than reduce, the level of squeal noise. The addition of only a moderate amount of damping can be sufficient to overcome this instability and eliminate squeal. Various forms of damping treatment which have been used successfully include: constrained layer damping applied to the wheel web; tuned absorbers attached to the inner edge of the tyre; friction damping by a ring in a groove in the inner edge of the tyre; and resilient wheels with a visco-elastic layer between the tyre and web.

Another control technique is friction management. The instability can be attributed in part to the difference between the static and dynamic coefficients of friction. Lubrication of the rail or wheel surface can reduce, or even eliminate, this difference and thereby overcome the instability. Grease is commonly used for this purpose, and water has also been used effectively. More recently solid lubricants have been developed that modify the friction characteristics beneficially. Steerable wheelsets can also be used effectively to eliminate squeal by removing the excitation. However, a compromise must be reached between adequate curving behaviour and stability at speed.

6.2.6   Aerodynamic noise

The sound power of aeroacoustic sources increases more rapidly with speed than that of mechanical sources (see also Section 6.3 and Chapter 7). For an aeroacoustic monopole source, such as the pulsating flow from an exhaust pipe, the sound power increases with flow speed according to the fourth power of the speed. This means that the sound power level increases at a rate of $40 \log_{10} V$. For a dipole type source, such as the tones generated by vortex shedding from a cylinder or turbulence acting on a rigid surface, the rate is $60\log_{10} V$, whereas for a quadrupole source such as free turbulent flow, the rate is $80\log_{10} V$.

As has already been shown in Figure 6.1, aerodynamic sources become dominant for exterior noise of trains above a speed of about $300\,\mathrm{km\,h}^{-1}$. Where noise barriers are placed alongside the track, the wheel/rail noise may be attenuated by 10–15 dB, while leaving the aerodynamic sources from the upper part of the train and pantograph exposed. This causes aerodynamic noise to become important at lower speeds. Aerodynamic sources are also important for interior noise in high-speed trains, particularly the upper deck of double-deck trains where rolling noise is less noticeable.

Turbulent air flow, which can be caused by many different parts of a rail vehicle, is an important source of aerodynamic noise (King III 1996). The locations of a number of sources have been identified and their strengths quantified in studies using specialised microphone arrays (Barsikow 1996). Important sources are found to fall into two main categories (Talotte 2000). The first category, which is dipole in nature, is generated by air flow over structural elements: the bogies, the recess at the inter-coach connections, the pantograph and electrical isolators on the roof and the recess in the roof in which the pantograph is mounted. In addition, the flow over the succession of cavities presented by louvred openings in the side of locomotives is a source of aerodynamic noise, the form of which depends on the length and depth of the cavity. In the second category, which may have a dipole or quadrupole nature, noise is created due to the turbulent boundary layer.

Empirically-based models for each source of aerodynamic noise from trains can be derived if the locations and source strengths are experimentally determined. Measurements may be complemented by the use of computational fluid dynamics (CFD) models. While a working theoretical model for the aerodynamic sources from trains is not yet available, it is the objective of current research (Talotte 2000).

6.2.7 Power unit noise

Power units on trains are mostly either electric or diesel. Noise from diesel locomotives is mostly dominated by the engine and its intake and exhaust. Space restrictions often limit the ability to silence the exhaust adequately, although in modern locomotives this has been given serious attention. On electrically powered stock, and on diesels with electric transmission, the electric traction motors and their associated cooling fans are a major source of noise. Most sources of noise from the power unit are largely independent of vehicle speed, depending rather on the tractive effort required. The whine due to traction motors is an exception to this.

A discussion of the sources of noise in internal combustion engines is presented in Section 6.3; the same considerations apply to railway diesel engines, although the scale is different. The mechanisms of transmission of noise from underfloor diesel engines to the interior of railway vehicles are also similar to those found in cars.

6.2.8 Noise inside trains

All the noise sources discussed above are also of relevance to interior noise in trains (Hardy and Jones 1989). Noise is transmitted from each of these sources to the interior by both airborne and structure-borne paths. The noise from the wheel–rail region is often the major source, with structure-borne transmission dominant at low frequencies and airborne transmission at high frequencies. Noise from the air-conditioning system can also require consideration in rolling stock where this is present. There is often very limited space in which to package the air-conditioning unit and ducts.

Passenger requirements for noise inside a train vary from one person to another (Hardy 1999). Clearly it is desirable that the noise should not interfere with conversation held between neighbours. However, particularly for a modern open saloon type vehicle, silence would not be the ideal. There needs to be sufficient background noise so that passengers talking do not disturb other passengers further along the vehicle (people talking loudly into mobile phones are a particular source of annoyance).

Due to the increase of the transmission loss of both airborne and structure-borne sound transmission with frequency, the spectrum of noise inside trains contains considerable energy at low frequency. An example spectrum is given in Figure 6.11, which was measured in a diesel multiple

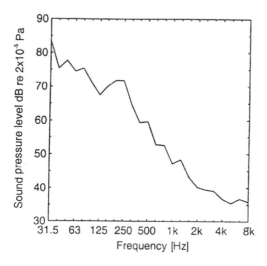

Figure 6.11 One-third octave band spectrum of noise inside a British class 158 diesel multiple unit train cruising at approximately 110 km h^{-1}.

unit with an underfloor engine. The A-weighted level in this case is 68 dB and is dominated by the peak around 250 Hz. This low-frequency sound energy can be a source of human fatigue, but is not effective in masking speech, for which noise in the range 200–6000 Hz is most effective.

6.3 Automotive noise

6.3.1 Overview of road vehicle noise

Road vehicle noise may be considered to fall into two very different categories, namely exterior and interior noise. Exterior noise is limited by legislation in many countries, and this has been the driving force behind many of the developments over the last three decades. Interior noise is purely a matter of customer comfort and preference, but is increasingly being used by manufacturers as a selling point, an indication of overall refinement or even a statement of individuality. In recent years this desire to portray quality through the vehicle's noise signature has also begun to influence exterior noise development, with manufacturers becoming aware that their brand image is now being examined (often subconsciously) by pedestrians and even neighbours of their customers.

6.3.1.1 Exterior noise

As an example of legislative limits, Table 6.1 shows how the required levels applying in European Union countries have been significantly reduced over the years. These noise levels correspond to a standard drive-by test (ISO 362 1994) which is both simple and easy to execute. The vehicle is required to

Table 6.1 Permissible sound levels of motor vehicles (EEC)

		1973	1981	1987	1989	1996
Passenger cars,	<9 seats	82	80		77*	74*
vans and minibuses	<2 tonnes	84	81		78*	76*
	>2 < 3.5 tonnes	84	81		79*	77*
	>150 kW	89	82		80	78
Coaches (>9 seats; 3.5 tonnes)	>150 kW	91	85		83	80
Trucks (>3.5 tonnes)	<75 kW	89	86		81	77
	>75 <150 kW	89	86		83	78
	>150 kW	91	88		84	80
Motor cycles	≤ 80 cc			77		75
	> 80 ≤ 175 cc			80		78
	>175 cc			82		80

Note
* 1 dB extra allowed for direct injection diesel engines.

accelerate between two microphone stations and the maximum A-weighted noise level is recorded (Figure 6.12). The entry speed into the test site is determined by various vehicle design factors, but never exceeds $50\,\mathrm{km\,h^{-1}}$. Although the test procedure is fairly robust, its transient nature can unfortunately make systematic noise reduction development difficult.

The sound levels given in Table 6.1 suggest that over a period of 25 years the overall A-weighted noise levels of passenger cars have decreased by 8 dB whilst the noise of commercial trucks has decreased by 11 dB. Much of these reductions have been achieved by attention to the contributions from the power unit, intake and exhaust. Unfortunately, tyre manufacturers have been unable to reduce the noise of their products significantly over this period, and they are therefore rapidly becoming the dominant source, even at the relatively slow speed specified by the current ISO 362. Owing to this relatively small reduction in tyre noise, high-speed vehicle noise, which is dominated by tyre noise, has seen only small reductions of around 1–2 dB which has been more than negated by the increase in traffic densities. In an attempt to place more pressure on the tyre manufacturers, an additional, or even totally alternative, vehicle test procedure is currently being discussed which will be more biased towards monitoring tyre noise emissions.

There are four primary sources of exterior noise during the drive-by test; namely the power unit, tyres, intake and exhaust. The characteristics of

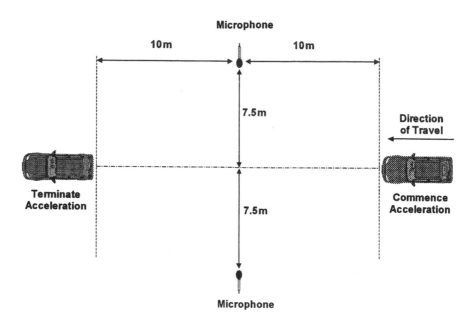

Figure 6.12 Exterior noise drive-by test.

these four sources are considered in greater detail in Sections 6.3.2–6.3.4. In addition there are at least a further four secondary sources; namely intake shell noise, exhaust shell noise, wind noise and body-radiated structure-borne engine noise. Noise radiated from the shell of the intake or exhaust system (as opposed to gas noise from the respective orifices) can be an issue; but robust solutions using traditional stiffening or damping measures are available. Wind noise is rarely significant in the drive-by test due to the low speed of the current test procedure. Body-radiated structure-borne engine noise is again insignificant, apart from that from some panel vans. These secondary sources clearly exist, but they are rarely considered as particularly challenging (with respect to the current test procedure), and few manufacturers make allowances for them when developing a new vehicle.

Although the drive-by test only requires overall noise to be measured in dB(A), a clear picture of the spectral content of the noise is most useful during the development of the vehicle. Most road vehicles generate noise of fairly similar spectral character (Figure 6.13), with a few low-frequency harmonics of engine firing generated by the intake and exhaust gas noise, and then a broad mid-frequency hump centred around 1–2 kHz generated by the power unit and tyres. A well-developed vehicle will usually have similar A-weighted levels in these two distinct frequency regions.

The various noise sources present in a vehicle fall into two categories; those related to engine speed (engine, intake and exhaust) and those related to road speed (tyres). The noise of the transmission falls into both categories,

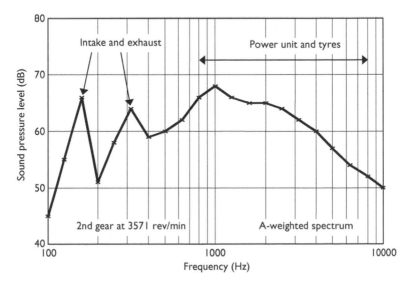

Figure 6.13 Typical light commercial vehicle drive-by spectrum at peak noise.

with some of its noise being related to engine speed and the rest being related to road speed. In most vehicles, engine speed and road speed are themselves related by the overall gear ratio selected. As the different gear ratios are selected (as required for most vehicles by the test procedure) so the balance between these two categories of noise source changes. In low gears the exterior noise typically is controlled by the power unit, while in the higher gears tyre noise becomes more dominant (Figure 6.14).

Unfortunately, little attenuation of these noise sources is provided by the vehicle. Even if partially shielded, the low-frequency character of the intake and exhaust noise ensures that their attenuation by the vehicle is small. The open architecture around the vehicle wheels again means the noise from the tyres also has minimal attenuation. Only power unit noise can benefit from the overall attenuation afforded by the vehicle body, and even then the degree of attenuation is still small ranging from just 1–2 dB for the open-architectured heavy commercial vehicle through to 5–6 dB for the more enclosed passenger car. Further attenuation can be achieved by the use of undershields, engine tunnels and engine bay absorption, but fully enclosing the engine is usually ruled out for practical reasons.

6.3.1.2 Interior noise

Interior noise has also fallen significantly over the years in response to customer expectation. In the past 40 years, the greatest improvements have been seen within the mass market with the once considerable refinement gap between the family saloon and the luxury car being rapidly closed. Unlike the case of exterior noise, it is believed that some interior noise is desirable so as to give character, feel and, perhaps most importantly, an indication of speed. It has been suggested that at low speeds some recent luxury vehicles may have even gone below this 'ideal noise' limit and are therefore too quiet (Goddard 1998). For an average passenger car, interior noise ranges from around 45 dB(A) at stationary idle up to about 70 dB(A) at $110 \, \text{km} \, \text{h}^{-1}$ under steady cruise conditions. At maximum acceleration, the noise may rise as high as 78 dB(A). The dramatic effect of both engine and vehicle speed on interior noise (~40 dB/decade) results in acceptably low levels at idle, close to that of a quiet office environment, through to potentially hazardous levels at high speed especially during acceleration and long hill climbs. Cruising at $130 \, \text{km} \, \text{h}^{-1}$ with full 'in-car entertainment', the driver may be exposed to continuous levels well in excess of 80 dB(A), especially in light commercial vehicles and on noisy road surfaces.

The noise source balance for interior noise varies greatly with speed and engine load conditions. At high speed cruise, the cabin noise is controlled by tyres and wind, whereas during acceleration through the gears, it has significant contributions from all four primary components. Clearly, static idle noise is dominated by engine noise and any operational ancillaries.

(a)

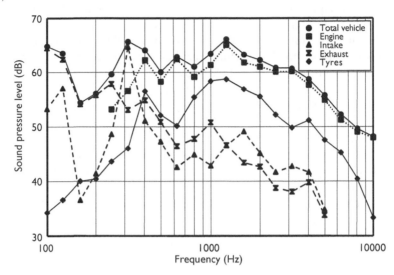

(b)

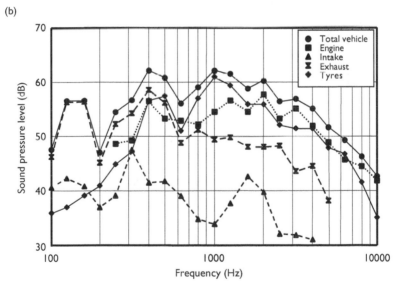

Figure 6.14 The effect of gear selection on drive-by A-weighted noise contributions for a light commercial vehicle: (a) 2nd gear at 47 km h^{-1}, 3200 rev/min and 9 m into test; (b) 3rd gear at 55 km h^{-1}, 2420 rev/min and 15 m into test.

Wind noise increases at a greater rate with speed than that of tyres and therefore becomes more dominant at high speeds. The speed at which they tend to be equal in a typical passenger car is around $110\,\mathrm{km\,h^{-1}}$ which corresponds to common motorway speed limits.

The interior noise spectrum of a road vehicle is dominated by low-frequency noise, much of which reaches the passenger compartment via structural routes (Figure 6.15). However, airborne contributions from intake and exhaust are also present at these low frequencies. As frequency increases, interior noise progressively becomes controlled by airborne routes. The frequency at which structure-borne and airborne noise are of similar levels typically occurs in the range 500–1000 Hz both for engine and tyre noise. Unlike exterior noise, the vehicle fortunately provides considerable attenuation of noise from the various sources before it reaches the interior of the vehicle.

Modern vehicles rely upon vibration isolation to reduce the structure-borne noise and vibration to an acceptable level. The body shell is isolated from the power unit and the road wheels by compliant engine mounts,

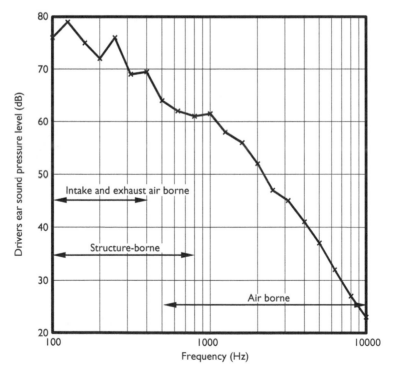

Figure 6.15 Typical interior noise spectrum of a passenger car ($110\,\mathrm{km\,h^{-1}}$).

road springs and rubber bushes. The force that is transmitted through these isolators to the vehicle body is, to the first approximation, the product of the spring element dynamic stiffness and the relative displacement across that spring element. Although this simple isolation concept can be designed to provide more than adequate performance over most of the speed range (Racca 1982), the necessarily low mounting resonances adopted (5–20 Hz) can result in low speed idle, pull away and general transient problems. In an attempt to overcome some of these problems, non-linear mount designs are being adopted, together with optimised damping performance. If cost allows, variable stiffness and variable damping mounts can further alleviate the problems, with strategic switching of performance for different engine (or road) conditions. Fully active mounts (Fursdon *et al*. 2000), where the mount contains a vibro-actuator, have great refinement potential, but as yet their cost and energy consumption continue to prevent them from being widely adopted in mass production vehicles.

The general response of a vehicle body to low-frequency forcing varies little with frequency. A typical Noise Transfer Function (NTF) from a major forcing point to the driver's ear ranges from around 50 dB *re* 2×10^{-5} paN^{-1} for a luxury passenger car through to 70 dB *re* 2×10^{-5} paN^{-1} for a light commercial vehicle. Heavy commercial vehicles can potentially benefit from the additional isolation provided by the separate chassis to cab mounts, but for various practical reasons, they rarely get close to the NTFs of luxury cars.

At low frequencies, both structural vibration and internal cavity acoustics of the vehicle are strongly modal, and so the principal refinement design guideline is to ensure that any resonant coupling between the forcing (for example, power unit driveline bending), the structure (for example, body panel modes) and the acoustics (for example, the first few cavity modes) is minimised. At higher frequencies, noise energy predominantly enters the vehicle via airborne routes. The attenuation of airborne noise paths increases with frequency at around 20 dB/decade over much of the audible range and is predominantly mass controlled. This increasing attenuation with frequency principally causes the domination of interior noise by low frequencies. Up to around 500 Hz the detail design of the vehicle can only have a small influence upon its attenuation performance unless a double panel construction for the bulkhead is adopted. At higher frequencies, the effect of the vehicle's trim becomes progressively more important and the differences between a 'body-in-white' with closures and a fully trimmed vehicle can be in excess of 20 dB (Figure 6.16a).

By defining specific transfer functions for the various sources (Dixon *et al*. 1993), considerable gains can be made by optimising the locations of the various high level sources. The attenuation from the forward facing side of a transverse power unit is often around 5 dB greater than that from the side facing the firewall or bulkhead (Figure 6.16b). Similarly, in

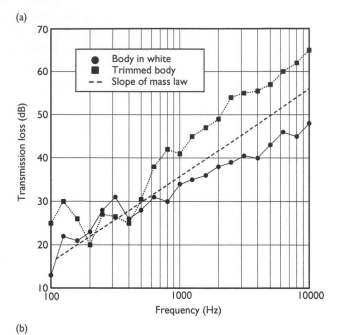

(a)

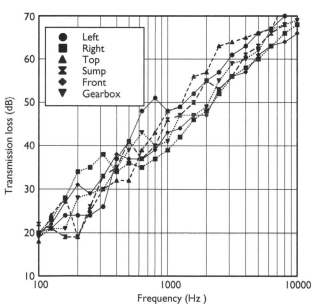

(b)

Figure 6.16 Typical engine bay to cabin transmission loss spectra for a passenger car: (a) comparison between body in white and fully trimmed body; (b) comparison between the 6 sides of a transverse engined passenger car.

a longitudinal installation, the attenuation from the transmission area can be some 8 dB lower than that from the front of the engine, thus often making the transmission the greatest power unit noise source in terms of vehicle interior noise.

In recent years, airborne noise reaching the passenger compartment has been significantly reduced through attention to detail of pass-throughs, seals, grommets, etc., as well as a general increase in trim performance. Unfortunately, in many cases this has resulted in an increase in vehicle weight. Statistical Energy Analysis (Steel 1998) has been adopted in an attempt to minimise this weight increase and reduce the quantity of redundant trim. A broad band model of the vehicle's structural acoustics is constructed (from measurement or theory) and then used to trade acoustic performance with added weight. Although far from being fully robust, such an approach shows considerable potential.

6.3.1.3 Sound quality

Controlling vehicle noise is no longer solely about reducing overall levels; it is now just as important to ensure the 'quality' of the vehicle's noise is up to customer expectation and conforms to the manufacturer's brand image. Considerable research effort has been applied over recent years to road vehicle sound quality (Williams and Balaam 1989), especially in the field of diesel-powered passenger cars (Russell 1993), but still no commonly agreed universal measure of sound quality has been forthcoming. Indeed, it is highly unlikely that any such universal measure will ever be achieved for such a subjective issue.

Arguably one of the most important components of sound quality is balance. Much of the art of sound engineering relies upon achieving the correct balance between different sounds or even different related components of the same sound. Regular readers of motoring columns or vehicle magazines cannot have failed to notice the increase over recent years in comments regarding refinement. However, seldom are the comments simply that a vehicle is too noisy. In nearly all cases, the comments fall into two categories; either that a particular noise source is too loud (which in reality could easily have implied that the other sources are too quiet), or the vehicle is 'refined'. The first category corresponds to unbalanced sound whilst the second category probably implies balanced sound.

Not everyone has the same view of ideal balance. Different ages (Fish 1998), different social profiles, and especially different nationalities (Hussain et al. 1998), appear to prefer different balances of sound. If the general balance preference for a targeted group follows a normal distribution, attempts can be made to set the balance such that the majority of the vehicle customers should be satisfied. However, even within the same age, social and national groups, some fairly fundamental balance preference distributions are far from normal. It has been shown that a straightforward balance between high-frequency and low-frequency noise is often too simplistic (Russell 1993). It would appear that a significant proportion of people are more sensitive to low-frequency noise than others, and are prepared to accept much higher levels of broad band, mid- and high-frequency noise so long as the low frequency is suppressed. Others prefer a more muted high-frequency content and are simply not aware of high levels of low-frequency sound. With two sections within the group preferring the two extreme ends of the preference scale, any attempt at a compromise can result in dissatisfaction for a sizeable proportion of the targeted group. Although these highly subjective issues make quantification and target setting difficult, most sound quality parameters, if understood, can be robustly engineered in exactly the same manner as the more straight-forward objective issue of level.

6.3.2   Engine noise

Of all the noise sources present within a road vehicle, engine noise has been the most intensively researched over the past 40 years. This is prob-ably due to its relative complexity as well as it originally being the major source of vehicle noise. For the first reason, engine noise is considered in the greatest depth within this chapter. Alternative sources of power are increasingly being investigated; however, well over 99 per cent of (four or more wheeled) road vehicles are currently powered by traditional recipro-cating, four-stroke, internal combustion engines. Such engines produce their power by igniting a fuel and air mixture and then harnessing the pressure generated by this combustion process through the use of a piston crank mechanism. This combustion process and the operation of the piston crank mechanism are the two primary sources of engine noise excitation.

6.3.2.1   Combustion induced noise

To satisfy performance and emission requirements, the combustion process takes place very rapidly over a relatively short period when the piston is close to the top of its travel. For the four stroke cycle this rapid pressure rise in each cylinder is repeated every other rotation of the crankshaft and if analysed provides a full spectrum with harmonics at 1/2 engine rotational frequency (i.e., at fundamental firing frequency of one cylinder). The forcing spectrum is dominated by the low-frequency orders, but when combined with the stiffness-controlled response function of an engine it results in a spectral character controlled by the mid frequencies (500–3000 Hz). The character of the combustion spectrum is controlled by three clearly defined features (Figure 6.17). At low frequencies the spectral level is closely related to the peak cylinder pressure level (P), at mid frequencies to the rate of rise

(a)

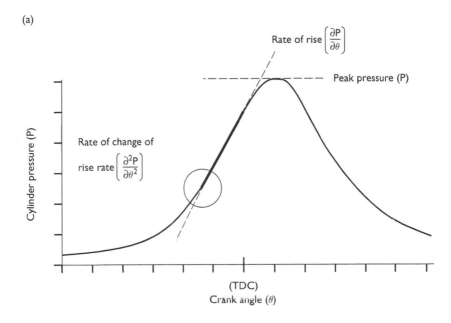

(b)

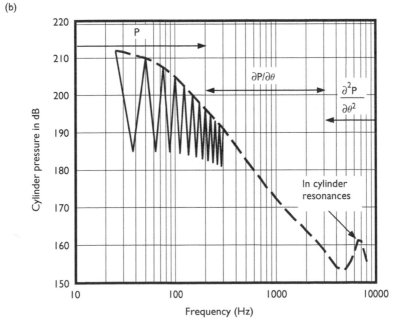

Figure 6.17 General relationship between the combustion diagram and its frequency spectrum: (a) typical fast burn combustion diagram; (b) typical fast burn combustion spectrum.

of cylinder pressure during combustion ($\partial P/\partial\theta$), whilst at high frequencies to the rate of change of pressure rise rate ($\partial^2 P/\partial\theta^2$), where θ is the crank angle. Of these three noise-controlling parameters, rate of pressure rise is arguably the most important due to its direct effect upon the characteristic mid-frequency hump that dominates most road vehicle engine noise spectra. The low-frequency region controlled by peak pressure should, however, not be forgotten, as these lower harmonics of excitation can be highly influential upon structure-borne engine noise.

As the rates of pressure rise increase, so the mid-frequency spectral slope decreases, and more high-frequency noise energy is generated (Figure 6.18). Normally aspirated direct injection diesel engines have the fastest rates of pressure rise that result in a spectral decay of around 30 dB per decade of frequency. The additional heat of turbocharging produces shorter combustion delay periods and therefore slower rates of pressure rise; this results in the spectral decay falling to around 40 dB per decade for turbocharged diesel engines. The slower, more progressive combustion of the petrol engine provides a spectral decay of around 50 dB per decade. For similar peak pressures, the steeper the spectral slope, the lower the combustion noise in the important mid-frequency region (Waters and Priede 1972).

The combustion pressure spectrum acts fairly uniformly over the full areas of the piston and cylinder head. The resulting forces impart vibrational energy into the cylinder block structure at the main bearings via the connecting rod and crankshaft and to a lesser extent through the relatively stiff cylinder head. As a large proportion of the combustion excitation occurs over a short period at around top dead centre, when the connecting rod is vertical and the bearing oil films are extremely stiff due to compression forces, the system can be reliably modelled using linear methods and solved in the frequency domain.

For reasons of economy and emissions, most modern diesel engines have adopted both exceedingly high pressure, electronically controlled injection systems as well as turbochargers. Fortunately, these two enforced trends are beneficial for combustion noise. The precise control given by modern fuel systems allows the injection to be tailored to give relatively slow pressure rise rates, and thus low combustion noise. The best injection shape for noise is seldom best for emissions, but modern injection systems allow considerable potential for trade-off optimisation. Although turbocharging can be highly beneficial for full load combustion noise, it can have detrimental side effects in terms of increased mechanical noise and in some cases turbocharger whine.

6.3.2.2 Mechanically induced noise

The other primary noise source within a reciprocating engine is that produced by the piston-crank mechanism as components traverse running

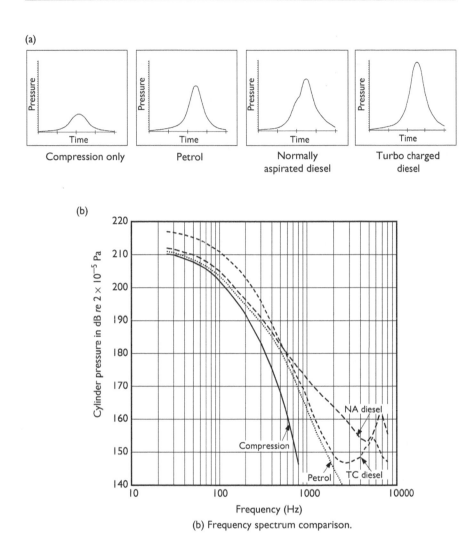

(a)

Compression only

Petrol

Normally
aspirated diesel

Turbo charged
diesel

(b) Frequency spectrum comparison.

Figure 6.18 Time and frequency comparison of various combustion processes: (a) pressure time history comparison; (b) frequency spectrum comparison.

clearances and create impacts (Lalor *et al.* 1980). Such impacts occur at main bearings, big end bearings and, more especially, between the pistons and their respective liners, commonly known as piston slap. Being of a sharp impulsive nature, most mechanical excitation has a far flatter spectral character than that of combustion, with considerably greater high-frequency content.

In the case of piston slap, the side force on the piston changes direction a number of times throughout its 720 degree operating cycle (Figure 6.19),

(a)

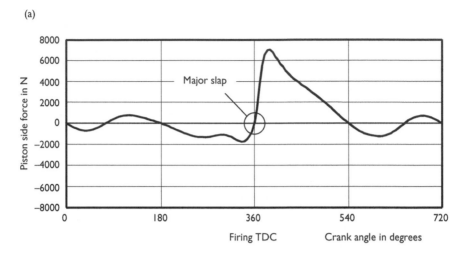

(b)

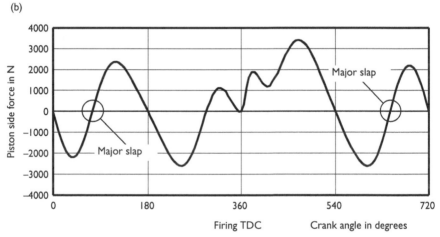

Figure 6.19 Comparison of piston side force diagrams for a large diesel and a medium-sized petrol engine: (a) a large diesel engine at 1500 rev/min; (b) a medium sized petrol engine at 5000 rev/min.

the number and location of the reversals being dependent upon speed and load. In large, slow-speed diesel engines, the biggest impact usually takes place around the firing top dead centre, and is principally controlled by combustion forces. In high-speed petrol engines the mechanical excitation is often more controlled by inertia forces and therefore the biggest impacts often occur at other locations in the cycle.

With the move towards turbocharging and the consequential increase in combustion pressures, mechanical forcing within larger diesel engines has increased in recent years. This has been somewhat countered by reduced clearances, small geometry changes and gudgeon pin offsets (Kamp and Spermann 1995); but in the majority of road vehicle applications, diesel engine noise is now mechanically controlled at full load. Smaller engines, especially petrol engines where mid-to-high-speed mechanical noise is more controlled by inertia forces, have seen recent improvements through significant weight reduction of moving parts and tighter control of clearances (Parker et al. 1986).

The incoming direct injection petrol (GDI) technology brings with it increased mechanical noise under conditions of light load and low speed due to its inherent high combustion pressures through lack of throttling (Bowden et al. 2000). Although seldom an absolute noise level issue, GDI engines can have a distinctly mechanical noise quality at low speed idle.

6.3.2.3 Secondary mechanical noise sources

There are a number of secondary mechanical noise sources within an engine which need to be considered, but they are seldom major contributors to overall noise if correct design strategies are adopted. These sources and their broad speed (N) or clearance (C) dependencies are listed in Table 6.2 in order of their common occurrence as noise concerns. Most of these secondary noise sources are principally speed controlled with little dependence upon engine load (with the exception of fuel injection systems). Although the speed dependency of broad band noise sources is clearly seen, those of many of the tonal sources are often masked by the resonant response of the engine structure and its covers.

Table 6.2 General characteristics of secondary mechanical noise sources with speed (N) and clearance (C)

Valve gear	Low speed rattle due to impacts ($\propto N^2$), high speed general noise sensitive to cam profile ($\propto N^{6-8}$)
Drive system	Whine at tooth pass frequency ($\propto N^{2-6}$), low speed rattle due to reversals across clearances ($\propto C^{2/3}$)
Diesel injection system	Low speed impulsiveness due to rapid pressure rise rates ($\propto N^2$), general noise of mechanical components ($\propto N^4$)
Turbocharger	Whine at shaft and blade pass frequency ($\sim \propto N^2$)
Oil pump	Whine at vane, gear or plunger pass frequency ($\propto N^{4-8}$)
Alternator	Whine at fan blade and pole pass frequency ($\propto N^6$)
Cooling fan	Hum at blade pass frequencies, high-speed air rush ($\propto N^6$)
Ancillary pumps	Whine at plunger frequency ($\propto N^{2-4}$)

6.3.2.4   Structural response

Internal combustion engines would be far noisier were it not for the considerable attenuation provided by the engine structure and its covers. Overall sound pressure levels in excess of 210 dB are typically generated within the combustion chamber of an engine, whereas noise measured at one metre from most modern engines is below 100 dB(A). All automotive engines have very similar structural response characteristics, with a steadily increasing accelerance (see Chapter 9) over much of the frequency range of interest (Priede *et al.* 1984). For both in-line and Vee configuration engines, the general mid-frequency slope of this accelerance is always close to 60 dB/decade (Figure 6.20).

Although an engine's structural response is complex, involving numerous modes, much of the modal activity over the frequency range of interest falls into two distinct categories. At low frequencies, the cylinder block and head respond as a beam having clear fundamental bending modes in the two planes, as well as a torsional mode. At higher frequencies the individual

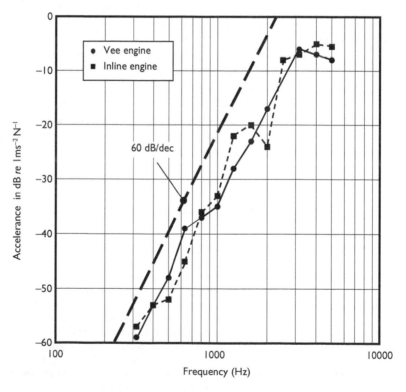

Figure 6.20 Examples of engine structural response to combustion forcing.

panel modes, first of the crankcase and then of the cylinder block, become dominant. Higher order beam modes also occur, but they are often lost in this broad family of panel modes.

As a major proportion of the forcing enters the engine structure through the main bearings, the greatest effort has been focused upon stiffening the crankcase or lower structure of the engine. Bed plates, ladder frames and bearing beams were being successfully demonstrated in the 1960s (Priede *et al.* 1964), but it is only fairly recently that these concepts have been commonly seen in production engines (Figure 6.21). An engine with a well-designed structural bottom end is around 2.5 dB quieter than a traditional deep skirt engine of similar size and configuration. Stiff bottom ends can also control the low-frequency coupling with the crankshaft modes, thereby reducing 1/2 orders and 'rumble' and thus resulting in improved sound quality (Teramoto *et al.* 1990).

In general, the covers attached to the main engine structure function as oil (and sometimes water) retainers and they rarely serve any structural purpose (with the possible exception of sumps). They are designed to be cheap, light and robust, but unfortunately they are often significant radiators of engine noise, and can contribute up to 50 per cent of overall engine sound power. Isolation of the covers can be a successful means of reducing their radiated noise; however, practical issues, especially sealing, often mean that the covers' natural mounting frequencies are too high, which can sometimes even result in a general amplification rather than attenuation. With recent developments in bonding processes which allow sealing and isolation mechanisms to be separated, isolated covers still hold considerable potential for noise control. The use of alternative materials for covers (for example, composites and magnesium) has demonstrated how difficult it is to beat the mass law, with a general increase in airborne transmitted noise through the cover being seen with the lighter designs of cover. The structure-borne noise control performance of these new material covers is not so clear cut, and appears to be very engine specific; but the general guideline of mismatching the impedances of the cover and its supporting structure still appears to be valid.

6.3.2.5 Parameters affecting engine noise

Although a reciprocating engine is a highly complex piece of machinery, which continues to be a challenge to all who attempt to model its noise-producing properties, it does appear to have only a few dominant noise-controlling parameters. The noise of all engines is dependent upon speed. This speed dependency ranges from around N^2 to N^6 and is closely linked to combustion process and operating conditions with the full load sound power of normally aspirated direct injection diesels being proportional to N^3, that of turbocharged direct injection diesels to N^4, whilst petrol engines are close to N^5.

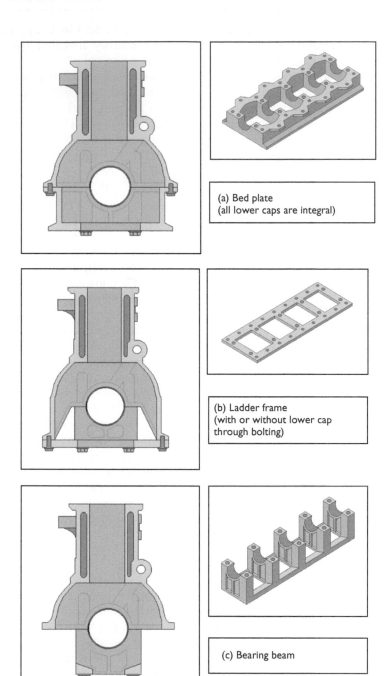

(a) Bed plate
(all lower caps are integral)

(b) Ladder frame
(with or without lower cap
through bolting)

(c) Bearing beam

Figure 6.21 Examples of stiff crankcase concepts.

Engine size is also a significant noise-controlling parameter, with sound power increasing with cylinder capacity to the power of 1.7 for large diesel engines through to 2.3 for passenger car size petrol engines. As sound power is proportional to the number of cylinders, it can be seen that, for a given overall engine capacity, the greater the number of cylinders (with correspondingly smaller individual capacity) the lower the engine noise. For example, for a petrol engine

$$\text{Sound power level} \propto 10 \log_{10} (\text{no. of cylinders})$$
$$+ 23 \log_{10} (\text{cylinder capacity}) \, \text{dB} \tag{6.5}$$

which indicates that the noise of a six-cylinder engine will be around 2.3 dB lower than that of a four-cylinder engine of the same overall capacity. Engine load has a surprisingly small effect upon radiated noise at rated speed. Petrol engines are typically only 2–3 dB noisier at full load, due to the high dominance of their inertia-controlled excitation. Conversely, turbocharged direct injection engines, are around 1–2 dB quieter at full load due to their potentially dominant combustion noise being smoothed out with the boosted hot air charge. Load effects can be very different for conditions away from rated speed; for example petrol engines at low speed can be as much as 10 dB noisier at full load than at no load due to the peak pressure controlling the dominant mechanical excitation.

6.3.2.6   Prediction of engine noise

Based upon the noise-controlling parameters outlined in Section 6.3.2.5, the following simple empirical noise prediction formulae for rated conditions have been developed for the various engine types (Anderton 1992; Lalor and Dixon 1994).

NA DI Diesel $\quad\quad$ $dB(A) = 30 \log_{10} N + 50 \log_{10} B - 108$ $\quad$ (6.6)

Large TC DI Diesel $\quad$ $dB(A) = 40 \log_{10} N + 50 \log_{10} B - 137.5$ $\quad$ (6.7)

Petrol $\quad\quad\quad$ $dB(A) = 50 \log_{10} N + 30 \log_{10} B$
$$+ 40 \log_{10} S - 223.5 \tag{6.8}$$

where $\quad$ $dB(A) =$ Average overall A-weighted sound pressure level of a four-cylinder engine measured at 1 m
$\quad\quad\quad\quad$ $N =$ engine speed in rev/min
$\quad\quad\quad\quad$ $B =$ bore in mm
$\quad\quad\quad\quad$ $S =$ stroke in mm

For engines with more or less than four cylinders, $10\log_{10}$ [number of cylinders/4] should be added to the prediction.

These empirical formulae represent the mean fits through large data sets, and therefore they will only predict the noise of the average engine, or what has been termed the 'noise potential' of an engine. A specific prediction may be further refined by the following additional design modifiers:

Structural bottom end	–	subtract up to 2.5 dB
Low-noise covers	–	subtract up to 3 dB
No intermediate main bearings	–	add 1.5 dB
Vee type engines	–	add up to 1.5 dB
Pressure rise rates above 2 bar/deg	–	add up to 5 dB

Although this simple empirically based prediction method can be of immense use at the design concept stage or for comparing competitor engines, it is of little benefit to the detailed development process.

To assist the refinement development of a design concept through to production and beyond, the near universally accepted approach is to use a full detailed model of the engine (Erotokritos *et al.* 1995). The structure, covers and moving components are modelled using finite element techniques; forces are either measured or predicted using dynamic models of the moving components; and the resulting noise is derived either from the average surface vibration or by using boundary element methods. Such structural modelling has, over the years, matured into a fairly robust and reliable process (Khan and Cook 1990), but predicting the dynamic performance of the numerous oil films present within an engine still provides the analyst with major challenges. The prediction of structural damping is again far from an exact science, and it is accepted that empirical values for damping are more reliable. The future of engine noise modelling may combine the design detail of the structural modelling, with the good absolute level accuracy of the empirical prediction approach, providing a hybrid method with immense development power (Dixon *et al.* 2000).

6.3.2.7 Low-frequency excitation

The low-frequency forcing of a powertrain up to around 500 Hz comprises a combination of a number of different mechanisms including simple component out-of-balance. However, the three main contributions are the inherent imbalance of the engine's crank mechanism, the combined inertia and gas force torque fluctuations of the crank mechanisms and the inertia imbalance of the valve train. As a crank rotates at constant angular velocity ω, so the piston reciprocates accordingly; but its motion is not perfectly sinusoidal and a Fourier analysis of the piston displacement reveals components at $\theta, 2\theta, 4\theta, 6\theta$, etc., where θ is the crank angle. Because these harmonic components decay exceedingly rapidly, orders above 2nd are rarely considered, and therefore the out-of-balance forcing F for a single piston may be approximated by

$$F = M\omega^2 r \cos\theta + \frac{M\omega^2\ r^2}{\ell}\cos 2\theta \qquad (6.9)$$

where $F.$ = Piston axis out-of-balance force (N)
M = Mass of piston and reciprocating component of connecting
rod (kg)
ω = Crankshaft angular velocity (rads/sec)
r = Crank radius (m)
l = Connecting rod length (m)
θ = Crank angle relative to top dead centre

The second term in equation (6.9), which is commonly called the secondary out-of-balance, is smaller than the first term, or primary out-of-balance, by the ratio of the crank radius to the connecting rod length (r/ℓ). For most automotive applications (r/ℓ) is in the range of 0.33–0.25. Thus the magnitude of the 2nd order forcing for a single piston is between 1/3 and 1/4 of the primary forcing, or some 10–12 dB lower. In multi-cylinder engines, the relative importance of the primary and secondary out-of-balance forces and couples is determined by the specific engine configuration (Heizler 1995). As can be seen in Table 6.3, the forcing due to inertia out-of-balance generally becomes less of an issue as the number of cylinders increases. Also, due to the inherent bank stagger, Vee configurations generally have more out-of-balance issues than in-line configurations. But these two statements are gross generalisations, and perhaps the engine with one of the

Table 6.3 Inertia forces and couples for the more common engine
configurations

Cylinder configuration	Primary force	Primary couple	Secondary force	Secondary couple
I2 (180°)	–	✓	✓	–
I3	–	✓	–	✓
I4	–	–	✓	–
I5	–	✓	–	✓
I6	–	–	–	–
V2 (90°)	✓	✓	✓	✓
V4 (90°)	–	✓	✓	✓
V6 (60°)	–	(✓)	–	✓
V8 (90°)	–	(✓)	–	–
V12 (60°)	–	–	–	–

Note
(✓) – Can be removed by crankshaft counterweighting.

biggest out-of-balance problems is the ever popular in-line four configuration that has a secondary out-of-balance force of considerable magnitude. A two-litre, in-line, four-cylinder, petrol engine at 6000 rev/min will produce secondary out-of-balance force of around 15 kN. To cancel in full, or in part, some of these inherent forces and couples, rotating out-of-balance weights are used as balancing devices. For second order issues, these balancers have to be driven at twice engine rotational speed. To cancel the second order vertical force of the in-line four-cylinder engine, two counter rotating balance weights may be used (as in the Lanchester balancer).

Although piston and crank inertia imbalance is totally independent of combustion forces, there is a second low-frequency excitation forcing due to the piston-crank mechanism that is considerably influenced by combustion, namely torque fluctuations or reversals (Rhodes *et al.* 1988). The reciprocating inertia forces combine with the combustion forces and, through their reaction with the connecting rod, create a piston side force and a corresponding equal and opposite main bearing force (Figure 6.22). As this applied couple varies and reverses throughout the engine cycle, a rolling motion is applied to the engine.

As with inertia imbalance so, in general, the couples or torque fluctuations decrease as the number of cylinders increases; but unlike inertia imbalance, the fluctuations are totally independent of cylinder configuration on the assumption of even firing intervals. Also, unlike inertia imbalance, the magnitude of these torque fluctuations is very load dependent especially at low speed due to the combustion term. At high engine speeds, the fluctuations are more controlled by the inertia-generated term. It is worth noting that, although the inertia-generated term of the torque fluctuations is directly related to the reciprocating force, it is far more harmonically dense due to the crank mechanism geometry.

In a similar manner to the piston-crank mechanism, the valve train can also produce inherent out-of-balance forces (Dixon 1990). As each valve is opened and closed, the force required to accelerate and decelerate its mass is reacted by the engine. When compared with the reacted forces due to piston inertia, these valve inertias may appear small, but they can combine in unfortunate ways to produce significant vibration. Six-cylinder engines suffer badly from this effect with in-line six-cylinder engines exhibiting an appreciable pitch at $1\frac{1}{2}$ order and the V6 a transverse motion again at $1\frac{1}{2}$ order. V8s have significant forces and couples at $\frac{1}{2}$, 1 and $1\frac{1}{2}$ order. There are a number of ways of reducing this forcing, but the best solution is to start with a balanced valve train concept that incorporates a rocker layout to counter balance the inertia of the valves.

Most, if not all, of this low-frequency forcing enters the vehicle via the engine mounts and other structure-borne routes (see Section 6.3.1.2). However, as mounting systems have dramatically improved over the past 40 years, so the once insignificant low-frequency airborne noise is now

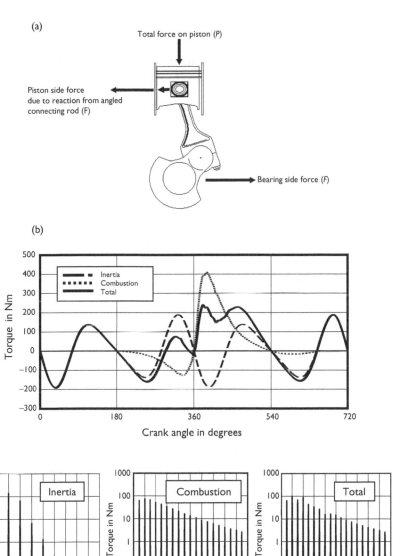

Figure 6.22 Typical torque fluctuations for a single cylinder engine: (a) generation of torque fluctuations through opposite forces at the liner and main bearings; (b) torque time histories; (c) torque spectra.

beginning to surface as a minor, yet measurable, source (Dixon and Phillips 1998). At these low frequencies, the power unit vibrates as a solid body on its soft mounts and may be considered as a simple dipole source. Clearly the radiation efficiency of the powertrain at these low frequencies is very poor, but so is the vehicle's acoustic transmission loss, so that for an in-line four-cylinder engine the typical net effect can be that the airborne noise comes within 5 dB of the structure-borne noise at certain speeds and loads. Unlike the complex mid- and high-frequency noise of an engine, this low-frequency excitation is relatively easy to model and accurately predict using simple mathematical representation; but its control is no less hard to implement and in many cases suffers more from trade-off compromise than the higher frequency mechanisms.

6.3.3   Intake and exhaust noise

Intake gas noise is generated when the fast moving incoming air charge is abruptly halted by the closing of the inlet valves. The resulting pressure pulse then propagates against the mean air flow back to the system's inlet orifice (or snorkel) where it radiates as noise. In a similar way, exhaust gas noise is generated when the exhaust valve opens and rapidly releases the residual pressure from the combustion process. The pressure pulse then propagates down the system with the flow, which can be relatively high at around Mach 0.3 at full load, and then radiates from the orifice (or tailpipe) as noise. Both intake and exhaust noise have predominantly low-frequency harmonic spectral characteristics, with exhaust noise often having additional broad-band-high-frequency-flow-induced noise at higher engine speeds and loads. Although base engine design features (e.g. valve size, cam profile) are very influential upon the generation of intake and exhaust noise, they are seldom allowed to be optimised for low noise; indeed they are generally optimised for high performance which often results in high noise levels. Therefore almost all of the required control of intake and exhaust noise is achieved by traditional attenuation methods.

6.3.3.1   System attenuation

Most intake and exhaust systems comprise a combination of various reactive and dispersive attenuation elements, the more common of which are listed in Table 6.4. With the pressure of legislation to reduce total vehicle noise, and with the ever increasing performance requirements from engines, intake and exhaust systems have become progressively more complex. The simple air cleaner expansion chamber that served the industry well for nearly a century, has now been replaced by multi-chamber systems often incorporating two or three quarter-wave or Helmholtz resonators. Exhaust

Table 6.4 Commonly used attenuation elements in both intake and exhaust systems

Attenuation element	Attenuation characteristics		Comment
Packed (dissipative)	Broad band	att / f	Low backpressure
Expansion chamber	Broad band	att / f	Max attenuation $\simeq \left(\dfrac{d_2^2}{2d_1^2} \right)^2$
1/4 Wave side branch	Narrow band (with odd order harmonics)	att / $\dfrac{c_0}{4L}$ f	More commonly used for higher frequencies
Helmholtz resonator S = area of neck L = length of neck	Narrow band	att / $\dfrac{c_0}{2}\sqrt{\dfrac{S}{LV}}$ f	More commonly used for lower frequencies
Cross flow and perforates	Various but generally broad band		Prone to high-frequency gas rush noise at high flows

systems have developed from a single silencer to two, or even three, separate boxes. In general, attenuation produces back pressure; and to minimise back pressure, but retain attenuation, system volume has to be increased. Thus the success or failure of a vehicle's intake or exhaust system is often determined at the vehicle concept stage when system packaging volume is being fought for.

Although the majority of the attenuation concepts used today have been around for many years, there are a few recent developments that have had an impact on system design. The exhaust catalytic converter that is now a feature on many classes of vehicles, is in itself a worthwhile attenuating element (Selamet *et al.* 2000), with an expansion and a contraction as well as through-matrix flow. The increasingly used turbocharger is also a useful attenuating element; however the physics behind its acoustic performance is not yet fully understood.

On the intake side, perhaps one of the more interesting developments has been the acoustically porous (dirty side) inlet duct. This concept originally met with scepticism, as it was felt to 'cheat' snorkel targets by releasing some of the noise energy prior to the orifice, but now it is becoming more accepted as a valid noise control measure. By releasing the noise energy progressively, the noise no longer comes from a point source, but from a line source, thus giving potential for acoustic interference noise cancellation. The porous duct can also be designed to be a reflective impedance in much the same way as traditional pulse holes, thus truly reducing intake levels. Active control of both intake and exhaust has been successfully demonstrated (Tanaka *et al.* 1995), but durability and cost issues remain a challenge.

The sound quality of a vehicle can be significantly influenced by its intake and exhaust noise (Naylor and Willats 2000). For example, the tailpipe length downstream from the silencer can greatly affect the sound quality of both the acceleration and overrun exhaust noise of a vehicle. Interior sound quality is best engineered using the induction system, with small spectral changes to the attenuation performance of the system greatly changing the noise character of the vehicle. Although often seen as a requirement, not all preferred intake or exhaust sound quality necessarily has to be maintained over the complete engine speed range. A 'sporty' sound cutting in at a certain speed provides both enjoyable anticipation as well as contrast.

Fortunately, most intake and exhaust noise problems occur at low frequencies (below 500 Hz), which, in most cases, are below the lowest acoustic cross-mode cut-off frequencies of the system's pipes and chambers. The acoustics of the intake and exhaust systems may therefore be success-fully modelled using one-dimensional theory (Davies and Harrison 1997). Despite the very high pressure levels present within an exhaust system ($\sim$185 dB) which would suggest considerable non-linearity, simple linear codes do provide surprisingly accurate predictions.

6.3.4 Wind noise

Wind noise is generated in a number of different ways (see Chapter 7) and, of all the vehicle noise sources considered, wind noise has the highest dependency on road speed V. Road vehicle wind noise has the character of a dipole source, with its sound power level increasing as V^6 over the speed range of interest (see Section 6.2.5). It is therefore a high speed phenomenon and only becomes an issue in passenger cars above about 110 km h^{-1}. Wind noise is not currently controlled by legislation, and for this reason probably varies considerably in level from vehicle to vehicle. Of all the major vehicle noise sources, it has received the least research effort.

Arguably the most fundamental and ultimately challenging of these aero-acoustic sources are those generated by the basic form of the vehicle moving

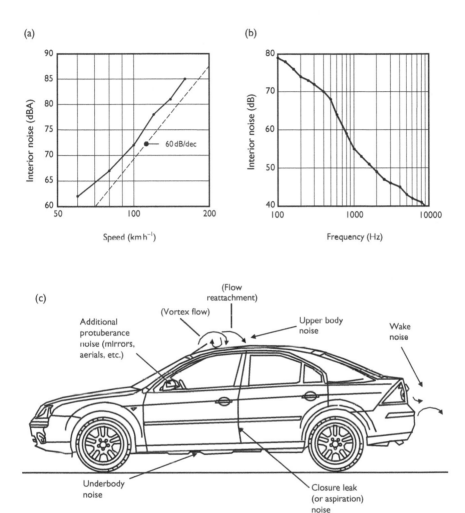

Figure 6.23 Characteristics and sources of wind noise: (a) typical effect of speed on interior wind noise for a family saloon; (b) typical spectrum of interior wind noise at 100 km h^{-1} for a family saloon; (c) sources of wind noise.

through the air (Figure 6.23). At mid-to-high vehicle speeds the airflow over the vehicle creates a turbulent boundary layer over much of its outer surface (George 1989). The inherent pressure fluctuations of this boundary layer impinge upon the body structure and thus generate noise within the vehicle. However, these pressure fluctuations can be greatly increased if the flow separates from the surface and then reattaches to the vehicle body downstream. The pressure fluctuations can be further enhanced if this separation induces strong vortex flow between separation and reattachment. Although all part of the same general mechanism, the vortex flow generation has greater high-frequency content than that due to reattachment (Stapleford and Carr 1970). The mechanism by which this pressure fluctuation energy enters the vehicle interior can be greatly affected by the local vehicle response. Although the general character of the attenuation closely follows the mass law above 200 Hz, there can be significant local differences, making the interior noise potentially sensitive to vortex locations and their corresponding reattachment lines.

The aeroacoustic generation mechanisms of turbulence, reattachment and vortex flow are the same over both the upperbody of the vehicle and its underbody; but in design terms, it is sometimes convenient to consider these two zones separately. The upperbody noise can be highly influenced by the style or fashion of the day, and the current trend for smooth body forms with faired trim is in most cases good for reducing wind noise. Styling rarely influences the underbody form of a vehicle, which, when compared with the upperbody, is a mass of abrupt shapes, cavities and close fitting protuberances. The increasing use of engine bay undershields slightly improves the aerodynamics, but the underbody of road vehicles is far from ideal. Unlike the upperbody, the underbody wind noise is generated in a semi-reverberant environment due to the reflectivity of the road surface. A clear resonance can often be seen in a vehicle's underbody wind noise at a frequency at which the mean body-ground clearance corresponds to half an acoustic wavelength. When the upperbody and underbody airflow meet at the rear of the vehicle, considerable wake-induced noise is generated. This wake noise has a significant low-frequency content, which can easily enter the vehicle via the rear window due to its relatively low acoustic transmission loss, when compared with that of the boot (trunk) zone (Dobrzynzki 1986).

Many wind noise problems are related to closure (or door) surrounds, and in general, door sealing design usually follows a route of compromise. Substantial, often compound, seals are required to minimise wind noise intrusion. Unfortunately such sealing arrangements can often result in heavy and poor door closure quality. Creative ways to overcome this compromise involve powered final closure and door geometry that allows gravity assistance during closure. The acoustic transmission performance of a well-sealed door at rest can become significantly degraded at high speed as the (often structurally weak) panels deflect under aerodynamic forces and distort sealing arrangements. Seal geometry can be distorted to such an extent that small leaks occur allowing air flow (or aspiration) noise to be generated (Peng and Morrey 1998). Closure surrounds can also be the generator of tonal noise as the flow across the small gap excites the resonance(s) of the gap cavity, the frequency of the sound produced being related to the dimensions of the cavity.

High levels of vortex noise can also be generated by additional protuberances such as wing mirrors, wiper blades and communication aerials. This noise then propagates through the air and enters the vehicle, being attenuated by the mass-governed transmission loss of the body and its trim. In addition, vortex shedding can excite the structure of the protuberances that are often attached solidly to the vehicle's body, thereby generating significant structure-borne interior noise. It is usually of a fairly narrow band nature, and for regular geometries such as a rod aerial can have a strong tonal character. As door seal design is improved, aerials are located within the vehicle's surface, wiper blades are parked in faired enclosures and even wing mirrors are replaced by video systems, so the major challenge continues to be the understanding and prediction of the wind noise created by both the 'clean' upperbody and the 'dirty' underbody of the vehicle.

6.3.5 Tyre noise

Tyre noise is a major source of both vehicle interior and exterior (drive-by) noise. Airborne tyre sound power increases at between V^3 and V^4, while structure-borne noise levels increase at between V and V^2. These rates of increase are somewhat sensitive to both tyre design and road surface (Underwood 1980). The rate is also affected by the presence of water on the road with an increase of up to 10 dB at low speed, but appreciably less at high speed. The absolute level of tyre noise at any one speed can range over 10 dB depending upon the tyre design and road surface type. Unfortunately there is, as yet, no tyre design concept that is universally quiet on all road surfaces.

Tyre noise is principally generated by air pumping from the tread cavities and by various vibration-inducing mechanisms at the contact patch (Figure 6.24a). These generation mechanisms may then be modified by the resonant characteristics of the tyre carcass and the air both inside and around the tyre (Figure 6.24b).

6.3.5.1 Air pumping

When a segment of a rolling tyre comes into contact with the road surface and takes weight, the tread deflects and air is squeezed out of the road and tyre tread voids. When the segment leaves the contact area the reverse occurs, the road and tread cavities return to their original dimensions, and air is now sucked back into the increasing volumes. This air pumping action has been shown to be a significant noise generation mechanism (Hayden 1971) and may be considered as a simple monopole source. The fourth power speed dependency associated with a monopole agrees well with the observed high-frequency characteristics of tyre noise. Tread pumping is dominant on a smooth road, while surface pumping, where the tyre enters and leaves the road cavities, is dominant with a smooth soft tyre (Kanaizumi 1971). The level

(a)

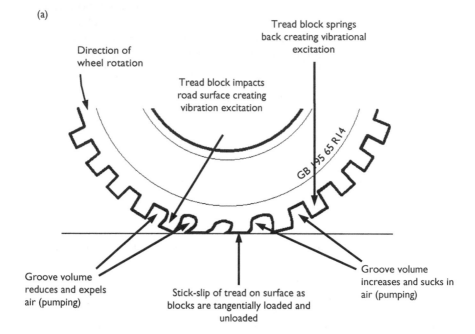

Direction of
wheel rotation

Tread block impacts
road surface creating
vibration excitation

Tread block springs
back creating vibrational
excitation

GB 195 65 R14

Groove volume
reduces and expels
air (pumping)

Stick-slip of tread on surface as
blocks are tangentially loaded and
unloaded

Groove volume
increases and sucks in
air (pumping)

(b)

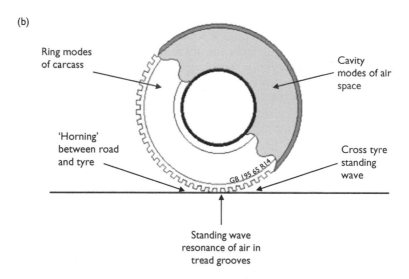

Ring modes
of carcass

Cavity
modes of air
space

'Horning'
between road
and tyre

Cross tyre
standing
wave

GB 195 65 R14

Standing wave
resonance of air in
tread grooves

Figure 6.24 Tyre noise generation and amplification: (a) tyre noise generation; (b) tyre
noise amplification.

of noise from this surface pumping mechanisms is related to the size of the cavities in both the tyre and the road, the load on the tyre, the pressure in the tyre and the smoothness of the road (Favre and Pachiaudi 1974).

6.3.5.2 Vibration of the tyre

Irregularities in both the tread pattern of the tyre and the texture of the road surface create impacts that excite the tyre. The resulting vibration of the carcass generates directly radiated airborne noise as well as entering the vehicle via the suspension as structure-borne noise. The tyre vibration is the result of both radial and tangential excitation. Radial impact excitation is principally controlled by road surface roughness (Desconet and Sandberg 1980) and is commonly believed to be the major cause of low-frequency excitation. Tangential excitation occurs through both impact and slip-stick mechanisms. As a tread element enters the contact patch it impacts with the road and is deflected. When that tread element is released from the contact patch it springs back to its relaxed position and again excites the tyre carcass. Especially at higher speeds and higher wheel torques, the rubber at the contact patch alternately sticks to the road surface and then slips. This mechanism has been shown to be a significant mid-to-high-frequency noise generator (Underwood 1981) and is termed 'stick-slip' noise.

6.3.5.3 Tyre noise amplifiers

The various tyre noise generation mechanisms can be amplified by a number of structural and acoustic resonances in and around the tyre. Low order ring modes of the tyre carcass can be excited and clear modal behaviour around the tyre may be seen up to around 300 Hz. At higher frequencies, the tyre response becomes progressively more localised around the contact patch. The cavity resonances of the toroidal air space within the tyre can also amplify tyre noise in the mid-to-high-frequency range. On the outside of the tyre, the closed and full width tread grooves can act as quarter-wave and half-wave resonators respectively, amplifying mid- and high-frequency sound. The geometry at the entrance and exit to the contact patch can provide both an acoustic horn effect (termed 'horning') and a cross tyre half-wave resonance behaviour.

6.3.5.4 Design parameters affecting tyre noise

Although the absolute level of tyre noise is controlled by its interaction with a particular road surface, there are a number of general tyre design parameters that are influential upon the noise generated at this interaction. Table 6.5 shows that a number of design features can have measurable impact on tyre noise; unfortunately, many of these noise-reducing guidelines

Table 6.5 Tyre design parameters and features affecting noise generation

Tyre feature	Requirement for low noise	Beneficial effect	Detrimental effect
Tread block	Randomised (or optimised pattern)	Reduces tonal noise	Small if any
Cross tread groove angle	Diagonal (with rearward exits) rather than transverse	Reduces pumping	Can reduce wet performance
Tread groove volume	Small volume	Reduces pumping	Can reduce wet performance
Tread groove layout	As circumferential as possible	Reduces groove cavity closures	Can reduce wet performance
Aspect ratio – $\frac{\text{section height}}{\text{section width}}$ (×100)	High aspect ratio	0.7 (70%) is 2 dB (approx.) quieter than 0.5 (50%)	Can reduce cornering performance. Contrary to current fashion
Sectional width	Narrow section	$L_w \propto 30 \log$ width	Can reduce cornering performance. Contrary to current fashion
Diameter	Large diameter	Approx. 1 dB reduction for 25% increase	Can compromise styling
Load (commercial vehicle)	Low load	Up to 4 dB quieter when unloaded	Can reduce ride quality

have detrimental effects upon other aspects of the tyre's performance. The primary trade-off partner to tyre noise is the emotive subject of safety. Significant reduction in tyre noise could be achieved if very small compromises on skid performance were to be made. This scenario is clearly never going to be acceptable. For this reason, perhaps the most promising future for tyre noise reduction continues to remain with the design of the road surface. Considerable advances have been made in this area over the past 10 years, but this topic is outside the scope of this chapter.

6.3.6   Future trends

Although apparently informed statements have regularly been made over the past 40 years to the effect that the reciprocating internal combustion engine has just 10 more years left, it continues to outperform anything on the horizon by a substantial margin. Undoubtedly, vehicles with alternative power sources will arrive, and whether it be a simple electrical battery or a fuel cell it is highly likely that it will be significantly quieter than the current combustion engines. However, broad band tyre and wind noise will still continue to challenge the NVH engineer, together with the rearguard action required to control the detrimental effects of the ever present desire to reduce vehicle weight and cost.

References

Anderton, D. (1992) Trends in IC engine noise generation and its effect on vehicle noise. *Proceedings of the 2nd International Conference of the Associazione Tecnica dell' Automobile on Vehicle Comfort*, Bologna, 2, 643–658.

Barsikow, B. (1996) Experiences with various configurations of microphone arrays used to locate sound sources on railway trains operated by DB-AG. *Journal of Sound and Vibration*, 193, 283–293.

Bowden, D., Derry, S. and Dixon, J. (2000) NVH characteristics of air-assisted direct injected (DI) spark ignition four-stroke engines. *Proceedings of the Institution of Mechanical Engineers*, London, C577/011/2000.

Davies, P. and Harrison, M. (1997) Predictive acoustic modelling applied to the control of intake/exhaust noise of internal combustion engines. *Journal of Sound and Vibration*, 202(2), 249–274.

de Beer, F., Janssens, M., Kooijman, P.P. and van Vliet, W.J. (2000) Curve squeal of railbound vehicles (part 1): frequency domain calculation model. *Proceedings of Internoise 2000*, Société Française d'Acoustique (SFA), Nice, 1560–1563.

Desconet, G. and Sandberg, U. (1980) Road surface influence on tire/road noise. *Proceedings of Internoise '80*, Noise Control Foundation, Miami.

Dixon, J. (1990) Six cylinder refinement – the cause and control of $1\frac{1}{2}$ order. *Proceedings of the Institution of Mechanical Engineers*, London, C420/002.

Dixon, J. and Phillips, A. (1998) Power unit low frequency airborne noise. *Institution of Mechanical Engineers, Transactions of European Conference on Vehicle Noise and Vibration*, London, C521/032/98, 71–75.

Dixon, J., Rhodes, D., Hughes, M. and Phillips, A.J. (1993) Recent developments in powertrain noise simulation. *Proceedings of Autotech 93*, Institution of Mechanical Engineers, C462/20.

Dixon, J., Baker, J. and Challen, B. (2000) A hybrid method for modelling engine noise. *Proceedings of the Institution of Mechanical Engineers*, London, C577/036/2000.

Dobrzynzki, W. (1986) Wind induced interior and farfield radiated exterior noise from automobiles. *Vehicle Aerodynamics – Von Karman Institute for Fluid Dynamics*, Lecture Series 1986-05.

Erotokritos, N., Dixon, J. and Lalor, N. (1995) Using computer models to design quiet engines. *Proceedings of the 3rd International Conference of the Associozione Tecnica dell' Automobile on Vehicle Comfort and Ergonomics*, Bologna, Paper 95A1004.

European Commission (1996) *Future Noise Policy*, European Commission Green Paper, COM(96) 540 final, Brussels, 4 November, CB-CO-96-548-EN-C.

Favre, B. and Pachiaudi, G. (1974) Bruit des pneumatiques, aspects théoriques et experimentaux. *Institut de Recherche des Transports*, Bron, France.

Fish, D. (1998) Evidence for the influence of presbycusis on the perception of vehicle noise. *Proceedings of Euro Noise 1998*, German Acoustical Society (DEGA), Munich, 509–514.

Fursdon, P., Harrison, A. and Stoten, D. (2000) The design and development of a self tuning active engine mount. *Proceedings of the Institution of Mechanical Engineers*, London, C577/018/2000.

George, A. (1989) Automobile Aeroacoustics. *Proceedings of the 12th Conference of the American Institute of Aeronautics and Astronautics on Aeroacoustics.* AIAA-89-1067, San Antonio.

Goddard, M. (1998) *Autocar*, 18 March, Road Test Number 4311.

Grassie, S.L. and Kalousek, J. (1993) Rail corrugation: Characteristics, causes and treatments. *Proceedings of the Institution of Mechanical Engineers*, 207F, 57–68.

Hardy, A.E.J. (1999) Railway passengers and noise. *Proceedings of the Institution of Mechanical Engineers*, 213F, 173–180.

Hardy, A.E.J. and Jones, R.R.K. (1989) Control of the noise environment for passengers in railway vehicles. *Proceedings of the Institution of Mechanical Engineers*, 203F, 79–85.

Hayden, R.E. (1971) Roadside noise from the interaction of a rolling tyre with the road surface. *Journal of the Aeronautical Society of America*, 50(1), Pt 1, 113(a).

Heckl, M.A. (2000) Curve squeal of train wheels, part 2: which wheel modes are prone to squeal? *Journal of Sound and Vibration*, 229, 695–707.

Heizler, H. (1995) *Advanced Engine Technology*. London: Arnold, ISBN 0340-568224, Chapter 3.

Hemsworth, B. (1979) Recent developments in wheel/rail noise research. *Journal of Sound and Vibration*, 66, 297–310.

Hemsworth, B., Gautier, P.E. and Jones, R. (2000) Silent freight and silent track projects. *Proceedings of Internoise 2000*, Société Française d'Acoustique (SFA), Nice, 714–719.

Hölzl, G. (1994) A quiet railway by noise optimised wheels (in German). *ZEV+DET Glas. Ann.*, 188, 20–23.

Hussain, M., Pflüger, M., Brandl, F. and Blermayer, W. (1998) Intercultral differences in annoyance response to vehicle interior noise. *Proceedings of Euro Noise 1998*, German Acoustical Society (DEGA) Munich, 521–526.

ISO 362 (1994) *Measurement of noise emitted by accelerating road vehicles – engineering method* (2nd edn), International Organization for Standardization, Geneva.

Jones, R.R.K. (1994) Bogie shrouds and low barriers could significantly reduce wheel/rail noise. *Railway Gazette International*, Reed Business Publishing, 459–462.

Kamp, H. and Spermann, J. (1995) New methods of evaluating and improving piston related noise in internal combustion engines. *Proceedings of SAE*, 951238.

Kanaizumi, A. (1971) Paved surface and driven motor vehicle noise. *Public Works Research Institute*, Ministry of Construction, Japan.

Khan, R. and Cook, G. (1990) Application for finite element analysis for improving powertrain noise and vibration. *Proceedings of the Institution of Mechanical Engineers*, London, C420/019.

King III, W.F. (1990) The components of wayside noise generated by tracked vehicles. *Proceedings of Internoise 1990*, Acoustic Society of Sweden, Gothenburg, 375–378.

King III, W.F. (1996) A precis of developments in the aeroacoustics of fast trains. *Journal of Sound and Vibration*, **193**, 349–358.

Lalor, N. and Dixon, J. (1994) Recent trends in I.C. engine noise control. *Proceedings of the 4th International Conference of the Associazione Tecnica dell' Automobile on new design frontiers for more efficient, reliable and economical vehicles*, 94A1118, Florence.

Lalor, N., Grover, E.C. and Priede, T. (1980) Engine noise due to mechanical impacts at pistons and bearings. *Proceedings of SAE*, 80402.

Mauclaire, B. (1990) Noise generated by high speed trains. New information acquired by SNCF in the field of acoustics owing to the high speed test programme. *Proceedings of Internoise 1990*, Acoustic Society of Sweden, Gothenburg, 371–374.

Meister, L. and Saurenman, H. (2000) Noise impacts from train whistles at highway/rail at-grade crossings. *Proceedings of Internoise 2000*, Société Française d'Acoustique (SFA), Nice, 1038–1043.

Naylor, S. and Willats, R. (2000) The development of a 'sports' tailpipe noise with predictions of its effect on interior vehicle sound quality. *Proceedings of the Institution of Mechanical Engineers, London*, C577/002/2000.

Newton, S.G. and Clark, R.A. (1979) An investigation into the dynamic effects on the track of wheelflats on railway vehicles. *Journal of Mechanical Engineering Science*, **21**, 287–297.

Parker, D., Richmond, J., Taylor, B., Auezou, J. and Bruni, L. (1986) The reduction of piston friction and noise. *AE Group Symposium*. Internal Publication.

Peng, C. and Morrey, D. (1998) An investigation into the effect of door seals on noise generated in the passenger compartment. *Proceedings of the Institution of Mechanical Engineers*, C521/020/98.

Priede, T., Austen, A. and Grover, E. (1964) Effect of engine structure on noise of diesel engines, *Proceedings of the Institution of Mechanical Engineers*, **179**(2A), 4, London.

Priede, T., Dixon, J., Grover, E. and Saleh, N. (1984) Experimental techniques leading to the better understanding of the origins of automotive engine noise. *Proceedings of the Institution of Mechanical Engineers*, C151/84.

Racca, R. (1982) How to select powertrain isolators for good performance and long service life. *Proceedings of SAE*, 821095.

Remington, P.J. (1987) Wheel/rail squeal and impact noise. What do we know? What don't we know? Where do we go from here? *Journal of Sound and Vibration*, **116**, 339–353.

Rhodes, D., Phillips, A. and Abbott, B. (1988) Prediction of engine vibration from combustion pressure measurements and examination of the effect of combustion variations in the frequency domain. *Proceedings of the Institution of Mechanical Engineers*, Birmingham, C28/88.

Russell, M.F. (1993) An objective approach to vehicle assessments. *Proceedings of Autotech '93, Recent Advances in NVH Technology*, Birmingham. Institution of Mechanical Engineers, C462/36/205.

Selamet, A., Kothamasu, V., Novak, J. and Kach, R. (2000) Experimental investigation of in-duct insertion loss of catalysts in internal combustion engines. *Applied Acoustics*, **60**, 451–487.

Stapleford, W. and Carr, G. (1970) Aerodynamic noise in road vehicles, Part 1 'The relationship between aerodynamic noise and the nature of airflow'. *Motor Industry Research Association*, Report 1971/2.

Steel, J. (1998) A study of engine noise transmission using statistical energy analysis. *Proceedings of the Institution of Mechanical Engineers*, London, **212**, Part D, 205–213.

Talotte, C. (2000) Aerodynamic noise, a critical survey. *Journal of Sound and Vibration*, **231**, 549–562.

Tanaka, K., Nishio, Y., Kohama, T. and Ohara, K. (1995) Technological development for active control of air induction noise. *Proceedings of SAE*, 951301.

Teramoto, T., Deguchi, H. and Shintani, H. (1990) Improvement in vehicle interior sound quality by newly developed power plant members. *Proceedings of the Institution of Mechanical Engineers*, London, C420/027.

Thompson, D.J. (1993a) Wheel–rail noise generation, Part I: Introduction and interaction model. *Journal of Sound and Vibration*, **161**, 387–400.

Thompson, D.J. (1993b) Wheel–rail noise generation, Part II: Wheel vibration. *Journal of Sound and Vibration*, **161**, 401–419.

Thompson, D.J. (1993c) Wheel–rail noise generation, Part V: Inclusion of wheel rotation. *Journal of Sound and Vibration*, **161**, 467–482.

Thompson, D.J. (1997) Experimental analysis of wave propagation in railway track. *Journal of Sound and Vibration*, **203**(5), 867–888.

Thompson, D.J., Hemsworth, B. and Vincent, N. (1996) Experimental validation of the TWINS prediction program for rolling noise, Part 1: Description of the model and method. *Journal of Sound and Vibration*, **193**, 123–135.

Thompson, D.J., Jones, C.J.C., Wu, T.X. and de France, G. (1999) The influence of the non-linear stiffness behaviour of rail pads on the track component of rolling noise. *Proceedings of the Institution of Mechanical Engineers*, 213F, 233–241.

Thompson, D.J., Jones, C.J.C. and Farrington, D. (2000) The development of a rail damping device for reducing noise from railway track. *Internoise 2000*, Société Française d'Acoustique (SFA), Nice, France, 685–690.

Underwood, M. (1980) *The Origins of Tyre Noise*. Institute of Sound and Vibration Research, University of Southampton, PhD Thesis.

Underwood, M.C.P. (1981) *Lorry Tyre Noise*, Transport and Road Research Laboratory, Report 974.

Vér, I.L., Ventres, C.S. and Myles, M.M. (1976) Wheel/rail noise, Part II: Impact noise generation by wheel and rail discontinuities, *Journal of Sound and Vibration*, **46**, 395–417.

Vincent, N. Bouvet, P., Thompson, D.J. and Gautier, P.E. (1996) Theoretical optimization of track components to reduce rolling noise. *Journal of Sound and Vibration*, **193**, 161–171.

Waters, P. and Priede, T. (1972) Origins of diesel truck noise and its control. *Proceedings of SAE*, 720636.

Wettschurek, R. and Hauck, G. (1994) Noise and ground vibration from railway traffic. In: M. Heckl and H.A. Müller (eds), *Taschenbuch der Technischen Akustik*, Springer-Verlag, 2nd edn (in German).

Williams, R. and Balaam, M. (1989) Understanding and solving noise quality problems. *Proceedings of Autotech 89*, Institution of Mechanical Engineers, C399/16.

Wu, T.X. and Thompson, D.J. (2001) A hybrid model for wheel/track dynamic interaction and noise generation due to wheel flats. *ISVR Technical Memorandum No. 859*, University of Southampton, January.

Chapter 7

Aircraft noise

P.F. Joseph and M.G. Smith

7.1 Introduction

Despite the best efforts of engineers over the last four decades, the problem of noise disturbance around airports continues to grow in importance. Restrictions are now regularly imposed on aircraft movements into and out of major airports, with obvious economic consequences. An understanding of the mechanisms of noise generation is clearly fundamental to the development and implementation of effective control measures. The principal aims of this chapter are to provide brief descriptions and explanations of the various mechanisms of aircraft noise generation, together with the associated controlling parameters, and to cite technical literature that may be consulted for greater detail.

The dominant contributor to overall aircraft noise is the power plant, although, at approach, a significant component of the noise comes from the airframe. The propulsion systems of the majority of commercial passenger aircraft comprise a number of high bypass ratio turbofan engines, each with many distinct noise sources. Each engine source has a particular dependence on operating conditions and a characteristic acoustic far-field frequency spectrum and directivity. Airframe noise is also generated by numerous sources, particularly on the wings (with high lift devices deployed) and the landing gear (Figure 7.1).

Achieving cost-effective and energy-efficient reductions in aircraft noise requires evaluation of the relative contributions to total radiated noise from the many engine and airframe sources. A typical component breakdown for an aircraft with acoustically treated engines was predicted by Owens (1979); the data are re-plotted in Figure 7.2, expressed in terms of perceived noise loudness.

For the generation of aircraft to which Figure 7.2 applies, the combined fan noise from the inlet and bypass duct dominates the overall noise of the aircraft at both takeoff and approach conditions. Next in importance is jet noise at takeoff, and airframe noise at approach. Since the time when this analysis was performed there has been considerable progress in

Figure 7.1 A commercial aircraft at takeoff.

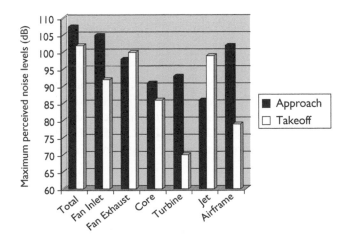

Figure 7.2 Predicted 'source' breakdown of perceived noise for an energy-efficient lined engine at approach and at takeoff, re-plotted from Owens (1979).

the reduction of engine noise; airframe noise is now comparatively more important.

This chapter presents a brief, mostly qualitative, description of the dominant noise sources present on a modern turbofan-powered aircraft, within a framework of the theory of aerodynamic noise generation. The issue of aircraft noise assessment is not addressed in this chapter; information on this subject can be found in specialist texts such as that of Smith (1989). The chapter is divided into four main sections: Section 7.2 outlines the fundamental theory of aerodynamic noise generation, Section 7.3 presents a qualitative description of the basic noise sources on an engine, Section 7.4 presents an overview of the noise sources on an airframe and Section 7.5 concludes the chapter with a brief overview of numerical methods for aerodynamic noise prediction.

7.2 Principles of aerodynamic noise generation

We begin with a brief outline of the Lighthill 'acoustic analogy', which provides the essential philosophical framework underpinning modern aeroacoustics theory.

7.2.1 Aerodynamic sound generation and the acoustic analogy

Aerodynamic sound generation is fundamentally different from most other sound-producing mechanisms in that it does not result from solid-body vibration. Sound produced by an aircraft in flight results from the presence of unsteady flow. Sound may be radiated from a region of free turbulence, as in the case of the jet emanating from the engine exhaust (Section 7.3.4), or may be produced by the interaction between unsteady flow and the components of the turbomachinery (Section 7.3.2), or by the lift and drag-inducing components of the airframe and undercarriage (Section 7.4). Much of the progress in the understanding of aerodynamic sound generation derives from the pioneering work of Lighthill (1952, 1954). He showed that by subtracting the time derivative of the exact mass conservation equation

$$\frac{\partial \rho}{\partial t} + \frac{\partial \rho u_i}{\partial x_i} = 0 \tag{7.1}$$

from the divergence of the exact mass conservation equation

$$\frac{\partial}{\partial t}\rho u_i + \frac{\partial}{\partial x_j}\left(p_{ij} + \rho u_i u_j\right) = 0 \tag{7.2}$$

a wave equation is produced of the form

$$\frac{\partial^2 \rho'}{\partial t^2} - c_0^2 \nabla^2 \rho' = \frac{\partial^2 T_{ij}}{\partial x_i \partial x_j} \tag{7.3}$$

where the Lighthill stress tensor, $T_{ij} = \rho u_i u_j + p_{ij} - c_0^2 \rho' \delta_{ij}$ and $\rho' = \rho - \rho_0$ is the fluctuating component of fluid density, where ρ_0 is the mean density. Here $p_{ij} = p' \delta_{ij} - \tau_{ij}$, where δ_{ij} is the Kronecker delta function, τ_{ij} is the viscous stress, u_i is the ith component of instantaneous flow velocity and ρ is the total density. Note that, for fluids of uniform entropy, $p' = \rho'/c_0^2$. The left-hand side of equation (7.3) (by design) is the familiar wave operator governing the propagation of density fluctuations though a uniform, stationary fluid, whilst the right-hand side represents the *equivalent* source distribution. The acoustic analogy therefore formally identifies aerodynamic sources with non-zero values of $q = c_0^{-2} \partial^2 p / \partial t^2 - \nabla^2 p$, where c_0 is the sound speed in the analogous fluid. The double divergence on the right-hand side of equation (7.3) suggests a distribution of quadrupole sources of strength T_{ij} per unit volume. Thus, the acoustic analogy treats the density fluctuations in the real fluid as if they were due to acoustic waves in a uniform stationary fluid driven by an externally applied fluctuating stress field. Here, the flow is assumed to be given and unaffected by the sound that it radiates. There is therefore an *exact* analogy between density fluctuations that occur in the actual flow and the small-amplitude fluctuations that would be produced by a quadrupole source distribution of strength T_{ij} in a stationary, fictitious fluid with sound speed c_0. The solution to equation (7.3) for the fluctuating (acoustic) density is

$$\rho'(\mathbf{x}, t) = \frac{\partial^2}{\partial x_i \partial x_j} \int_V \frac{T_{ij}\left(\mathbf{y}, t - \dfrac{r}{c_0}\right)}{4\pi c_0^2 r} \, d\mathbf{y} \tag{7.4}$$

where $r = |\mathbf{x} - \mathbf{y}|$, with $\mathbf{y}$ taken over the volume V occupying the turbulence and T_{ij} evaluated at the retarded time $t - r/c_0$. Evaluation of the acoustic field radiated by the turbulence volume therefore requires knowledge of the turbulence velocity at all positions in space at all instants in time.

7.2.2   The effect of solid boundaries

For aeroacoustic sound sources, other than a turbulent jet, a surface is usually present and the wave equation of equation (7.3) must be supplied with the associated boundary conditions. The presence of solid boundaries plays a direct role in the sound generation process, and often causes a significant amplification of the radiated sound power compared with that produced without boundaries. With a solid surface S present, the solution for ρ' becomes modified as in equation (7.5) (for example, Goldstein 1976)

$$4\pi c_0^2 \rho'(\mathbf{x}, t) = \underbrace{\frac{\partial^2}{\partial x_i \partial x_j} \int_V \left[\frac{T_{ij}}{r}\right] \mathrm{d}V}_{\text{I}} + \underbrace{\frac{\partial}{\partial x_i} \int_S \left[\frac{n_j}{r}(\rho u_i u_j + p_{ij})\right] \mathrm{d}S}_{\text{II}}$$

$$-\underbrace{\frac{\partial}{\partial t} \int_S \left[\frac{n_j}{r}\frac{\partial \rho u_j}{\partial t}\right] \mathrm{d}S}_{\text{III}} \tag{7.5}$$

where n_i is the unit vector normal pointing outwards from the surface and the square brackets indicate that the integrands are to be evaluated at the retarded time $t - r/c_0$. The form of equation (7.5) explicitly indicates that the acoustic field is the result of three distinct contributions: I, a volume distribution of quadrupole sources as given by equation (7.4); II, a surface distribution of dipole sources caused by local fluctuating stresses exerted *by* the surface *on* the adjacent fluid; and III, a surface distribution of monopole sources due to the mass flux crossing the surface, for example, due to permeability of the surface or that created by flow inducing vibration. The presence of solid surfaces in the unsteady flow may therefore fundamentally alter the character of the aerodynamic sound sources by virtue of the forces and fluid motion on the surface caused by the unsteady flow. Ffowcs-Williams and Hawkings (1969a) subsequently extended this approach to include the effects of moving boundaries. The FWH formulation may therefore be regarded as the most general representation of the acoustic analogy, and is also the most widely used for making turbomachinery noise predictions.

7.2.3   Scaling laws for elementary aerodynamic sources

Turbulence is rarely known in sufficient detail to allow accurate predictions of the acoustic pressure over all space and all time. However, dimensional analysis applied to the terms in equation (7.5) can provide the fundamental scaling laws for the basic aerodynamic source distributions in terms of mean aerodynamic quantities. For example, in a jet with mean flow velocity U (Mach Number $M = U/c_0$) and diameter D, the characteristic frequency of the noise is known to scale as $f \sim U/D$, and hence with wavelength $\lambda \sim DM^{-1}$. At suitably low jet velocities $\lambda \gg D$ (i.e. a compact jet), retarded time variations over the source region may be ignored and the following scaling relationships may be written: $\partial/\partial x_i \equiv M/D$, $\int_V \mathrm{d}V \equiv D^3$ and $T_{ij} \equiv \rho_0 U^2$

(obtained by assuming $u_i \propto U$). Substitution of these relationships in term I of equation (7.5) gives

$$I \sim \overline{p^2} \sim \rho_0^2 M^8 \frac{D^2}{|\mathbf{x}|^2} \qquad (M \ll 1) \tag{7.6}$$

Equation (7.6) is the famous Lighthill 'eighth power law' relating the far-field acoustic intensity I radiated from a cold compact subsonic jet to the eighth power of its mean velocity. It shows that the most efficient means of reducing jet noise is to reduce the mean jet velocity. This is one of the main factors motivating the current trend towards having increasingly high bypass ratio engines. See Section 7.3.4 for a more detailed account of jet noise. A similar scaling argument applied to terms II and III in equation (7.5) predicts $I \sim M^6$ for the dipole sources (Curle 1955) and $I \sim M^4$ for the monopole sources.

7.2.4 Trailing-edge effects

Sound produced by aerofoil–flow interaction is generally dominated by the surface dipole sources in equation (7.5), for which an M^6 scaling law is predicted following the argument in Section 7.2.3 (Curle 1955). However, in trailing-edge noise experiments a scaling law much closer to M^5 is generally observed (for example, Brooks *et al.* 1989). The trailing edge is found to enhance the effectiveness of acoustic radiation. The turbulent boundary layer induces a surface pressure that is supported by the blade but cannot be supported in the wake downstream of the blade. The sudden adjustment between these situations generates acoustic waves, which then propagate to the far field more efficiently than when the trailing edge is absent. Ffowcs-Williams and Hall (1970) investigated this problem by modelling a typical turbulent eddy as a quadrupole point source near the edge of a half-plane. They showed that the important parameter controlling sound radiation is kr_0, where k is the acoustic wavenumber and r_0 is the distance of the quadrupole to the edge of the half-plane. For quadrupoles close to the edge $(kr_0 \ll 1)$ having fluid motion perpendicular to the edge, the sound output from the associated quadrupoles decreases by a factor of $(kr_0)^3$. By assuming that the eddy perturbation velocity and the acoustic frequency scale linearly with the characteristic velocity U, Ffowcs-Williams and Hall showed that the acoustic intensity varies by a factor of M^3 relative to quadrupole radiation in free space. Quadrupole noise sources in the vicinity of a sharp trailing edge therefore radiate noise sound power proportional to M^5, which for low Mach number flows $(M < 1)$ is greater than the M^6 scaling predicted in the case of Curle's surface dipoles situated well away from sharp edges.

A comprehensive review of trailing-edge noise theories is presented by Howe (1978). The theoretical predictions of Ffowcs-Williams and Hall are largely supported by the trailing-edge experimental results reported by Brooks and Hodgson (1981) (see Figure 7.14). Pressure measurements were made in an anechoic, quiet-flow facility by sensors located near the trailing edge, and by microphones placed at various angular positions θ with respect to the chord, at distances of between two and four chords from its trailing edge. Consistent with the Ffowcs-Williams and Hall predictions, the radiation intensity was found to scale as M^5 with far-field directivity proportional to $\sin^2(\theta/2)$, where θ is the polar angle of the observer to the plane parallel to the chord; the trailing-edge effects in relation to airframe noise are discussed further in Section 7.4.1.1.

7.2.5   Sound radiation by an aerofoil in a gust

The dominant sound generation mechanism on an aircraft involves the interaction of a region of unsteady flow and an aerofoil; for example, a rotor blade, a stator vane or an aircraft wing. A simple model of the unsteady forces exerted by the aerofoil on the adjacent fluid, which leads to sound radiation, is obtained at very low (non-dimensional) frequency by invoking the quasi-steady approximation. Here, the fluctuating lift and drag forces are attributed to an unsteady angle-of-attack caused by a fluctuating inflow velocity. The monopole source contribution ('thickness' noise) may be neglected for thin aerofoils in uniform motion, and the contribution of the volume distribution of quadrupole sources has been shown to be negligible, except at high frequency and at very high loading conditions. At higher frequencies, the model for unsteady lift forces on the aerofoil is unsatisfactory. In addition to the aforementioned mechanism, there are also contributions to the total fluctuating lift from the wake vorticity acting back on the aerofoil, and to 'apparent mass' effects due to the force required to accelerate the surrounding fluid.

The starting point of most analyses of aerofoil sound radiation is the assumption of an incident gust in the form of a small-amplitude vortical velocity $\mathbf{u}$ that is convected over the blade at the flow velocity U_1 (Taylor's hypothesis), as sketched in Figure 7.3.

The velocity $\mathbf{u}$ due to a harmonic vortical gust of wavenumber vector $\mathbf{k}$ and amplitude $\mathbf{u}_0$, convecting in the x_1-direction with mean velocity U_1, is of the form

$$\mathbf{u} = \mathbf{u}_0 \exp\left[j\left(\mathbf{k}\cdot\mathbf{x} - k_1 U_1 t\right)\right] \tag{7.7}$$

where k_1 is the gust wavenumber vector component in the chordwise direction and $\mathbf{x}$ denotes a coordinate system attached to the blade. Turbulence

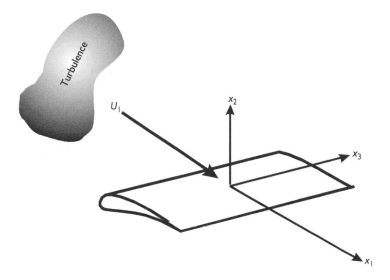

Figure 7.3 Turbulence interacting with an aerofoil and associated coordinate system.

comprising harmonic wave components of the form of equation (7.7) takes the form of a velocity field that convects as an unchanging, 'frozen' pattern with constant velocity U_1. This 'frozen gust' assumption is generally valid for aerofoil–flow interaction problems because turbulence velocities are usually only a few per cent of the mean flow speed. Significant changes to the turbulence field therefore only occur over distances much larger than a blade chord. The assumption of a small fluctuating velocity allows for linearisation, and hence simplification, of the governing equations. Solutions obtained in the linear approximation are appropriate to lightly loaded aerofoils with small camber and thickness at small angle-of-attack ($\pm 10°$). Under these assumptions, the distribution of fluctuating lift force per unit span F_2, acting normal to the chord of an aerofoil idealised as a flat plate due to interaction with the harmonic gust of equation (7.7), can be expressed in the form

$$F_2 = \pi \rho_0 c U_1 u_{02} \exp\left[j(k_3 x_3 - k_1 U_1 t)\right] \Gamma(k_1, k_3, M_1) \qquad (7.8)$$

where Γ is a non-dimensional aerofoil response function, c is the blade chord and u_{02} is the velocity amplitude of the incident just normal to the chord (upwash velocity). The derivation of this expression can be found in, for example, Goldstein (1976). For a turbulent inflow with continuous frequency and wavenumber spectrum, the corresponding form of equation (7.8) for the fluctuating lift, expressed as a power spectral density (mean square force per unit length per unit radian frequency), is given (Amiet 1975a; Goldstein 1976) as

$$S_{FF}(\omega, k_3) = \lim_{T \to \infty} \frac{\pi}{T} E \left\{ |F_2(\omega, k_3)|^2 \right\}$$

$$= \pi^2 \rho_0^2 c^2 U_1 |\Gamma(K_1, k_3, M_1)|^2 \int_{-\infty}^{\infty} \Phi_{22}(K_1, k_2, k_3) \, dk_2 \qquad (7.9)$$

Here, $\Phi_{22}(k_1, k_2, k_3)$ is the wavenumber spectral density of u_2 evaluated in a frame of reference, $\mathbf{x}' = \mathbf{x} + U_1 t$, moving with the mean convection speed U_1. It may be deduced from the relation $\Phi_{22}(\mathbf{k}) = (2\pi)^{-3} \int_V R_{22}(\Delta \mathbf{x}') e^{-i\mathbf{k} \cdot \Delta \mathbf{x}'} d(\Delta \mathbf{x}')$, where $R_{22}(\Delta \mathbf{x}') = \langle u_2(\mathbf{x}') u_2^*(\mathbf{x}' + \Delta \mathbf{x}') \rangle$ is the upwash velocity spatial correlation function for homogenous turbulence measured in the moving reference frame. This formulation is appropriate, for example, to the prediction of fan broadband and airframe noise. A consequence of making the 'frozen gust' assumption of equation (7.7) is that only the single streamwise wavenumber $K_1 = \omega/U_1$ contributes to the blade response in a narrow frequency band centred on ω.

Unlike the idealised flat-plate aerofoils assumed in the derivation of equations (7.8) and (7.9), realistic aerofoil geometries act to distort both the potential flow field around the aerofoil and the oncoming vorticity wave. The inclusion of these effects, which are generally coupled phenomena, requires analysis to second and higher order. For example, one of the main effects of finite blade camber and thickness has been shown to allow chordwise velocity fluctuations to generate unsteady lift whereas, to first order, this velocity component has negligible effect. Lorence and Hall (1995) have demonstrated that, compared with the flat-plate calculations, introducing blade thickness, camber and stagger angle produces a redistribution of sound power levels between the inlet duct and the aft duct. Overall sound power levels were roughly unchanged, however. Evers and Peake (2002) have also studied the effects of camber and thickness on the unsteady lift, and hence on the radiated sound. They conclude that, whilst these effects are important for pure tone generation, they are substantially less important for broadband noise generation.

7.2.6   Isolated flat-plate aerofoils

Von Karman and Sears (Von Karman and Sears 1938; Sears 1941) were the first to obtain analytic closed-form solutions for the unsteady lift produced by a single flat-plate aerofoil at small angles of attack, subject to a two-dimensional harmonic gust ($k_3 = 0$) convecting with the free stream. The solutions are completely defined by the non-dimensional (reduced) frequency $\sigma_1 = \omega c / 2u_1$ and the mean flow Mach number over the aerofoil

M_1 ($\ll 1$). These early analyses neglected the effects of fluid compressibility, whose influence is only important at high frequencies by enhancing the unsteady lift though the back-reaction of the generated sound on the acoustic source strength.

A three-dimensional turbulent flow field has velocity contributions across all three wavenumber components, thus leading to variations in aerodynamic parameters along the span. The simplest way of predicting the aerofoil response to these spanwise gust variations, that cannot be made using the normal-incidence Sears functions, is by dividing the span into a number of strips parallel to the flow. Each strip is then treated as if it were a two-dimensional plate subject to a fictitious two-dimensional gust whose amplitude equals that of the actual gust at that point. In a numerical study aimed at comparing a strip theory calculation with an exact three-dimensional lift calculation, Kobayashi (1978) has concluded that, 'two-dimensional calculations of the unsteady forces are adequate for fan noise prediction'. Estimates of the modal sound power levels obtained using the two approaches were shown to be within 2 dB. One explanation for this finding is provided by Amiet (1975a). He showed that, due to phase cancellation of the blade response along the blade span, a harmonic gust arriving obliquely at the aerofoil leading edge produces significantly less lift compared with a gust of the same frequency arriving close to normal incidence. Analytic expressions for the oblique gust aerodynamic response function on a flat-plate aerofoil, subject to a two-dimensional normally incident mean flow, have been obtained by Filotas (1969) and Graham (1970).

The effects of compressibility on the aerodynamic response function Γ become important when $\sigma_1 M_1 > 1$ (Figure 7.4) corresponding to when the time taken for an acoustic disturbance to traverse the chord c/c_0 is long compared with the period of oscillation $2\pi/\omega$. Closed-form solutions to the compressible skewed-gust problem were obtained by, for example, Graham (1970) and Adamczyk and Brand (1972). Graham showed that when the trace velocity of the gust along the leading edge moves slower than the sound speed, the solution for the unsteady lift could be expressed in terms of the three-dimensional, incompressible, lift response function. When the trace velocity along the leading edge is supersonic, he showed that the lift response function could be expressed in terms of that for the two-dimensional compressible response function. These results have become known as the 'Graham similarity principle'.

7.2.7 Cascade effects

Rotor and stator blades present in turbomachines do not exist in isolation but as part of a cascade. Mutual interference effects between neighbouring blades can significantly alter the unsteady loading, and hence the sound

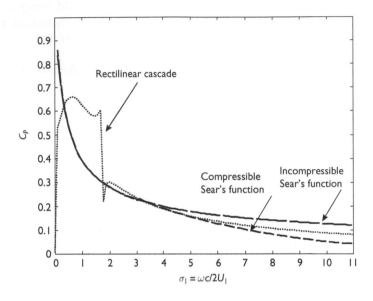

Figure 7.4 Predicted unsteady lift coefficient versus reduced frequency obtained using the incompressible and compressible Sears function for $M_1 = 0.3$, and from the LINSUB code, the cascade solution of Smith (1972).

radiation, from that which would occur on a blade existing in isolation (Figure 7.4). One of the main features of the cascade response, that is not present in the isolated aerofoil response, is the occurrence of acoustic modes in the space between adjacent blades. Single aerofoils radiate sound spherically whereas the sound from cascades radiate as modes that are bounded and which propagate unattenuated. Smith (1972) was the first to develop the theory for predicting the unsteady blade loading, the vortical field and the acoustic field upstream and downstream of a two-dimensional unloaded blade row due an incoming harmonic gust that may be either acoustical or vortical.

Whitehead (1970, 1987) subsequently implemented this theory in the well-known computer code LINSUB, which is still in widespread use today for fan noise prediction and for validating cascade predictions obtained using more sophisticated prediction techniques (see Section 7.5). Examples of more recent work on unsteady aerofoil cascade aerodynamics for acoustic prediction are the works of Glegg (1999), who has recently solved the general problem of the unsteady lift on a rectilinear cascade of flat plates at zero angle-of-attack in a uniform cross-flow for an oblique gust, and of Evers and Peake (2002), who have considered real aerofoil effects such as camber and thickness. Glegg's solution is useful for predicting the response of swept blades in which the effects of spanwise unsteady blade loading are important. Evers and Peake conclude that, whilst camber and thickness effects are important in influencing the unsteady aerofoil response function for single-frequency plane wave gusts (up to $+10\,\text{dB}$ compared with flat-plate solutions), these effects are comparatively unimportant for broadband turbulent gusts. Estimates of sound power level, involving a cascade of aerofoils with non-zero camber and thickness effects, were shown to differ by up to $2\,\text{dB}$ from the equivalent flat-plate predictions.

Examples of unsteady lift coefficients, defined by $C_p = F_2/\pi\rho_0 U_1 u_{02} c$, are shown in Figure 7.4 plotted against non-dimensional reduced frequency $\sigma_1 = \omega c/2u_1$. The smooth curves are computed from the incompressible and compressible Sears functions (Sears 1941; Adamczyk and Brand 1972) for $M_1 = 0.3$. The dashed curve is a typical response function obtained using the LINSUB code for a rectilinear cascade of unloaded flat-plate aerofoils. The calculation was performed for $M_1 = 0.3$, a space–chord ratio of 0.5, a stagger angle of $\pi/4$ radians and an inter-blade phase angle equal to 3 radians. These quantities completely define the flat-plate cascade response. All solutions tend to zero as σ_1 tends to infinity, implying that single aerofoils and aerofoil cascades are relatively unaffected by gusts of high reduced frequency.

The compressible and incompressible isolated aerofoil solutions plotted in Figure 7.4 are in close agreement for $\sigma_1 M_1 \le 1$ ($M_1 = 0.3$), but deviate for values above this. As explained earlier, this condition identifies the frequency above which the time taken for sound to propagate across the blade chord roughly equals the period of oscillation. At these frequencies, the effects of fluid compressibility become important, the main effect being to produce a significant back-reaction of the generated sound on the acoustic source strength. Cascade effects are most significant near their modal cut-off frequencies (for example, Goldstein 1976). The sound field exerts a significant back-reaction on its source that causes the blade forces to vanish. The radiated power is therefore prevented from becoming infinite, as would be the case for predictions of blade forces based on isolated aerofoil theory.

The unsteady aerodynamic response of realistic three-dimensional cascade geometries operating in typical flow environments is currently being predicted on the basis of the linearised Euler equations. For example, Rangwalla and Rai (1993) have modelled the complete problem of the interaction tones produced by rotor wakes interacting with a downstream blade row. A time-marching solution, requiring many hundreds of CPU hours, was used to obtain accurate three-dimensional solutions for a single fan speed/configuration. The availability of three-dimensional codes, that can be used in the compressor design stage, in which many hundreds of configurations are tested, is therefore unlikely in the foreseeable future. Computational techniques for aeroacoustic predictions are discussed further in Sections 7.3.4.4 and 7.5.

7.3   Engine noise

This section briefly describes the main noise sources present on a modern
high bypass ratio engine. The sound produced by ducted sources is most
conveniently described in terms of the natural acoustic duct modes. Before
discussing the different source mechanisms, we begin with the basic theory
of duct acoustics.

7.3.1   Sound transmission and radiation from hard-walled cylindrical ducts

A complication in the prediction of turbofan engine noise is the presence
of the duct surrounding the turbomachinery. The duct imposes its own
particular characteristics on both the interior and the radiated sound fields
that are absent with the duct removed. However, the effect of the (hard-
walled) duct on the transmitted, and hence radiated, sound power steadily
diminishes as non-dimensional frequency ka is increased (ka is Helmholtz
number, where a is the duct radius). In this section we investigate briefly
some aspects of sound transmission and radiation from hard-walled, cylindrical ducts. We will adopt the modal model of sound transmission and
radiation, although there are alternative models, such as ray theory, that
are equally revealing.

7.3.1.1   Sound transmission

The acoustic pressure $p(x, r, \phi, t)$ in a source-free region in the duct satisfies the wave equation of equation (7.3) with zero right-hand side. For
simplicity, and without any loss of generality, we shall consider single-
frequency solutions to the wave equation in ducts in the absence of flow.
A family of solutions may be found, which when appropriately weighted and
summed, describe the acoustic pressure field in the duct for any arbitrary
source distribution. Each of these solutions p_{mn} are referred to as duct
modes, which for the cylindrical duct may be written as separable solutions
of radius r and circumferential angle ϕ, of the form

$$p_{mn} = \overline{a}_{mn} e^{j(\omega t - \alpha_{mn}kx)} G_{mn}(r) H_m(\phi) \tag{7.10}$$

where $\alpha_{mn}k$ is the axial wavenumber component and $\overline{a}_{mn}$ is the pressure
mode amplitude of the (m, n)th mode. Substituting equation (7.10) into
the homogenous wave equation produces separate second-order differential
equations in ϕ and r, the latter being Bessel's equation. Solutions to these
equations are

$$H_m(\phi) = e^{jm\phi}, \qquad G_{mn}(r) = J_m(\kappa_{mn}r) \tag{7.11a,b}$$

where J_m is the Bessel function of the first kind of order m (the Bessel function of the second kind can be discounted as it does not satisfy the hard wall boundary condition). Transverse mode wavenumbers κ_{mn} must comply with the condition of vanishing radial component of particle velocity at the duct wall, i.e. $J'_m(\kappa_{mn}a) = 0$, where the prime signifies differentiation with respect to its argument. This boundary condition restricts their values to an infinite set of discrete solutions $\kappa_{mn} = j'_{mn}/a$, where j'_{mn} represents the nth stationary value of the Bessel function of order m.

Modal solutions of equation (7.11) only satisfy the wave equation when the mode wavenumbers satisfy $(\kappa_{mn}/k)^2 + \alpha_{mn}^2 = 1$. This so-called dispersion equation between the discrete transverse mode wavenumbers κ_{mn}, the axial propagation wavenumber $\alpha_{mn}k$ and the excitation frequency though k, has significant consequences for sound generation by turbomachinery. It indicates that an essential condition for modal propagation, whereby α_{mn} is real, is that wavenumber components k_i in any direction i must be less than, or equal to, its value k in the direction of propagation. Equivalently, it suggests that the modal phase speeds in these directions must exceed the sound speed c_0, i.e. $c_i = \omega/k_i \geq c_0$. Thus, an essential requirement for efficient sound generation by ducted rotating sources is that their speed of rotation, either actual or effective (phase speed), is supersonic. However, it should be noted that the corresponding group velocity (the velocity with which energy is transmitted) is always less than, or equal to, c_0.

The dispersion relationship may be rearranged to give

$$\alpha_{mn} = \sqrt{1 - \left(\frac{j'_{mn}}{ka}\right)^2} \tag{7.12}$$

For a particular mode, equation (7.12) makes explicit the existence of a critical, or cut-off, frequency $\omega_{mn} = c_0\kappa_{mn}$, below which, α_{mn} is purely imaginary and the mode decays exponentially with distance along the duct. At frequencies above ω_{mn}, the mode is 'cut off' or evanescent. In an infinite (but not a semi-infinite) duct, the pressure and particle velocities of single-frequency cut-off modes are in quadrature. The time-averaged axial intensity flow is therefore zero, and hence no acoustic energy is transmitted along the duct. The significance of the cut-off frequency in determining the phase speed of propagating modes and the decay rate of cut-off modes can be made explicit by expressing the axial propagation wavenumber α_{mn} and axial phase speed c_{mn} in terms of its cut-off frequency. The results are $\alpha_{mn} = \sqrt{1 - (\omega_{mn}/\omega)^2}$ and $c_{mn}/c_0 = \left(1 - (\omega_{mn}/\omega)^2\right)^{-1/2}$. The rate at which cut-off modes decay with distance therefore increases with frequency ratios ω_{mn}/ω greater than unity.

At a single frequency, equation (7.12) indicates that only a finite number of modes $N(ka)$ are able to propagate unattenuated along the duct. At high normalised frequencies $ka \gg 1$, N approximates to $N(ka) = \frac{1}{2}ka + \left(\frac{1}{2}ka\right)^2$ (Rice 1976); all other higher-order modes decay exponentially along the duct. Thus, in seeking a solution for the pressure field in a duct we obtain, not a single unique solution but, a family of solutions. The resultant acoustic pressure in the duct is the weighted sum of these modal solutions, the weighting factor being dependent upon the source distribution. Each mode may be regarded as a stationary pressure pattern across the duct cross-section, which propagates axially along the duct at its own characteristic axial phase speed (modal dispersion) $c_{mn} = c_0/\alpha_{mn}$ (the group velocity is $c_0\alpha_{mn}$). At a single frequency, the general solution for the acoustic pressure can therefore be written as

$$p(r, \theta, x, t) = \sum_{m=-\infty}^{\infty} \sum_{n=1}^{\infty} \bar{p}_{mn} \Phi_{mn}(r, \phi) \, e^{j(\omega t - \alpha_{mn}kx)};$$

$$\Phi_{mn}(r, \phi) = J_m(\kappa_{mn}r) \frac{e^{jm\phi}}{N_{mn}} \qquad (7.13a,b)$$

where the constant N_{mn} is introduced to satisfy the normalisation condition $S^{-1} \int_S \Phi_{mn}^2(\mathbf{y}) \, d\mathbf{y} = 1$, where S is the duct cross-sectional area. The mode amplitude $\bar{p}_{mn}$ is determined from the source distribution $q(\mathbf{y})$ by means of the mode-coupling integral $\bar{p}_{mn} = S^{-1} \int_S \Phi_{mn}^*(\mathbf{y}) \, q(\mathbf{y}) d\mathbf{y}$. For fan tone noise, the sound field usually comprises of a few dominant modes, unlike fan broadband noise, which generally comprise all the propagating modes of various amplitude. Equation (7.13) shows that a point of constant phase $\phi_{mn}(x, \phi, t)$ on the modal wavefront is described by $m\phi + \omega t - \alpha_{mn}kx = \phi_{mn}$, which represents a spiral pattern winding around the duct axis. These modal solutions are often referred to as spinning modes, and their circumferential phase speeds at the duct wall at fixed axial position x is obtained as $V_{p_{mn}} = a\,(\partial\phi/\partial t) = a\omega/m$. The lowest (non-dimensional) cut-off frequency of a spinning mode of order m is approximately given by $j'_{m,1} \approx m > ka$. Recalling that $k = \omega/c_0$, the previous two results may be combined to produce the cut-off condition, $V_{p_{mn}} > c_0$. Thus, a necessary condition for modal propagation is that its circumferential phase speed at the duct wall should exceed the sound speed.

Equation (7.12) may be re-written as $\alpha_{mn} = c_0/c_{mn}$, where c_{mn} is the mode axial phase speed. By geometry, $\alpha_{mn} = \cos\theta_{mn}$, where θ_{mn} is the propagation angle between the local modal wavefront and the duct axis. The parameter may therefore be interpreted as a measure of the degree to which the mode is cut-on, and is often called the cut-off ratio. Different modes with the same cut-off ratio have very similar transmission and radiation characteristics. Cut-off (or the reciprocal quantity, cut-on) ratio is the single most important parameter for characterising modal behaviour in ducts. Well cut-on modes are characterised by α-values close to unity. Their axial propagation angle θ_{mn} is small and hence points on the wavefront only encounter a small number of interactions with the duct wall before radiation to the far field. Duct liners therefore have little effect on well cut-on modes (and also because the theoretical reflectivity of locally reacting liners tend to unity in the zero-grazing angle limit). Modes close to cut-off, however, are most susceptible to suppression by acoustic liner treatments since they interact more frequently with the duct wall.

An alternative normalised, non-dimensional measure of modal cut-on is the cut-off ratio $\xi_{mn} = k/\kappa_{mn}$, which has the advantage of taking real values for both cut-on and cut-off modes. It falls into the three ranges of values corresponding to the following categories of modal transmission:

$\xi_{mn} > 1$ Mode is cut-on and propagates at an angle $\sin^{-1}(1/\zeta_{mn})$ to the duct axis. Cut-on modes have an axial phase speed greater than c_0, and group velocity less than c_0.

$\xi_{mn} = 0$ Mode is just cut-on (or just cut-off) and propagates with infinite phase speed (and zero-group velocity). No modal power is transmitted.

$\xi_{mn} < 1$ Mode is cut-off and decays exponentially along the duct. In an infinitely long duct, pressure and particle velocity are in-quadrature and zero power is transmitted.

7.3.1.2 Modal radiation

The far-field acoustic pressure $p_{mn}(R, \theta, \phi)$ due to the (m, n)th mode incident upon the open termination of a semi-infinite, cylindrical, hard-walled duct may be expressed in terms of a non-dimensional modal directivity factor H_{mn} defined by Joseph and Morfey (1999)

$$p_{mn}(R, \theta, \phi) = \overline{p}_{mn} H_{mn}(ka, \theta) \left(\frac{a}{R}\right) e^{jm\phi - jkR + j\omega t}; \quad kR \gg 1 \quad\quad (7.14)$$

where θ and ϕ are the polar and azimuthal angles, respectively. Tyler and Sofrin (1962) derived closed-form expressions for H_{mn} for a duct terminated by an infinite rigid flange. In this flanged duct case, $\overline{p}_{mn}$ refers to the mode amplitudes at the duct termination *after* reflection, i.e. it includes the reflected wave. For the more realistic case of the unflanged duct, the exact solution for H_{mn} is more complicated and involves the solution to a Wiener-Hopf integral equation. This was first solved by Levine and Schwinger (1948) for the plane wave mode, and by Weinstein (1969) for the higher-order modes. The solution includes the reflection process at the open end so that $\overline{p}_{mn}$ in equation 7.14 refers to the amplitude of the incident mode. The flanged and unflanged solutions are generally in close agreement in the forward arc (not too close to sideline directions) at normalised frequencies ka greater than about 5 (not too close to the cut-off frequencies). Some representative unflanged-duct directivity plots (in decibels) for the $(m, n) = (40, 3)$ mode at three cut-off ratios are presented in Figure 7.5.

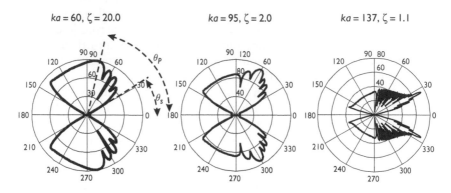

Figure 7.5 Theoretical modal directivity functions of the mode $(m, n) = (40, 3)$ for a semi-infinite hard-walled duct for three representative frequencies above cut-off.

Whilst the detailed behaviour of these directivity patterns is complicated, a few simple rules govern their broad characteristics:

- The angle θ_{Pmn} of the principal radiation lobe equals $\theta_{Pmn} = \sin^{-1}(1/\zeta_{mn})$, which in the absence of flow is identical to the axial propagation angle within the duct.
- Modal radiation becomes progressively weaker, as the frequency approaches cut-off from above, tending to zero, exactly at cut-off. Correspondingly, the modal pressure reflection coefficient at the duct open end tends to unity in the same limit.
- No major or minor lobes occur in the rear arc.
- Zeros (or nulls) in the radiation pattern occur at angles = $\sin^{-1}(1/\zeta_{mj}), j \neq n$. Angles of the minor lobes occur roughly mid-way between the angles of the zeros. The number of zeros and minor lobes increase roughly as the frequency squared.
- Symmetrical angles θ_s exist, beyond which, modal radiation is extremely weak. These are referred to as shadow zones (or cones of silence) and occur at $\theta_s = \sin^{-1}(m/ka)$.

A single mode incident upon the open end of a circular duct scatters into reflected modes of different radial mode indices but with the same m index. Theoretical predictions by Zorumski (1973) for flanged ducts, and by Lansing *et al.* (1970) for unflanged ducts, demonstrate that at frequencies well above cut-off, modes suffer negligible reflection back along the duct. An equivalent result has been obtained by Morfey (1969) who shows that, in this high-frequency limit, modal radiation efficiency tends asymptotically to unity. Below the cut-off frequency, modal radiation is finite, but weak compared with that of the propagating modes (note that it is zero in an infinite duct). Modal radiation efficiency falls sharply with decreasing frequency below cut-off. Modes that are only just cut-off, such as those present at some frequencies in buzz-saw noise, may therefore contribute significantly to the far-field radiation. Modes precisely at cut-off represent a singular condition in which incident modal power is completely reflected back along the duct, with zero power being radiated to the far field.

Single-frequency, coherent excitation of the modes, as for example by the mechanisms responsible for fan tone generation described in Section 7.3.2, causes the individual modal radiation fields to sum coherently, usually leading to complicated interference patterns in both the polar and the azimuthal far-field angles. Broadband sources generally excite a much larger number of modes. These modes are mutually incoherent, and lead to a far-field mean square pressure variation that is axisymmetric and slowly varying in polar angle. Based on prescribed mode amplitude distribution functions of α, Rice (1978) and Joseph and Morfey (1999) have obtained comparatively simple expressions for the multi-mode far-field radiation by exploiting the approximate data collapse which exists between cut-off ratio and the polar angle of far-field main radiation lobe.

7.3.2 Fan tones

This section presents a qualitative description of the mechanisms of pure tone generation by supersonic and subsonic rotors. Typical sound pressure level spectra measured in the far field of a modern turbofan engine with 28 rotor blades are presented in Figure 7.6 plotted against EO (Engine Order, defined as frequency non-dimensionalised on the shaft rotation frequency). The measurements were made at fan tip speeds less than, and greater than, the speed of sound.

The sound power spectrum at subsonic tip speeds is characterised by the presence of a number of peaks centred on the blade passing frequency and its harmonics superimposed on a broadband noise floor. Not all frequency harmonics are present, however. Certain tones have intentionally not been excited as a result of the judicious choice that is usually made of the number of rotor blades and stator vanes (Section 7.3.2.2). For rotor tip speeds exceeding the sound speed, the radiation spectrum is markedly different. Spectral peaks, called 'buzz-saw' tones, are now present at all harmonics of the shaft rotation frequency (Section 7.3.2.1).

7.3.2.1 'Buzz-saw' tones

Buzz-saw noise is particularly noticeable during takeoff and climb, and affects both community noise levels and cabin noise in regions ahead of the engine. The overall subjective impression, first documented by Sofrin and McCann (1966), was that it is similar to the noise produced by a sawmill.

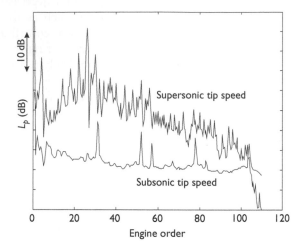

Figure 7.6 Typical measured far-field sound pressure level spectra of the noise radiated from the inlet of a ducted fan rotating with tip speed less than, and greater than, the speed of sound.

Buzz-saw noise occurs exclusively at relative rotor tip speeds exceeding the speed of sound where the region of the air at the blade tip moves supersonically. Tyler and Sofrin (1962) express this condition for the onset of buzz-saw noise as:

$$M_x^2 + \left(\frac{B\Omega}{c_0 \kappa_{B,1}}\right)^2 > 1 \qquad (7.15)$$

where $\kappa_{B,1}$ is the transverse wavenumber of the lowest order mode associated with the rotor-excited spinning mode $m = B$, M_x is the axial flow Mach number and Ω is the shaft rotational frequency. Equation (7.15) was derived entirely from kinematic considerations and consequently provides little insight into the physical processes involved in buzz-saw tone interaction. For hub-tip ratios close to unity (i.e. a hollow duct), equation (7.15) reduces to $M_x^2 + M_t^2 = M_r^2 > 1$, where M_t denotes the blade tip Mach number and M_r is the Mach number of the flow relative to the blade.

Ahead of each blade a shock-wave is produced, analogous to the sonic boom produced by an aircraft travelling at supersonic speeds. The pressure signature due to a hypothetical fan with identical rotor blades has the shape of a regular 'saw-tooth', whose energy is confined to the blade passing frequency and its harmonics. In practice, small differences in blade geometry and in the high-speed aero-elastic response between different blades cause the pressure signature due to real fans operating at supersonic tip speeds to be an irregular saw-tooth, whose spectral content is distributed amongst all

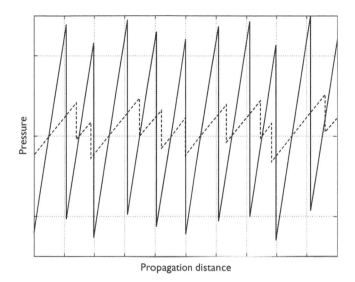

Figure 7.7 Theoretical pressure variation due to buzz-saw noise close to the fan face (solid line) and upstream of the fan (dashed line).

EOs. Stagger angle variations between the blades of the order of 0.1° are sufficient to lead to this effect.

Figure 7.7 illustrates the pressure variation due to buzz-saw noise close to the fan face (solid line) and upstream of the fan (dashed line). At the fan face, the blade-to-blade periodicity causes most of the energy to be concentrated at the blade passing frequency and its harmonics. However, the shocks in an irregular saw-tooth propagate at different speeds, and the saw-tooth becomes increasingly irregular with distance upstream of the fan, as shown by the dashed line in Figure 7.7. The non-linear propagation of an initially high-amplitude regular saw-tooth causes energy to be redistributed amongst all EOs. Energy transferred to higher frequencies is more readily dissipated by the shocks. By the end of the inlet duct, most of the acoustic energy is at EOs, other than the blade passing frequency. The occurrence of buzz-saw noise is therefore inextricably linked with the generation and subsequent non-linear propagation of a high-amplitude irregular saw-tooth pressure signature.

7.3.2.2 Rotor–stator interaction

In Section 7.3.1.1, it was indicated, although not shown explicitly, that modes phase-locked to the rotor blades at subsonic rotor tip speeds are cut-off and do not radiate efficiently to the far field. However, for certain combinations of rotor blades and stator vanes, the wakes impinging onto the downstream stator vanes from subsonic rotors can produce a disturbance, at the blade passing frequency and its harmonics, whose *effective* speed of rotation around the stator is supersonic. This disturbance is able to couple directly to propagating duct modes, which then radiate to the far field. By an identical process, noise is also produced by the interaction between the potential near field of the rotor and the stator. The rapid exponential decay with distance x from a blade of the potential flow field, compared with the slower $x^{-1/2}$ decay of the viscous wake, suggests that potential field interaction can usually be neglected except for rotors and stators very close together compared to the acoustic wavelength, as for example in the case of a turbine at low to mid-frequencies. In a classic study of compressor noise by Tyler and Sofrin (1962), it was shown that a number of rotor blades B adjacent to a number of stator vanes V produce a pattern occurring at s times the blade passing frequency in m circumferential wavelengths. Spinning mode orders m produced by this interaction process are given by their well-known formula

$$m = sB \pm kV \tag{7.16}$$

where k is any positive integer. The sign of m may be positive or negative indicating that the modes may spin in either direction relative to the stator (or rotor). However, only those modes satisfying the cut-off condition implied by equation (7.12) propagate. This formula assumes identical, equally spaced blades, so that the pressure pattern due to each blade is out of phase with those of the remaining blades. All harmonic frequencies that are not multiples of B cancel exactly. Small irregularities between blades therefore generate small tonal contributions at all harmonics of the shaft rotation frequency; these are often observed in practice.

The angular phase speed Ω_{mns} of the (m, n)th mode at the sth harmonic frequency of the blade passing frequency (*bpf*) is given by

$$\frac{\Omega_{mns}}{\Omega} = \frac{sB}{m} = \frac{sB}{sB \pm kV} \tag{7.17}$$

In Section 7.2.5.1 it was demonstrated that a necessary condition for modal propagation is that modal phase speed in the circumferential direction should exceed, or be equal to, the free-space sound speed. Applying this principle to the modal angular phase speed yields the cut-on condition, $\Omega_{mns}/\Omega \geq 1$. This condition is more usefully expressed in terms of the ratio of the number of stator vanes to rotor blades V/B necessary to achieve cut-off of the modes at the sth harmonic frequency of *bpf*

$$\frac{V}{B} \geq 1.1 \, (1 + M_t) \, s \tag{7.18}$$

For rotor tip speeds close to sonic $M_t \approx 1$, all modes at $1bpf$ and $2bpf$ ($s = 1$ and 2), are cut-off for numbers of stator vanes approximately equal to $2B+1$ and $4B+1$, respectively.

The detailed mechanism of rotor–stator interaction is complex. Stator vanes are usually highly loaded, and their interaction with the viscous wake shed by the rotors is complex. The presence of swirl distorts the wakes, which then sweep along the stator vanes, rather than arriving at all points on the stator leading edge simultaneously. Another complication in the prediction of fan tones produced by rotor–stator interaction is the effects of blockage by the rotor. Sound produced downstream of the rotor must traverse the rotor before being radiated from the duct inlet. In addition, there is blockage caused by the presence of high-speed flow in the space between the rotor blades opposing the direction of propagation. In general, modes whose directions of propagation are closely aligned with the chord-line suffer less blockage through the rotor than for modes which are not (the so-called 'venetian blind effect'). Two of the first models of sound transmission and reflection by a linear cascade of flat-plate aerofoils were presented by Kaji and Okazaki (1970) and Koch (1971).

Smith and House (1967) suggested that the tone level produced by rotor–stator interaction varies as $10 \log_{10}(\Delta x/c)$, where Δx is the axial gap between blade rows. Sutcliff *et al.* (1997) have compared acoustic predictions obtained using measured fan wake data and compared the results with the prediction obtained using empirical wake correlations. Differences in sound power level predictions by up to a few dB were observed, although trends with the subsonic fan speed were similar.

The mean-velocity profile of the wake is the main factor in determining the distribution of tone amplitudes across the harmonics of the blade passing frequency. An idealised wake profile is sketched in Figure 7.8. It depicts the main wake parameters of free stream velocity U_0, maximum, centre-line wake velocity deficit u_0, and the wake width W. Superimposed on this mean-velocity profile is a random component due to the turbulence generated within the wake, which is responsible for one of the main components of the radiated broadband noise (Section 7.3.3.4).

Wygnanski *et al.* (1986) and Gliebe *et al.* (2000) have studied the variation of u_0^2 and W with mean aerodynamic parameters in the wake generated by an *isolated* aerofoil in a laboratory wind tunnel and in the wakes generated by the blades of a fan simulator, respectively. In both cases, scaling laws were obtained of the form $(U_\infty/u_0)^2 = A\overline{X}$ and $(L/\theta)^2 = B\overline{X}$, where $\overline{X} = (X - X_0)/2\theta$, X_0 is the virtual centre of the wake, $\theta = C_d c/2$ is the wake momentum thickness, where C_d is the blade profile drag coefficient. Here, A and B are empirical constants determined from measurement. For thin aerofoils at small angles of attack, Wygnanski *et al.* have found that $A = 1.56^2/2$, $B = 2.032^2$ and $X_0 = -380\theta$.

The wake profile sketched in Figure 7.8 arrives at the stator periodically at the blade passing frequency. Fourier series decomposition of this periodic

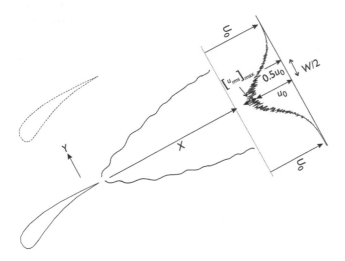

Figure 7.8 Parameters defining the mean and turbulence characteristics of an aerofoil wake.

train of transient signals determines the relative distribution of wake velocity between the various harmonic frequencies of *bpf*. Equation (7.8) suggests that the centre-line velocity deficit u_0 in the upwash direction determines the strength of the unsteady blade loading at the stator vanes. The wake width W determines the distribution of the unsteady blade loading between the harmonics of *bpf*. At small (non-dimensional) rotor–stator separation distances $\overline{X}$, the wake width W is comparatively small and the wake profile is impulse-like. The discrete spectrum of harmonic frequencies is therefore comparatively flat. At greater separation distances between rotor and stator, the wake scaling laws above indicate that the maximum wake deficit is smaller but the wake width greater. Most of the spectral energy is now concentrated at $1bpf$, with the levels of the higher harmonic frequencies falling-off faster with frequency. In general, however, the number of rotor blades and stator vanes is chosen so that the tone at $1bpf$ is cut-off.

7.3.2.3   Inlet steady flow distortion

Another mechanism by which a subsonic rotor can excite propagating duct modes is through its interaction with distortions of the mean intake-flow velocity at the fan face. This problem is analysed in, for example, Goldstein (1976). Intake-flow distortions occur naturally in the atmosphere, or may be produced by non axisymmetric duct features such as droop and scarfing. A number of rotor blades B, interacting with a circumferential variation in axial flow speed of spatial harmonic p, produce a pattern occurring at s times the *bpf* in m circumferential wavelengths given by,

$$m = sB + p \qquad\qquad (7.19)$$

This interaction mechanism is particularly important at the blade passing frequency since the contribution to the sound field due to rotor–stator at this frequency has usually been designed to be cut-off (Section 7.3.2.2). Slowly varying circumferential flow variations around the fan face comprise low spatial harmonic frequencies p. In accordance with equation (7.19), subsequently excited modal orders m are high, with correspondingly high cut-off frequencies. These low-p contributions are therefore unimportant over much of the frequency range, or may be masked by rotor-alone modes at takeoff. Higher levels of flow distortion contain higher spatial harmonics that produce smaller mode orders, which cut-on at much lower frequencies. Not only are these modal contributions present over a wider frequency range, but their amplitudes may also be much higher.

7.3.3 Fan broadband noise

Fan broadband noise is generated by the interaction between the various sources of turbulence present in a turbofan engine, and the rotor and stator. The importance of broadband noise to perceived annoyance of overall aircraft noise was highlighted in a recent study into turbomachinery noise by Gliebe (1996). He states that, 'If one could completely eliminate all fan tones, total system noise would be reduced by 0.5 to 1.5 EPNdB, depending on the operating conditions'. The growing importance of broadband noise in relation to the fan tones is mostly due to advances in liner technology. Liners, which may be tuned to a particular operating condition at a particular frequency, are becoming increasingly effective in attenuating fan tones. Narrow-band liner performance is therefore less effective in attenuating broadband noise, which is present over a wide frequency bandwidth.

Controlling fan broadband noise 'at source' is also extremely difficult. The number of distinct source mechanisms is large, and their identification is not possible using conventional source location techniques. Whilst the origin and mechanism of fan tone generation can be deduced from measurement of the far-field radiation, and by decomposition of the in-duct sound field into its constituent duct modes (particularly in m-orders which can be traced to its excitation mechanism though $m = sB \pm kV$), mode detection is not useful for broadband noise. This is because, in general, all broadband sources excite all modes. A further complication in the use of source location techniques for broadband noise sources is that the number of effectively uncorrelated sources is extremely large since these sources tend to be distributed over the large areas of the rotor and the stator but possess small length scale on the scale of the acoustic wavelength. A systematic technique for locating the sources of broadband noise therefore remains elusive. The recent source decomposition study by Ganz *et al.* (1998) on a model fan appears to give plausible results but relies on the removal of the stator vanes, thereby modifying the aerodynamics of the rotor and hence its noise generation.

A typical turbofan engine broadband noise spectrum falls off slowly with increasing frequency (Figures 7.6 and 7.9). The dominant broadband 'sources' in the forward and rear arcs are listed below, although all sources are present to some degree in both forward and rear arcs.

(i) Forward arc

- Ingested turbulent flow on to the rotor;
- Blade tip interaction with the turbulent boundary at the casing wall.

(ii) Rear arc

- Turbulence generated in the blade boundary layer and scattered from the rotor trailing edge;
- Turbulent wakes shed from the rotor impinging onto the stator.

Less-important contributions to broadband noise radiation arise from rotor blade interactions with the turbulent boundary at the hub, and sound produced by secondary flows caused by vortices generated at the blade tip. These sources will not be considered here. Ganz *et al.* (1998) have performed extensive measurements on an 18-inch diameter fan rig aimed at decomposing the total radiated broadband noise power into the constituent sources. A typical 'source' decomposition measurement (with tones removed) is reproduced in Figure 7.9.

The source decomposition was accomplished by comparing measurements of the transmitted sound power before and after the removal of stator vanes, before and after bleeding off the duct-wall boundary layer, and by taking precautions to smooth the flow entering the engine so as to reduce the noise contribution arising from ingested turbulence. This study concludes that there is not one single dominant source, but rather a number of competing mechanisms, which are more or less of equal importance, depending on fan design and flow conditions. Thus, not only is fan broadband noise poorly attenuated by conventional narrow-band liner treatments, but the multiplicity of sources present suggest that reductions in total broadband noise can only be achieved by the attenuation of *all* the dominant broadband noise sources simultaneously.

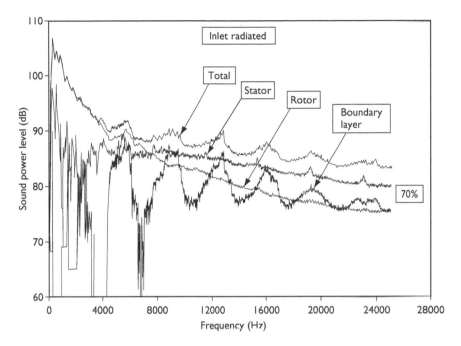

Figure 7.9 Breakdown of fan broadband noise at 70% fan speed radiated from the duct inlet into its constituent components of rotor self-noise, rotor–stator interaction and rotor–boundary layer interaction. (Reproduced with permission from Ganz *et al.*)

7.3.3.1   *Characteristics of broadband noise generation*

The following sections briefly describe the various characteristics of the broadband source mechanisms and their dependence on aerodynamic parameters. Modern analytical approaches to the prediction of fan broadband noise begin with a form of equation (7.9) to compute the spectral density of the unsteady blade loading, which is then integrated across a blade chord to compute the coupling to the acoustic duct modes. An analytic model is usually used to represent the moving-frame frequency-wavenumber spectral density $\Phi_{22}(k_1, k_2, k_3)$ for the upwash velocity or pressure (see Section 7.2.5).

7.3.3.2   *Self (trailing-edge) noise*

A rotor operating in a perfectly uniform oncoming flow generates broadband noise due to the turbulence that develops in the boundary layer over the blade surface and then scatters as it is convected past the trailing edge. The incompressible, hydrodynamic component of pressure associated with the convected turbulence is typically 20 dB greater than the component of acoustic pressure beneath the boundary layer of a thin aerofoil, far upstream of the trailing edge. At approach conditions, for example, the hydrodynamic component of pressure on the blade surface, far upstream of the trailing edge, convects subsonically with a velocity U_c and therefore does not radiate sound to the far field. Conversion of this energy to sound occurs at the trailing edge by a scattering process in which the high stream-wise wavenumber components associated with the incident hydrodynamic vortical motion in the boundary layer $(\sim\omega/U_c)$ are converted to the low wavenumber components of acoustic pressure $(\leq\omega/c_0)$.

Rotor-alone noise, alternatively referred to as self-noise, was for a long time believed to be the dominant broadband noise source. The noise source decomposition results of Ganz *et al.* (1998) suggest that, depending on operating conditions, this is not necessarily the case. Rotor self-noise sound power spectral levels were found to be between 2 and 3 dB below that due to rotor–stator interaction (even in the forward arc for which sound radiated from the stator must overcome rotor blockage). Measured spectra of sound power were found to vary smoothly and monotonically with frequency, and to increase in level with increasing loading and tip clearance.

Morfey (1970) has observed that the shapes of the broadband noise spectra produced by different engines at different operating conditions collapsed when plotted against the non-dimensional Helmholtz frequency $He_l = fl_e/c_0$, defined with respect to a length scale l_e. Here, l_e is related to the trailing-edge wake momentum thickness, which characterises the size of the eddies leaving the trailing edge. This observation is unusual in the sense that spectral collapse of flow-acoustic problems is usually obtained on the basis of Strouhal frequency $St_l = fl_e/U$ (Brooks and Hodgson 1981; Brooks *et al.* 1989). Since $He_l = MSt_l$, a Mach number of dependence of M^5 is predicted, and not the M^6 dependence usually associated with aero-dynamic dipole sources (see Section 7.2.3). This power law is consistent with the M^5 predictions of Ffowcs-Williams and Hall (1970) and with the measurements by Brooks *et al.* (1989) (Section 7.2.4). A comprehensive review of trailing-edge noise theories for isolated aerofoils is presented by Howe (1978).

Brooks and coworkers (Brooks and Hodgson 1981; Brooks *et al.* 1989) have performed detailed far-field and near-field acoustic pressure measurements of aerofoil self-noise for a variety of aerofoil trailing-edge geometries, blade chords, flow speeds and angles of attack. Five distinct mechanisms of self-noise were identified, each with different dependencies on mean aero-dynamic parameters. A list of these mechanisms, their description and the conditions under which they occur are presented in Table 7.1.

Glegg and Jochault (1997) have formulated a model for predicating rotor self-noise in a duct. Boundary layer surface pressures on the blades were obtained by inverting predictions of the trailing-edge noise spectra obtained using the model by Brooks and Hodgson (1981) for a single aerofoil in

Table 7.1 Aerofoil self-noise mechanisms: description and the conditions under which they occur

Types of self-noise	Description	Conditions
Turbulent boundary layer TE noise	Boundary layer turbulence convected past TE	High Reynolds numbers (with respect to chord) Small angles of attack
Laminar boundary layer – vortex shedding noise	Vortex shedding at the TE coupled to acoustically excited aerodynamic feedback loops	Laminar boundary layer over at least one surface Low Reynolds numbers Small angles of attack
Separation-stall noise	Boundary layer separation	Moderate to high angles of attack
Trailing-edge bluntness – vortex shedding noise	Vortex shedding at blunt TE	Blunt trailing edges
Tip vortex shedding noise	Turbulence present in the locally separated flow convected past the tip region	Formation of a tip vortex

an open jet, with the trailing-edge directivity functions derived by Amiet (1976). This unsteady surface pressure field was then input to a cascade model and, by using strip theory to treat the spanwise loading variations, rotor self-noise was computed for a ducted rotor. A fifth power law for variation of sound power with fan speed was predicted for low fan speeds, and a sixth power law at higher fan speeds. Rotor self-noise was predicted to increase by 2.4 dB per degree of incidence angle for an unstalled fan. This prediction is consistent with the forward arc measurements of Gliebe *et al*. (2000) who reported an increase in radiated sound power at a rate of 2.5 dB per degree increase in stagger angle.

A complicating feature of making predictions of rotor-alone sound power radiated from the inlet is the effect of blockage. As mentioned previously, the presence of high-speed flow between adjacent rotor blades, in the direction opposite to the direction of acoustic propagation, acts to impede the flow of acoustic energy through the rotor. Cascade effects also complicate the prediction of rotor self-noise, which affects both the steady and the unsteady aerodynamics (see also Section 7.2.8). Measurements of the self-noise radiated from a cascade of loaded aerofoils has recently been reported by Sabah and Roger (2001). The radiated sound power was observed to vary as $U^{5.8}$ over a 10-degree range of incidence angles. In his work on cascades, Glegg (1998) has shown theoretically that the presence of adjacent blades increases the trailing-edge noise of an aerofoil by up to 6 dB at the resonance frequencies and produces nulls in the radiation at anti-resonances. Nevertheless, the comparatively simple trailing-edge noise theory due to

Amiet (1976) for a single aerofoil, using as input data the pressure spectra measured at the wall (well away from the trailing edge), was found to reproduce the measured radiation spectra and the mean-velocity scaling law extremely well.

7.3.3.3   Rotor interaction with inflow turbulence

Another potentially important contributor to fan broadband noise arises from the interaction between inflow turbulence and the rotating fan. The turbulence may originate from that ingested from the atmosphere or from the boundary layer on the duct walls (casing and hub). The characteristics of the two sources of turbulence, and hence in their radiation spectra, differ significantly (for example, Majumdar and Peake 1998; Joseph and Parry 2001). The nearly isotropic turbulence of the atmosphere in free air becomes strongly deformed as it is ingested into the engine. Turbulent eddies become elongated into long narrow filaments due to strong streamtube contraction experienced by the steady, non-uniform mean flow generated by the fan. It has been estimated that length scales in the streamwise direction are typically 100–200 times those in the transverse directions (Hanson 1997). By contrast, a number of recent measurements of boundary layer turbulence at the wall of duct inlets have indicated that, in the moving reference frame, it does not deviate too far from isotropy (Ganz *et al.* 1998; Gliebe *et al.* 2000). Streamwise length scales measured for the three velocity components were found to be of the order of the boundary layer thickness.

Eddy length scales in the streamwise direction play an important role in determining the characteristics of the unsteady blade-loading spectrum, and hence of the radiated noise. Eddies passing through the fan become periodically sliced at the blade passing frequency. The result is that the surface pressures between the different blades are partially coherent, which gives rise to a broadband noise spectrum that contains a series of 'humps' of bandwidth $\Delta\omega$ centred on the blade passing frequency and its harmonics $\omega = sB\Omega$. The degree of inter-blade coherence is related to the time taken for a characteristic eddy to traverse the fan. By making the assumption that an eddy loses its identity when it travels a distance greater than its own diameter, Ffowcs-Williams and Hawkings (1969b) showed that spectral spreading of the tones $\Delta\omega$ relative to its centre frequency ω could be related to the number of times N an eddy is chopped by a blade,

$$\frac{\Delta\omega}{\omega} = \frac{1}{N} \tag{7.20}$$

Here, N may be calculated from the eddy size, the convection speed and the fan speed. Unlike during static testing, when noise due to inflow turbulence is found to be significant, during flight where the extent of streamtube contraction is much less severe, noise due to inflow turbulence from the atmosphere has been shown to be comparatively unimportant (Majumdar and Peake 1998). To reduce this effect in ground tests a turbulence control screen in the form of a 'golf ball' covered by a honeycomb mesh is usually introduced over the inlet to smooth the ingested flow. By making noise measurement with and without the boundary layer, Ganz *et al.* (1998) were able to isolate and quantify its contribution to overall fan broadband noise. The spectrum was shown to have the characteristic humped form, as shown in Figure 7.9. Broadband levels were found generally to be 10 dB below the overall broadband level, rising to a few decibels below the overall level at frequencies close to the blade passing frequencies.

It may be shown that sound radiation from ducted rotor blades at a frequency ω, is produced by an unsteady loading over the blades at frequencies $\omega - m\Omega$ measured in the reference frame rotating with the fan (Mani 1971; Homicz and George 1974). This has a number of important consequences for the spectrum of turbulence–rotor interaction noise. Since the aerodynamic response function and the turbulence intensity spectrum both generally peak close to $\omega = 0$ (Figure 7.3), the unsteady blade-loading spectrum also peaks at $\omega = m\Omega$. For B identical blades, however, there is perfect cancellation of the modal sound field for all $m \neq sB$ so that the sound radiation spectrum peaks at harmonic frequencies of *bpf*, i.e., $\omega = sB\Omega$. A further consequence of this frequency scattering phenomenon is an asymmetry of the mode amplitude distribution with respect to the spinning mode index m, with most sound power being transmitted by the modes spinning in the direction of the fan rotation. This behaviour is observed in the modal analysis results of rotor–turbulence interaction noise by Ganz *et al.* (1998).

7.3.3.4   Rotor–stator interaction

Another important mechanism of broadband noise generation is that arising from the interaction between the turbulence in the wake shed from the rotor and the stator (Ganz *et al.* 1998). They showed with an 18-inch diameter fan rig that the stator is the dominant source contributor in both the rear and the forward arcs. The radiated power was found to increase roughly in proportional to the number of stator vanes V (suggesting that the unsteady loading on different blades are uncorrelated), except at high solidity ($V = 60$) and low frequency, when the radiated power is roughly independent of V. The effects of loading on radiated power were also measured (by changing throttle settings on the fan rig). They found that, whilst turbulence intensities present at the stators generally increased with increasing loading, particularly close to the duct wall, the mean flow velocity over the stator vanes decreases (except close to the hub when the trend is reversed). Since radiated power increases with increasing turbulence intensity and mean flow speed, the dependence on radiated power radiation on loading is a balance of these opposing influences. In this particular fan rig, the net result of changing from low to high loading was found to be a small reduction in downstream power of between 1 and 2 dB.

The space-time characteristics of the three-dimensional turbulent field in a rotor wake incident upon a stator are highly complex. Nevertheless, useful insights into the evolution of turbulent rotor wakes, as they convect towards the stator, are provided in a recent report by Gliebe *et al.* (2000). Hot wire measurements were made in a low-speed compressor fan simulator with the objective of establishing generalised wake-flow correlations in the form $a + b\overline{X}$ between turbulence quantities and mean wake quantities, where $\overline{X}$ is the non-dimensional convection distance defined in Figure 7.8. Specifically, correlations were obtained between the mean square centre-line maximum turbulence velocities $\left[u_{i,rms}\right]^2_{\max}$ and the square of the maximum velocity deficit, u_0^2, and between the streamwise integral length scales L_i and the wake width W (Figure 7.8). The results confirm earlier aerofoil wake-turbulence measurements by Wygnanski *et al.* (1986) who have made measurements under laboratory conditions of the two-dimensional turbulent wake generated by a stationary aerofoil. They showed that turbulence generated by the rotor, measured in the stationary frame, is nearly isotropic.

Correlations of the streamwise length scales have also been obtained experimentally by Ganz *et al.* (1998) at a number of positions along the span of an 18-inch diameter fan. The measured length scales were shown to collapse on the wake width W according to $L_{2,3}/W \approx 0.2, L_1/W \approx 0.35$. Large deviations from this behaviour were observed close to the hub and tip regions where tip-leakage is the dominant turbulence generation mechanism. A similar result for turbulence wake length scales was obtained by Wygnanski *et al.* (1986). These relationships establish a link between the 'size' of the wake turbulence, and hence its radiation efficiency, and the more easily predicted and measured quantity of mean wake width W. The measurements by Wygnanski *et al.*, taken at more than 70-momentum thicknesses downstream of the trailing edge, suggest the existence of a self-preserving state for two-dimensional fully developed wakes of the form $\overline{u_i^2}\left(\overline{Y}\right)/u_0^2 = g_i\left(\overline{Y}\right)$, where $\overline{Y} = Y/W, i = 1, 2$. Here, g_i are universal functions that depend only upon the characteristics of the wake generator. Joseph *et al.* (2003) have used these wake-turbulence correlation measurements in a model aimed at predicting broadband noise due to rotor–stator interaction.

A modified form of the Tyler and Sofrin equations (7.16) and (7.19), for determining the spinning mode orders present in the tones generated by rotor–stator interaction and rotor–inflow distortion interaction, governs the modal composition of the broadband noise (Morfey 1970; Mani 1971; Homicz and George 1974). The result is $m = \mu \pm kV$, where k is any integer and μ is the spatial Fourier component of the turbulence in the circumferential direction, which takes integer values owing to the periodicity of turbulence in this direction. Thus, very high azimuthal turbulence wavenumbers $k_\theta \approx \mu/r$, upon interaction with the stator cascade, may scatter into lower wavenumbers lying within the acoustic region, which are then able to excite propagating acoustic waves directly.

7.3.4 Jet noise

The term 'jet noise' is sometimes used to describe the total noise radiated by the engine exhaust system. It is dominated by the noise produced by the turbulent mixing between the jet exhaust and the ambient fluid downstream of the nozzle exit. Other sources of jet noise arise from incorrectly expanded jet exhaust flows, the resulting shock having a tonal component called 'screech' and a broadband component associated with the shock. We shall limit our attention to turbulent mixing noise since this is the most important, and certainly the most difficult to reduce by means other than reducing the jet velocity. Excellent review articles dealing with jet noise can be found in Lilley (1995) and Tam (1998).

Despite more than 50 years of research, our ability to predict jet-mixing noise remains poor. The difficulty lies, not with the basic understanding of sound generation by turbulent flows, which has its basis in the established Lighthill Acoustic Analogy, but in our tenuous understanding of turbulence, which constitutes the basic aeroacoustic sources. In the acoustic analogy, the fluid flow dynamics, including the generation of noise within the flow and its interaction with the flow, is included in the strength and distribution of the equivalent acoustic source field. However, these equivalent sources are notoriously difficult to quantify since they involve predictions of the turbulence to a degree of approximation that is beyond current computational capabilities. Computational power has increased by a factor of a thousand in the last 20 years and so there is every hope that today's computational tools may be feasible for jet noise prediction in the near future. Current jet noise predictions contain some element of empiricism and include simplifying assumptions about the statistical properties of the turbulence sources.

A typical jet noise spectrum is shown in Figure 7.10 plotted against Strouhal number St_D defined with respect to the jet diameter D. It was obtained from a semi-empirical jet noise model and relates to the field at a far-field observer located at 90° to the jet axis at a distance of 7.5 m from the jet of 75 cm diameter and 0.3 Mach number.

The pioneering work of Lighthill (1952, 1954) was complemented by a number of experimental studies that confirmed, amongst other predicted features of jet noise, the now-famous 'eighth power law' for the sound power radiated by cold subsonic, very low Mach number jets. This power law suggests that (fortunately!) free-space turbulence is a highly inefficient radiator of sound energy. Goldstein (1976) has showed that the ratio

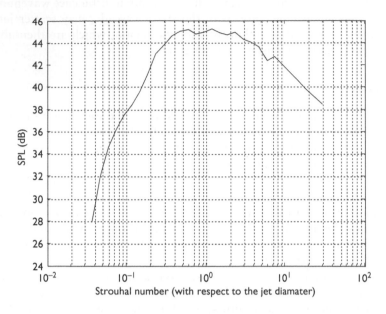

Figure 7.10 Semi-empirical prediction of mixing noise radiated from a jet of 75 cm diameter, 0.3 Mach number, at 90° to the jet axis at a distance of 7.5 m.

of radiated acoustic power W to the total available hydrodynamic power $W_{kin} = \frac{\pi}{8}\rho_0 U^3 D^2$ associated with the kinetic energy flux is given by $W/W_{kin} = 8 \times 10^{-5} M^5$. Thus, at $M = 0.3$, only 10^{-8} of the available hydrodynamic power is converted to sound.

Although predictions obtained from acoustic analogy match experimental results reasonably well, modifications are required to take account of important effects present in a jet engine. The three most important effects are: source convection effects, flow–acoustic interaction effects and the effects of temperature. These are discussed briefly in Sections 7.3.4.1–7.3.4.3.

7.3.4.1   Source convection effects

The compactness assumption made in the derivation of the Lighthill U^8 power law (see Section 7.2.3) begins to break down for jet velocities approaching the sound speed. Lighthill showed that the restriction on Mach number could be relaxed by allowing the quadrupole sources T_{ij} to convect at an effective velocity M_c. With source convection taken into account, the far-field intensity at angle θ to the jet axis becomes

$$I(\mathbf{x}, \theta) \sim \rho_0^2 M^8 \frac{D^2}{|\mathbf{x}|^2} (1 - M_c \cos\theta)^{-5} \quad (M \ll 1) \tag{7.21}$$

where, typically, $M_c \approx 0.7M$. Convection effects are therefore predicted to be greatest along the jet axis with the radiation being preferentially directed in the downstream direction $\theta = 0°$, with zero change being predicted for $\theta = 90°$. Enhanced sound power radiation due to source motion is referred to as convective amplification. At higher jet speeds this effect is more pronounced. By extending Lighthill's dimensional argument to include the effect of source convection, Ffowcs-Williams (1963) found that for very high-speed jets, the power of the radiated noise should vary as the third power of the jet velocity, i.e. $W \propto U^3$. This and the Lighthill U^8 power law are the two most important and celebrated results of classical acoustic analogy theory.

7.3.4.2   Mean flow–acoustic interaction

Sound generated by the quadrupole sources must first propagate through the jet flow before reaching the far field. In a simplified model of jet flow, in which the velocity is uniform over the cross-section, a wavefront in the jet propagates at the sound speed c_s plus the local flow velocity Mc_0. Outside of the jet, it propagates at the sound speed c_0. Application of Snell's law at the fluid boundary shows that the range of angles θ_0, over which rays emerge from the jet unattenuated, is confined to angles above a critical angle θ_c given by:

$$\cos \theta_c = \frac{1}{\dfrac{c_s}{c_0} + M} \tag{7.22}$$

Below the critical angle, $0 \leq \theta_0 \leq \theta_c$, the intensity is much weaker. This range of emergence angles is commonly referred to as the cone of silence. Note that for typical flow conditions, $\theta_c \approx 60°$. In practice, the jet has a non-uniform profile, which varies with distance along the jet axis. Leakage of sound into the cone of silence is therefore observed. Nevertheless, experimentally, the intensity has been observed to drop by more than 20 dB in the cone of silence. Lilley (1974) has proposed a formal solution to this problem for parallel shear flows, in which the operator on the left-hand side of the wave equation is the linearised Euler equation. Further information on the Lilley acoustic analogy can be found in Goldstein (1976) and Tester and Morfey (1976).

7.3.4.3   Temperature effects on jet-mixing noise

Jet noise measurements by, for example, Hoch *et al.* (1973) and Lush *et al.* (1973) have shown that increasing jet temperature increases noise radiation at low jet velocities, but reduces the noise at high velocities. Lush was the first to propose a physical model for this observation. He suggested that in a source region in which the speed of sound differs from that in the ambient fluid, it was no longer permissible to ignore the term $p' - c_0^{-2}\rho'$ in the Lighthill stress tensor, as was done in Section 7.2.1 for isothermal jets. Morfey's (1973) analysis of this problem indicated that the corresponding source term on the right-hand side of equation (7.3) was of the form $-\partial^2(p' - c_0^{-2}\rho')/\partial t^2$, which he showed radiated as a volume distribution of dipole source. Dimensional analysis applied to this source term suggests that a heated volume of fluid with density ρ_s, with a temperature difference of ΔT compared to the ambient fluid of temperature T_0, predicts the following scaling law for the far-field intensity:

$$\bar{I} \sim \frac{\rho_s^2 U^6 D^2}{\rho_0 c_0^3 |\mathbf{x}|^2} \left(\frac{\Delta T}{T_0}\right)^2 \tag{7.23}$$

Thus, sound is radiated whenever a temperature exists between the jet and the ambient fluid. Whilst equation (7.23) explains the observed increase in noise with temperature at low jet velocities, it cannot explain the converse behaviour at higher velocities. In this case, the quadrupole term of equation (7.6) might be expected to dominate, and this is predicted to be temperature-independent. One explanation for the failure of this model has been proposed by Morfey (1973). He suggested that at higher Mach numbers the wavelength becomes comparable to the shear layer thickness, and therefore, rather than the sources radiating into the ambient fluid, they may be considered to be radiating into a fluid whose properties are similar to those of the source region. By replacing in the analysis the ambient values (ρ_0, c_0) by those of the source region (ρ_s, c_s), the observed high-velocity dependence is correctly predicted.

7.3.4.4 Modern developments in jet noise

Although the acoustic analogy has formed the principal theoretical basis of our understanding of jet noise for the last 50 years, the legitimacy of its basic philosophy has recently been questioned, most notably by Tam and Aurault (1999). However, Morris and Farassat (2002) have shown that, whilst the Tam and Aurault model yields better agreement with experimental noise measurements than that given by the acoustic analogy for the same level of empiricism, it is not because of some basic flaw in the acoustic analogy. Instead, it is because of the assumptions made in the approximation of the turbulent source statistics. When consistent assumptions are made, the Tam and Aurault model and the acoustic analogy yield identical noise predictions.

With the acoustic analogy firmly established as the basis of aeroacoustic theory, current jet noise research has mainly focused on gaining clearer understanding and better predictions of turbulence in jets. Experimental techniques, such as laser doppler anenometry (LDA) and particle velocity imaging (PVI), are beginning to be used to image the turbulent flow in jets and hence provide greater insight into the complex behaviour of jet turbulence.

Computational fluid dynamics (CFD) techniques are also increasingly being used for making jet noise predictions based on direct calculations of turbulence velocity fields. Previously, predictions of turbulence statistics in jets were inferred from mean flow similarity laws (for example, Tam *et al.* 1996). The computation, however, is challenging, mainly due to the large differences in amplitude and scale between the turbulence and the sound fields. Nevertheless, recent work has indicated that, providing sufficient care is taken with the boundary conditions, and in the construction of the mesh, modern CFD techniques offer the potential to provide considerable insight into the mechanisms of sound generation by turbulent. Some recent examples of the use of CFD for jet noise prediction are discussed.

Mitchell *et al.* (1999) have used direct numerical simulation (DNS) to compute sound generated by vortex pairing in an axisymmetric jet. Here, the compressible Navier-Stokes equation is solved directly to compute the vorticity and the acoustic fields simultaneously. These noise predictions are in good agreement with predictions obtained from the Kirchhoff surface method (see Section 7.5.3) and the acoustic analogy. Although the jet in this case was not fully turbulent, the application of DNS to this problem provides valuable guidelines of its potential application to more challenging problems, once the necessary computing power becomes available. Self and Bassetti (2003) have used Reynolds-Averaged Navier-Stokes (RANS) models to compute the equivalent aerodynamic sources in an isothermal jet. RANS models solve the Reynolds equations for the mean-velocity field, from which the turbulence quantities are inferred from a turbulence model. The results obtained are shown to be in close agreement with those from an empirical database over a wide range of frequencies and radiation angles.

Large Eddy simulation (LES) is another promising technique for making jet noise predictions. Here, the larger scales of turbulence motion are represented whilst the effects of the smaller scales are modelled. Hu *et al.* (2003) have used this approach, coupled with a Ffowcs-Williams–Hawkings solver to compute the sound radiation from a subsonic plane jet at a Mach number of 0.9 and a Reynolds number of 2000. Different subgrid scale models were used to model the smaller scales of motion, and the results compared with predictions obtained using DNS, in which all scales are represented 'exactly'. A two-dimensional slice through the DNS prediction is plotted in Figure 7.11. The lower half depicts contours of vorticity magnitude (11 equally spaced levels); the upper plot shows contours of dilatation (10 equally spaced levels). Dilatation is the divergence of velocity and is used here as a representation of sound radiation. It relates to the irrotational part of the velocity field, which in the far field is due entirely to acoustic

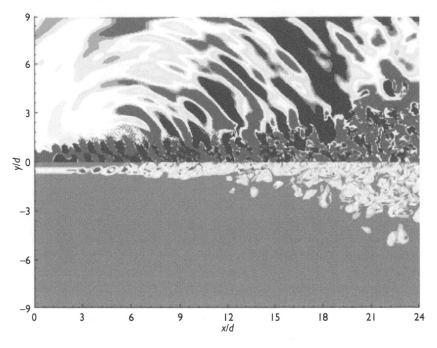

Figure 7.11 DNS prediction of vorticity (lower half) and dilatation (upper half) contours through a two-dimensional slice of a plane jet with 0.9 Mach number and Reynolds number of 2000 (after Hu *et al.* 2003).

radiation. Consequently, the variation in dilatation shown in Figure 7.11 exhibits wave-like behaviour, which appears to originate roughly six jet diameters down stream of the jet nozzle, corresponding to the end of the potential core.

One of the main conclusions from this study is that jet noise calculations derived from LES turbulence predictions are sensitive to the choice of subgrid model. RANS approaches are also dependent upon a choice turbulence model, and are therefore ultimately unsatisfactory. In view of this, and the fact that DNS predictions for full three-dimensional geometries at realistic Reynolds numbers require computational power significantly beyond that which is currently available in terms of processor speed and memory, CFD is, therefore, still a long way off from being a reliable and practicable tool for making jet noise predictions. Nevertheless, with the relentless improvement in computing power, these and many other preliminary studies suggest that CFD, combined with a form of the acoustic analogy, will one day provide an indispensable tool for the accurate prediction of not only jet noise, but also of the many other engine noise sources already described.

7.4 Airframe noise

With increasingly effective design and treatment on the power plant, a significant component of the overall noise of modern aircraft now originates from flow over the airframe. In this section we present a review of these airframe noise sources within the general context of the theories of aero-dynamic noise generation outlined in the Section 7.2. Figure 7.12 shows a typical noise map of 'sources' on an aircraft in its landing configuration with the engines idling. Such maps are generated by data recorded from an array of microphones on the ground during an aircraft flypast (see, for example, Piet *et al.* 2002). Some care is needed in the interpretation of noise maps since, because of the finite spatial resolution of the array, they tend to emphasise localised sources and may underestimate the total integrated noise from other large areas of wing or fuselage.

Data from such maps, together with evidence from other flight and wind tunnel tests, indicate, nonetheless, that the dominant sources of airframe noise are the landing gears and the high-lift devices (leading-edge slats and trailing-edge flaps) that are deployed to increase the lift coefficient of

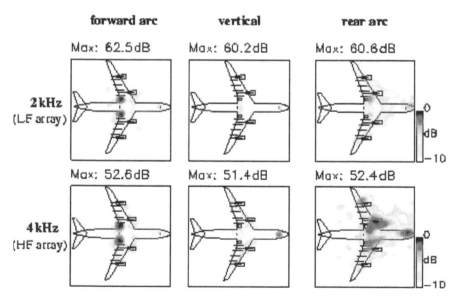

Figure 7.12 Noise map of an aircraft in flight with flaps, slats and landing gear deployed and with the engine at idle.

the wings during takeoff and landing. These components are considered separately in the following sections.

7.4.1 Noise from wings and high-lift devices

There is a significant number of identifiable noise source mechanisms associated with the wings, slats and flaps, the main ones being depicted in Figure 7.13. The sources shown are located at leading, trailing or side edges. Because they generally obey approximately M^5 power laws they are the most efficient radiators of sound at low speeds. Other sources arising from noise radiation by the turbulent flow over the wing surfaces themselves obey the M^6 power law and are therefore of secondary importance at low speeds.

7.4.1.1 Trailing-edge noise sources

Noise radiated due to flow past a trailing edge was the subject of considerable research into propeller and helicopter rotor noise in the 1970s. Its relevance for airframe noise was recognised early on, and the review by Crighton (1991) includes a section on the subject with a detailed description of the approaches adopted by various researchers. Experimental confirmation of theoretical results is presented. A deficiency of the early rotor noise research, when applied to wings and high-lift devices, is that these generally comprise isolated airfoils, which are compact on the wavelength scale. One or both of these assumptions are breached for all of the trailing-edge sources shown in Figure 7.13.

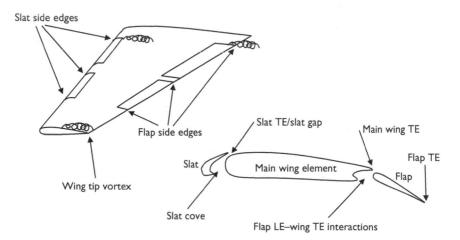

Figure 7.13 Sources of aerodynamic noise associated with wings, flaps and slats.

Current understanding of trailing-edge noise, which is described in Section 7.2.5, originates from the work of Ffowcs-Williams and Hall (1970) who derived the following expression for the noise due to turbulent flow convected past a trailing edge with sweep angle β as:

$$\overline{p^2} \approx \rho_0^2 \overline{u_2^2} V^2 M_c \left(\frac{Ll_3}{R^2}\right) \frac{\sin \alpha \sin^2\left(\dfrac{\theta}{2}\right) \cos^3 \beta}{(1 + M \cos \theta)^4} \tag{7.24}$$

Here $\overline{u_2^2}$ is the mean square turbulence velocity in the frequency band-width of interest, V is a typical mean flow velocity and M_c is the convection Mach number of the turbulence and α is the azimuthal angle to the observer. Assuming that the three velocity terms scale with M, they combine to give the M^5 dependence discussed in Section 7.2.4. The turbulence velocity u_2 is Strouhal number-dependent so that the spectrum of trailing edge noise varies with the flow speed. The angles α and β define the orientation of the edge relative to the flow. In practice, $\sin \alpha$ and $\cos \beta$ are close to unity for the airframe noise problem and can therefore be neglected.

In equation 7.24 the directivity of the trailing-edge source is controlled by two terms: $\sin^2(\theta/2)$ and $(1 + M \cos \theta)^{-4}$. The first term is the characteristic, high-frequency directivity of trailing-edge noise and indicates that sound is directed predominantly into the forward arc. The second factor is due to convective amplification, which further increases the level in the forward arc and decreases its importance in the rear arc. Confirmation of this behaviour is observed in the experimental results of Brooks and Hodgson, as shown in Figure 7.14. Results are compared with the theoretical directivity $\sin^2(\theta/2)$ and $\sin^2(\theta)$ for an infinite flat plat and a rod, respectively.

The work of Ffowcs-Williams and Hall was supplemented by Amiet (1976) and Howe (1978), with experimental confirmation by Kambe (1986) for the case of a vortex flowing past a trailing edge. An outline of the Amiet trailing-edge noise model, which is formulated in terms of the surface pressure spectrum upstream of the edge and the frozen gust assumption, is given in Section 7.2.5.

Of particular importance in Howe's theory is the question of whether or not to apply the Kutta condition, an aerodynamic boundary condition that ensures that the flow leaves the edge of the plate smoothly and which thus determines the strength of the vorticity that is spread into the wake. Howe showed that the far-field sound pressure level predicted when no Kutta condition was applied, was up to 10 dB above the level predicted when it was applied. He also established the conditions under which it is admissible to use the surface pressure close to the trailing edge to predict the far-field noise. A major factor, which may be expected to enhance the importance of several of the trailing-edge sources identified in Figure 7.13, is the high-flow speed through the slat and the flap gaps. Depending on

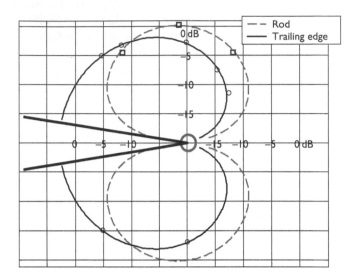

Figure 7.14 Comparison of theoretical and measured directivities of trailing-edge noise at $M = 0.1$ (O) compared with a dipole source on a rod (□) (after Brooks and Hodgson 1981).

the angle of slat and flap deployment, this may be up to 50 per cent higher than the flight speed of the aircraft.

The discussion so far concerns airfoils with sharp trailing edges, whereas for practical reasons, the edges are often blunt. Although this is unlikely to alter the acoustic scattering of the edge, it does modify the spectrum of turbulence shed into the wake, and hence the resultant noise. The spectrum of trailing-edge noise is fairly narrow band and centred approximately on a Strouhal number of 0.2, defined with respect to trailing-edge thickness. The trailing edge of aircraft wings may be up to 10 mm thick; so for a flow of about 80 m/s the peak frequency in a frame of reference moving with the aircraft would be approximately 1600 Hz. On the ground, in the forward arc, the peak frequency will be Doppler-shifted to higher frequencies. Trailing edge-bluntness noise has been identified in studies on wind turbines (for example, Grosveld 1985), but the lack of mention of it in publications on airframe noise means either that, it is not a major contributor to overall airframe noise, or that it has not yet been recognised on aircraft.

7.4.1.2   Leading-edge noise sources

Noise radiated by an airfoil in a turbulent flow was described in Section 7.2.5. For the airframe noise sources identified in Figure 7.13, this mechanism is likely to be particularly important for noise from the leading edge of the flaps due to the impinging wake from the main airfoil. It is also a possibility for noise to be generated on the main wing element by turbulence shed by the slats. Inflow-turbulence noise was also of interest for helicopter rotors; one generally accepted model is that developed by Amiet (1975a). The mean square pressure predicted by the Amiet model is:

$$S_{pp}(x, R, \omega) \approx \left(\frac{\rho_0 \omega c M \sin \theta}{R(1 + M \sin \theta)} \right)^2 d \, |L(x, K_x, 0)|^2 \, l_3(\omega) \Phi_{22}(\omega) \qquad (7.25)$$

where d is the semi-length of the aerofoil, $L(x, K_x, k_y)$ is an aerofoil response function, which specifies the far-field pressure at a position with streamwise coordinate x, due to an incident harmonic upwash velocity with component wavenumbers $(K_x = \omega/U_c, k_y)$, $l_3(\omega)$ is the spanwise correlation length and $\Phi_{22}(\omega)$ is the frequency spectrum of the upwash velocity fluctuations. The Amiet model was applied by Molin and Roger (2000) who, with appropriate assumptions about the turbulence spectrum $\Phi_{22}(\omega)$, found reasonable agreement with scale model measurements.

The directivity of leading-edge noise is not discussed explicitly by Amiet, whose analysis focused on the sound field at 90° to the plane of a compact rotor blade. It seems plausible from physical considerations, however, that for non-compact airfoils the trailing-edge $\sin^2(\theta/2)$ directivity might also apply to the leading-edge noise, though with the angle taken from the airfoil leading edge. This would suggest that noise from the leading edge of flaps would be radiated into the rear arc of the aircraft.

7.4.1.3 Side-edge noise sources

The side edges of slats and flaps were identified as sources of intense noise in the experimental work of Ahtye et al. (1979), Fink and Schinkler (1979) and Kendall and Ahtye (1980). More recently, these findings have been qualitatively confirmed for aircraft in flight using noise maps, such as Figure 7.11, though the importance of side edge noise relative to that from other sources is yet to be firmly established.

Models for side-edge flap noise have been proposed by Hardin (1980), Howe (1982), Sen (1996) and Guo (1997). The fundamental mechanism arises from the discontinuity in the aerodynamic lift force, which generates a vortex, but the modelling of this process has been tackled differently by the various researchers. The Hardin model is based on a physical picture of the boundary layer vorticity being swept around the edge by spanwise flow on the flap – a chordwise oriented line vortex. The Howe model, however, considers the flap edge as a slot in an otherwise plane trailing edge. Sen's is similar to that of Hardin's in that it uses a vortex to represent the side-edge flow, but models the noise generation as being due to perturbation of this main vortex by secondary vortices or upstream turbulence disturbing the equilibrium. The model was used as the basis of the study by Martin (1996), who calculated the far-field noise using the equation developed by Ffowcs-Williams and Hawkings (1969b).

7.4.1.4   Noise from cavities

Aircraft wing surfaces often have cavities, such as fuel vents, and these can be significant sources of tonal noise. These sources may be associated with Helmholtz volume resonator effects, or with 1/4 wave effects in pipes exposed to the flow. Such sources are quite easy to deal with by careful design of the orifice.

7.4.2   Noise from landing gears

Whereas aerofoil noise was a major area of fundamental research in the 1970s, because of its importance for propellers and rotors, the landing gear noise problem has received far less attention. Amongst the earliest significant works was that of Heller and Dobrzynski (1978) who carried out tests on model scale landing gears. It is now realised that such tests often give inaccurate results because of the lack of detail in the scale models.

For many years, the standard model of landing gear noise was that produced originally by Fink (1977), and later implemented in ESDU (1990). Variants of this model have been used by airframe manufacturers to scale landing gear noise data from flight test, rather than making absolute predictions. The Fink model is quite basic in its scaling laws, and more recently, there has been a major drive for much more accurate predictions of the noise source distribution on a landing gear noise with a view to optimising the design of noise control fairings. In practice, however, the geometrical and aerodynamic complexity of the problem has meant that there is a continued reliance on semi-empirical noise models, though with the benefit of far more extensive databases. The empirical model developed by Smith and Chow (1998, 2002) was derived from the results of a comprehensive test programme using full-scale landing gears installed in a wind tunnel (Dobrzynski and Buckholz 1997), as shown in Figure 7.15.

Tests on full-scale landing gear in a wind tunnel are complex and expensive but are necessary in order to achieve the correct physical detail on the gear at the correct Reynolds number. Data obtained during these tests have to be corrected for a range of wind tunnel effects, for example the refraction of sound through the shear layer of the wind tunnel core flow, as described by Amiet (1975b). Even so, the tests are only approximations to a gear actually installed on an aircraft, since the distribution of flow over the gear is not the same as for a gear installed under a lifting wing surface.

The primary structure of a landing gear comprises a main strut with integral shock absorber and stays to take the lateral and the drag forces. There are also a considerable number of secondary components associated with the hydraulic folding and stowing mechanisms, the steering system on the nose gear and the braking system on the main gear. Small secondary components are often referred to as 'dressings' which, together with the joints and bluff shape of the main structural components, are believed to

Figure 7.15 Landing gear noise tests in the DNW anechoic open-jet wind tunnel.

be important sources of high-frequency noise. The wheels and hatch doors or fairings are also potential noise sources.

7.4.2.1 Noise from simple bluff bodies: fluctuating lift and drag forces

The basic scaling law for the aerodynamic noise generated by a compact solid body in a flow was derived by Curle (1955). Using this theory, the essence of recent noise models has therefore been to refine the geometrical description of the landing gear to take better account of the size of individual components and the local flow distribution over the gear. The noise sources of individual components may be modelled as a distribution of compact dipoles whose strength is determined by the fluctuating drag or lift force on each component. Thus the far-field acoustic pressure may be written as

$$\overline{p^2} \quad = \frac{(\rho_0 c_0^2)^2}{R^2} M^6 L D\, f(\text{St}_D) \qquad\qquad (7.26)$$

where L is the characteristic length of the component, D is its width or diameter, R is observer radius and $f(\mathrm{St}_D)$ is a non-dimensional function of the Strouhal number, defined with respect to D, which represents the dipole strength. The noise is predicted to follow an M^6 power law, and this has been confirmed by the experimental results.

The non-dimensional spectrum $f(\mathrm{St}_D)$ incorporates a range of physical phenomena which are as yet poorly understood, such as, for example, the dependence of the shed turbulence spectrum on the detailed geometry of the component, and the efficiency with which sound is radiated from the surface dipoles when there is scattering by edges, joints and dressings. It may also be expected that $f(\mathrm{St}_D)$ depends on Reynolds number. This simple scaling method therefore provides a means of comparing noise radiation from geometrically similar components under similar flow conditions.

Uniform cylindrical rods shed vortices at a frequency corresponding to a Strouhal number of 0.2 over a wide range of Reynolds numbers (Blevins 1977). When a feedback mechanism is present, such as a resonant dynamic response of the structure, the result is a discrete frequency noise (aeolian tone). In the absence of a feedback mechanism the noise radiation is broadband, though peaking at the expected Strouhal number. Such a spectrum is shown in the first curve (clean strut) of Figure 7.16.

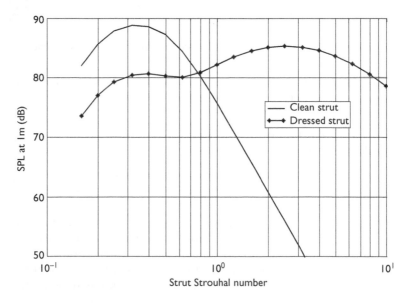

Figure 7.16 Non-dimensional source spectra for clean and dressed struts as proposed by Smith and Chow (1998).

7.4.2.2   Noise breakdown by component

The struts of a landing gear are rarely sufficiently regular to produce a narrow spectral peak. Smith and Chow (1998) postulated that the spectrum broadens with the increasing irregularity of the strut profile and density of small components and joints attached to it. This effect is shown in the second curve (dressed struts) of Figure 7.16, where some noise is radiated at the original frequency, but there is also a broad spectrum of high-frequency noise produced by the dressings. The high-frequency noise is expected to scale with the Strouhal number of the small components, based on a nominal component diameter, and the broad spectral shape function at high frequencies reflects the expected variation in component sizes. These non-dimensionalised spectra form the basis of a semi-empirical noise model from which detailed noise source breakdowns may be obtained, an example of which is shown in Figure 7.17. This figure also includes the spectrum of noise that was associated with the wheels of the landing gear.

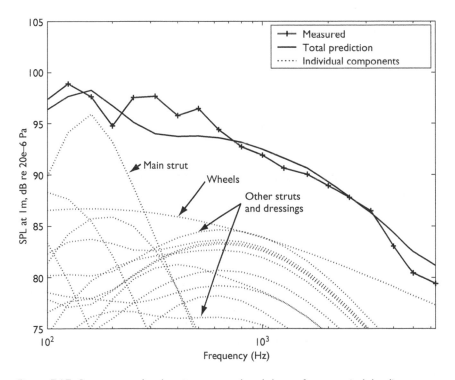

Figure 7.17 Component level noise source breakdown for a typical landing gear as proposed by Smith and Chow (1998).

A complete model of aircraft noise must also take account of a variety of other effects such as atmospheric attenuation and installation effects. One installation effect that has already been mentioned is the altered flow over the main landing gear due to the circulating flow around the wing. A second effect, suggested by Crighton (1991), is that, as the gear is installed under a large wing, there will be a modified directivity due to the image source. This type of effect is again impossible to calculate accurately for a large distributed source such as a landing gear and is a candidate for an empirical directivity model.

7.5 Numerical methods for predicting aerodynamic noise

Since 1990 there has been a big increase in the research effort aimed at numerical prediction of aerodynamic noise, which is the principal objective of the new field of computational aeroacoustics (CAA). This section provides an overview of its current status. More detailed reviews of the subject are provided in Hardin and Hussaini (1993), which contains a number of papers on various aspects of CAA, Lyrintzis (1994), which considers Kirchhoff methods in CAA, and Tam (1995), which reviews the computational methods of CFD and their extension to CAA.

CAA is capable, in principle, of modelling both the aerodynamic sound source and the propagation to the far field. The principle constraint is that the computing resources required to solve the full problem are beyond current technology in terms of both computer storage and numerical accuracy. Limiting the computational domain control to the required resources introduces a number of secondary problems concerning numerical methods and the modelling of boundary conditions.

The general equation governing unsteady fluid flow, including acoustic wave propagation, is the Navier-Stokes equation

$$\rho \frac{Dv_i}{Dt} + \frac{\partial T_{ij}}{\partial x_j} = 0 \tag{7.27}$$

where the T_{ij} is the Lighhill stress tensor defined in Section 7.2.1. When viscosity is neglected, the Navier-Stokes equation reduces to the Euler equation, which is also one of the bases of the linearised acoustic wave equation, but inherently excludes sources of aerodynamic noise:

$$\rho \frac{Dv_i}{Dt} = -\frac{\partial p}{\partial x_i} \tag{7.28}$$

Direct solution of the non-linear Navier-Stokes equation is very difficult. Lyrintzis (1994) outlines three alternative methods by which a compromise solution of the sound propagation problem can be achieved:

1   Direct numerical solution of the wave equation using the source terms (surface pressures) derived from a CFD calculation, as required by Lighthill's acoustic analogy (equation (7.4)).
2   Use non-linear CFD (Navier-Stokes, equation (7.26)) to calculate the source region and a near-field domain, and use linearised non-viscous CFD (Euler, equation (7.27)) to compute the far field.
3   Use CFD to calculate the near field, particularly on a surface enclosing the source region, and use the Kirchhoff integral over this surface to compute the far-field radiation.

At present there is much debate about the merits of each method. A number of researchers have compared the results from the various methods. Compromises for computing solutions in mixed domains are quite common in acoustics (for example, finite element/boundary element methods for structure–fluid interaction), and it seems likely that one of these hybrid methods will provide the most effective solution for the aerodynamic noise problem.

7.5.1   Acoustic analogy methods

A review of the acoustic analogy approach is presented by Farassat (1993), who is one of the main proponents of this method. One of the reasons he gives for preferring the acoustic analogy to the Kirchhoff method is that the precision of the aerodynamic calculation usually falls off in the region where the boundary surface is located. It should be noted that Farassat's particular interest is engine and helicopter rotors where the moving sources add to the complexity of the Kirchhoff method (Farassat and Myers 1988). The acoustic analogy and Kirchhoff methods were compared by Brentner et al. (1997) who found that the Kirchhoff method was somewhat sensitive to the location of the Kirchhoff surface if it was positioned too close to the rotor blade, and that the acoustic analogy method was not as sensitive to the extent of volume included in the quadrupole calculation. The computational requirements of both methods were comparable.

The fact that there are difficulties with quadrupoles in the acoustic analogy method is perhaps the reason why Farassat (1993), in listing the successful applications of the method, concentrates on the prediction of deterministic sources such as periodic blade thickness noise, blade–vortex interaction noise and high-speed impulsive noise.

Zhang et al. (1995) used the acoustic analogy to predict the far-field noise radiation from an unsteady supersonic flow over a cavity. The near-field model was based on a solution of the RANS on a rectangular grid of

160×200 cells with appropriate boundary conditions at the solid surface and at the edge of the computational domain. Instabilities were induced by applying a turbulence model. It was found that the calculated near-field pressure fluctuations compared well with experimental results, but no data were available to validate the far-field noise predictions.

A more relevant example for the problem of airframe noise is provided by the companion studies of Khorrami *et al.* (1998) who solved the RANS equation and Streett (1998) who predicted the pressure fluctuations in a flap-edge flow field using a simplified flow model. The model of Khorrami used a mesh of 19 million grid points, requiring 50 h of a Cray C-90 CPU; even so, Streett argued that the resolution of the unsteady flow field is inadequate for noise predictions. Instead, he used three approximations to reduce the complexity of the problem: the acoustic analogy to predict noise radiation, the incompressible Navier-Stokes equations to simulate the near field and a temporal method of reducing the full three-dimensional flow to a series of two-dimensional calculations.

7.5.2   CFD plus the linearised Euler equation

In this method the source of aerodynamic noise is modelled directly using non-linear CFD methods in an interior region, and then a linear equation is used to model propagation to the far field. Particular problems with the method are the specification of the boundary conditions at the interface between the interior and the exterior regions and the adequate meshing of the far field. These problems were examined by Freund *et al.* (1993) and Hardin and Pope (1995) who used the method to predict sound radiation from flow over a cavity. The general problems of direct simulation of turbulence using CFD are considered in a number of papers in the book edited by Hardin and Hussaini (1993). Some aspects of the numerical implementation of the Euler method (particularly the boundary conditions) are discussed in Goodrich (1997) and Goodrich and Hagstrom (1997).

7.5.3   Kirchhoff method

In the Kirchhoff method the pressure and velocity on a surface bounding the source region are calculated using CFD as in the previous method, and then the far field is calculated from an integral over the surface. The same integral is the basis of the boundary element method in acoustics. The paper by Lyrintzis (1994) provides a review of the use of the Kirchhoff method. Much of the research using this method has concerned noise radiation from rotors, particularly with a transonic blade. These problems require special treatment because of the motion of the source, and Farassat and Myers (1988) developed appropriate theory to cover this case.

A major problem with the Kirchhoff method is that the surface should entirely enclose the source region but lie outside the turbulent flow. In the case of a body with a turbulent wake, this can be a very extended region. As a result, it is commonly assumed that the function to be integrated is zero in regions where it cannot be evaluated. This problem has been addressed by Freund *et al.* (1993) who developed a method of correcting the surface integral to account for missing portions. They claim that this significantly reduced errors in a model problem. Lyrintzis *et al.* (1997) compared the Kirchhoff and the acoustic analogy methods with experimental data for test cases with both high-speed impulsive and blade–vortex interaction noise. Computed results showed that both computational methods gave similar results for cases where there is no transonic flow and required similar computational resources (which were in any case small when compared to those of the CFD calculation).

A boundary element method for predicting sound radiation from a body in a low Mach number flow was developed by Astley and Bain (1986). The integration is taken over the surface of the body, and this is achieved by using linearised equations in the near-field domain. This method was used by Guidati *et al.* (1997) to predict inflow-turbulence noise on airfoils. Unlike the combined CFD–Kirchhoff method, the source terms are not predicted using the numerical model, but are assumed to be represented by some incident vorticity. It was found that, although the absolute levels of noise radiation were not reliably predicted, changes in noise due to modification of the airfoil were predicted with reasonable accuracy. The influence of the wavenumber spectrum of the inflow-turbulence has been examined by Devenport *et al.* (1997).

7.5.4 Validation of numerical studies

The book by Hardin and Hussaini (1993) contains a section entitled 'Validation Methodology', and the section summary includes some useful comments on the problem of validation, particularly the observation that there is much experimental information available that can be used to validate models, and also relatively simple aeroacoustic phenomena for which there are theoretical solutions available. An indication of the current status of the various numerical models and calculation procedures may be gleaned from the fact that the most common comparisons in recent papers are between two forms of numerical modelling rather than with experimental data.

Acknowledgements

The authors would like to express their gratitude to Drs Mike Fisher, Alan McAlpine, Rod Self and Zhiwei Hu for helpful discussions during the preparation of this chapter.

References

Adamczyk, J.J. and Brand, R.S. (1972) Scattering of sound by an aerofoil of finite span in a compressible stream. *Journal of Sound and Vibration*, 25, 139–156.

Ahtye, W.F., Miller, W.R. and Meecham, W.C. (1979) Wing and flap noise measured by near and far-field cross-correlation techniques. AIAA-79-0667.

Amiet, R.K. (1975a) Acoustic radiation from an airfoil in a turbulent stream. *Journal of Sound and Vibration*, 41, 407–420.

Amiet, R.K. (1975b) Correction of open jet wind tunnel measurements for shear layer refraction. AIAA-75-532.

Amiet, R.K. (1976) Noise due to turbulent flow past a trailing edge. *Journal of Sound and Vibration*, 47, 387–393.

Astley, R.J. and Bain, J.G. (1986) A 3-dimensional boundary element scheme for acoustic radiation in low mach number flows. *Journal of Sound and Vibration*, 109(3), 445–465.

Blevins, R.D. (1977) *Flow Induced Vibration* (2nd edn) (Van Nostrand Reinhold), ISBN 0-442-20651-8.

Brentner, K.S., Lyrintzis, A.S. and Koutsavdis, E.K. (1997) Comparison of computational aeroacoustic prediction methods for transonic rotor noise. *Journal of Aircraft*, 34(4), 531–538.

Brooks, T.F. and Hodgson, T.H. (1981) Trailing edge noise prediction from measured surface pressures. *Journal of Sound and Vibration*, 78(1), 69–117.

Brooks, T.F., Pope, D.S. and Marcolini, M.A. (1989) Airfoil self-noise and prediction. *NASA reference publication 1218*.

Crighton, D.G. (1991) Aeroacoustics of Flight Vehicles, Theory and Practice, Volume 1 Noise Sources, Chapter 7, Airframe Noise. *NASA Technical Report 90-3052*.

Curle, N. (1955) The influence of solid boundaries upon aerodynamic sound. *Proceedings of the Royal Society*, A231, 505–514.

Devenport, W.J., Wenger, C.W., Glegg, S.A.L. and Miranda, J.A. (1997) Wavenumber frequency spectra of turbulence in a lifting wake for broadband noise prediction. *3rd AIAA/CEAS Aeroacoustics Conference*, AIAA-97-1699-cp.

Dobrzynski, W.M. and Buckholz, H. (1997) Full scale noise testing on airbus landing gears in the German Dutch Wind Tunnel. AIAA-97-1597.

Engineering Sciences Data Unit (ESDU), 1990, Airframe noise prediction. *ESDU Report 90023*, amended 1992.

Evers, I. and Peake, N. (2002) On sound generation by the interaction between turbulence and a cascade of airfoils with non-uniform mean flow. *Journal of Fluid Mechanics*, 463, 25–52.

Farassat, F. (1993) The acoustic analogy as a tool of computational aeroacoustics. In Hardin and Hussaini (eds), pp. 133–155.

Farassat, F. and Myers, M.K. (1988) Extension of Kirchhoffs formula to radiation from moving surfaces. *Journal of Sound and Vibration*, 123(3), 451–460.

Ffowcs-Williams, J.E. (1963) The noise from turbulence convected at high speeds. *Philosophical Transactions of the Royal Society of London, Series A*, 255, 469–503.

Ffowcs-Williams, J.E. and Hall, L.H. (1970) Aerodynamic sound generation by turbulent flow in the vicinity of a scattering half plane. *Journal of Fluid Mechanics*, 40(4), 657–670.

Ffowcs-Williams, J.E. and Hawkings, D.L. (1969a) Sound generation by turbulence and surfaces in arbitrary motion. *Philosophical Transactions of the Royal Society of London, Series A*, **264**, 321–342.

Ffowcs-Williams, J.E. and Hawkings, D.L. (1969b) Theory relating to the noise of rotating machinery. *Journal of Sound and Vibration*, 10, 10–21.

Filotas, L.T. (1969) Theory of airfoil response in a gust atmosphere. Part I – Aerodynamic transfer function. UTIAS-139, Toronto University.

Fink, M.R. (1977) Airframe noise prediction method. FAA-RD-77-29.

Fink, M.R. (1979) Noise component method for airframe noise. *Journal of Aircraft*, 16(10), 659–665.

Fink, M.R. and Schinkler, R.H. (1979) Airframe noise component interaction studies. *NASA Report 3110*.

Freund, J.B., Lele, S.K. and Moin, P. (1993) Matching of near/far-field equation sets for direct computations of aerodynamic sound. AIAA-93-4326.

Ganz, U.W., Joppa, P.D., Patten, J.T. and Scharpf, D. (1998) Boeing 18-inch fan rig broadband noise test. NASA CR-1998-208704.

Glegg, S.A.L. (1998) Airfoil self-noise generated in a cascade. *AIAA Journal*, 36(9), 1575–1582.

Glegg, S.A.L. (1999) The response of a swept blade row to a three dimensional gust. *Journal of Sound and Vibration*, 227(1), 29–64.

Glegg, S.A.L. and Jochault, C. (1997) Broadband self noise from a ducted fan. AIAA-97-1612.

Gliebe, P.R. (1996) Fan broadband noise – the floor to high bypass ratio engine noise reduction. *Noise Con 96*, Seattle, Washington.

Gliebe, P.R., Mani, R., Shin, H., Mitchell, B., Ashford, G., Salamah, S. and Connell, S. (2000) Aero prediction codes. NASA/CR-2000-210244.

Goldstein, M.E. (1976) *Aeroacoustics* (New York: McGraw-Hill International Book Company).

Goodrich, J.W. (1997) High accuracy finite difference algorithms for computational aeroacoustics. *3rd AIAA/CEAS Aeroacoustics Conference*, AIAA-97-1584.

Goodrich, J.W. and Hagstrom, T. (1997) A comparison of two accurate boundary treatments for computational aeroacoustics. *3rd AIAA/CEAS Aeroacoustics Conference*, AIAA-97-1585.

Graham, J.M.R. (1970) Similarity rules for thin airfoils in non-stationary flows. *Journal of Fluid Mechanics*, 43, 753–766.

Grosveld, F.W. (1985) Prediction of broadband noise from horizontal axis wind turbines. *Journal of Propulsion*, 1(4), 292–299.

Guidati, G., Bareiss, R., Wagner, S., Dassen, T. and Parchen, R. (1997) Simulation and measurement of inflow-turbulence noise on airfoils. AIAA-97-1698-cp.

Guo, Y. (1997) On sound generation by unsteady flow separation at airfoil sharp edges. *3rd AIAA/CEAS Aeroacoustics Conference*, AIAA-97-1697.

Hanson, D.B. (1997) Quantification of in-flow turbulence for prediction of cascade broadband noise. *Fifth International Congress on Sound and Vibration*, Australia.

Hardin, J.C. (1980) Noise radiation from the side edges of flaps. *AIAA Journal*, 18(5), 549–562.

Hardin, J.C. and Hussaini, M.Y. (1993) *Computational Aeroacoustics* (Springer-Verlag).

Hardin, J.C. and Pope, D.S. (1995) Sound generation by flow over a 2-dimensional cavity. *AIAA Journal*, 33(3), 407–412.

Heller, H.H. and Dobrzynski, W.M. (1978) *ICAS* Paper GL-03.

Hoch, R.G., Duponchel, J.P., Cocking B.J. and Bryce, W.D. (1973) Studies of the influence of density on jet noise. *Journal of Sound and Vibration*, 28, 649–668.

Homicz, G.F. and George, A.R. (1974) Broadband and discrete frequency radiation from subsonic rotors. *Journal of Sound and Vibration*, 36(2), 151–177.

Howe, M.S. (1978) A review of the theory of trailing edge noise. *Journal of Sound and Vibration*, 61(3), 437–465.

Howe, M.S. (1982) On the generation of side-edge flap noise. *Journal of Sound and Vibration*, 80(4), 555–573.

Hu, Z.W., Morfey, C.L. and Sandham, N.D. (2003) *9th AIAA/CEAS Conference*, Hilton Head, South Carolina, AIAA-2003-3166.

Joseph, P. and Morfey, C.L. (1999) Multi-mode radiation from an unflanged circular duct. *Journal of the Acoustical Society of America*, 105, 2593–2600.

Joseph, P. and Parry, A. (2001) Rotor/wall boundary layer interaction broadband noise in turbofan engines. *Seventh AIAA/CEAS Aeroacoustics Conference*, Maastricht, The Netherlands, Paper AIAA-2001-2244.

Joseph, P., Britchford, K. and Loheac, P. (2003) A model of fan broadband noise due to rotor–stator interaction. *5th European Conference on Turbomachinery*, Prague, Czech Republic.

Kaji, S. and Okazaki, T. (1970) Propagation of sound waves through a blade row, II. Analysis based on acceleration potential method. *Journal of Sound and Vibration*, 11, 355–375.

Kambe, T. (1986) Acoustic emissions by vortex motions. *Journal of Fluid Mechanics*, 173, 643–666.

Kendall, J.M. and Ahtye, W.F. (1980) Noise generation by a lifting wing/flap combination at Reynolds numbers 2.8 e6. AIAA-80-0035.

Khorrami, M.R., Singer, B.A. and Radeztsky, R.H. (1998) Reynolds averaged Navier-Stokes computations of a flap-edge flow field. AIAA-98-0768.

Kobayashi, H. (1978) Three-dimensional effects on pure tone fan noise due to inflow distortion. AIAA-78-1120.

Koch, W. (1971) On the transmission of sound through a blade row. *Journal of Sound and Vibration*, 18, 111–128.

Lansing, D.L., Driscler, J.A. and Pusey, C.G. (1970) Radiation of sound from an unflanged duct with flow. *Journal of the Acoustical Society of America*, 48(1), 75.

Levine, H. and Schwinger, J. (1948) On the radiation from an unflanged circular pipe. *Physical Review*, 73(4).

Lighthill, M.J. (1952) On sound generated aerodynamically. I. General Theory. *Proceedings of the Royal Society of London*, 211A, 1107, 564–587.

Lighthill, M.J. (1954) On sound generated aerodynamically. II. Turbulence as a source of sound. *Proceedings of the Royal Society of London*, 222A, 1148, 1–32.

Lilley, G.M. (1974) On the noise from jets. Noise mechanisms, AGARD-CP-131, pp. 13.1–13.12.

Lilley, G.M. (1995) Jet noise classical theory and experiments, Chapter 4. In: H.H. Hubbard (ed.), *Aeroacoustics of Flight Vehicles* (Acoustical Society of America).

Lorence, C.B. and Hall, K.C. (1995) Sensitivity analysis of the aerodynamic response of turbomachinery blade rows. AIAA-95-0166.

Lush, P.A., Fisher, M.J. and Ahuja, K. (1973) Noise from hot jets *Proceedings of the British Acoustical Society Spring Meeting*, London, Paper 73ANA2.

Lyrintzis, A.S. (1994) Review – the use of Kirchhoffs method in computational aeroacoustics. *Journal of Fluids Engineering-transactions of the ASME*, **116**(4), 665–676.

Lyrintzis, A.S., Koutsavdis, E.K. and Strawn, R.C. (1997) A comparison of computational aeroacoustic prediction methods. *Journal of the American Helicopter Society*, **42**(1), 54–57.

Majumdar, S.J. and Peake, N. (1998) Noise generation by the interaction between ingested turbulence and a rotating fan. *Journal of Fluid Mechanics*, **359**, 181–216.

Mani, R. (1971) Noise due to interaction between inlet turbulence with isolated stators and rotors. *Journal of Sound and Vibration*, **17**, 251–260.

Martin, J.E. (1996) A computational investigation of flap side-edge flow noise. *NASA Technical Report* ID: 19970025364 N 97N24918.

Mitchell, B.E., Lele, S.K. and Moin, P. (1999) Direct computation of the sound generated by vortex pairing in an axisymmetric jet. *Journal of Fluid Mechanics*, **383**, 113.

Molin, N. and Roger, M. (2000) Use of Amiet's methods in predicting the noise from 2D high lift devices, AIAA-2000-2064.

Morfey, C.L. (1969) A note on the radiation efficiency of acoustic duct modes. *Journal of Sound and Vibration*, **9**, 367–372.

Morfey, C.L. (1970) Sound generation in subsonic turbomachinery. *Journal of Basic Engineering, Trans ASM*, 461–491.

Morfey, C.L. (1973) Amplification of aerodynamic noise by convected flow inhomogeneities. *Journal of Sound and Vibration*, **31**, 391–397.

Morris, P.J. and Farassat, F. (2002) Acoustic analogy and alternative theories for jet noise prediction. *AIAA Journal*, **40**(4), 671–680.

Owens, R.E. (1979) Energy efficient engine propulsion system – aircraft integration evaluation. *NASA* CR-159488.

Piet, J.F., Michel, U. and Böhning, P. (2002) Localisation of the acoustic sources of the A340 with a large phased microphone array during flight tests. AIAA-2002-506.

Rangwalla, A.A and Rai, M.M. (1993) A numerical analysis of tonal acoustics in rotor–stator interactions. *Journal of Fluids and Structures*, **7**, 611–637.

Rice, E.J. (1976) Modal density functions and number of propagating modes in ducts. NASA TM-73839.

Rice, E.J. (1978) Multimodal far-field acoustic radiation pattern using mode cutoff ratio. *AIAA Journal*, **16**(9).

Sabah, M. and Roger, M. (2001) Experimental study and model predictions of cascade broadband noise. *7th AIAA/CEAS Aeroacoustics Conference*, Maastricht.

Sears, W.R. (1941) Some aspects of non-stationary airfoil theory and its practical applications. *Journal of Aeronautical Science*, **8**(3), 104–188.

Self, R.H. and Bassetti, A. (2003) A RANS based jet noise prediction scheme. *9th AIAA/CEAS Conference*, Hilton Head, South Carolina, AIAA-2003-3325.

Sen, R. (1996) Local dynamics and acoustics in a simple 2-d model of airfoil lateral-edge noise. *AIAA Aeroacoustics Conference*, AIAA-96-1673.

Smith, M.G. and Chow, L.C. (1998) Prediction method for aerodynamic noise from aircraft landing gears. *AIAA Aeroacoustics Conference*, AIAA-98-2228.

Smith, M.G. and Chow, L.C. (2002) Validation of a prediction model for aerodynamic noise from aircraft landing gear. AIAA-2002-2581.

Smith, M.J.T. (1989) *Aircraft Noise* (Cambridge University Press).

Smith, M.J.T. and House, M.E. (1967) Internally generated noise from gas turbines: measurement and prediction. *Journal of Engineering for Power: Transactions of the American Society of Mechanical Engineers, Series A*, 89, 117–190.

Smith, S.N. (1972) Discrete frequency sound generation in axial flow turbomachines, *British Aeronautical research Council, England*, R&M 3709.

Sofrin, T.G. and McCann, J.C. (1966) Pratt and Whitney experience in compressor noise reduction. *Journal of the Acoustical Society of America*, 40(5), 1248–1259.

Streett, C.L. (1998) Numerical simulation of fluctuations leading to noise in a flap-edge flowfield. *36th Aerospace Sciences Meeting and Exhibition*, AIAA-98-0628.

Sutcliff, D.L., Bridges, J. and Envia, E. (1997) Comparison of predicted low speed fan rotor/stator interaction. AIAA-97-1609.

Tam, C.K.W. (1995) Computational aeroacoustics – issues and methods. *AIAA Journal*, 33(10), 1788–1796.

Tam, C.K.W. (1998) Jet noise since 1952. *Theoretical Computation of Fluid Dynamics*, 10, 393–405.

Tam, C.K.W. and Aurault, L. (1999) Jet mixing noise from fine-scale turbulence. *AIAA Journal*, 37(2), 145–153.

Tam, C.K.W., Golebiowski, M. and Seiner, J.M. (1996) On the two components of turbulent mixing noise from supersonic jets. AIAA-96-1716.

Tester, B.J. and Morfey, C.L. (1976) Developments in jet noise modeling, Theoretical predictions and comparisons with measured data. *Journal of Sound and Vibration*, 46, 79–103.

Tyler, J.M. and Sofrin, T.G. (1962) Axial flow compressor noise studies. *Transactions of SAE*, 70, 209–332.

Von Karman, Th. and Sears, W.R. (1938) Airfoil theory for non-uniform motion. *Journal of Aeronautical Science*, 5(10), 379–390.

Weinstein, L.A. (1969) *The Theory of Diffraction and the Factorisation Method* (Golem Press).

Whitehead, D.S. (1970) Vibration and sound generation in a cascade of flat plates in subsonic flow. Department of Engineering, University of Cambridge, CUED/A-Turbo/TR 15.

Whitehead, D.S. (1987) Classical two-dimensional methods. Chapter 3, AGARD Manual on aeroelasticity in axial flow machines.

Wygnanski, I., Champagne, F. and Marasli, B. (1986) On the structures in two-dimensional, small definite turbulent wakes. *Journal of Fluid Mechanics*, 168, 31–71.

Zhang, X., Rona, A. and Lilley, G.M. (1995) Far-field noise radiation from an unsteady supersonic cavity flow. CEAS/AIAA-95-040.

Zorumski, W.E. (1973) Generalised radiation impedances and reflection coefficients of circular and annular ducts. *Journal of the Acoustical Society of America*, 54(6), 1667–1673.

Chapter 8

Active noise control

S.J. Elliott and P.A. Nelson

8.1 Introduction

Active sound control exploits the destructive interference which can exist between the sound fields generated by an original 'primary' acoustic source and a controllable 'secondary' acoustic source. In order for this destructive interference to be effective, the waveform of the sound field generated by the secondary source must be very close to that of the primary source, and also the spatial distributions of the two sound fields must be well matched in the region where the sound is to be controlled.

The need for close spatial matching between the interfering sound fields means that active sound control is best suited to the control of low-frequency sounds, for which the acoustic wavelength is large. Conventional passive methods of sound control can struggle to attenuate the noise in this frequency region and so the two methods of sound control can be complementary in many applications.

Although the principle of active sound control has been known since the 1930s (Lueg 1936; Figure 8.1), and single-channel analogue control systems were developed in the 1950s (Olsen and May 1953; Conover 1956), it was not until the development of modern digital signal processing (DSP) devices in the 1980s that adaptive digital controllers enabled the technique to be used in many practical problems (Chaplin 1983; Roure 1985). Since then, there has been considerable interest in the commercial application of active sound control, and this has led to a more detailed investigation of its fundamental acoustic limitations (Nelson and Elliott 1992).

At the time of writing, the most successful applications of active sound control are probably in active headsets, which are widely used in both military and civil applications (Wheeler 1987; Rafaely 2000) and in the control of sound inside aircraft (Billout *et al.* 1995; Ross and Purver 1997). Although the active control of both engine noise and road noise at low frequencies in cars was demonstrated some time ago (Elliott *et al.* 1988b; Sutton *et al.* 1994), the automotive application has taken some time to reach mass production, mostly because of cost constraints. Similarly, the

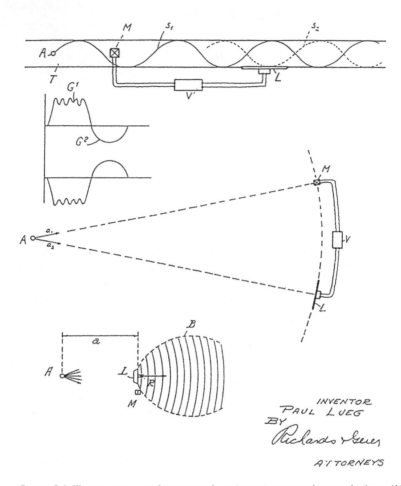

Figure 8.1 Illustrations page from an early active noise control patent by Lueg (1936).

active control of low-frequency sound in air-conditioning ducts has been well studied in the laboratory, but it has been difficult for this technology to compete economically with passive absorbers, except in specialist applications where weight or size are of prime consideration.

In this chapter, the physical mechanisms and limitations of active sound control will be illustrated, initially for plane, freely propagating waves in a duct, and then for three-dimensional sound fields in free space. The different strategies of controlling enclosed sound fields will then be illustrated in a one-dimensional duct before the more realistic problem of controlling sound in a three-dimensional enclosure is considered. Finally, the local control of sound is described, which can generate zones of quietness in specific regions of a space.

8.2 Control of wave transmission

In this section, we consider the active control of sound that is transmitted as plane propagating waves in a uniform duct. For sufficiently low-frequency excitation only *plane waves* of sound can propagate in such a duct. In that case, the waves have a uniform pressure distribution across any section of the duct and obey the one-dimensional wave equation. The fundamental approaches to active wave control can be illustrated using this simple example, but some of the complications involved in controlling higher-order modes in ducts are also briefly discussed at the end of the section.

8.2.1 Single secondary actuator

To begin with, it will be assumed that an incident harmonic sound wave, travelling in the positive-x direction along a duct, is controlled by a single acoustic secondary source such as a loudspeaker mounted in the wall of the duct, as illustrated in Figure 8.2. The duct is assumed to be infinite in length, to have rigid walls and to have a uniform cross-section. The complex pressure of the incident primary wave is expressed as

$$p_{p+}(x) = Ae^{-jkx} \quad \text{for all } x \tag{8.1}$$

where the subscript p+ denotes the primary wave travelling in a positive-x direction or *downstream*. An acoustic source, such as a loudspeaker driven at the same frequency as that of the incident wave, will produce acoustic waves travelling both in the downstream direction and in the *upstream* or negative-x direction, whose complex pressures can be written as

$$p_{s+}(x) = Be^{-jkx} \quad \text{for } x > 0, \quad p_{s-}(x) = Be^{+jkx} \quad \text{for } x < 0 \tag{8.2a,b}$$

where the secondary source has been assumed to be at the position corresponding to $x = 0$, and B is a complex amplitude which is linearly dependent

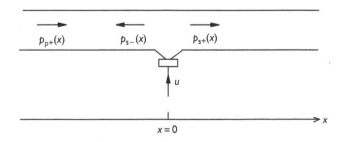

Figure 8.2 Active control of plane sound waves in an infinite duct using a single secondary source.

on the electrical input to the secondary source u in Figure 8.2. If this electrical input is adjusted in amplitude and phase so that $B = -A$, the total downstream pressure will be

$$p_{p+}(x) + p_{s+}(x) = 0 \quad \text{for } x > 0 \tag{8.3}$$

indicating that the pressure will be perfectly cancelled at all points downstream of the secondary source. This suggests that a practical way in which the control input could be adapted is by monitoring the tonal pressure at any point downstream of the secondary source and adjusting the amplitude and phase of the control input until this pressure is zero. However, we are mainly interested here in the physical consequences of such a control strategy and so we calculate the total pressure to the left, on the upstream side of the secondary source, which in general will be

$$p_{p+}(x) + p_{s-}(x) = Ae^{-jkx} + Be^{+jkx}, \quad x < 0 \tag{8.4}$$

If the secondary source is adjusted to cancel the pressure on the downstream side, then $B = -A$, and the pressure on the upstream side becomes

$$p_{p+}(x) + p_{s-}(x) = -2jA \sin kx, \quad x < 0 \tag{8.5}$$

since $e^{jkl} - e^{-jkl} = 2j \sin kl$. Thus a perfect acoustic *standing wave* is generated by interference between the positive-going primary wave and the negative-going wave generated by the secondary source. Notice that this standing wave has nodes of pressure at the position of the secondary source, $x = 0$, and at $x = -\lambda/2$, $x = -\lambda$, etc., where λ is the acoustic wavelength, and that when $x = -\lambda/4$, $x = -3\lambda/4$, etc., its amplitude is exactly twice the amplitude of the incident primary wave. The distribution of the pressure amplitude in the duct under these circumstances is shown in Figure 8.3.

In cancelling the pressure downstream of the secondary source, the pressure at the secondary source location has been driven to zero by the effect of the active control system. The secondary source thus acts to create a pressure-release boundary condition as far as the incident wave is concerned, and effectively *reflects* this wave back up the duct with equal amplitude and inverted phase, which gives rise to the standing wave observed in Figure 8.3. The acoustic power generated by a loudspeaker is equal to the time-averaged product of its volume velocity and the acoustic pressure on the cone surface. The fact that the acoustic pressure at the secondary source location is zero means that the secondary source can generate no acoustic power when operating to cancel the incident wave, and it acts as a purely reactive element.

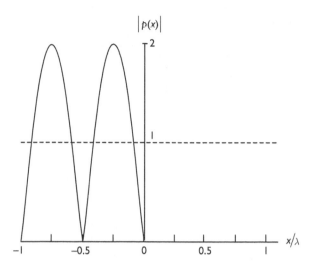

Figure 8.3 Amplitude of the pressure distribution in an infinite duct after a single secondary source at $x = 0$ has been adjusted to cancel an incident plane wave of unit amplitude travelling in the positive-x direction.

8.2.2 Two secondary actuators

It is also possible to use an active control system to *absorb* the whole of the incident primary wave, instead of reflecting it back upstream, but such a strategy requires a pair of secondary sources. This is illustrated in Figure 8.4, where two loudspeakers act as secondary sources, at positions corresponding to $x = 0$ and $x = l$, which can individually generate complex pressures on their upstream and downstream sides which are given by

$$p_{s1+}(x) = Be^{-jkx} \quad \text{for } x > 0, \qquad p_{s1-} = Be^{+jkx} \quad \text{for } x < 0 \qquad (8.6\text{a,b})$$

and

$$p_{s2+} = Ce^{-jk(x-l)} \quad \text{for } x > l, \qquad p_{s2-} = Ce^{+jk(x-l)} \quad \text{for } x < l \qquad (8.7\text{a,b})$$

where the complex amplitudes B and C are proportional to the complex inputs to the two secondary sources u_1 and u_2 in Figure 8.4.

The two secondary sources can be driven so that they only affect the downstream wave if the net upstream pressure they produce in the duct is zero, so that

$$p_{s1-}(x) + p_{s2-}(x) = 0, \quad x < 0 \qquad (8.8)$$

This requires that the control input u_1 is adjusted relative to u_2 so that

$$B = -Ce^{-jkl} \qquad (8.9)$$

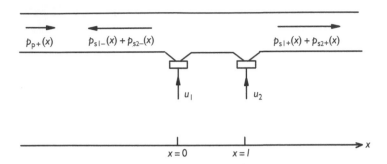

Figure 8.4 Active control of plane waves in an infinite duct using a pair of secondary sources.

that is to say that the loudspeaker at $x = 0$ produces a delayed and inverted version of that produced by the loudspeaker at $x = l$ (Swinbanks 1973), and the two loudspeakers act as a secondary source array.

The total pressure produced downstream of the two secondary sources is then

$$p_{\mathrm{p}+}(x) + p_{\mathrm{s}1+}(x) + p_{\mathrm{s}2+}(x) = \left[A + C(e^{jkl} - e^{-jkl}) \right] e^{jkx}, \quad x > l \qquad (8.10)$$

The loudspeaker array can thus be arranged to cancel the primary incident wave, provided u_2 is adjusted to ensure that

$$C = \frac{-A}{2j\sin kl} \qquad (8.11)$$

It should be noted that, at some frequencies, C will have to be very large compared with A. Both secondary sources will have to drive very hard if $\sin kl \approx 0$, which occurs when $l \ll \lambda$, $l \approx \lambda/2$, etc. The frequency range over which such a secondary source array can be operated is thus fundamentally limited (Swinbanks 1973), although this does not cause problems in many practical implementations that only operate over a limited bandwidth (Winkler and Elliott 1995).

The pressure distribution between the two secondary sources can be calculated using equations (8.6) and (8.7) with the conditions given by equations (8.9) and (8.11), and turns out to be part of another standing wave. The total pressure distribution after the absorption of an incident primary wave by a pair of secondary sources is shown in Figure 8.5 for the case in which $l = \lambda/10$. The pressure amplitude upstream of the secondary source pair is now unaffected by the cancellation of the downstream incident wave. In this case, the secondary source furthest downstream, driven by signal u_2, still generates no acoustic power, because it has zero acoustic pressure on it, but the other secondary source, driven by signal u_1, must absorb the power carried by the incident primary wave.

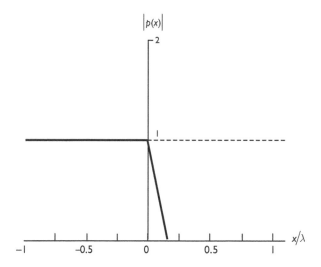

Figure 8.5 The amplitude of the pressure distribution in an infinite duct after a pair of secondary sources at $x = 0$ and $x = \lambda/10$ has been adjusted to cancel an incident plane wave of unit amplitude travelling in the positive-x direction, and also to suppress the wave generated by the secondary sources travelling in the negative-x direction.

8.2.3   Control of multiple modes

As the excitation frequency is increased, and the acoustic wavelength becomes comparable with the dimensions of the duct cross-section, it becomes possible for more than just plane acoustic waves to propagate in the duct. The other types of acoustic field that can propagate in the duct have pressure distributions that are not uniform across the cross-section of the duct and are referred to as higher-order modes. For a rectangular duct with a height L_y which is greater than its depth L_z, the first higher-order mode can propagate in the duct when it is excited above its first cut-on frequency (also known as cut-off frequency) given by

$$f_1 = \frac{c_0}{2L_y} \tag{8.12}$$

The pressure distribution of the first higher-order mode in the y direction, that is measured across the height of a rectangular duct, is illustrated in Figure 8.6, together with the pressure distribution of the plane wave or zeroth-order mode. If two identical secondary sources are used, one on either side of the duct, which are driven at the same amplitude, a plane wave will be produced when they are driven in phase; when they are driven out of phase they will only excite the higher-order mode. These two secondary

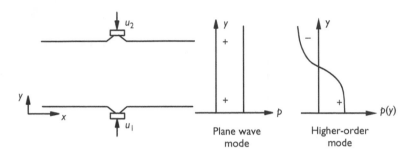

Figure 8.6 Cross-section of the part of a duct driven by two loudspeakers mounted
in the same plane, which can be driven in phase to produce a plane wave
mode, and out of phase to produce a higher-order mode. The pressure
distribution of these two modes across the duct are also shown.

sources can thus be used to excite any combination of amplitudes of these
two modes and so could be used to actively control both modes. In general,
it will require N secondary sources to control N modes, provided the
combination of secondary sources is able to independently excite each of
these modes.

Although it is, in principle, possible to actively control higher-order modes
using multiple secondary sources, there are several additional problems
which are not encountered in the active control of plane acoustic waves, as
discussed in more detail by, for example, Fedorynk (1975), Tichy (1988),
Eriksson *et al.* (1989), Silcox and Elliott (1990), Stell and Bernhard (1991)
and Zander and Hansen (1992). There is an increasing interest in the
active control of higher-order acoustic modes in short ducts because of
the potential application in controlling the fan tones radiated from the
inlet of aircraft engines, particularly as the aircraft is coming in to land
(see, for example, Burdisso *et al.* 1993; Joseph *et al.* 1999). It should be
noted, however, that the number of higher-order modes which are able to
propagate in a duct increases significantly with the excitation frequency, as
illustrated in Figure 8.7 for both a rectangular duct and a circular duct. The
number of propagating modes is proportional to the square of the excitation
frequency when it is well above the lowest cut-on frequency. If a significant
amount of energy is being transmitted in each of these modes, then an active
control system with a very large number of channels would be required to
attenuate the overall pressure level at high frequencies. However, if it is only
required to reduce the sound radiating at particular angles from the end of
the duct, such as that which causes significant sound on the ground from an
aeroengine inlet duct, this could be achieved by controlling a much smaller
number of modes. It is possible to detect the amplitude of these modes using
an axial array of sensors placed inside the duct (Joseph *et al.* 1996).

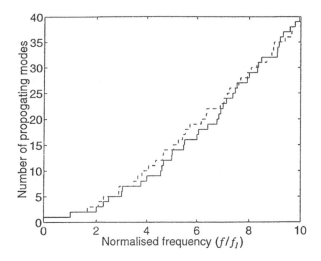

Figure 8.7 The number of acoustic modes that can propagate in a rectangular duct (solid line) or a circular duct (dashed line) as a function of excitation frequency, which has been normalised by the lowest cut-on frequency of the duct f_1.

8.3 Control of sound radiation in the free field

In the previous section, we saw that waves travelling in one direction could be either reflected using a single secondary source, or absorbed using a pair of secondary sources. In this section we consider the active control of disturbances propagating as waves in three dimensions. The physical interpretation is clearest if we initially restrict ourselves to the control of waves propagating only away from the sources, that is to say in infinite systems with no reflections. The waves propagating in such three-dimensional systems cannot be perfectly cancelled unless the secondary source is physically collocated with the primary source. Significant reductions in wave amplitude can, however, still be achieved if the separation between primary and secondary sources is not too great compared with a wavelength. The total acoustic power radiated by the combination of sources provides a convenient way of quantifying the space-average or *global* effect of various control strategies. In this section we will introduce the idea of minimising such a global measure of performance by adjusting the secondary source strengths, rather than arranging for the secondary sources to cancel the primary field perfectly.

In a mechanical system, the power supplied by a mechanical point force is equal to the time-averaged product of the force multiplied by the resultant velocity, but we must be more careful with acoustic sources since the force is a distributed quantity, determined by the acoustic pressure, and an acoustic source can also have a non-uniform distribution of velocity. However, if we assume that the source is vibrating with equal amplitude in all directions, such as a pulsating sphere, and that this sphere is small compared with the acoustic wavelength, so that the pressure on its surface is reasonably uniform, its power output can be simply calculated by taking the time-average product of the acoustic pressure at the surface and the volume velocity of the source (Nelson and Elliott 1992). The volume velocity q is equal to the radial surface velocity multiplied by the surface area. Such an acoustic source is known as a monopole. A practical example of such an acoustic source might be the end of an engine's exhaust pipe radiating sound at low frequencies, which could be the primary source in an active control system.

The relation between the complex acoustic pressure at a distance r from an acoustic monopole source operating at a single frequency may be expressed as

$$p(r) = Z(r)q \tag{8.13}$$

where q is the complex volume velocity of the monopole source and $Z(r)$ is the complex acoustic transfer impedance. In an infinite medium with no acoustic reflections, that is to say a free field, the acoustic transfer impedance is given by Morse (1948) (see also Nelson and Elliott 1992)

$$Z(r) = \frac{\omega^2 \rho_0}{4\pi c_0} \left(\frac{je^{-jkr}}{kr} \right) \tag{8.14}$$

where ρ_0 and c_0 are the density and the speed of sound in the medium, and k is the acoustic wavenumber which is equal to ω/c_0 or $2\pi/\lambda$, where λ is the acoustic wavelength.

The time-average acoustic power generated by a monopole source is thus equal to

$$\Pi = \frac{1}{2} \text{Re}\{p^*(0)q\} \tag{8.15}$$

Using equation (8.13), this can be written as

$$\Pi = \frac{1}{2} |q|^2 \text{Re}\{Z(0)\} \tag{8.16}$$

where $\text{Re}\{Z(0)\}$ is the real part of the acoustic input impedance, which in a free field is given from equation (8.14) by

$$\text{Re}\{Z(0)\} = \frac{\omega^2 \rho_0}{4\pi c_0} \tag{8.17}$$

If an array of monopole sources is present, whose complex volume velocities are represented by the vector

$$q = [q_1, q_2, \ldots, q_N]^T \qquad (8.18)$$

which generates complex pressures at each of these sources, represented by the elements of the vector

$$p = [p_1, p_2, \ldots, p_N]^T \qquad (8.19)$$

then the total power radiated by the array of monopoles can be written as

$$\Pi = \frac{1}{2} \operatorname{Re}\{p^H q\} \qquad (8.20)$$

where ^{H} denotes the Hermitian or complex conjugate transpose.

The pressure at each of the source positions depends on the volume velocity of each of the sources in a way which can be represented in matrix form as

$$p = Z_p q \qquad (8.21)$$

where Z is the matrix of input and transfer acoustic impedances between the sources, which is symmetric since the system is assumed to be reciprocal.

The total radiated power can now be written as

$$\Pi = \frac{1}{2} \operatorname{Re}[q^H Z^H q] = \frac{1}{2} q^H R q \qquad (8.22)$$

where $R = \operatorname{Re}\{Z\}$.

The vectors of source strengths and pressures can now be partitioned into those due to primary sources q_p and p_p and those due to secondary sources q_s and p_s so that

$$\begin{bmatrix} p_p \\ p_s \end{bmatrix} = \begin{bmatrix} Z_{pp} & Z_{ps} \\ Z_{sp} & Z_{ss} \end{bmatrix} \begin{bmatrix} q_p \\ q_s \end{bmatrix} = \qquad (8.23)$$

If we similarly partition the matrix R as

$$R = \operatorname{Re} \begin{bmatrix} Z_{pp} & Z_{ps} \\ Z_{sp} & Z_{ss} \end{bmatrix} = \begin{bmatrix} R_{pp} & R_{ps} \\ R_{sp} & R_{ss} \end{bmatrix} \qquad (8.24)$$

then the total radiated power can be written as (Nelson and Elliott 1992)

$$\Pi = \frac{1}{2} \left(q_s^H R_{ss} q_s + q_s^H R_{sp} q_p + q_z^H R_{sp}^T q_s + q_p^H R_{pp} q_p \right) \qquad (8.25)$$

where $R_{ps} = R_{sp}^T$, because of reciprocity.

Equation (8.25) is of Hermitian quadratic form, as described by Nelson and Elliott (1992), so that the power is a quadratic function of the real and imaginary parts of each element of the vector q_s. This quadratic function must always have a minimum rather than a maximum associated with it, since otherwise for very large secondary sources, the total radiated power would become negative, which would correspond to the impossible situation of an array of sources absorbing power from the otherwise passive medium. It is shown by Nelson and Elliott (1992) that the minimum possible value of power is given by a unique set of secondary sources, provided the matrix $\mathbf{R}_{ss}$ is positive definite, and that this optimum set of secondary sources is given by

$$q_{s,opt} = -\mathbf{R}_{ss}^{-1}\mathbf{R}_{sp}q_p \qquad (8.26)$$

The positive definiteness of $\mathbf{R}_{ss}$ is guaranteed on physical grounds in this case, since the power supplied by the secondary forces acting alone is equal to $q_s^H \mathbf{R}_{ss} q_s$ which must be positive for all q_s, provided they are not collocated. The minimum value of the total power that results from this optimum set of secondary sources is given by

$$\Pi_{min} = \frac{1}{2}q_p \left[\mathbf{R}_{pp} - \mathbf{R}_{sp}^T\mathbf{R}_{ss}^{-1}\mathbf{R}_{sp}\right] q_p \qquad (8.27)$$

Each of the elements in the matrix $\mathbf{Z}$ in equation (8.23) can be calculated by using equation (8.16), using the geometric arrangement of the primary and secondary forces to compute the distance between each of them. The maximum possible attenuation of the input power for this geometric arrangement of primary and secondary sources can thus be calculated by taking the ratio of the power before control, given by equation (8.25) with q_s set to zero, and after control, by equation (8.27).

The maximum attenuation in input power has been calculated for the arrangements of primary and secondary sources shown in Figure 8.8, in which a single primary source is controlled by one, two, four or eight

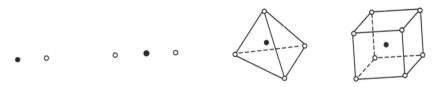

Figure 8.8 Arrangement of primary (solid sphere) and secondary acoustic monopole sources (open spheres) in free space, used in the calculation of the attenuation of total power output. Each of the secondary sources is positioned at a distance d from the primary source.

secondary sources, each uniformly spaced at a distance d from the primary source. The results of such a calculation are shown in Figure 8.9, in which the attenuation for an optimally adjusted set of secondary sources is plotted against the separation distance normalised by the acoustic wavelength. A practical arrangement corresponding to this model problem may be the use of an array of loudspeakers used as secondary sources placed round the end of an engine's exhaust pipe in the open air.

When the separation distance is small compared with the acoustic wavelength, the attenuation achieved in radiated power for a single secondary source is large. The decrease of attenuation, as the normalised separation distance increases, is rapid, however. With a larger number of secondary sources, attenuation can be achieved for greater separation distances. The general trend is that the number of secondary sources required to achieve a given attenuation increases as the *square* of the normalised separation distance d/λ. The maximum separation distance allowable between each of the multiple secondary sources to achieve 10 dB attenuation in total power output is about half a wavelength.

With a single layer of acoustic monopoles, all at equal distances from the primary source, the power output of the primary source can be reduced by reflecting the outward-going wave back towards the primary source. It can be shown that the power output of all the secondary sources is identically zero when controlling a single primary source (Elliott *et al.* 1991),

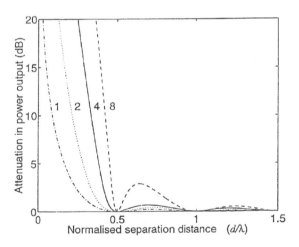

Figure 8.9 Attenuation in the total acoustic power radiated into free space by a primary monopole source when optimally controlled by one ($-\cdot-\cdot-\cdot$), two ($\cdots\cdots$), four (——) or eight (- - - -) secondary monopole sources, as a function of the distance from the primary to the secondary sources d, normalised by the acoustic wavelength.

emphasising their entirely reactive role; this case is thus directly analogous to the one-dimensional case whose pressure distribution is shown in Figure 8.3. It is also possible to use a double layer of acoustic monopoles, at unequal distances from the primary source, to *absorb* the waves radiated by the primary source, by direct analogy with the one-dimensional acoustic case, whose pressure distribution is shown in Figure 8.5. When operating in this manner, the pairs of monopole sources can be shown to be synthesising a single-point monopole/dipole source (Nelson and Elliott 1992), which is sometimes called a tripole (Jessel 1968; Mangiante 1994). The absorption of sound in two- and three-dimensional systems have been studied by Zavadskaya *et al.* (1976) and Konaev *et al.* (1977), who showed that the number of such sources required for a given level of control was again proportional to $(d/\lambda)^2$, where d is the distance from the primary source to the secondary source.

In practical applications, the sum of the squared pressures measured at microphones around the source array can be used as a measure of the radiated sound power. Multichannel control systems can be implemented to adjust the real and imaginary parts of the secondary source strengths to minimise this cost function, as described by Nelson and Elliott (1992) and in more detail by Elliott (2001). One practical application of this method of controlling sound radiation is illustrated in Figure 8.10. This shows the results of some experiments undertaken by Hesselman (1978), who used two secondary loudspeakers on either side of an electrical transformer to control its sound power output at a frequency of 100 Hz.

8.4   Strategies of control in enclosed sound fields

A source driving an infinite system will only generate waves moving away from the source position. In practice, most systems have boundaries which reflect the outgoing waves from a source within them, and the interference between the incident and reflected waves creates *resonances* at certain excitation frequencies. In the following section, the response of finite two- and three-dimensional systems will be described entirely in terms of the *modes* that are associated with these resonances. For one-dimensional systems, however, it is sometimes clearer to retain a model of the system in terms of the constituent waves, since this provides a more direct way of calculating its response.

There are many different active control strategies that can be implemented in a finite system. In this section, the physical effect of several of these strategies is illustrated using the simple one-dimensional acoustic system shown in Figure 8.11. This consists of a long duct with rigid walls, in which only plane acoustic waves can propagate, with a primary source of volume velocity q_p at the left-hand end $x = 0$, and a secondary source of

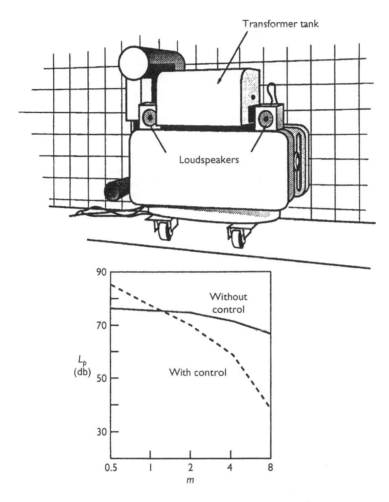

Figure 8.10 Results of the experiment reported by Hesselman (1978), using two loudspeakers to control the sound radiated by a 100 kVA transformer tank. A sketch of the experimental arrangement is shown, together with the variation of sound pressure level at 100 Hz (averaged over several directions) with distance from the transformer tank. Note the increase in sound pressure in the near field of the tank.

volume velocity q_s at the right-hand end $x = L$. This simple system was originally used by Curtis *et al.* (1987) to illustrate the effect of different control strategies on the energy in the duct. The modification presented here, based on input power, was originally presented by Elliott (1994), who also considered the equivalent mechanical system of a thin beam excited by

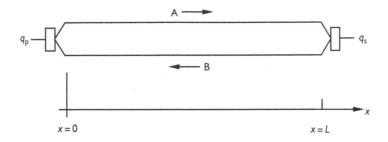

Figure 8.11 A long duct, driven at one end by a primary source of volume velocity q_p and at the other end by a secondary source of volume velocity q_s, which is used to illustrate the effect of different control strategies on a finite system.

moments at the two ends. A more comprehensive treatment of the structural case is provided by Brennan *et al.* (1995).

8.4.1 Acoustic impedances in a finite duct

In this section, the acoustic input and transfer impedances in the duct are derived, and the acoustic power supplied by the primary source is defined. The complex pressure $p(x)$ at position x along such a duct can be written as the sum of two travelling waves, one of amplitude A travelling in the positive-x direction and one of amplitude B travelling in the negative direction, so that

$$p(x) = Ae^{-jkx} + Be^{+jkx} \qquad (8.28)$$

The complex particle velocity associated with a plane acoustic wave is $+1/\rho_0 c_0$ times the pressure for waves travelling in the positive-x direction, and $-1/\rho_0 c_0$ times the pressure for wave travelling in the negative-x direction, where ρ_0 and c_0 are the density and the speed of sound in the medium, respectively (Nelson and Elliott 1992; Kinsler *et al.* 1999). The complex particle velocity in the duct is thus

$$u(x) = \frac{1}{\rho_0 c_0} \left[Ae^{-jkx} - Be^{+jkx} \right] \qquad (8.29)$$

Expressions for the acoustic input and transfer impedance of the duct can be derived from these equations, which are

$$Z(0) = \frac{p(0)}{q_p} = Z_c \frac{1 + e^{-j2kL}}{1 - e^{-j2kL}} = -jZ_c \cot kL \qquad (8.30)$$

and

$$Z(L) = \frac{p(L)}{q_p} = Z_c \frac{2e^{-jkL}}{1 - e^{-j2kL}} = \frac{-jZ_c}{\sin kL} \qquad (8.31)$$

where Z_c is the characteristic impedance $\rho_0 c_0 / S$ and S is the duct cross-sectional area. With these definitions of the input and transfer impedances, and using the symmetry of the duct, expressions for the total pressure at the two ends of the duct, with both primary and secondary sources operating, can be readily deduced as

$$p_p = Z(0)q_p + Z(L)q_s \qquad (8.32)$$

and

$$p_s = Z(L)q_p + Z(0)q_s \qquad (8.33)$$

where, in this case, $p_p = p(0)$ and $p_s = p(L)$. Note that $Z(L)$ is equal to the transfer impedance from the secondary source to the primary source position, as well as that from the primary source to the secondary source position. These expressions can now be used to calculate the effect of various active control strategies. The strategies are compared by calculating their effect on the acoustic power output of the primary source, which, from equation (8.16), is found to be

$$\Pi_p = \frac{1}{2} \text{Re} \{p_p^* q_p\} \qquad (8.34)$$

With the primary source operating alone in the duct, this quantity is plotted as the solid line in Figure 8.12 and in the subsequent figures, normalised with respect to the power output of the primary source in an infinite duct which is equal to

$$\Pi_{p,\text{infinite}} = \frac{1}{2} |q_p|^2 Z_c \qquad (8.35)$$

The normalised power output is plotted as a function of the normalised frequency, which is equal to $L/\lambda = kL/2\pi = \omega L/2\pi c$, where k is the wavenumber and L the length of the duct. It is important to include some dissipation or loss in the duct, since otherwise the input impedance, equation (8.30), is entirely reactive and no power is ever supplied by the primary or the secondary source. This dissipation is provided here by assuming a complex wavenumber

$$k = \frac{\omega}{c_0} - j\alpha \qquad (8.36)$$

where α is a positive number that represents the attenuation of a wave propagating in the duct, and is assumed to be small compared with ω/c_0. The value of α was chosen in these simulations to give an effective damping ratio of about 1 per cent.

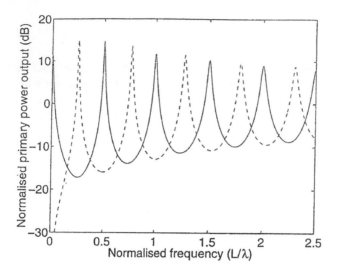

Figure 8.12 Power output of the primary source in the finite duct before control (solid line) and after the secondary source has been adjusted at each frequency to cancel the pressure in front of the secondary source (dashed line).

8.4.2 Cancellation of pressure

The first control strategy considered is that of cancellation of the pressure at the secondary source position at the right-hand end of the duct. This pressure is equal to p_s in equation (8.33), which can be set to zero if the secondary source is driven so that

$$q_{s1} = -\frac{Z(L)}{Z(0)} q_p \tag{8.37}$$

Using equations (8.30) and (8.31), this expression reduces to

$$q_{s1} = -\frac{q_p}{\cos kL} \tag{8.38}$$

The effect of this control strategy on the power output of the primary source is shown in Figure 8.12. It can be seen that the power supplied by the primary source has been reduced at the uncontrolled resonant excitation frequencies, which occurred when $L/\lambda = 0, 1/2, 1$, etc., but is now a maximum when $L/\lambda = 1/4, 3/4$, etc. This is because the secondary source is acting to create a pressure-release boundary at the right-hand end of the duct, which now exhibits the resonance frequencies associated with a closed/open duct, rather than a closed/closed duct. The fact that the duct can still be resonant after the pressure has been cancelled at the secondary source position means that the power supplied to the duct by the primary source is significantly increased by the active control action at the new resonance frequencies.

8.4.3 Absorption of incident wave

The second strategy we will consider is that in which the secondary source is driven to suppress the wave reflected from the right-hand end of the duct; this corresponds to absorption of the incident wave. The required condition for the secondary source in this case can be obtained by calculating the amplitude B of the reflected wave as a function of the primary and secondary source strengths, which gives

$$B = \frac{Z_c \left(q_p e^{-jkL} + q_s \right)}{\left(e^{+jkL} - e^{-jkL} \right)} \tag{8.39}$$

The reflected wave amplitude is zero if the secondary source volume velocity is equal to

$$q_{s2} = -q_p e^{-jkL} \tag{8.40}$$

see, for example, Section 5.15 of Nelson and Elliott (1992), where this is referred to as the absorbing termination, and the earlier reference of Beatty (1964). Practical realisations of such absorbing terminations have also been developed by Guicking and Karcher (1984), Orduña-Bustamante and Nelson (1991) and Darlington and Nicholson (1992). The effect of such a termination on the power output of the primary source is shown in Figure 8.15, from which it can be seen that after control the power output is the same as that produced by the primary source in an infinite duct, equation (8.35), as expected.

The control strategies of cancelling the pressure at the secondary source position and of absorbing the plane wave incident upon the secondary sources can be considered as two special cases of the general approach of local pressure feedback. In the general case the secondary source volume velocity q_s is arranged to be equal to a real gain factor $-g$, times the pressure on p_s on it so that

$$q_s = -g p_s \tag{8.41}$$

The acoustic impedance presented by the secondary source to the duct using local pressure feedback is thus equal to

$$Z_s = \frac{p_s}{-q_s} = \frac{1}{g} \tag{8.42}$$

If the gain factor is zero, the volume velocity of the secondary source is also zero, and Z_s then corresponds to a rigid boundary, as assumed for the duct with no control. If the gain factor is very high, the pressure in front of the secondary source is almost cancelled by the effect of the feedback and Z_s tends to zero, with the result shown in Figure 8.14. If the gain factor is set equal to the reciprocal of the characteristic impedance $g = 1/Z_c$, the impedance Z_s presented to the duct by the secondary source Z_s is equal to the characteristic impedance of the duct, which creates the absorbing termination whose effects are shown in Figure 8.13. The broken line in Figure 8.13 can thus be considered as being midway between the two extremes of a continuous set of responses that can be obtained by varying the gain in a local pressure feedback system. The limits of these responses are shown by the solid and broken lines in Figure 8.12.

In principle, local pressure feedback is stable for any gain factor g, because the acoustic response of the system under control, equation (8.30), has a strictly positive real part (Elliott 2001). In practice, however, the volume velocity of a loudspeaker will not be a frequency-independent function of its electrical input, because of its dynamic properties. Thus, the response of the system under control is not as simple as the acoustic input impedance given by equation (8.30), and some care must be taken to maintain the stability of such a feedback system, as explained by Guicking and Karcher (1984) and Darlington and Nicholson (1992), for example.

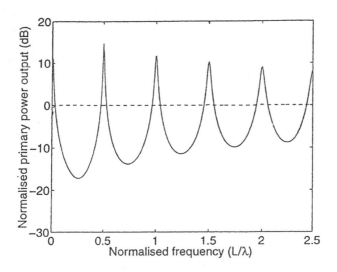

Figure 8.13 Power output of the primary source in the finite duct before control (solid line) and after the secondary source has been adjusted at each frequency to cancel the sound wave reflected by the secondary source (dashed line).

8.4.4   Maximisation of secondary power absorption

The final two control strategies that will be considered in this section are based on the sound power output of the sources. In the first of these we naively attempt to do an even better job than just absorbing the incident wave, as in Figure 8.13, by *maximising* the power absorbed by the secondary source (Elliott *et al.* 1991). The power output of the secondary source can be written as

$$\Pi_s = \frac{1}{2}\text{Re}\{p_s^* q_s\} = \frac{1}{4}\left(p_s^* q_s + q_s^* p_s\right) \tag{8.43}$$

and, using equation (8.39), this can be expressed in the Hermitian quadratic form

$$\Pi_s = \frac{1}{4}\left(2q_s^*\text{Re}\{Z(0)\}q_s + q_s^* Z(L)q_p + q_p^* Z^*(L)q_s\right) \tag{8.44}$$

When the secondary volume velocity corresponds to the global minimum of this Hermitian quadratic form, which is

$$q_{s3} = -\frac{Z(L)q_p}{2\text{Re}\left(Z(0)\right)} \tag{8.45}$$

the power output of the secondary source is minimised, so that the power absorbed by the secondary source is maximised.

The results of generating a secondary source strength corresponding to equation (8.45) in the duct are shown as the broken line in Figure 8.14. It can be seen that at most excitation frequencies, the power output of the primary source has been *increased* by this control strategy. In general, the maximisation of the power absorbed by the secondary source is a rather dangerous control strategy, particularly for narrowband disturbances in resonant systems, because of the ability of the secondary source to increase the power output of the primary source. The secondary source achieves this increase by altering the impedance 'seen' by the primary source and making the physical system appear to be resonant over a much broader range of frequencies than would naturally be the case. A more detailed analysis of the power balance within the duct reveals that only a fraction of the power that is induced out of the primary source is actually absorbed by the secondary source, and the remaining power is dissipated in the duct, which consequently has a significantly increased level of energy stored within it. It should be pointed out, however, that the effectiveness of a maximum power absorbing strategy will also depend on the nature of the primary excitation and, if this has a random waveform, on the bandwidth of the excitation (Nelson 1996).

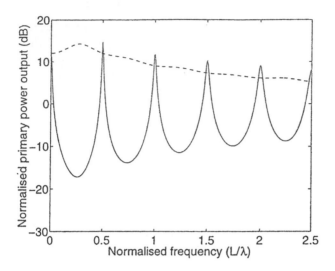

Figure 8.14 Power output of the primary source in the finite duct before control (solid line) and after the secondary source has been adjusted at each frequency to maximise the power absorption by the secondary source (dashed line).

8.4.5 Minimisation of total power output

The final control strategy we consider in this section is that of minimising the total power output of the primary and the secondary sources. The total power output can be written, by analogy with equation (8.26), as

$$\Pi_T = \frac{1}{2} \left(q_s^* R_0 q_s + q_s^* R_L q_p + q_p^* R_L q_s + q_p^* R_0 q_p \right) \tag{8.46}$$

where $R_0 = \text{Re}\{Z(0)\}$ and $R_L = \text{Re}\{Z(L)\}$. This Hermitian quadratic form is minimised by the secondary source volume velocity

$$q_{s4} = -\frac{R_L q_p}{R_0} \tag{8.47}$$

The resulting power output of the primary source is shown in Figure 8.15, which reveals that the power output of the primary source is always reduced by this control strategy and that the residual level of output power at each frequency is then similar to that produced under anti-resonant conditions. The action of the secondary source in this case is to alter the impedance seen by the primary source so that the physical system appears to be anti-resonant over a much broader range of frequencies than would naturally be the case. In order to implement this control strategy in practice, the power

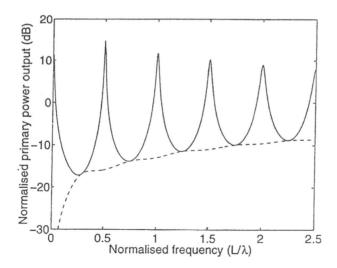

Figure 8.15 Power output of the primary source in the finite duct before control (solid line) and after the secondary source has been adjusted at each frequency to minimise the total power output of both the primary and secondary sources (dashed line).

output of the two acoustic sources could be deduced from measurements of their source strength and the pressure on them, or from the amplitudes of the waves propagating in the two directions in the duct. However, we will see in the next section that minimising the total power supplied to a system is very nearly the same as minimising the energy stored in the system, and that this energy can be readily estimated using an array of sensors distributed within the system under control.

8.5 Control of energy in enclosed sound fields

One measure of the overall or *global* response of a finite system is the energy stored within it. In this section we discuss the active minimisation of the acoustical energy in an enclosure. Before we consider this specific example, it is useful to establish the relationship between the energy stored in a system and the power supplied to it. These two quantities are related by the dissipation or damping in the system, since under steady state conditions the power supplied to a system must increase its stored energy until the power lost through dissipation is the same as the power supplied by the sources. We would thus expect that if an active control system were adjusted to minimise the total power supplied to a system, the energy stored by the system would also be minimised. This turns out to be true in the majority of cases of practical interest, and although the minimisation of total input power and total stored energy does give slightly different analytical results, the differences are generally of academic interest rather than of practical importance.

8.5.1   Power input and total energy

The connection between input power and stored energy can be illustrated by considering the response of an acoustic enclosure to a single monopole source (Elliott 2001). This analysis is also used to introduce the modal model for finite systems. The response of a finite linear distributed system can always be represented by a summation in terms of its natural modes (Meirovitch 1990). The steady state complex pressure at a point x, y, z when excited at a frequency ω can thus be written as

$$p(x, y, z, \omega) = \sum_{n=0}^{\infty} a_n(\omega)\psi_n(x, y, z) \tag{8.48}$$

where $a_n(\omega)$ is amplitude of the nth acoustic mode, which has a mode shape given by $\psi_n(x, y, z)$. The mode shapes are orthogonal and will be assumed here to be entirely real and normalised so that

$$\frac{1}{V} \int_v \psi_n(x, y, z)\psi_m(x, y, z)\mathrm{d}x\,\mathrm{d}y\,\mathrm{d}z = \begin{cases} 1 & \text{if } n = m \\ 0 & \text{if } n \neq m \end{cases} \tag{8.49}$$

where V is the volume of the enclosure. For example, the acoustic mode shapes for a rigid rectangular enclosure of dimensions L_x by L_y by L_z are given by

$$\psi_n(x, y, z) = \varepsilon_{n_1}\varepsilon_{n_2}\varepsilon_{n_3}\cos\left(\frac{n_1\pi x}{L_x}\right)\cos\left(\frac{n_2\pi y}{L_y}\right)\cos\left(\frac{n_3\pi z}{L_z}\right) \tag{8.50}$$

where $\varepsilon_{ni} = 1$ if $n_{i=0}$ and $\varepsilon_{ni} = 2$ if $n_i > 0$ and n_1, n_2 and n_3 are the three modal integers denoted by the single index n in equation (8.48).

If the enclosure is excited by a point monopole of source strength q located at position (x_0, y_0, z_0) the mode amplitude can be written as

$$a_n(\omega) = \rho_0 c^2 A_n(\omega)\psi_n(x_0, y_0, z_0)q \tag{8.51}$$

The term $A_n(\omega)$ in equation (8.51) denotes the modal resonance term which can be written as

$$A_n(\omega) = \frac{\omega}{B_n\omega + j(\omega^2 - \omega_n^2)} \tag{8.52}$$

where B_n is the modal bandwidth and ω_n is the natural frequency of the nth mode. If viscous damping is assumed, $B_n = 2\omega_n\zeta_n$, where ζ_n is the modal damping ratio. The power radiated into the enclosure by this monopole source is equal to

$$\Pi(\omega) = \frac{1}{2} \operatorname{Re} \{ p^*(x_0, y_0, z_0) \, q \} \tag{8.53}$$

and using the modal expansion for $p(x, y, z)$ this can be written as

$$\Pi(\omega) = \frac{\rho_0 c^2}{2V} |q|^2 \sum_{n=0}^{\infty} \operatorname{Re}\{A_n(\omega)\} \, \psi_n^2(x_0, y_0, z_0) \tag{8.54}$$

The *total potential energy* stored in the enclosure is proportional to the volume integral of the mean-square pressure, and can be written as (Nelson and Elliott 1992)

$$E_p(\omega) = \frac{1}{4\rho_0 c_0^2} \int_v |p(x, y, z)|^2 \, dv \tag{8.55}$$

When the pressure distribution is expressed in terms of the modal expansion, equation (8.48), and using the orthonormal properties of the modes, equation (8.49), the total energy potential can be written as

$$E_p(\omega) = \frac{V}{4\rho_0 c^2} \sum_{n=0}^{\infty} |a_n|^2 \tag{8.56}$$

and so is proportional to the sum of the squared mode amplitudes. Using equation (8.51) for $a_n(\omega)$, and the fact that the mode resonance term (equation (8.52)) has the interesting property that

$$|A_n(\omega)|^2 = \frac{\operatorname{Re}\{A_n(\omega)\}}{B_n} \tag{8.57}$$

then the total potential energy can be written as

$$E_p(\omega) = \frac{\rho_0 c_0^2}{2V} |q|^2 \sum_{n=0}^{\infty} , \operatorname{Re}\{A_n(\omega)\} \frac{\psi_n^2(x_0, y_0, z_0)}{2B_n} \tag{8.58}$$

Comparing the terms in equation (8.58) for the total potential energy with those in equation (8.54) for the power radiated, we can see that they differ by a factor of twice the modal bandwidth. If the enclosure is lightly damped and excited close to the resonance frequency ω_m of the mth mode, then only the mth term in the modal summations for the energy and power will be significant. In this case we can express the total potential energy as,

$$E_p(\omega_m) \approx \frac{\pi(\omega_m)}{2B_m} \tag{8.59}$$

so that the stored energy is equal to the supplied power multiplied by a time constant, which depends on the damping in the system (Elliott *et al.* 1991). If the modal bandwidth is similar for a number of modes adjacent to the mth mode, then equation (8.59) will continue to be approximately true over a frequency band that includes several modes.

8.5.2 Control of acoustic energy in an enclosure

The total acoustic potential energy provides a convenient cost function for evaluating the effect of global active control of sound in an enclosure. Because of the assumed orthonormality of the acoustic modes, E_p is proportional to the sum of the squared mode amplitudes, and these mode amplitudes can be expressed in terms of the contributions from the primary and secondary sources, as in equation (8.44). Thus the total acoustic potential energy is a Hermitian quadratic function of the complex strengths of the secondary acoustic sources, which can be minimised in exactly the same way as described in Section 8.3.

The result of simulation of minimising the total acoustic potential energy in an enclosure of dimensions $1.9 \times 1.1 \times 1.0\,\mathrm{m}$ is illustrated in Figure 8.16. The acoustic modes have an assumed damping ratio of 10 per cent, which is fairly typical for a reasonably well-damped acoustic enclosure such as a car interior at low frequencies. The acoustic mode shapes in a rigid-walled rectangular enclosure are proportional to the product of three cosine functions in the three coordinates. The lowest order mode, with all ns set to zero in equation (8.3), has a uniform mode amplitude throughout the enclosure and corresponds to a uniform compression or expansion of the air at all points. The mode with the next highest natural frequency corresponds to fitting a half wavelength into the longest dimension of the enclosure. This first axial mode has a natural frequency of about 90 Hz for the enclosure shown in Figure 8.16, which is similar in size to the interior of a small car.

Figure 8.17 shows the total acoustic potential energy in the enclosure when driven only by the primary source placed in one corner of the enclosure and when the total acoustic potential energy is minimised by a single secondary acoustic source in the opposite corner (broken line) or by seven secondary acoustic sources positioned at each of the corners of the enclosure

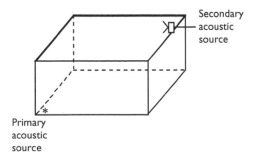

Figure 8.16 Physical arrangement for the simulation of the active control of tonal sound in a rectangular enclosure, which is about the size of a car interior, excited by a primary acoustic source in one corner and a secondary acoustic source in the opposite corner.

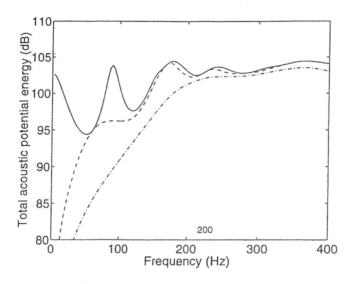

Figure 8.17 Total acoustic potential energy in the enclosure when driven by the primary acoustic source alone at discrete frequencies (solid line) and when the total potential energy has been minimised using either the single secondary source shown in Figure 8.16, optimally adjusted at each excitation frequency (dashed line), or seven secondary acoustic sources placed in all the corners of the enclosure not occupied by the primary source (dash-dot line).

not occupied by the primary source (dash-dot line). Considerable attenuations in the total acoustic potential energy are achieved with a single secondary source below about 20 Hz, where only the zeroth-order acoustic mode is significantly excited, and for excitation frequencies close to the natural frequency of the first longitudinal mode at about 90 Hz.

The response of the system does not, however, show clear modal behaviour for excitation frequencies above about 150 Hz, and very little attenuation can be achieved with a single secondary source above this frequency. This is because the spacing of the natural frequencies of the acoustic modes in a three-dimensional enclosure becomes smaller with higher mode order. Even introducing seven secondary sources into the enclosure does not allow global control to be achieved at frequencies above about 250 Hz in this case.

The amplitude of an individual mode of a physical system could be completely controlled using a single secondary source, provided it was not placed on a nodal line. In controlling this mode, however, the secondary source will tend to increase the excitation of other modes of the system. The minimisation of total energy generally involves a balance between cancelling the dominant modes and not overly exciting the other, residual, modes of the system. This balance is automatically maintained when the total energy in the system is minimised. The attenuation which can be obtained in the energy at any one excitation frequency will generally depend on the number of modes that contribute substantially to the response.

8.5.3   The effect of the modal overlap

The number of modes that are significantly excited in a system at any one excitation frequency can be quantified by a dimensionless parameter known as the *modal overlap* M(ω). This is defined to be the average number of modes whose natural frequencies fall within the half-power bandwidth of any one mode at a given excitation frequency ω. M(ω) is equal to the product of the modal density (average number of modal natural frequencies per Hz) and the modal half-power bandwidth (in Hz), and both of these quantities can be calculated for the structural modes in a panel and the acoustic modes in the enclosure used in the simulations in Section 8.5.2.

For a three-dimensional rectangular enclosure, an approximate expression for the acoustic modal overlap can be calculated from that for the modal densities (Morse 1948). It is given by Nelson and Elliott (1992) as

$$M(\omega) = \frac{2\zeta\omega L}{\pi c_0} + \frac{\zeta\omega^2 S}{\pi c_0^2} + \frac{\zeta\omega^3 V}{\pi^2 c_0^3} \tag{8.60}$$

where L is the sum of the linear dimensions of the enclosure, S is its total surface area, V is the volume of the enclosure, ζ is the damping ratio and c_0 is the speed of sound. At high frequencies the acoustic modal overlap increases as the cube of the excitation frequency. The modal overlap calculated for the acoustic modes in the enclosure shown in Figure 8.16 is plotted in Figure 8.18. In this case the modal overlap is less than unity for excitation frequencies below about 150 Hz, which was the limit of global control with one source in Figure 8.17, and is under seven for excitation frequencies below about 250 Hz, which was the limit of global control with seven sources in Figure 8.17.

The modal overlap can thus be seen to be a useful method of characterising the complexity of the modal structure in a system at a given excitation frequency. It can also be used as a very approximate guide to the number of secondary sources required to achieve a given level of global control; that is, a given reduction in energy level in a system. The difference between the variation of modal overlap with frequency of a typical structure and of an acoustically excited enclosure explains the very significant difference in the physical limitations of active control in these two cases.

The total acoustic potential energy in an enclosure is proportional to the volume integral of the mean-square pressure. In a practical active control system, the total potential energy can be estimated using the sum of the

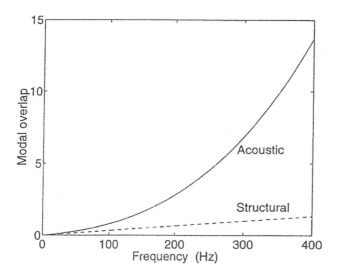

Figure 8.18 The modal overlap $M(\omega)$, for the structural modes on a plate and for the acoustic modes in the enclosure used for the simulations shown in Figure 8.17 (solid line).

squared outputs of a number of pressure microphones. The number of microphones required to obtain an accurate estimate of the total acoustic potential energy is proportional to the number of substantially excited acoustic modes within the enclosure and thus increases sharply at higher excitation frequencies. Microphones can be placed at a smaller number of locations if they simultaneously measure pressure and pressure gradient in the three directions (Sommerfeldt and Nasif 1994). The active control of sound in the passenger cabin of a propeller aircraft is used as an example of the application of a multichannel system for tonal noise in Section 8.6. If control over only a part of the enclosure is required, then the microphones can be concentrated in this region. The zone of quiet generated in the limiting case where only one microphone is used is discussed in Section 8.7.

8.6 Control of propeller noise inside fixed-wing aircraft

Many short-haul aircraft, having up to about 50 seats, are driven by propellers instead of jet engines because they can be considerably more efficient at speeds below about 300 mph (480 kph). The development of prop-fan engines to power larger faster aircraft has also been suggested because of their improved fuel efficiency. The spectrum of the sound pressure in the passenger cabins of such aircraft contain strong tonal components at the *blade passage frequency* (BPF) of the propellers, which are difficult to attenuate using passive absorption (Wilby *et al.* 1980; Metzger 1981). Active control of these tones has been considered since the early 1980s (Chaplin 1983; Ffowcs-Williams 1984; Bullmore *et al.* 1987) and appears to be a good solution to this problem since active sound control works particularly well at low frequencies, where the acoustic modal overlap is relatively small, as described in Section 8.4. Also, with lightweight loudspeakers, active control potentially carries a significantly smaller weight penalty than passive methods of noise control. In this section we will briefly describe the results of some flight trials of a practical active control system operating in the passenger cabin of a British Aerospace 748 turboprop aircraft during a series of flight trials undertaken in early 1988 (Dorling *et al.* 1989; Elliott *et al.* 1989, 1990) and the subsequent development of this technology.

The 50-seat aircraft used for the trials had a fully trimmed passenger cabin about 9 m long by 2.6 m diameter, and was flown at an altitude of about 10,000 feet under straight and level cruise conditions with the engines running at a nominal 14,200 rpm. As a result of the gearing in the engine and the number of propeller blades, this produces a BPF of 88 Hz. The control system used in the experiments discussed here employed a tachometer on one of the engines to generate reference signals at the BPF and its second and third harmonics (88 Hz, 176 Hz and 264 Hz). These reference signals were passed through an array of adaptive digital filters, and their outputs were used to drive 16 loudspeakers. The coefficients of the digital filters were adjusted to implement the steepest-descent algorithm, in order to minimise the sum of the square values of the 32 microphone signals (Elliott 2001). Many different configurations of loudspeaker and microphone positions were investigated. The results presented here are for the distribution illustrated in Figure 8.19, with the sixteen 200 mm diameter loudspeakers and thirty-two microphones distributed reasonably uniformly on the floor and on the luggage racks throughout the cabin, although there is a greater concentration at the front of the aircraft (seat row 1) since this is close to the plane of the propellers.

The levels of reduction achieved in the sum of the mean-squared pressures at all 32 control microphones ΔJ(dB), are listed in Table 8.1 at the three control frequencies. These were achieved by using the control system with all the loudspeakers operating on the pressure field from either the port or starboard propeller alone. It is clear that significant reductions are achieved at the BPF with smaller reductions at the second and third harmonics. The normalised acoustic pressures measured at the error microphones at 88 Hz, before and after control, are plotted in terms of their physical position in the cabin in Figure 8.20. Not only has the control system given substantial reductions in the sum of the squared error signals, as listed in Table 8.1, but the individual mean-square pressures at each of the microphones has also been reduced in this case. The maximum pressure level at any one

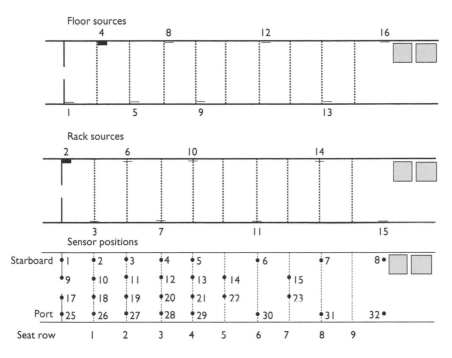

Figure 8.19 Locations of loudspeakers in the aircraft cabin, on the floor (upper) and in the hat racks (middle), and of the microphones (lower), which are all placed at seated head height. The shaded blocks at the rear of the cabin (right side of the figure), show the position of the control system.

Table 8.1 Changes in the level of the sum of the squared output of 32 control microphones measured during flight trials of a 16-loudspeaker-32-microphone control system in a B.Ae. 748 aircraft (Elliott *et al.* 1990)

Configuration	Propeller	88 Hz ΔJ(dB)	176 Hz ΔJ(dB)	264 Hz ΔJ(dB)
Loudspeakers distributed throughout the cabin	Port	−13.8	−7.1	−4.0
	Starboard	−10.9	−4.9	−4.8

microphone position was reduced by about 11 dB for the results shown in Figure 8.20.

At the second and third harmonics, the attenuations in the sum of squared microphone signals are somewhat smaller than at the BPF. This reflects the greater complexity of the acoustic field within the aircraft cabin at these

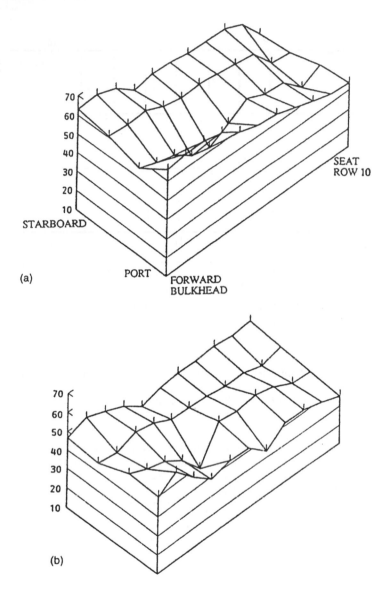

Figure 8.20 Distribution of normalised sound pressure level at 88 Hz, measured at the 32 control microphones illustrated in Figure 8.19 with the control system off (a) and on (b).

higher frequencies, which makes the sound field more difficult to control. By moving some of the loudspeakers to a circumferential array in the plane of the propellers, at the front of the aircraft, somewhat greater reductions than those shown in Table 8.1 could be achieved at the second and third harmonics. This improvement is achieved by the ability of the secondary loudspeakers to more closely match the spatial distribution of the fuselage vibration caused by the propellers, and thus control the sound at source without it being radiated into the cabin. Some care must be taken in the interpretation of the results from the error microphones at these higher frequencies, however, since the greater complexity in the sound field makes it more likely that the pressure level has been substantially reduced only at the locations of the error microphones, and that the average pressure level in the rest of the cabin has remained substantially unaltered. More recent flight trials have included a separate array of monitoring microphones, which are not used in the cost function minimised by the active control system, so that an independent measure of attenuation can be achieved (Borchers *et al.* 1994).

Although the aircraft control system did not have to respond very quickly under straight and level cruise conditions if the speeds of the two propellers were locked together, this was not possible on the aircraft used for the flight trials. The results presented in Table 8.1 were obtained with only one engine operating at 14,200 rpm and the other detuned to 12,700 rpm, so that its contribution to the sound field was ignored by the control system. As the speeds of the two engines are brought closer together, the sound fields due to the two propellers beating together and the amplitude tracking properties of the adaptive algorithm allow the control system to follow these beats, provided they are not quicker than about 2 beats per second. Reductions in overall A-weighted sound pressure level, of up to 7 dB(A) were measured with the active control system operating at all three harmonics (Elliott *et al.* 1990). These measurements include a frequency-dependent A-weighting function to account for the response of the ear (Kinsler *et al.* 1999).

Since this early work, which demonstrated the feasibility of actively controlling the internal propeller noise using loudspeakers, a number of commercial systems have been developed (Emborg and Ross 1993; Billout *et al.* 1995) and are now in service on a number of aircraft. Instead of using loudspeakers as secondary sources it is also possible to use structural actuators attached to the fuselage to generate the secondary sound field. These have a number of potential advantages over loudspeakers, since they are capable of reducing both the cabin vibration as well as the sound inside the aircraft. It may also be possible to integrate structural actuators more easily into the aircraft manufacturing process, since no loudspeakers would need to be mounted on the trim panels. Early work in this area using piezoceramic actuators on the aircraft fuselage (Fuller 1985) demonstrated that relatively few secondary actuators may be needed for efficient acoustic control. More recent systems have used inertial electromagnetic actuators mounted on the aircraft frame (Ross and Purver 1997) or actively tuned resonant mechanical systems (Fuller *et al.* 1995; Fuller 1997) to achieve a more efficient mechanical excitation of the fuselage at the BPF and its harmonics.

8.7 Local control of sound

Apart from minimising a global cost function, such as radiated power or total energy, an active control system can also be designed to minimise the local response of a system. The physical effect of such a local control system will be illustrated in this section by considering the cancellation of the pressure at a point in a room.

8.7.1 Cancellation of pressure in a large room

Cross-sections through the acoustic zones of quiet, generated by cancelling the pressure at a point in a large three-dimensional space, are shown in Figure 8.21. The sound field in the room is assumed to be diffuse, which occurs when the excitation frequency is such that the modal overlap in equation (8.60) is well above unity. In this case, when the distance from the monopole secondary source to the cancellation point is very small, compared with the acoustic wavelength, the zone of quiet forms a *shell* around the secondary source, as indicated by the shaded area in the two-dimensional cross-section shown in Figure 8.21a. If the pressure is cancelled in the nearfield of an acoustic monopole source, the secondary field will only be equal and opposite to the primary field at distances from the secondary source that

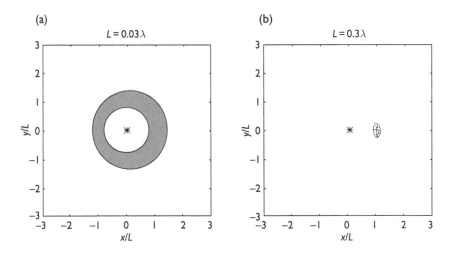

Figure 8.21 The spatial extent of the acoustic 'zone of quiet' generated by cancelling the pressure at $x = L$ in a three-dimensional free field using a point monopole acoustic secondary source at the origin of the coordinate system for (a) $L = 0.03\lambda$ and (b) $L = 0.3\lambda$, where λ is the acoustic wavelength. The shaded area within the solid line corresponds to a 10 dB attenuation in the diffuse primary field.

are about the same as the distance to the cancellation point, thus generating a shell of cancellation. At higher frequencies when, the distance from the secondary source to the cancellation point is not small compared with the wavelength, then the zone of quiet does not form a complete shell around the secondary source but is now concentrated in a sphere centred on the cancellation point, whose diameter is about $\lambda/10$ (Elliott *et al.* 1988a), as shown in Figure 8.21b.

The advantage of a local control system is that the secondary source does not have to drive very hard to achieve control, because it is very well coupled to the pressure at the cancellation point. Thus local zones of quiet can often be generated without greatly affecting the overall energy in the system. Local active control systems also have the advantage that, because the secondary actuator and the error sensor are close together, there is relatively little delay between them, which can improve the performance of both feedforward and feedback control systems. One of the earliest designs for a local active control system was proposed by Olsen and May (1953), who suggested that a loudspeaker on the back of a seat could be used to generate a zone of quiet around the head of a passenger in a car or an aircraft, as illustrated by Figure 8.22. Olson and May had the idea of using a feedback control system to practically implement such an arrangement, although in the 1950s most of the design effort had to be

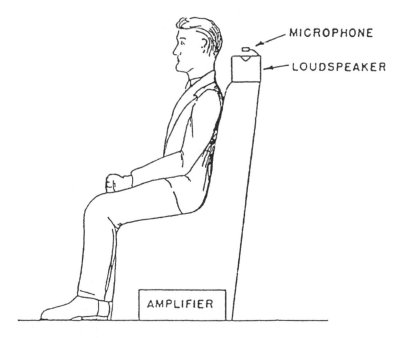

Figure 8.22 Illustration of a local system for the active control of sound near the head of a seated passenger (taken from Olson and May 1953).

spent in reducing the phase lag in the audio amplifier to ensure stability of the feedback loop. More recent investigations of such systems (Rafaely *et al.* 1999) have demonstrated the trade-off in such a system between good acoustic performance and robust stability of the feedback loop. The upper frequency of control in such a system will be fundamentally determined by the acoustic considerations and the extent of the listener's head movements. Zones of quiet of a useful size can be achieved in practice up to several hundred Hertz. The obvious extension to such a local control system would be to arrange for the loudspeakers to move with the ears, as in an active headset. Under these conditions the entrance of the ear canal is kept very close to the position of the loudspeaker, and the fundamental acoustic limitations can generally be avoided up to a frequency of about 1 kHz.

8.8   Conclusions

The physical principles of active sound control are most easily illustrated by the one-dimensional problem of plane waves propagating in a duct. A single secondary source can be used to reflect an incident primary plane wave, and a pair of secondary sources can be used to absorb an incident plane wave. At higher frequencies, multiple waves can propagate in the duct, and multiple actuators then need to be used to control the transmission of acoustic energy.

The problem of spatial matching becomes more severe when secondary acoustic sources are used to attenuate the sound power output of a primary source in a three-dimensional space. If the sources are in a free field, a single monopole secondary source must be spaced within about one-tenth of an acoustic wavelength of a monopole primary source if reductions of 10 dB are to be achieved in sound power output. If multiple secondary sources are used, the number of secondary sources necessary to obtain a given level of performance increases with the square of their distance from the primary source divided by the acoustic wavelength.

When the primary source drives an enclosed sound field a number of different active control strategies may be used. These strategies can be illustrated most easily by returning to the one-dimensional field in a duct, which is closed at the two ends by the primary and secondary sources. Cancelling the pressure in front of the secondary source effectively changes the end condition in the duct from open to closed; resonances are still apparent but are shifted in frequency. Driving the secondary source so as to absorb the incident soundwave suppresses these resonances. However, the most effective strategy for reducing the stored energy in the enclosure is to drive the secondary source so as to minimise the total power input to the duct from both the primary and secondary sources. This strategy maintains the sound field in an anti-resonant condition at each frequency and can give significant reductions in the stored energy at each of the resonant frequencies, which are equally spaced in this one-dimensional case.

The number of modes of a three-dimensional enclosure which are significantly excited at a given driving frequency increases with the cube of that frequency, and a single secondary source is not sufficient to control such a sound field above the first resonance frequency. Multiple secondary sources can be used to achieve control in this case, provided there are at least as many secondary sources as significantly excited acoustic modes. Minimising the total acoustic potential energy in the enclosure has almost the same effect as minimising the total acoustic power input. By using an array of microphones, a practical approximation to the total acoustic potential energy can be measured in an enclosure, which can be used to automatically adjust the amplitude and phase of the multiple secondary sources to minimise this quantity.

Such a control system with 16 loudspeakers and 32 microphones was used in the original demonstration of the control of propeller noise in a passenger aircraft. Commercial systems are now available for this purpose, although more recent systems use electromagnetic shakers attached to the fuselage as actuators, instead of loudspeakers, which can control the vibration in the passenger cabin as well as the noise.

Finally, the possibility of operating an active control system at higher frequencies is considered, by controlling only the local sound field around a person's head instead of the global sound field throughout an enclosure. It is shown that such a strategy can generate a zone of quiet which forms a shell around the secondary source at low frequencies. At high frequencies the quiet zone reduces, on average, to a sphere centred on the cancellation microphone, whose diameter is one-tenth of an acoustic wavelength. Unless this zone of quiet can always be maintained close to the ear, as in the case of an active headset, degradation due to the natural movement of a listener's head will restrict the frequency range of a headrest system using such local active control to frequencies below about 500 Hz.

Active control techniques are thus seen to complement conventional passive control methods since they are most effective at low frequencies. Current applications are mainly in the aerospace field, where the reduced weight of an active system for low-frequency noise control, compared with a passive one, is worth the additional cost and complexity of an electronic system.

References

Beatty, L.G. (1964) Acoustic impedance in a rigid-walled cylindrical sound channel terminated at both ends with active transducers. *Journal of the Acoustic Society of America*, 36, 1081–1089.

Billout, G., Norris, M.A. and Rossetti, D.J. (1995) Systéme de controle actif de bruit Lord NVX pour avions d'affaire Beechcraft Kingair, un concept devanu produit. *Active Control Conference*, Senlis, France.

Borchers, I.U., Tougard, D. and Klöppel, V. (1994) Advanced study of active noise control in aircraft (ASANCA), in *Advances in Acoustics Technology*, J.M.M. Hernandez (ed.) (John Wiley & Sons).

Brennan, M.J., Elliott, S.J. and Pinnington, R.J. (1995) Strategies for the active control of flexural vibration on a beam. *Journal of Sound and Vibration*, **186**, 657–688.

Bullmore, A.J., Nelson, P.A., Elliott, S.J., Evers, J.F. and Chidley, B. (1987) Models for evaluating the performance of propeller aircraft active noise control systems. *AIAA 11th Aeroacoustics Conference*, Palo Alto, C.A., Paper AIAA-87-2704.

Burdisso, R.A., Thomas, R.H., Fuller, C.R. and O'Brien, W.F. (1993) Active control of radiated inlet noise from turbofan engines. *Second Conference on Recent Advances in the Active Control of Sound and Vibration*, Blacksburg, Virginia, USA, 848–860.

Chaplin, G.B.B. (1983) Anti-sound – The Essex breakthrough. *Chartered Mechanical Engineer*. **30**, 41–47.

Conover, W.B. (1956) Fighting noise with noise. *Noise Control*, **2**, 78–82.

Curtis, A.R.D., Nelson, P.A., Elliott, S.J. and Bullmore, A.J. (1987) Active suppression of acoustic resonance. *Journal of the Acoustical Society of America*, **8**, 624–231.

Darlington, P. and Nicholson, G.C. (1992) Theoretical and practical constraints on the implementation of active acoustic boundary elements. *2nd Inc. Congress on Recent Developments in Air and Structure-borne Sound and Vibration*.

Dorling, C.M., Eatwell, G.P., Hutchins, S.M., Ross, C.F. and Sutcliffe, S.G.C. (1989) A demonstration of active noise reduction in an aircraft cabin. *Journal of Sound and Vibration*, **128**, 358–360.

Elliott, S.J. (1994) Active control of sound and vibration. Keynote lecture, *ISMA19*, Katholic University of Leuven, Belgium.

Elliott, S.J. (2001) *Signal Processing for Active Control* (London: Academic press).

Elliott, S.J., Joseph, P., Bullmore, A.J. and Nelson, P.A. (1988a) Active cancellation at a point in a pure tone diffuse field. *Journal of Sound and Vibration*, **120**, 183–189.

Elliott, S.J., Stothers, I.M., Nelson, P.A., McDonald, A.M., Quinn, D.C. and Saunders, T. (1988b) The active control of engine noise inside cars. *InterNoise '88*, Avignon, 987–990 (INCE).

Elliott, S.J., Nelson, P.A., Stothers, I.M. and Boucher, C.C. (1989) Preliminary results of in-flight experiments on the active control of propeller-induced cabin noise. *Journal of Sound and Vibration*, **128**, 355–357.

Elliott, S.J., Nelson, P.A., Stothers, I.M. and Boucher, C.C. (1990) In-flight experiments on the active control of propeller-induced cabin noise. *Journal of Sound and Vibration*, **140**, 219–238.

Elliott, S.J., Joseph, P., Nelson, P.A. and Johnson, M.E. (1991) Power output minimisation and power absorption in the active control of sound. *Journal of the Acoustical Society of America*, **90**, 2501–2512.

Emborg, U. and Ross, C.F. (1993) Active control in the SAAB 340, *Recent Advances in the Active Control of Sound and Vibration*, 567–573.

Eriksson, L.J., Allie, M.C., Hoops, R.H. and Warner, J.V. (1989) Higher order mode cancellation in ducts using active noise control. *InterNoise 89*, 495–500 (INCE).

Fedorynk, M.V. (1975) The suppression of sound in acoustic waveguides. *Soviet Physics Acoustics*, **21**, 174–176.

Ffowcs-Williams, J.E. (1984) Review lecture: anti-sound. *Royal Society of London*, **A395**, 63–88.

Fuller, C.R. (1985) Experiments on reduction of aircraft interior noise using active control of fuselage vibration. *Journal of the Acoustical Society of America*, **78** (S1), S88.

Fuller, C.R. (1997) Active control of cabin noise – lessons learned? *Fifth International Congress on Sound and Vibration*, Adelaide.

Fuller, C.R., Maillard, J.P., Meradal, M. and Von Flotow, A.H. (1995) Control of aircraft interior noise using globally detuned vibration absorbers. *The First Joint CEAS/AIAA Aeroacoustics Conference*, Munich, Germany, Paper CEAS/AIAA-95-082, 615–623.

Guicking, D. and Karcher, K. (1984) Active impedance control for one-dimensional sound. *American Society of Mechanical Engineers Journal of Vibration, Acoustics, Stress and Reliability in Design*, **106**, 393–396.

Hesselman, M. (1978) Investigation of noise reduction on a 100 kVA transformer tank by means of active methods. *Applied Acoustics*, **11**, 27–34.

Jessel, M.J.M. (1968) Sur les absorbeurs actifs. *6th International Conference on Acoustics*, Tokyo. Paper F-5-6, 82.

Joseph, P., Nelson, P.A. and Fisher, M.J. (1996) An in-duct sensor array for the active control of sound radiated by circular flow ducts. *InterNoise 96*, 1035–1040 (INCE).

Joseph, P., Nelson, P.A. and Fisher, M.J. (1999) Active control of fan tones radiated from turbofan engines. I: External error sensors, and, II: In-duct error sensors. *Journal of the Acoustical Society of America*, **106**, 766–778 and 779–786.

Kinsler, L.E., Frey, A.R., Coppens, A.B. and Sanders, J.V. (1999) *Fundamentals of Acoustics* (4th edn) (New York: John Wiley).

Konaev, S.I., Levedev, V.I. and Fedorynk, M.V. (1977) Discrete approximations of a spherical Huggens surface. *Soviet Physics Acoustics*, **23**, 373–374.

Lueg, P. (1936) Process of silencing sound oscillations. U.S. Patent, No. 2,043,416.

Mangiante, G. (1994) The JMC method for 3D active absorption: A numerical simulation. *Noise Control Engineering*, **41**, 1293–1298.

Meirovitch, L. (1990) *Dynamics and Control of Structures* (New York: John Wiley).

Metzger, F.B. (1981) Strategies for reducing propeller aircraft cabin noise. *Automotive Engineering*, **89**, 107–113.

Morse, P.M. (1948) *Vibration and Sound* (2nd edn) (New York: McGraw-Hill) (reprinted in 1981 by the Acoustical Society of America).

Nelson, P.A. (1996) Acoustical prediction. *Internoise 96*, 11–50, (INCE).

Nelson, P.A. and Elliott, S.J. (1992) *Active Control of Sound* (London: Academic Press).

Olson, H.F. and May, E.G., (1953) Electronic sound absorber. *Journal of the Acoustical Society of America*, **25**, 1130–1136.

Orduña-Bustamante, F. and Nelson, P.A. (1991) An adaptive controller for the active absorption of sound. *Journal of the Acoustical Society of America*, **91**, 2740–2747.

Rafaely, B. (2000) Active noise reducing headset. *First Online Symposium for Electronics Engineers (OSEE)*.

Rafaely, B., Elliott, S.J. and Garcia-Bonito, J. (1999) Broadband performance of an active headrest. *Journal of the Acoustical Society of America*, 102, 787–793.

Ross, C.F. and Purver, M.R.J. (1997) Active cabin noise control. *ACTIVE 97*, Budapest, Hungary, xxxix–xlvi (Publishing Company of the Technical University of Budapest).

Roure, A. (1985) Self adaptive broadband active noise control system. *Journal of Sound and Vibration*, 101, 429–441.

Silcox, R.J. and Elliott, S.J. (1990) Active control of multi-dimensional random sound in ducts. *NASA Technical Memorandum 102653*.

Sommerfeldt, S.D. and Nasif, P.J. (1994) An adaptive filtered-*x* algorithm for energy-based active control. *Journal of the Acoustical Society of America*, 96, 300–306.

Stell, J.D. and Bernhard, R.J. (1991) Active control of higher-order acoustic modes in semi-infinite waveguides. *Transactions of the American Society of Mechanical Engineering Journal of Vibration and Acoustics*, 113, 523–531.

Sutton, T.J., Elliott, S.J. and McDonald, A.M. (1994) Active control of road noise inside vehicles. *Noise Control Engineering Journal*, 42, 137–147.

Swinbanks, M.A. (1973) The active control of sound propagating in long ducts. *Journal of Sound and Vibration*, 27, 411–436.

Tichy, J. (1988) Active systems for sound attenuation in ducts. *International Conference on Acoustics, Speech and Signal Processing ICASSP88*, 2602–2605 (IEEE Press).

Wheeler, P.D. (1987) The role of noise cancellation techniques in aircrew voice communications systems. *Proceedings of the Royal Aeronautical Society Symposium on Helmets and Helmet-mounted Displays*.

Wilby, J.F., Rennison, D.C., Wilby, E.G. and Marsh, A.H. (1980) Noise control prediction for high speed propeller-driven aircraft. *American Institute of Aeronautics and Astronautics 6th Aeroacoustic Conference*, Paper AIAA-80-0999.

Winkler, J. and Elliott, S.J. (1995) Adaptive control of broadband sound in ducts using a pair of loudspeakers. *Acustica*, 81, 475–488.

Zander, A.C. and Hansen, C.H. (1992) Active control of higher-order acoustic modes in ducts. *Journal of the Acoustical Society of America*, 92, 244–257.

Zavadskaya, M.P., Popov, A.V. and Egelskii, B.L. (1976) Approximations of wave potentials in the active suppression of sound fields by the Malyuzhinets method. *Soviet Physics Acoustics*, 33, 622–625.

Part IV

Vibration

Chapter 9

Mobility and impedance methods in structural dynamics

P. Gardonio and M.J. Brennan

9.1 Introduction

The mobility and impedance representation is a very powerful tool in structural dynamics. It allows structures to be subdivided but does not have the high-frequency limitations of the finite element method. Problems can be formulated in terms of variables that are easily measurable, and the representation is particularly useful in the training of young researchers and engineers as it forces them to think of the physics of the problem in hand rather than simply the mathematics.

In this chapter, mobility–impedance methods for both lumped parameter and distributed one- and two-dimensional systems are described. Following a historical review in Section 9.2, Section 9.3 describes the way in which mobilities can be used to study a variety of vibration problems. In Section 9.4 the mobility and impedance approach to describing lumped parameter systems is discussed. This section is based on the chapter written by Hixson in the *Shock and Vibration Handbook* (Hixson 1976). The following three sections extend the mobility–impedance to structural elements such as beams and plates that have distributed mass and stiffness.

9.2 Historical background

Mobility and impedance methods have been used for many years to study either coupled electrical–mechanical or simply mechanical systems (Gardonio and Brennan 2002). Professor Arthur G. Webster was the first to realise the possibility of using the electric impedance concept in the study of vibrating mechanical systems. During a meeting of the Physical Society in Philadelphia held in December 1914, he read a paper (Webster 1919) in which he first defined the acoustic impedance terms for single degree of freedom (sdof) acoustic and mechanical oscillating systems. Once the similarity between mechanical and electrical systems was realised, network theory developed for purely electrical systems was used to analyse electro-mechanical transducers (Guillemin 1953). The method eventually became known as the 'direct analogy' (Nickle 1925; Maxfield and Harrison 1926; Herbert 1932; Pawley 1937; Mason 1942; Olson 1943; Bloch 1945), based on the following methodology: first, the electrical circuit analogous to the mechanical problem is drawn; second, the electrical problem is solved using the electric network theory; and third, the answer is reworked into mechanical terms. Velocity and force parameters of the mechanical system are represented by current and electromotive force (emf) of the equivalent electrical circuit respectively (Maxfield and Harrison 1926), and equivalent network impedances are derived from the corresponding mechanical impedances by assuming appropriate conversion factors and connecting rules (Bloch 1945).

Since the early studies on analogies it had been apparent that the variables used in the differential equations of electrodynamic or electromagnetic systems were not consistent with those of purely electrical systems (Wegel 1921; le Corbeiller 1929). It was concluded that electrodynamic and electromagnetic systems' network diagrams could not be directly transposed to an equivalent purely electrical network (Firestone 1933). Darrieus (1929) first mentioned the possibility of defining the analogy in a different way where force is equivalent to current rather than emf. A new analogy was subsequently published by Hähnle (1932) and Firestone (1933), who defined the term 'bar impedance'. This was the ratio of a kinematic variable to a kinetic variable and gave a new electromechanical analogy for solving vibroacoustic problems. The new analogy was free from the problems of consistency highlighted by Firestone and became known as the 'inverse analogy'. The transformation of the mechanical bar impedances to electric impedances used new conversion factors and connecting rules as described by Bloch (1945).

Table 9.1 summarises the main features of the direct and inverse electromechanical analogies. The first page shows two sdof mass-spring-damper systems. The system denoted by M_f is a parallel connected system and is excited by a force. The system denoted by M_v is a series connected system and is excited by a velocity source. The following two pages of Table 9.1 show the analogous, or inversely analogous, electrical networks. The two mechanical systems depicted, and their direct or inverse analogue electric network pairs, are dual. The equation of motion of the mechanical system M_f can be derived using d'Alembert principle directly: the applied force is equal to the sum of the inertial force (which is proportional to the absolute acceleration of the mass), the damping force (which is proportional to the relative velocity across the viscous damper) and the elastic force (which is proportional to the relative displacement across the spring). For this system it is straightforward to derive the equivalent electrical circuit using the direct analogy where the forces from the three mechanical elements are

Table 9.1 Electromechanical analogies. The factors ε_f, ε_v, δ_f, δ_v are the conversion factors from force or velocity variables to voltage or current variables

Mechanical system

m = mass (kg)

k = stiffness (N/m)

c = viscous damping coefficient (N/ms^{-1})

$f(t) = F\exp(j\omega t)$ harmonic force (N)

$v(t) = V\exp(j\omega t)$ harmonic velocity (m/s)

M$_f$

$$m\frac{dv}{dt} + cv + k\int v\,dt = f(t)$$

Dual of

M$_v$

$$\frac{1}{k}\frac{df}{dt} + \frac{f}{c} + \frac{1}{m}\int f\,dt = v(t)$$

Table 9.1 (Continued)

Equivalent electric circuit from

Direct analogy

E_v

Dual of

E_i

$f(t) \leftrightarrow e(t)$

$v(t) \leftrightarrow i(t)$

L = inductance (H)

C = capacitance (F)

R = resistance (Ω)

$e(t) = E \exp(j\omega t)$ Voltage (V)

$i(t) = I \exp(j\omega t)$ Current (A)

$$L\frac{di}{dt} + Ri + \frac{1}{C}\int i\,dt = e(t) \qquad C\frac{de}{dt} + \frac{1}{R}e + \frac{1}{L}\int e\,dt = i(t)$$

where $L = m/\alpha^2$ $R = c/\alpha^2$ $C = \alpha^2/k$ with $\alpha^2 = \varepsilon_v/\varepsilon_f$ $E = \varepsilon_f F$ $I = \varepsilon_v V$

Table 9.1 (Continued)

Equivalent electric circuit from

Inverse analogy

$f(t) \leftrightarrow i(t)$

$v(t) \leftrightarrow e(t)$

$L =$ inductance (H)

$C =$ capacitance (F)

$R =$ resistance (Ω)

$e(t) = E \exp(j\omega t)$ Voltage (V)

$i(t) = I \exp(j\omega t)$ Current (A)

E_i

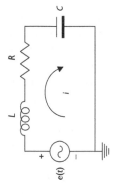

Dual of

E_v

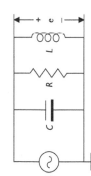

$$C\frac{de}{dt} + \frac{1}{R}e + \frac{1}{L}\int edt = i(t)$$

$$L\frac{di}{dt} + Ri + \frac{1}{C}\int idt = e(t)$$

where $\quad L = \beta^2/k \quad R = \beta^2/c \quad C = m/\beta^2 \quad$ with $\quad \beta^2 = \delta_v/\delta_f \quad I = \delta_f F \quad E = \delta_v V$

represented by emf differences across an inductance, a resistor and a capacitor connected in series such that they balance the voltage generator emf: E_v (second page left).

Alternatively the equation of motion of the mechanical system M_v (first page right) can be derived using the principle of compatibility. The sum of the absolute displacement of the mass and the relative displacements of the dash-pot and spring elements is equivalent to that generated by the scotch-yoke motion generator. In this case it is easier to derive the equivalent electrical circuit by using the inverse analogy where the displacements of the three mechanical elements are represented by the emf differences across an inductance, a resistor and a capacitor connected in series in such a way to balance the voltage generator emf: E_v (third page right). Thus the choice of using the direct or inverse analogy depends on the topology of the mechanical system. Nevertheless, Table 9.1 clearly shows that, as advocated by Firestone, the electrical networks derived with the inverse analogy have the same topological features as the mechanical systems from which they are derived. By contrast, the electric networks derived with the direct analogy have the node pair and meshes exchanged with reference to the mechanical system from which they are derived. Detailed introductions to electromechanical analogies can be found in books such as Mason (1942), Olson (1943), Beranek (1954), Koenig and Blackwell (1961), Hurty and Rubinstein (1964) and Skudrzyk (1968).

Because Firestone's definition of 'bar impedance' was not widely adopted, he introduced a new parameter called mobility (ease of motion), which is the ratio of a kinematic and a kinetic parameters (Firestone 1938). A problem could now be formulated in terms of mechanical variables, and all the electric circuit laws, theorems and principles were applied to a purely mechanical network. In the mobility formulation, force and velocity laws are equivalent to Kirchhoff's current and voltage laws. There are equivalent Thévenin and Norton theorems for force or velocity sources together with mechanical forms of the principles of reciprocity, superposition and compensation. Thus the mobility method proposed by Firestone had two main advantages: first, a problem could be formulated directly in terms of mechanical variables and second, a more intuitive formulation of mechanical network was possible. In a later paper, Firestone (1956) reworked the direct and inverse analogies so that instead of drawing an analogous electrical network the acoustical or mechanical network diagrams were drawn directly. The solution of a network problem was then carried out with reference to acoustical or mechanical units.

As with the electromechanical analogies, the choice of either the impedance or mobility methods is dependent upon the system of interest. Many mechanical systems can be modelled by rigidly connected, lumped elements, where the adjacent elements have the same velocity, or volume velocity in the acoustic case (M_v system in Table 9.1). The mobility approach is convenient for this type of system since the mechanical network can be drawn by inspection. When adjacent elements of a system have the same forces at their terminals, or pressures in the acoustic case (M_f system in Table 9.1), it is easier to derive the impedance network. In certain cases it could be desirable to draw part of the system scheme with the mobility method and part with the impedance method. These two different networks are then linked by means of specific couplers (Bauer 1953).

Table 9.2 shows the impedance (second page top) and mobility (second page bottom) schematic diagrams of the two mechanical systems M_f and M_v considered in Table 9.1. The equivalent impedance and mobility schematic diagrams have been derived from the appropriate mechanical diagrams for either the impedance or mobility method. The lines connecting the lumped elements have a different significance in the two diagrams. With the impedance method the connecting lines represent a 'force junction' where there is equal force at the terminals of two adjacent elements. When the mobility method is employed the lines connecting the elements represent a 'velocity junction' that ensures equal velocity at the terminals of adjacent elements. The diagrams obtained with the mobility method keep the same topological features of the mechanical systems from which they are derived; so the mobility approach, in general, is more intuitive. This is true except in a few cases where the system is such that the transmission of force is imposed at the junction between the elements (M_v in Table 9.2). In many cases, acoustical filters, with or without side branches, are better represented in terms of impedances. However it is now common practice to draw a mobility schematic diagram for the system under study and then to solve the problem using either mobility or impedance function using the appropriate network laws as described in Section 9.4.

In 1958 the American Society of Mechanical Engineers (ASME) organised a colloquium on mechanical impedance and mobility methods, and this is one of the key milestones in the theory of impedance and mobility methods (Plunket 1958). In 1963, The American National Standards Institute was the first to introduce a standard nomenclature: *Nomenclature and Symbols for Specifying the Mechanical Impedance of Structures* (ANSI, S2.6-1963 R 1971). There are few books that introduce the main principles of impedance or mobility methods. Perhaps the most complete presentation is by Hixson (1976).

9.3 Use of mobility and impedance in vibration problems

Impedance and mobility methods can be used in a straightforward manner in the analysis of *linear* mechanical systems subject to three types of excitation: periodic, transient and stationary random.

Table 9.2 Impedance and mobility diagrams

Mechanical system

m = mass (kg)

k = stiffness (N/m)

c = viscous damping coefficient (N/ms^{-1})

$f(t) = F\exp(j\omega t)$ harmonic force (N)

$v(t) = V\exp(j\omega t)$ harmonic velocity (m/s)

M$_f$

$$m\frac{dv}{dt} + cv + k\int v\,dt = f(t)$$

Dual of

M$_v$

$$\frac{1}{k}\frac{df}{dt} + \frac{f}{c} + \frac{1}{m}\int f\,dt = v(t)$$

Table 9.2 (Continued)

Impedance schematic or tubing diagram from the

Impedance method

Z_T total impedance

$Z_m = j\omega m$ impedance of the mass

$Z_c = c$ impedance of the damper

$Z_k = \dfrac{k}{j\omega}$ impedance of the spring

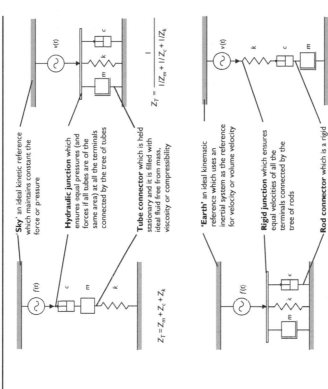

'Sky', an ideal kinetic reference which maintains constant the force or pressure

Hydraulic junction which ensures equal pressures (and forces if all tubes are of the same area) at all the terminals connected by the tree of tubes

Tube connector which is held stationary and it is filled with ideal fluid free from mass, viscosity or compressibility

$$Z_T = Z_m + Z_c + Z_k$$

$$Z_T = \dfrac{1}{1/Z_m + 1/Z_c + 1/Z_k}$$

Mobility schematic or wiring diagram from the

Mobility method

Y_T total mobility

$Y_m = \dfrac{1}{j\omega m}$ mobility of the mass

$Y_c = \dfrac{1}{c}$ mobility of the damper

$Y_k = \dfrac{j\omega}{k}$ mobility of the spring

'Earth' an ideal kinematic reference which uses an inertial system as the reference for velocity or volume velocity

Rigid junction which ensures equal velocities of all the terminals connected by the tree of rods

Rod connector which is a rigid element free from mass, friction or compliance

$$Y_T = \dfrac{1}{1/Y_m + 1/Y_c + 1/Y_k}$$

$$Y_T = Y_m + Y_c + Y_k$$

9.3.1   Time-varying response to harmonic, periodic and transient excitation

The instantaneous response $x(P_1, t)$ at a point P_1 of a linear mechanical system excited by a force $f(P_2, t)$ acting at point P_2 of the system is given by the convolution or Duhamel integral

$$x(P_1, t) = \int_{-\infty}^{t} f(P_2, \tau) h(t - \tau) d\tau \tag{9.1}$$

where $h(t)$ is the impulse response of the system (Thomson 1993). This expression is valid for any form of excitation history and can be used to study systems composed of both lumped and flexible elements. The response $x(P_1, t)$ could be a displacement, velocity or acceleration variable as long as $h(t)$ is the appropriate impulse response. If the excitation is harmonic, with the form $f(t) = \text{Re}\{F(\omega)e^{j\omega t}\}$, where ω is circular frequency and $F(\omega)$ is complex, the steady-state displacement response is given by $x(t) = \text{Re}\{X(\omega)e^{j\omega t}\}$, where $X(\omega)$ is complex. The displacement can also be written as (Thomson 1993)

$$x(t) = \text{Re}\{\alpha(\omega)F(\omega)e^{j\omega t}\} \tag{9.2}$$

where $\alpha(\omega) = X(\omega)/F(\omega)$ is the receptance frequency response function (FRF) of the system (Bishop and Johnson 1960; Ewins 2000). The relation between receptance and mobility is described in Section 9.3.3.

In many cases the excitation is periodic, with period T, but not harmonic. In this case the excitation can be regarded as the sum of an infinite number of harmonic functions, whose circular frequency is an integer multiple of the fundamental $\omega_o = 2\pi/T$; that is, $f(t) = \sum_{n=-\infty}^{n=+\infty} F(\omega_n)e^{j\omega_n t}$, where $\omega_n = n\omega_o$ (Wylie and Barrett 1982). Because the system is linear, the steady-state displacement response can be calculated by superposing the harmonic responses due to each of the exciting harmonic forces $x(\omega_n, t) = \text{Re}\{\alpha(\omega_n) \cdot F(\omega_n)e^{j\omega_n t}\}$, so that (Wylie and Barrett 1982; Ewins 2000)

$$x(t) = \sum_{n=-\infty}^{n=+\infty} \alpha(\omega_n)F(\omega_n)e^{jn\omega_o t} \tag{9.3}$$

If the excitation is not periodic, the *inverse Fourier transform* can be used and the excitation can be represented by $f(t) = (1/2\pi)\int_{-\infty}^{+\infty} F(\omega)e^{j\omega t}d\omega$ (Wylie and Barrett 1982). The response can also be expressed using the *inverse Fourier transform* and the system receptance as (Wylie and Barrett 1982; Ewins 2000)

$$x(t) = \frac{1}{2\pi}\int_{-\infty}^{+\infty} \alpha(\omega)F(\omega)e^{j\omega t}d\omega \tag{9.4}$$

In the remaining part of this chapter the dependence of time or frequency will be omitted in the formulae and lower case symbols will indicate time-dependent functions while upper case symbols will indicate frequency-dependent functions.

9.3.2 Time-average response to random stationary excitation

The response of a system to a random excitation is also random, but filtered by the dynamics of the system. In the study of random phenomena no attempt is made to specify the amplitude and phase of the response instant by instant. For stationary random excitation, that is, an excitation whose probability distribution is invariant with reference to time, the mean value of the response of a linear system is given by (Crandall and Mark 1963)

$$\overline{x(P_1, t)} = E\left[\int_{-\infty}^{+\infty} f(P_2, t - \tau)h(\tau)d\tau\right] = \overline{F(P_2, t)}\alpha(0) \tag{9.5}$$

where τ is the integration time variable which is defined with reference to the time t at which the response is evaluated and E is the expectation operator. The mean square response of the system can be written in terms of the power spectral density of the random excitation, $S_{FF}(\omega)$, as

$$\overline{x^2(P_1, t)} = \int_{-\infty}^{+\infty} |\alpha(\omega)|^2 \, S_{FF}(\omega)\, d\omega = \int_{-\infty}^{+\infty} S_{XX}(\omega)d\omega \tag{9.6}$$

where

$$S_{XX}(\omega) = |\alpha(\omega)|^2 \, S_{FF}(\omega) \tag{9.7}$$

is the spectral density of displacement response of the system. In this chapter, mobility and impedance functions are given for beams and plates using the hysteretic damping model. It should be noted that this type of damping model is only applicable for steady-state harmonic excitation and that if transient or random vibration is of interest, a viscous damping model should be used because hysteretic damping gives an a-causal impulse response.

9.3.3 Forms of FRF

In the previous sub-sections it has been shown that the receptance FRF plays an important role in the prediction of the vibration response of linear systems to various types of excitation. The FRF (kinetic variable/kinematic variable) assumes different forms depending on the kinematic variable used to describe the response of the mechanical system: displacement, velocity or acceleration. For each FRF there is an inverse function (kinetic variable/kinematic variable), which is also an FRF.

The three standard definitions of FRFs and the equivalent inverse functions are summarised in Table 9.3. When the kinematic and kinetic variables are collocated, the FRFs listed in Table 9.3 have the prefix 'driving point' or simply 'point', and if the two parameters correspond to two different positions on the structure then the terms listed in Table 9.3 have the prefix 'transfer'. For example the ratio between velocity and force can be a 'driving point mobility' or a 'transfer mobility'.

9.4 Lumped parameter systems

In this section the impedance and mobility relations for mass, stiffness and viscous damper elements are presented together with the rules for combining them. The reader is referred to the chapter by Hixson (1976) for a comprehensive treatment of the subject. An sdof system is used as an example to illustrate the way in which the elements are combined, and to show how the impedance/mobility concepts are useful in the interpretation of dynamic behaviour. The mobility approach is then used to describe a vibration source and to couple it to a dynamic load.

9.4.1 Simple lumped elements

Figure 9.1 shows a linear system (element) with two connections. If this element is a spring, a viscous damper or a mass, the respective equations for these elements are

$$f_1 = -f_2 = k(x_1 - x_2), \qquad f_1 = -f_2 = c(\dot{x}_1 - \dot{x}_2), \qquad f_1 + f_2 = m\ddot{x}_1$$

$$(9.8a,b,c)$$

where k is the spring constant, c is the damping coefficient, m is mass and the dot denotes differentiation with respect to time. Assuming harmonic force excitation of the form $f = Fe^{j\omega t}$ with displacement response of the form $x = Xe^{j\omega t}$, where ω is circular frequency and $j = \sqrt{-1}$, then fixing one point of the spring and of the damper, and setting one force on the mass to zero, yields expressions for the mechanical impedances of the elements

$$Z_k = \frac{F_1}{\dot{X}_1} = -j\frac{k}{\omega}, \qquad Z_c = \frac{F_1}{\dot{X}_1} = c, \qquad Z_m = \frac{F_1}{\dot{X}_1} = j\omega m \qquad (9.9a,b,c)$$

where the following notation is used in this chapter: $\dot{X} = j\omega X$ and $\ddot{X} = -\omega^2 X$. The moduli of the impedances are plotted in Figure 9.2a on a log scale as a function of log frequency, and the impedances are plotted in the complex plane in Figure 9.2b. It is worthwhile examining these impedances

Table 9.3 Definitions of frequency response functions with $\dot{X} = j\omega X$ and $\ddot{X} = -\omega^2 X$

Response variable	Direct or standard frequency response function		Inverse frequency response function	
	Formula	Name	Formula	Name
Displacement	$\alpha(\omega) = \dfrac{X(P_1, \omega)}{F(P_2, \omega)}\bigg\|_{F(P \neq P_2)=0}$	Receptance Admittance Compliance Dynamic flexibility	$K(\omega) = \dfrac{F(P_1, \omega)}{X(P_2, \omega)}\bigg\|_{X(P \neq P_2)=0}$	Dynamic stiffness
Velocity	$Y(\omega) = \dfrac{\dot{X}(P_1, \omega)}{F(P_2, \omega)}\bigg\|_{F(P \neq P_2)=0}$	Mobility	$Z(\omega) = \dfrac{F(P_1, \omega)}{\dot{X}(P_2, \omega)}\bigg\|_{\dot{X}(P \neq P_2)=0}$	Mechanical impedance
Acceleration	$A(\omega) = \dfrac{\ddot{X}(P_1, \omega)}{F(P_2, \omega)}\bigg\|_{F(P \neq P_2)=0}$	Inertance Accelerance	$M(\omega) = \dfrac{F(P_1, \omega)}{\ddot{X}(P_2, \omega)}\bigg\|_{\ddot{X}(P \neq P_2)=0}$	Apparent mass

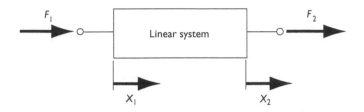

Figure 9.1 Linear system with two forces applied.

in some detail, because their characteristics are often sought in both measured and simulated frequency response functions to aid interpretation and give physical insight into the behaviour of a system. From Figure 9.2a it can be seen that the impedance of a spring decreases with frequency, the impedance of a mass increases with frequency and the impedance of a damper is independent of frequency. Referring to Figure 9.2b and equations (9.9a,b,c) it can be seen that the impedances of a spring and a mass are both imaginary, but they are of opposite sign. For these two elements the applied harmonic force and the resultant velocity are in quadrature and hence these elements do not dissipate power: they are reactive. However, in the case of a damper, the force and the velocity are in phase and hence this element is dissipative.

The mobilities of spring, damper and mass-like elements are simply the reciprocals of equations (9.9a,b,c) and are thus given by

$$Y_k = \frac{\dot{X}_1}{F_1} = \frac{j\omega}{k}, \qquad Y_c = \frac{\dot{X}_1}{F_1} = \frac{1}{c}, \qquad Y_m = \frac{\dot{X}_1}{F_1} = -j\frac{1}{\omega m} \qquad (9.10a,b,c)$$

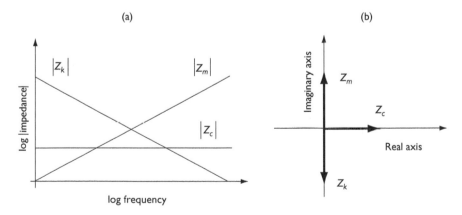

Figure 9.2 Impedance characteristics: (a) impedances as a function of frequency; (b) complex plane representation.

The moduli of the mobilities are plotted in Figure 9.3a on a log scale as a function of log frequency, and the mobilities are plotted in the complex plane in Figure 9.3b.

When examining measured frequency response functions of a passive structure a useful check on the validity of the measurements is to see whether or not the real part of the driving point impedance or mobility is greater than zero. This must be so, as energy cannot flow from the structure to the exciter; and if it is not, then there must be an instrumentation phase error. The reactive character of the structure can also be determined. If the imaginary part of the driving point mobility is negative at a particular frequency then the structure is behaving as a mass at this frequency: if it is positive, the structure is spring-like.

A hysteretically damped spring of which the complex stiffness is given by $k' = k(1 + j\eta)$, where η is the loss factor, can also be described using the impedance and mobility approach. The impedance is given by $Z_k' = Z_k(1 + j\eta)$, and provided that $\eta \ll 1$, the mobility is given by $Y_k' = Y_k(1 - j\eta)$. As discussed in the Introduction, an advantage of the impedance and mobility approach is that it is relatively easy to combine mass, spring and damper elements using simple rules without having to write down the equations of motion.

If a system is represented by a network of mobility elements arranged in parallel, the force applied to the system is split between the elements, which all respond with the same velocity. In this case the impedances are simply summed, and the mobilities combine as a reciprocal sum; that is to say, the total impedance and mobility of N elements Z_T and Y_T are respectively given by

Mobility scheme with elements in parallel
$$Z_T = \sum_{n=1}^{N} Z_n, \qquad \frac{1}{Y_T} = \sum_{n=1}^{N} \frac{1}{Y_n} \qquad (9.11a,b)$$

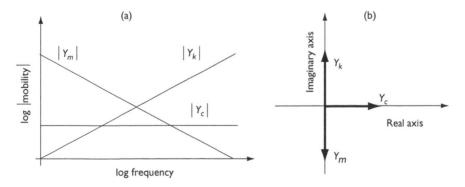

Figure 9.3 Mobility characteristics: (a) mobilities as a function of frequency; (b) complex plane representation.

If the mobility network elements are arranged in series, then provided that any mass elements are connected to an inertial system of reference, the velocity at the point where the force is applied is the sum of the velocities of the individual elements. This means that the sum of the impedances is a reciprocal sum and the total mobility is the sum of the individual mobilities, that is

Mobility scheme with elements in series
$$\frac{1}{Z_T} = \sum_{n=1}^{N} \frac{1}{Z_n}, \qquad Y_T = \sum_{n=1}^{N} Y_n \qquad (9.12a,b)$$

As an example of the addition of mobility or impedance elements in parallel, consider the sdof system shown in Figure 9.4a(i). The mobility representation of this system is shown in Figure 9.4a(ii). Note the convention used to represent a mass in this type of diagram (Hixson 1976) and that the mass has been 'connected' to an inertial system of reference. According to Table 9.2

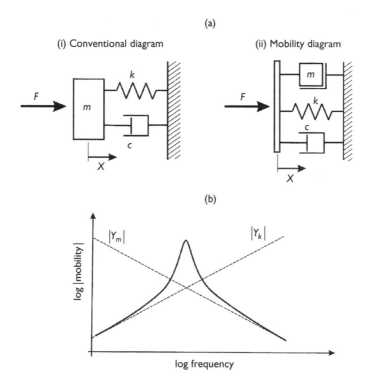

Figure 9.4 Combining lumped parameter elements connected in parallel: (a) mobility representation of a single-degree-of-freedom system; (b) driving point mobility of the single-degree-of-freedom system shown in (a) above.

the total (driving point) impedance is given by $Z_T = Z_k + Z_m + Z_c$ and the total (driving point) mobility by $Y_T = 1/Z_T$. The modulus of the mobility is plotted in Figure 9.4b, where the moduli of the mass and stiffness mobilities are also plotted to show which element dominates the dynamic behaviour of the system at any particular frequency. At the resonance frequency of the system, the moduli of the spring and mass impedances are equal, and $Z_k = -Z_m$, which means that the impedance of the system at this frequency is simply Z_c: thus the damping alone controls the vibration response.

If the sdof system is excited at the base as shown in Figure 9.5a(i) then the mobility diagram is as shown in Figure 9.5a(ii). The spring and damper impedances are in parallel, which are connected to the impedance of the mass in series. According to Table 9.2 the total (driving) impedance is given by $1/Z_T = 1/Z_m + 1/(Z_k + Z_c)$ and the mobility is given by $Y_T = 1/Z_T$ as before. The input mobility of this system is plotted in Figure 9.5b, where the moduli of the mass and damper mobilities are also plotted to aid interpretation. When impedances are connected in parallel, the largest

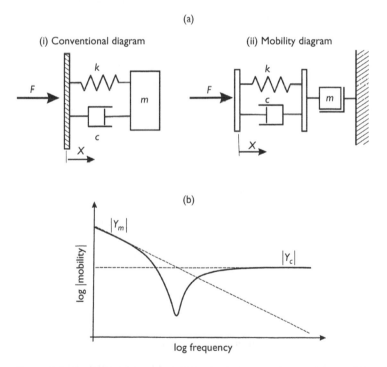

Figure 9.5 Combining lumped parameter elements connected in parallel and series. (a) Mobility representation of a base-excited single-degree-of-freedom system and (b) Driving point mobility of the single-degree-of-freedom system shown in (a) above.

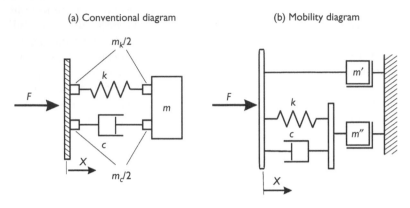

Figure 9.6 Accounting for the masses of springs and dampers.

impedance controls the dynamic behaviour and when impedances are connected in series, the smallest impedance controls the dynamic behaviour. Referring to Figure 9.5b, it can be seen that the mobility exhibits mass-like behaviour at low frequencies. This is because the spring has a large impedance at low frequencies and hence short-circuits the damper. There is, in effect, a mass in series with a spring and because the mass has the smaller impedance it then controls the response. At high frequencies, the impedance of the spring becomes very small and thus the damper short-circuits the spring. Because the impedance of the damper is much smaller than the mass, it controls the response. At the natural frequency of the system both the damper and the mass control the minimum in the response.

If the masses of springs and dampers are to be included in the model then care must be taken in drawing the mobility diagram prior to analysis. First the mass is split equally and added to each end of the spring or damper element as shown in Figure 9.6(a), where m_k and m_c are the masses of the spring and damper respectively. When the diagram is converted to a mobility diagram as shown in Figure 9.6(b), all masses must be connected to a stationary inertial reference point. Thus in this diagram $m' = (m_k/2) + (m_c/2)$ and $m'' = m + (m_k/2) + (m_c/2)$. Once the mobility diagram is drawn the impedance can be calculated as before to give $Z_T = Z_{m'} + 1/[1/Z_{m''} + 1/(Z_k + Z_c)]$.

9.4.2   Connection to a vibration source

It has been shown above that it is quite straight-forward to combine lumped parameter elements to give a single impedance at a selected point. When excited by a vibrating source at that point, this could be called the 'load impedance' Z_l. The impedance approach provides a convenient way to describe a vibration source. This is illustrated here for a single-point connected system, but it can be generalised for a multi-point source. The source can either be represented as a blocked force F_b in parallel with an internal impedance Z_i (a Thévenin equivalent source), or a free velocity $\dot{X}_f$ in series with an internal impedance (a Norton equivalent source). These sources are shown in Figure 9.7 connected to the load impedance. The choice of source representation depends upon personal preference and/or available data. The relation between the blocked force, the free velocity and the internal impedance of the source is given by

$$F_b = Z_i \dot{X}_f \tag{9.13}$$

In many practical situations it is relatively straight-forward to measure the free velocity of a source but difficult to measure the blocked force, especially over a wide frequency range. However, either Thévenin or the Norton equivalent system can be used to represent the source using the relationship given in equation (9.13), provided that the internal impedance of the source can be measured.

(a) A Thévenin equivalent system

(b) A Norton equivalent system

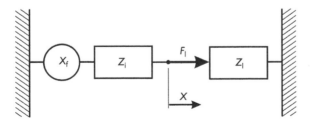

Figure 9.7 Vibration sources of internal impedance Z_i connected to a load impedance Z_l.

Referring to the upper system in Figure 9.7, the force applied to the load is related to the blocked force of the source by

$$F_1 = \frac{F_b}{1 + \dfrac{Z_i}{Z_l}} \tag{9.14}$$

If the modulus of the load impedance is much greater than the modulus of the internal impedance of the source ($|Z_l| \gg |Z_i|$), then $F_1 \approx F_b$, which means that the source behaves as a *force source* (Ungar and Dietrich 1966). *This means that the force applied to the load is insensitive to the dynamic behaviour of the load.* Referring to the lower system in Figure 9.7, the velocity of the load is related to the free velocity of the source by

$$\dot{X}_1 = \frac{\dot{X}_f}{1 + \dfrac{Z_l}{Z_i}} \tag{9.15}$$

If the load impedance is much less than the impedance of the source ($|Z_i| \gg |Z_l|$), then $\dot{X}_1 \approx \dot{X}_f$, which means that the source behaves as a *velocity source* (Ungar and Dietrich 1966). *This means that the velocity input to the load is insensitive to the dynamic behaviour of the load.*

9.5 General linear systems

If the system in Figure 9.1 is not a simple lumped parameter element then it has to be represented by an impedance matrix. The impedance relationship is given by

$$\begin{bmatrix} F_1 \\ F_2 \end{bmatrix} = \begin{bmatrix} Z_{11} & Z_{12} \\ Z_{21} & Z_{22} \end{bmatrix} \begin{bmatrix} \dot{X}_1 \\ \dot{X}_2 \end{bmatrix} \tag{9.16a}$$

which can be written in vector form as

$$\mathbf{f} = \mathbf{Z}\dot{\mathbf{x}} \tag{9.16b}$$

The corresponding mobility matrix is given by

$$\begin{bmatrix} \dot{X}_1 \\ \dot{X}_2 \end{bmatrix} = \begin{bmatrix} Y_{11} & Y_{12} \\ Y_{21} & Y_{22} \end{bmatrix} \begin{bmatrix} F_1 \\ F_2 \end{bmatrix} \tag{9.17a}$$

which can be written in vector form as

$$\dot{\mathbf{x}} = \mathbf{Y}\mathbf{f} \tag{9.17b}$$

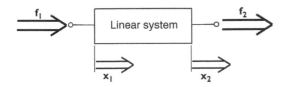

Figure 9.8 A multi-dimensional linear system.

These relationships can be extended to any number of inputs, and the system representing this situation is shown in Figure 9.8. The relevant matrix equations can be written in partitioned form as

$$\begin{bmatrix} f_1 \\ f_2 \end{bmatrix} = \begin{bmatrix} Z_{11} & Z_{12} \\ Z_{12}^T & Z_{22} \end{bmatrix} \begin{bmatrix} \dot{x}_1 \\ \dot{x}_2 \end{bmatrix}, \qquad \begin{bmatrix} \dot{x}_1 \\ \dot{x}_2 \end{bmatrix} = \begin{bmatrix} Y_{11} & Y_{12} \\ Y_{12}^T & Y_{22} \end{bmatrix} \begin{bmatrix} f_1 \\ f_2 \end{bmatrix} \qquad (9.18a,b)$$

where the superscript ^{T} denotes the transpose. Equations (9.18a,b) can also be represented by vector equations similar to equations (9.16b) and (9.17b).

It is important to note that impedance and mobility matrices are derived by applying different boundary conditions. To determine an element of the impedance matrix, the point at which the force is applied is unconstrained, but *all* other points are constrained, so that all other velocities are set to zero, thus $Z_{ij} = (F_i/\dot{X}_j)\big|_{\dot{X}_{k \neq j} = 0}$. Conversely when deriving an element of the mobility matrix, the point at which a force is applied is constrained, but *all* other points are set free, i.e., $Y_{ij} = (\dot{X}_i/F_j)\big|_{F_{k \neq j} = 0}$. O'Hara (1966) made the important point that if another force is added to an existing system, then the new mobility matrix is simply extended by adding another row and column to the existing mobility matrix. This, however, is not the case with the impedance matrix. The addition of a force means that a new row and column have to be added to the matrix, but in this case *all* the elements of the existing impedance matrix have to be changed. This is because of the different boundary conditions employed when determining the elements of the matrices. An interesting example of this problem is discussed in Appendix B of Gardonio and Elliott (2000).

If we combine equations (9.16b) and (9.17b) then we get the relationship between the impedance and the mobility matrices

$$Z = Y^{-1} \qquad (9.19)$$

It can be seen for a general linear system that $Z_{ij} \neq 1/Y_{ij}$; this is only true for single input systems similar to the lumped parameter elements discussed in the previous section. There are two important properties of impedance and mobility matrices that are worth noting here. The first is that the matrices are symmetric because of the principle of reciprocity (O'Hara 1966), and the second is that the real parts of the matrices are either positive definite or positive semi-definite. The second condition is a result of passivity, which means that structure vibrational power can only flow *into* the structure, which means that the real parts of the diagonal elements of the matrices must be greater than or equal to zero.

In the impedance and mobility formulations, the applied forces and resultant velocities are grouped together into force and velocity vectors as discussed above. An alternative formulation is the transfer matrix formulation in which the applied forces and resultant velocities at a particular point are grouped into a state-vector. The transfer matrix formulation for the system described by equations (9.16a,b) is given by

$$
\begin{bmatrix} \dot{x}_2 \\ f_2 \end{bmatrix} = \begin{bmatrix} T_{11} & T_{12} \\ T_{21} & T_{22} \end{bmatrix} \begin{bmatrix} \dot{x}_1 \\ -f_1 \end{bmatrix}
\tag{9.20a}
$$

which can be written as

$$
a_2 = T a_1
\tag{9.20b}
$$

and the elements in the transfer matrix are given by the impedance formulae $T_{11} = -Z_{12}^{-1} Z_{11}$, $T_{12} = -Z_{12}^{-1}$, $T_{21} = Z_{21} - Z_{22} Z_{12}^{-1} Z_{11}$ and $T_{22} = -Z_{22} Z_{12}^{-1}$ or by the mobility formulae $T_{11} = Y_{22} Y_{12}^{-1}$, $T_{12} = Y_{22} Y_{12}^{-1} Y_{11} - Y_{21}$, $T_{21} = Y_{12}^{-1}$ and $T_{22} = Y_{12}^{-1} Y_{11}$.

The reason for the negative sign in front of the force vector f_1 is so that structural elements in tandem can be connected together by simple multiplication (Rubin 1967). The impedance and mobility matrices can be written in terms of the transfer sub-matrices as

$$
\begin{bmatrix} f_1 \\ f_2 \end{bmatrix} = \begin{bmatrix} T_{12}^{-1} T_{11} & -T_{12}^{-1} \\ \left(-T_{12}^{-1}\right)^T & T_{22} T_{12}^{-1} \end{bmatrix} \begin{bmatrix} \dot{x}_1 \\ \dot{x}_2 \end{bmatrix} \quad \text{or}
$$

$$
\begin{bmatrix} \dot{x}_1 \\ \dot{x}_2 \end{bmatrix} = \begin{bmatrix} T_{21}^{-1} T_{22} & T_{21}^{-1} \\ \left(T_{21}^{-1}\right)^T & T_{11} T_{21}^{-1} \end{bmatrix} \begin{bmatrix} f_1 \\ f_2 \end{bmatrix}
\tag{9.21a,b}
$$

It can be seen that the impedance, mobility and transfer matrices all contain the same information; they are just different representations. The advantage of the mobility matrix is that the elements can generally be related to free space measurements where test structures are suspended on soft elastic chords. The advantage of the impedance matrix is demonstrated in Section 9.7 when a beam is connected to lumped parameter elements. As mentioned above, it is sometimes advantageous to reformulate the problem in terms of transfer matrices to aid computation using modern matrix manipulation packages. This is particularly useful for one-dimensional elements connected together end to end, and where the response is required at the ends of the structural chain.

9.6   Impedance and mobility matrices for rod, shaft and beam

This section describes the impedance and mobility matrices for one-dimensional distributed parameter elements such as rods, shafts and beams (Mead 1986). Distributed one-dimensional elements can be described in terms of waves or modes; both approaches are discussed here. One advantage of the wave approach is that it is possible to relate the dynamic behaviour of finite structures to infinite structures and it allows exact mobility and impedance expressions to be derived. Expressions for infinite structures are simpler than those for finite structures. They facilitate physical insight and, in fact, relate to frequency-average dynamic behaviour (Skudrzyk 1968). This approach is particularly appropriate for highly damped systems and for high frequencies, where the wavelength is much smaller than the length of the structure and the modal overlap factor is high (see Chapter 11).

The structures considered are infinite, semi-infinite and finite rods, shafts and beams. The point and transfer mobilities for the infinite and semi-infinite elements are given in Tables 9.4 and 9.5, respectively. S, E, ρ, G and I are the cross-sectional area, Young's modulus, density, shear modulus and second moment of area about the neutral axis of the elements, respectively. J is the polar second moment of area of the shaft, and $k_1 = \sqrt{(\rho/E)}\omega$, $k_s = \sqrt{(\rho/G)}\omega$, $k_b = \sqrt[4]{(\rho S/EI)}\sqrt{\omega}$ are the *quasi*-longitudinal, shear and bending (flexural) wave numbers respectively. It should be noted that the longitudinal and shear wavenumbers are proportional to frequency and the bending wavenumber is proportional to the square-root of frequency. This means that the longitudinal and shear wave speeds ($c = \omega/k$) are independent of frequency (non-dispersive) but the bending wave speed is proportional to the square root of frequency (dispersive). It can be seen from Tables 9.4 and 9.5 that the mobilities for rods and shafts are independent of frequency; but for the beam in flexure, the mobilities have differing dependencies on frequency, and the mobilities have different units. Expressions for the mobilities for the rod and the beam are derived below as examples, where the $\exp(j\omega t)$ time dependency is assumed but omitted for simplicity. The other mobilities can be derived using similar principles.

9.6.1   Infinite and semi-infinite rod

To determine the waves that can exist in an element, its equation of forced vibration is considered. The equation of motion of axial vibration of a rod due to an axial distributed force per unit length $f_x(x, t)$ is given by Bishop and Johnson (1960) as

$$ES\frac{\partial^2 u(x)}{\partial x^2} - \rho S\frac{\partial^2 u(x, t)}{\partial t^2} = -f_x(x, t) \tag{9.22}$$

Table 9.4 Mobility functions of infinite one-dimensional elements

Element	Mobilities		Units
	Point	Transfer ($x \geq 0$)	
Infinite rod	$\dfrac{\dot{U}(0)}{F} = \dfrac{1}{2S\sqrt{E\rho}}$	$\dfrac{\dot{U}(x)}{F} = \dfrac{1}{2S\sqrt{E\rho}}e^{-jk_lx}$	$mN^{-1}s^{-1}$
Infinite shaft	$\dfrac{\dot{\Phi}(0)}{T} = \dfrac{1}{2J\sqrt{G\rho}}$	$\dfrac{\dot{\Phi}(x)}{T} = \dfrac{1}{2J\sqrt{G\rho}}e^{-jk_sx}$	$radN^{-1}m^{-1}s^{-1}$
Infinite beam	$\dfrac{\dot{W}(0)}{F} = \dfrac{\omega(1-j)}{4EIk_b^3}$	$\dfrac{\dot{W}(x)}{F} = \dfrac{-\omega}{4EIk_b^3}\left(je^{-k_bx} - e^{-jk_bx}\right)$	$mN^{-1}s^{-1}$
	$\dfrac{\dot{W}(0)}{M} = 0$	$\dfrac{\dot{W}(x)}{M} = \dfrac{-j\omega}{4EIk_b^2}\left(e^{-k_bx} - e^{-jk_bx}\right)$	$N^{-1}s^{-1}$
	$\dfrac{\dot{\Theta}(0)}{F} = 0$	$\dfrac{\dot{\Theta}(x)}{F} = \dfrac{j\omega}{4EIk_b^2}\left(e^{-k_bx} - e^{-jk_bx}\right)$	$radN^{-1}s^{-1}$
	$\dfrac{\dot{\Theta}(0)}{M} = \dfrac{\omega(1+j)}{4EIk_b}$	$\dfrac{\dot{\Theta}(x)}{M} = \dfrac{j\omega}{4EIk_b}\left(e^{-k_bx} - je^{-jk_bx}\right)$	$radN^{-1}m^{-1}s^{-1}$

which has a solution to harmonic excitation

$$U(x) = A_R e^{-jk_lx} + A_L e^{jk_lx} \tag{9.23}$$

where A_R is the complex axial displacement of a right-going propagating wave, A_L is the complex axial displacement of a left-going propagating

Table 9.5 Mobility functions of semi-infinite one-dimensional elements

Element	Mobilities		Units
	Point	Transfer $(x \geq 0)$	
Semi-infinite rod	$\dfrac{\dot{U}(0)}{F} = \dfrac{1}{S\sqrt{E\rho}}$	$\dfrac{\dot{U}(x)}{F} = \dfrac{1}{S\sqrt{E\rho}}e^{-jk_l x}$	$m\,N^{-1}\,s^{-1}$
Semi-infinite shaft	$\dfrac{\dot{\Phi}(0)}{T} = \dfrac{1}{J\sqrt{G\rho}}$	$\dfrac{\dot{\Phi}(x)}{T} = \dfrac{1}{J\sqrt{G\rho}}e^{-jk_s x}$	$rad\,N^{-1}\,m^{-1}\,s^{-1}$
Semi-infinite beam	$\dfrac{\dot{W}(0)}{F} = \dfrac{\omega(1-j)}{EIk_b^3}$	$\dfrac{\dot{W}(x)}{F} = \dfrac{\omega(1-j)}{2EIk_b^3}\left(e^{-k_b x}+e^{-jk_b x}\right)$	$m\,N^{-1}\,s^{-1}$
	$\dfrac{\dot{W}(0)}{M} = \dfrac{-\omega}{EIk_b^2}$	$\dfrac{\dot{W}(x)}{M} = \dfrac{-\omega(1-j)}{2EIk_b^2}\left(je^{-k_b x}+e^{-jk_b x}\right)$	$N^{-1}\,s^{-1}$
	$\dfrac{\dot{\Theta}(0)}{F} = \dfrac{-\omega}{EIk_b^2}$	$\dfrac{\dot{\Theta}(x)}{F} = \dfrac{-\omega(1-j)}{2EIk_b^2}\left(e^{-k_b x}+je^{-jk_b x}\right)$	$rad\,N^{-1}\,s^{-1}$
	$\dfrac{\dot{\Theta}(0)}{M} = \dfrac{\omega(1+j)}{EIk_b}$	$\dfrac{\dot{\Theta}(x)}{M} = \dfrac{\omega(1+j)}{2EIk_b}\left(e^{-k_b x}+e^{-jk_b x}\right)$	$rad\,N^{-1}\,m^{-1}\,s^{-1}$

wave, and x is the distance along the rod. The amplitudes of these waves are dependent upon the geometrical configuration and the amplitude and phase of the excitation force. They can be determined by applying the appropriate boundary conditions.

For an infinite rod such as that shown in Table 9.4, excited at $x = 0$, one of the boundary conditions is $F = -2ESU'(0)$, where the prime denotes differentiation with respect to x. The right-going wave amplitude is determined to be $A_R = F/j2k_1ES$. Applying the condition of continuity of displacement gives $A_L = -F/j2k_1ES$. To determine the mobility of the rod for $x \geq 0$, A_R is substituted into equation (9.23), A_L is set to zero, and equation (9.23) is differentiated with respect to time ($\times j\omega$) and rearranged to give the transfer mobility

$$\frac{\dot{U}(x)}{F} = \frac{\omega}{2ESk_1} = \frac{1}{2S\sqrt{E\rho}}e^{-jk_1x} \tag{9.24}$$

To determine the driving point mobility, x is simply set to zero. Because this is entirely real, the rod has the characteristics of a damper. For a semi-infinite rod with the free end being at $x = 0$, the boundary condition is simply $F = -ESU'(0)$, which means that the point and transfer mobilities are twice that of an infinite rod.

A shaft has very similar dynamic behaviour to that of a rod, but the motion is torsional. The similarity between the mobility expressions for a rod and a shaft can be seen in Tables 9.4 and 9.5.

9.6.2 Infinite and semi-infinite Euler–Bernoulli beam

The equation of motion per unit length of flexural vibration of a Euler–Bernoulli beam due to a transverse distributed force per unit length $f_z(x, t)$ is given by Bishop and Johnson (1960) as

$$EI\frac{\partial^4 w(x)}{\partial x^4} + \rho S\frac{\partial^2 w(x, t)}{\partial t^2} = f_z(x, t) \tag{9.25}$$

which has a solution to harmonic excitation

$$W(x) = C_R e^{-k_b x} + D_R e^{-jk_b x} + C_L e^{k_b x} + D_L e^{jk_b x} \tag{9.26}$$

where C_R and D_R are the wave amplitudes of the transverse displacement near-field in the region $x > 0$ and propagating right-going waves and C_L and D_L are the wave amplitudes of the near-field in the region $x > 0$ and propagating left-going waves, respectively.

For an infinite beam with a lateral force applied at $x = 0$, the boundary conditions are $F = 2EIW'''(0)$ (shear force = applied force), $EIW''_-(0) = EIW''_+(0)$ (bending moment to the left of the applied force = bending moment to the right of the applied force), and $W_-(0) = W_+(0)$ (the lateral displacement to the left of the applied force = lateral displacement to the right of the applied force). The resulting wave amplitudes are $C_R = C_L = -F/4EIk_b^3$ and $D_R = D_L = -jF/4EIk_b^3$. To determine the mobility of the beam for

$x \geq 0$, C_R and D_R are substituted into equation (9.26) with $C_L = D_L = 0$, it is differentiated with respect to time ($\times j\omega$) and rearranged to give the transfer mobility

$$\frac{\dot{W}(x)}{F} = \frac{-\omega}{4EIk_b^3} \left(je^{-k_b x} - e^{-jk_b x} \right) \tag{9.27}$$

To obtain the driving point mobility, x is simply set to zero to give $\dot{W}(0)/F = \omega(1-j)/4EIk_b^3$. For a semi-infinite beam excited at the free end as shown in Table 9.5, the left-going waves are zero. Applying the boundary conditions of $EIW''(0) = 0$ (zero bending moment) and $EIW'''(0) = F$ (shear force = applied force), gives the wave amplitudes as $C_R = D_R = -(1+j)F/2EIk_b^3$. Substituting this into equation (9.26), differentiating with respect to time and rearranging gives the transfer mobility as

$$\frac{\dot{W}(x)}{F} = \frac{\omega(1-j)}{2EIk_b^3} \left(e^{-k_b x} + e^{-jk_b x} \right) \tag{9.28}$$

Noting that $k_b \propto \sqrt{\omega}$ in equations (9.27) and (9.28), it can be seen that the point and transfer mobilities of an infinite and semi-infinite beam are proportional to $1/\sqrt{\omega}$. It can also be seen from Tables 9.4 and 9.5 that the point mobilities consist of a positive real part and a negative imaginary part. By substituting for k_b and considering the behaviour of the lumped parameter elements discussed in Section 9.4, it can be seen that infinite and semi- infinite beams behave as if they were a frequency-dependent damper in parallel with a frequency-dependent mass. For example, for a semi-infinite beam, the equivalent damping coefficient is $c_{eq} = (EI)^{3/4}(\rho S)^{1/4}\sqrt{\omega}$ and equivalent mass is $m_{eq} = (EI)^{3/4}(\rho S)^{1/4}/\sqrt{\omega}$. The inertia force of this mass can counteract the stiffness of pipe supports to produce strong resonant response to support vibration.

9.6.3   Finite rod

Figure 9.9 shows the finite, one-dimensional elements considered in this section: they are all of length l. The mobilites are calculated using the wave approach for free–free rods and are given for three common boundary conditions for the modal formulation. When connecting elements together in line, it is the free–free mobilities that are required.

The impedance and mobility matrices for the rod are of order 2×2 and are given by equations (9.16a) and (9.17a) respectively, but with X replaced with U. The elements in these matrices can be determined in a similar manner to that for the semi-infinite structures, by applying the appropriate boundary conditions and solving for the wave amplitudes in equation (9.23). The boundary conditions are determined by applying constraints

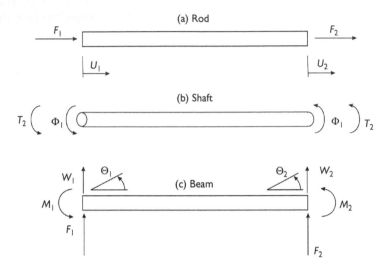

Figure 9.9 Finite one-dimensional distributed parameter elements.

as discussed in the previous section, for example $Z_{11} = (F_1/\dot{U}_1)\big|_{\dot{U}_2=0}$, $Z_{12} = (F_1/\dot{U}_2)\big|_{\dot{U}_1=0}$, $Z_{21} = (F_2/\dot{U}_1)\big|_{\dot{U}_2=0}$, $Z_{22} = (F_2/\dot{U}_2)\big|_{\dot{U}_1=0}$; $Z_{11} = Z_{22}$ because of symmetry and $Z_{12} = Z_{21}$ because of reciprocity. The elements of the impedance matrix are thus given by

$$Z_{11} = Z_{22} = -jS\sqrt{E\rho}\cot(k_l l), \qquad Z_{12} = Z_{21} = j\frac{S\sqrt{E\rho}}{\sin(k_l l)} \qquad \text{(9.29a,b,c,d)}$$

where k_l is the longitudinal wavenumber. At low frequencies, when the wavelength is much greater than the length of the element, the rod reduces to a simple spring, with elements

$$Z_{11} = Z_{22} = \frac{K_L}{j\omega}, \qquad Z_{12} = Z_{21} = \frac{-K_L}{j\omega} \qquad \text{(9.30a,b,c,d)}$$

where $K_L = EA/l$ is the longitudinal stiffness of the rod. The elements of the mobility matrix are given by

$$Y_{11} = Y_{22} = -j\frac{\cot(k_l l)}{S\sqrt{E\rho}}, \qquad Y_{12} = Y_{21} = \frac{-j}{S\sqrt{E\rho}\sin(k_l l)} \qquad \text{(9.31a,b,c,d)}$$

At low frequencies the mobilities of the rod become mass-like because of the unconstrained boundary conditions and the elements are given by

$$Y_{11} = Y_{22} = Y_{12} = Y_{21} = \frac{1}{j\omega m} \qquad (9.32a,b,c,d)$$

where $m = \rho S l$ is the mass of the rod.

The mobilities of a rod can also be written in terms of its natural frequencies and natural modes. The expression for the transfer mobility between response at point x_j and force at x_i is given by the summation of an infinite number of modal responses as

$$Y_{ij} = j\omega \sum_{n=1}^{\infty} \frac{\psi_n(x_i)\psi_n(x_j)}{\rho S l\left(\omega_n^2(1+j\eta) - \omega^2\right)} \qquad (9.33)$$

where ω_n is the nth undamped natural frequency, $\psi_n(x_i)$ is the natural mode shape evaluated at x_i and η is the loss factor which, for simplicity, can generally be taken to be a constant that is dependent upon the material of the rod, except for the rigid body mode ($n = 0$) for the free–free condition, when it is zero. The natural frequencies and natural modes of a rod are dependent upon the geometry, material properties and boundary conditions; they are given in Table 9.6 for three common conditions.

9.6.4   Finite shaft

To describe the torsional dynamic behaviour of the shaft, the force vector has to be replaced with a torsion vector, and the velocity vector by a rotational velocity vector in equations 9.16a and 9.17a to give

$$\begin{bmatrix} T_1 \\ T_2 \end{bmatrix} = \begin{bmatrix} Z_{11} & Z_{12} \\ Z_{21} & Z_{22} \end{bmatrix} \begin{bmatrix} \dot{\Phi}_1 \\ \dot{\Phi}_2 \end{bmatrix}, \qquad \begin{bmatrix} \dot{\Phi}_1 \\ \dot{\Phi}_2 \end{bmatrix} = \begin{bmatrix} Y_{11} & Y_{12} \\ Y_{21} & Y_{22} \end{bmatrix} \begin{bmatrix} T_1 \\ T_2 \end{bmatrix} \qquad (9.34a,b)$$

Table 9.6 Natural frequencies and natural modes for a rod (with $\int_0^l \psi^2(x)dx = 1$)

Boundary conditions	ω_n	$\psi_n(x)$
Clamped–clamped $U(0) = 0$ $U(l) = 0$	$\dfrac{n\pi}{l}\sqrt{\dfrac{E}{\rho}}$	$\sqrt{2}\sin\dfrac{n\pi x}{l}$
Free–free $U'(0) = 0$ $U'(l) = 0$	$\dfrac{n\pi}{l}\sqrt{\dfrac{E}{\rho}}$	$\sqrt{2}\cos\dfrac{n\pi x}{l}$
Clamped–free $U(0) = 0$ $U'(l) = 0$	$\dfrac{(2n-1)\pi}{2l}\sqrt{\dfrac{E}{\rho}}$	$\sqrt{2}\sin\dfrac{(2n-1)\pi x}{2l}$

The impedance matrix elements are similar to those for a rod. The area S is replaced by the polar second moment of area J, Young's modulus E is replaced by the shear modulus G, and the longitudinal wave number k_l is replaced by the shear wave number $k_s = \sqrt{(\rho/G)}\omega$. Thus we get

$$Z_{11} = Z_{22} = -jJ\sqrt{G\rho}\cot(k_s l), \qquad Z_{12} = Z_{21} = j\frac{J\sqrt{G\rho}}{\sin(k_s l)}$$

$$(9.35a,b,c,d)$$

At low frequencies the shaft impedance matrix reduces to a simple stiffness-like impedance, with elements

$$Z_{11} = Z_{22} = \frac{K_T}{j\omega}, \qquad Z_{12} = Z_{21} = \frac{-K_T}{j\omega} \qquad (9.36a,b,c,d)$$

where $K_T = GJ/l$ is the torsional stiffness of the rod. The elements of the mobility matrix are given by

$$Y_{11} = Y_{22} = -j\frac{\cot(k_s l)}{J\sqrt{G\rho}}, \qquad Y_{12} = Y_{21} = \frac{-j}{J\sqrt{G\rho}\sin(k_s l)} \qquad (9.37a,b,c,d)$$

At low frequencies the mobilities of the rod become inertia-like because of the unconstrained boundary conditions, and the elements are given by:

$$Y_{11} = Y_{22} = Y_{12} = Y_{21} = \frac{1}{j\omega J\rho l} \qquad (9.38a,b,c,d)$$

As described in Section 9.5, if the transfer matrices are required then they can be formed using equation (9.21).

As with a rod, the mobilities of a shaft can also be written in terms of its natural frequencies and natural modes. The expression for the mobility between response point x_j and input point x_i is given by the summation of an infinite number of modal responses as

$$Y_{ij} = j\omega \sum_{n=1}^{\infty} \frac{\psi_n(x_i)\psi_n(x_j)}{\rho Jl(\omega_n^2(1+j\eta) - \omega^2)} \qquad (9.39)$$

This expression is similar to that for a rod given in equation (9.33), but with the area S replaced with the polar moment of area J. The natural frequencies and natural modes for three common boundary conditions are given in Table 9.7. As with the rod the loss factor η is zero for the rigid body mode ($n = 0$) for the free–free condition.

Table 9.7 Natural frequencies and natural modes for a shaft (with $\int_0^l \psi^2(x)dx = 1$)

Boundary conditions	ω_n	$\psi_n(x)$
Clamped–clamped $\Phi(0) = 0$ $\Phi(l) = 0$	$\dfrac{n\pi}{l}\sqrt{\dfrac{G}{\rho}}$	$\sqrt{2}\sin\dfrac{n\pi x}{l}$
Free–free $\Phi'(0) = 0$ $\Phi'(l) = 0$	$\dfrac{n\pi}{l}\sqrt{\dfrac{G}{\rho}}$	$\sqrt{2}\cos\dfrac{n\pi x}{l}$
Clamped–free $\Phi(0) = 0$ $\Phi'(l) = 0$	$\dfrac{(2n-1)\pi}{2l}\sqrt{\dfrac{G}{\rho}}$	$\sqrt{2}\sin\dfrac{(2n-1)\pi x}{2l}$

9.6.5 Finite Euler–Bernoulli beam

The solution to the equation of motion can be written down in terms of near-field and propagating waves; but for a finite beam it is more convenient to write it down in terms of trigonometric and hyperbolic functions as

$$W(x) = A\sin(k_b x) + B\cos(k_b x) + C\sinh(k_b x) + D\cosh(k_b x) \tag{9.40}$$

where A, B, C, and D are constants. By applying geometric and natural boundary conditions at each end of the beam the impedance and the mobility matrices can be determined. For example, to determine the $(1,1)$ element in the impedance matrix given in equation (9.41), a unit velocity $\dot{W}_1$ is imposed and the force F_1 determined whilst $\dot{W}_2$, $\dot{\Theta}_1$ and $\dot{\Theta}_2$ are all set to zero. Both the impedance and mobility matrices are of dimension 4×4. The impedance matrix is given by

$$\begin{bmatrix} F_1 \\ M_1 \\ F_2 \\ M_2 \end{bmatrix} = \frac{EIk_b^3}{j\omega N} \begin{bmatrix} -K_{11} & -P & K_{12} & V \\ -P & Q_{11} & -V & Q_{12} \\ K_{12} & -V & -K_{11} & P \\ V & Q_{12} & P & Q_{11} \end{bmatrix} \begin{bmatrix} \dot{W}_1 \\ \dot{\Theta}_1 \\ \dot{W}_2 \\ \dot{\Theta}_2 \end{bmatrix} \tag{9.41}$$

where

$$K_{11} = \cos(k_b l)\sinh(k_b l) + \sin(k_b l)\cosh(k_b l)$$

$$K_{12} = \sin(k_b l) + \sinh(k_b l)$$

$$P = \frac{\sin(k_b l)\sinh(k_b l)}{k_b}$$

$$V = \frac{\cos(k_b l) - \cosh(k_b l)}{k_b}$$

$$Q_{11} = \frac{\cos(k_b l) \sinh(k_b l) - \sin(k_b l) \cosh(k_b l)}{k_b^2}$$

$$Q_{12} = \frac{\sin(k_b l) - \sinh(k_b l)}{k_b^2}$$

$$N = \cos(k_b l) \cosh(k_b l) - 1$$

At low frequencies, when the length of the beam is much less than a bending wavelength, then the impedance matrix reduces to:

$$\begin{bmatrix} F_1 \\ M_1 \\ F_2 \\ M_2 \end{bmatrix} = \frac{EI}{j\omega l^3} \begin{bmatrix} 12 & 6l & -12 & 6l \\ 6l & 4l^2 & -6l & 2l^2 \\ -12 & -6l & 12 & -6l \\ 6l & 2l^2 & -6l & 4l^2 \end{bmatrix} \begin{bmatrix} \dot{W}_1 \\ \dot{\Theta}_1 \\ \dot{W}_2 \\ \dot{\Theta}_2 \end{bmatrix} \tag{9.42}$$

It can be seen that there are no mass terms in this matrix. Thus at low frequencies all elements in the impedance matrix are stiffness-like.

The mobility matrix can be obtained by inverting the impedance matrix or by applying appropriate boundary conditions. To determine the $(1,1)$ element, for example, a unit force F_1 is applied and velocity $\dot{W}_1$ determined whilst F_2, M_1 and M_2 are all set to zero. The mobility matrix is given by

$$\begin{bmatrix} \dot{W}_1 \\ \dot{\Theta}_1 \\ \dot{W}_2 \\ \dot{\Theta}_2 \end{bmatrix} = \frac{j\omega}{EI k_b N} \begin{bmatrix} -Q_{11} & -P & Q_{12} & V \\ -P & K_{11} & -V & K_{12} \\ Q_{12} & -V & -Q_{11} & P \\ V & K_{12} & P & K_{11} \end{bmatrix} \begin{bmatrix} F_1 \\ M_1 \\ F_2 \\ M_2 \end{bmatrix} \tag{9.43}$$

At low frequencies, when the length of the beam is much smaller than a bending wavelength, this matrix reduces to:

$$\begin{bmatrix} \dot{W}_1 \\ \dot{\Theta}_1 \\ \dot{W}_2 \\ \dot{\Theta}_2 \end{bmatrix} = \frac{1}{j\omega m} \begin{bmatrix} 4 & \dfrac{-6}{l} & -2 & \dfrac{-6}{l} \\ \dfrac{-6}{l} & \dfrac{12}{l^2} & \dfrac{6}{l} & \dfrac{12}{l^2} \\ -2 & \dfrac{6}{l} & 4 & \dfrac{6}{l} \\ \dfrac{-6}{l} & \dfrac{12}{l^2} & \dfrac{6}{l} & \dfrac{12}{l^2} \end{bmatrix} \begin{bmatrix} F_1 \\ M_1 \\ F_2 \\ M_2 \end{bmatrix} \tag{9.44}$$

It can be seen that there are no stiffness-like terms in this matrix. Thus at low frequencies all elements in the mobility matrix are mass-like. It should be noted that it is not possible to invert the low-frequency impedance matrix,

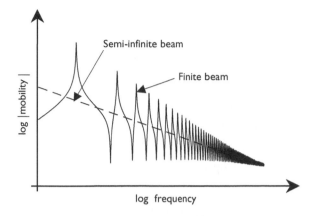

Figure 9.10 Driving point mobility of a beam.

equation (9.42), to give the low-frequency mobility matrix, equation (9.44), as it is singular. This means that it is not possible to have a mobility matrix for a stiffness-like element and it is not possible to have an impedance matrix for a mass-like element.

The moduli of the point mobilities of a finite and a semi-infinite beam are shown in Figure 9.10. It can be seen that the input mobility at one end of the finite beam tends to that of a semi-infinite beam at high frequencies. In fact the frequency-averaged mobility of the finite beam is the same as that of the semi-infinite beam.

The transfer matrix for a beam can be determined by partitioning the impedance matrix as follows

$$
\begin{bmatrix} F_1 \\ M_1 \\ \cdots \\ F_2 \\ M_2 \end{bmatrix} = \frac{EIk_b^3}{j\omega N} \left[\begin{array}{cc:cc} -K_{11} & -P & K_{12} & V \\ -P & Q_{11} & -V & Q_{12} \\ \hdashline K_{12} & -V & -K_{11} & P \\ V & Q_{12} & P & Q_{11} \end{array} \right] \begin{bmatrix} \dot{W}_1 \\ \dot{\Theta}_1 \\ \cdots \\ \dot{W}_2 \\ \dot{\Theta}_2 \end{bmatrix}
\tag{9.45}
$$

and then applying equation (9.21).

As with a rod and a shaft a modal approach may also be used to determine the elements of the mobility matrix, which may be written as

$$
\begin{bmatrix} \dot{W}_1 \\ \dot{\Theta}_1 \\ \dot{W}_2 \\ \dot{\Theta}_2 \end{bmatrix} = j\omega \begin{bmatrix} \alpha_{11} & \alpha_{11'} & \alpha_{12} & \alpha_{12'} \\ \alpha_{1'1} & \alpha_{1'1'} & \alpha_{1'2} & \alpha_{1'2'} \\ \alpha_{21} & \alpha_{21'} & \alpha_{22} & \alpha_{22'} \\ \alpha_{2'1} & \alpha_{2'1'} & \alpha_{2'2} & \alpha_{2'2'} \end{bmatrix} \begin{bmatrix} F_1 \\ M_1 \\ F_2 \\ M_2 \end{bmatrix}
\tag{9.46}
$$

where

$$\alpha_{11} = \alpha_{22} = \sum_{n=-1}^{\infty} \frac{\psi_n^2(0)}{\rho Sl(\omega_n^2(1+j\eta) - \omega^2)}$$

$$\alpha_{1'1} = \alpha_{11'} = -\alpha_{2'2} = -\alpha_{22'} = \sum_{n=-1}^{\infty} \frac{\psi_n'(0)\psi_n(l)}{\rho Sl(\omega_n^2(1+j\eta) - \omega^2)}$$

$$\alpha_{12} = \alpha_{21} = \sum_{n=-1}^{\infty} \frac{\psi_n(0)\psi_n(l)}{\rho Sl(\omega_n^2(1+j\eta) - \omega^2)}$$

$$\alpha_{1'1'} = \alpha_{2'2'} = \sum_{n=-1}^{\infty} \frac{\psi_n'^2(0)}{\rho Sl(\omega_n^2(1+j\eta) - \omega^2)}$$

$$\alpha_{12'} = \alpha_{2'1} = -\alpha_{1'2} = -\alpha_{21'} = \sum_{n=-1}^{\infty} \frac{\psi_n(0)\psi_n'(l)}{\rho Sl(\omega_n^2(1+j\eta) - \omega^2)}$$

$$\alpha_{1'2'} = \alpha_{2'1'} = \sum_{n=-1}^{\infty} \frac{\psi_n'(0)\psi_n'(l)}{\rho Sl(\omega_n^2(1+j\eta) - \omega^2)}$$

where $\psi_n(x)$ is the nth natural mode of a free–free beam given by Bishop and Johnson (1960)

$$n = -1 \quad \psi_{-1}(x) = 1 \qquad \text{(even rigid body mode)}$$

$$n = 0 \quad \psi_0(x) = \sqrt{3}\left(1 - \frac{2x}{l}\right) \qquad \text{(rocking rigid body mode)}$$

$$n \geq 1 \quad \psi_n(x) = (\cosh k_{nb}x + \cos k_{nb}x)$$

$$- \frac{\cosh k_{nb}l - \cos k_{nb}l}{\sinh k_{nb}l - \sin k_{nb}l} (\sinh k_{nb}x + \sin k_{nb}x)$$

and k_{nb} is the bending wavenumber at the nth natural frequency. The prime denotes differentiation with respect to x: $\psi_n'(x) = \partial\psi_n(x)/\partial x$. As with the rod and the shaft, the loss factor η is zero for the rigid body mode, $n = 0$ and $n = 1$, and is taken to be a constant for all the other modes.

The mobilities for a variety of end conditions can be determined by using the natural frequencies and natural modes given in Table 9.8. For force excitation at x_i and a transverse velocity response at x_j, the mobility is given by

$$Y_{ij} = j\omega \sum_{n=1}^{\infty} \frac{\psi_n(x_i)\psi_n(x_j)}{\rho Sl(\omega_n^2(1+j\eta) - \omega^2)} \tag{9.47a}$$

For force excitation at x_i and rotational velocity at x_j, or moment excitation at x_j and transverse velocity response at x_i,

$$Y_{ij'} = j\omega \sum_{n=1}^{\infty} \frac{\psi_n(x_i)\psi_n'(x_j)}{\rho Sl(\omega_n^2(1+j\eta) - \omega^2)} \tag{9.47b}$$

Table 9.8 Natural frequencies and natural modes for beams (with $\int_0^l \psi^2(x)dx = 1$)

Boundary conditions	$\omega_n = \sqrt{\frac{EI}{\rho S}} k_{nb}^2$		$\psi_n(x)$
	n	$k_{nb}l$	
Pinned–pinned $\quad W(0)=0 \quad W''(0)=0$ $\quad W(l)=0 \quad W''(l)=0$	$1, 2, \ldots$	$n\pi$	$\sqrt{2}\sin k_{nb}x$
Clamped–clamped $\quad W(0)=0 \quad W'(0)=0$ $\quad W(l)=0 \quad W'(l)=0$	1 2 3 4 5 $6, 7, \ldots$	4.73004 7.85320 10.9956 14.1372 17.2788 $(2n+1)\pi/2$	$(\cosh k_{nb}x - \cos k_{nb}x) - \sigma_n (\sinh k_{nb}x - \sin k_{nb}x)$ $\sigma_n = \dfrac{\cosh k_{nb}l - \cos k_{nb}l}{\sinh k_{nb}l - \sin k_{nb}l}$
Free–free $\quad W''(0)=0 \quad W'''(0)=0$ $\quad W''(l)=0 \quad W'''(l)=0$	even rigid body mode rocking rigid body mode 1 2 3 4 5 $6, 7, \ldots$	4.73004 7.85320 10.9956 14.1372 17.2788 $(2n+1)\pi/2$	1 $\sqrt{3}(1 - 2x/l)$ $(\cosh k_{nb}x + \cos k_{nb}x) - \sigma_n (\sinh k_{nb}x + \sin k_{nb}x)$ $\sigma_n = \dfrac{\cosh k_{nb}l - \cos k_{nb}l}{\sinh k_{nb}l - \sin k_{nb}l}$

Table 9.8 (Continued)

Boundary conditions	$\omega_n = \sqrt{\dfrac{EI}{\rho S}}\, k_{nb}^2$		$\psi_n(x)$
	n	$k_{nb}l$	
Clamped–free $W(0)=0 \quad W'(0)=0$ $W''(l)=0 \quad W'''(l)=0$	1 2 3 4 5 6, 7, …	1.87510 4.69409 7.85476 10.9955 14.1372 $(2n-1)\pi/2$	$(\cosh k_{nb}x - \cos k_{nb}x) - \sigma_n\,(\sinh k_{nb}x - \sin k_{nb}x)$ $\sigma_n = \dfrac{\sinh k_{nb}l - \sin k_{nb}l}{\cosh k_{nb}l + \cos k_{nb}l}$
Clamped–pinned $W(0)=0 \quad W'(0)=0$ $W(l)=0 \quad W''(l)=0$	1 2 3 4 5 6, 7, …	3.92660 7.06858 10.2102 13.3518 16.4934 $(4n+1)\pi/4$	$(\cosh k_{nb}x - \cos k_{nb}x) - \sigma_n\,(\sinh k_{nb}x - \sin k_{nb}x)$ $\sigma_n = \dfrac{\cosh k_{nb}l}{\sinh k_{nb}l}$
Free–pinned $W''(0)=0 \quad W'''(0)=0$ $W(l)=0 \quad W''(l)=0$	rocking rigid body mode 1 2 3 4 5 6, 7, …	 3.92660 7.06858 10.2102 13.3518 16.4934 $(4n+1)\pi/4$	$\sqrt{3}\,(l-x)/l$ $(\cosh k_{nb}x + \cos k_{nb}x) - \sigma_n\,(\sinh k_{nb}x + \sin k_{nb}x)$ $\sigma_n = \dfrac{\cosh k_{nb}l}{\sinh k_{nb}l}$

For moment excitation and rotational velocity response the mobility is given by

$$Y_{i'j'} = j\omega \sum_{n=1}^{\infty} \frac{\psi_n'(x_i)\psi_n'(x_j)}{\rho Sl(\omega_n^2(1+j\eta) - \omega^2)} \tag{9.47c}$$

where, again, the prime denotes differentiation with respect to x.

9.7 Connecting distributed and lumped parameter one-dimensional systems together

The procedure for connecting lumped parameter elements to a distributed parameter system, such as a beam or a rod is illustrated for two impedances attached to a beam as shown in Figure 9.11. The equation describing the system can be written down in a form where the distributed parameter system is described separately from the matrix containing the attached impedances. The equation describing the beam alone is given by

$$\begin{bmatrix} \dot{W}_1 \\ \dot{W}_2 \\ \dot{W}_3 \end{bmatrix} = \begin{bmatrix} Y_{11} & Y_{12} & Y_{13} \\ Y_{21} & Y_{22} & Y_{23} \\ Y_{31} & Y_{32} & Y_{33} \end{bmatrix} \begin{bmatrix} F_1 \\ 0 \\ 0 \end{bmatrix} + \begin{bmatrix} Y_{11} & Y_{12} & Y_{13} \\ Y_{21} & Y_{22} & Y_{23} \\ Y_{31} & Y_{32} & Y_{33} \end{bmatrix} \begin{bmatrix} 0 \\ F_2 \\ F_3 \end{bmatrix} \tag{9.48a}$$

which can be written in vector matrix form as

$$\mathbf{w} = \mathbf{Y}\mathbf{f}_1 + \mathbf{Y}\mathbf{f} \tag{9.48b}$$

The attached impedances can be described by the equation

$$\begin{bmatrix} 0 \\ F_2 \\ F_3 \end{bmatrix} = \begin{bmatrix} 0 & 0 & 0 \\ 0 & Z_2 & 0 \\ 0 & 0 & Z_3 \end{bmatrix} \begin{bmatrix} \dot{W}_1 \\ \dot{W}_2 \\ \dot{W}_3 \end{bmatrix} \tag{9.49a}$$

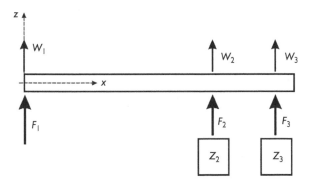

Figure 9.11 Attachment of two impedances to a beam.

which can also be written in vector matrix form as

$$\mathbf{f} = \mathbf{Zw} \tag{9.49b}$$

Equations (9.48b) and (9.49b) can be combined to give

$$\mathbf{w} = [\mathbf{I} - \mathbf{YZ}]^{-1} \mathbf{Yf}_1 \tag{9.50}$$

Equation (9.50) describes the dynamic behaviour of the coupled system in terms of the dynamics of the component parts. It is, in fact, a general equation for systems described in terms of their component mobilities and impedances. This approach has been used by the authors to study an active isolation system (Gardonio *et al.* 1997), to couple vibration absorbers to a beam (Brennan and Dayou 2000), to describe the vibration transmission between two plates connected by a set of mounts (Gardonio and Elliott 2000), to study a structural-acoustic system (Gardonio and Elliott 1999; Kim and Brennan 1999), and to describe a rotor-dynamics system (Bonello and Brennan 2001).

9.8 Mobility of thin flat plates

9.8.1 Mobility matrices of thin plates

This section gives expressions for the out-of-plane and in-plane mobility matrices for infinite and finite thin plates. As with the beam, the infinite plate mobilities are given in terms of bending and longitudinal shear waves generated by either a point force or a point moment excitation. A modal approach only is used for finite plates. The mobilities have been derived for the particular case of a rectangular plate, although the formulation can be used for any type of boundary conditions for which it is possible to derive analytical or numerical expressions of the natural frequencies and natural modes.

9.8.2 Out-of-plane vibration of an infinite plate

Out-of-plane bending waves in a thin flat plate are uncoupled from the in-plane longitudinal and shear waves so that out-of- plane vibration can be treated separately (Reddy 1984). Assuming 'classical plate theory' for thin plates, the equation of motion for the transverse displacement of an infinite thin plate subject to a distributed transverse force per unit area $p_z(x, y, t)$ is (Reddy 1984; Cremer *et al.* 1988)

$$B\left(\frac{\partial^4 w(x, y)}{\partial x^4} + 2\frac{\partial^4 w(x, y)}{\partial x^2 \partial y^2} + \frac{\partial^4 w(x, y)}{\partial y^4}\right) + m\frac{\partial^2 w(x, y, t)}{\partial t^2} = p_z(x, y, t)$$

$$\tag{9.51}$$

where $B = EI/(1 - \nu^2)$ is the bending stiffness, E is Young's modulus, ν is Poisson's ratio, $I = h^3/12$ is the cross-sectional second moment of area per unit width, h is the thickness of the plate, $m = \rho h$ is the mass per unit area and ρ is the density of the material. This can be solved in a similar way to that for a beam to give the characteristics of the wave motion in the plate. The particular wave pattern that is excited is dependent upon the nature of the excitation.

Consider the infinite plate shown in Figure 9.12. The relationship between the applied forces and moments at position 1, (x_1, y_1), and the resulting transverse and rotational velocities at position 2, (x_2, y_2), is given by

$$
\begin{bmatrix} \dot{W} \\ \dot{\Theta}_x \\ \dot{\Theta}_y \end{bmatrix}_2 = \begin{bmatrix} Y_{W,F_z} & Y_{W,M_x} & Y_{W,M_y} \\ Y_{\Theta_x,F_z} & Y_{\Theta_x,M_x} & Y_{\Theta_x,M_y} \\ Y_{\Theta_y,F_z} & Y_{\Theta_y,M_x} & Y_{\Theta_y,M_y} \end{bmatrix} \begin{bmatrix} F_z \\ M_x \\ M_y \end{bmatrix}_1
\tag{9.52}
$$

where

$$
Y_{W,F_z} = \left.\frac{\dot{W}}{F_z}\right|_{M_x,M_y=0} \qquad Y_{W,M_x} = \left.\frac{\dot{W}}{M_x}\right|_{F_z,M_y=0} \qquad Y_{W,M_y} = \left.\frac{\dot{W}}{M_y}\right|_{F_z,M_x=0}
$$

$$
Y_{\Theta_x,F_z} = \left.\frac{\dot{\Theta}_x}{F_z}\right|_{M_x,M_y=0} \qquad Y_{\Theta_x,M_x} = \left.\frac{\dot{\Theta}_x}{M_x}\right|_{F_z,M_y=0} \qquad Y_{\Theta_x,M_y} = \left.\frac{\dot{\Theta}_x}{M_y}\right|_{F_z,M_x=0}
$$

$$
Y_{\Theta_y,F_z} = \left.\frac{\dot{\Theta}_y}{F_z}\right|_{M_x,M_y=0} \qquad Y_{\Theta_y,M_x} = \left.\frac{\dot{\Theta}_y}{M_x}\right|_{F_z,M_y=0} \qquad Y_{\Theta_y,M_y} = \left.\frac{\dot{\Theta}_y}{M_y}\right|_{F_z,M_x=0}
$$

Considering the notation shown in Figure 9.13a, the out-of-plane velocity at any position (r, α) in response to a harmonic point force of amplitude F_z acting on a massless, rigid indenter is given by Cremer *et al.* (1988) and Ljunggren (1983) as

$$
\dot{W}(r, \alpha) = \frac{\omega F_z}{8 B k_B^2} \left[H_0^{(2)}(k_B r) - j\frac{2}{\pi} K_0(k_B r) \right]
\tag{9.53}
$$

(a)   (b)

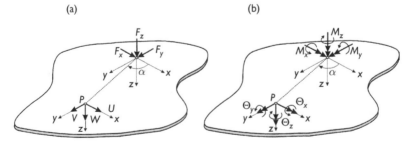

Figure 9.12 Notation for linear displacements and force excitations (a) and for angular displacements and moment excitations (b) acting on a thin infinite plate.

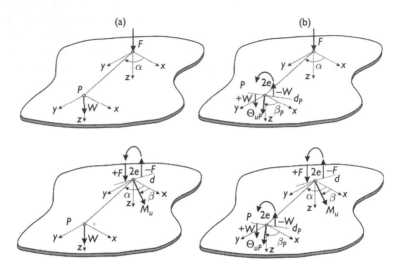

Figure 9.13 Notation for linear-angular velocities and force-moment excitations related to out-of-plane flexural waves in a plate.

where $H_i^{(2)}(k_B r)$ is an ith order Hankel function of the second kind, $K_i(k_B r)$ is an ith order modified Bessel function of the second kind, and $k_B = \sqrt[4]{m/B}\sqrt{\omega}$ is the bending (flexural) wave number. Figure 9.14 shows the real and imaginary parts of the plate velocity $\dot{W}(r, \alpha)$ per unit input force for $k_B r = 0{-}5$ and $\alpha = 90°{-}360°$. It can be seen that for small values of $k_B r$ the real part of the velocity response is dominated by a strong near-field

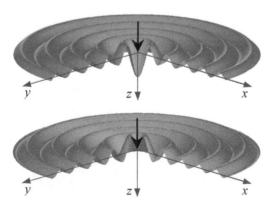

Figure 9.14 Real (top) and Imaginary (bottom) part of the out-of-plane velocity $\dot{W}$ when an infinite plate is excited by a unit out-of-plane force F_z ($k_B r = 0{-}5$).

component, which decays rapidly as $k_B r$ increases while the imaginary part of the velocity response tends to zero for $k_B r \to 0$.

The angular velocity $\dot{\Theta}_{uP}$, with orientation u_P, at position (r, α) generated by the out-of-plane force of amplitude F_z acting on a small rigid indenter fixed to the plate is derived by calculating the gradient of the transverse velocity $\dot{W}$ along a line d_P which, as shown in Figure 9.13b, is orthogonal to the direction u_P (Gardonio and Elliott 1998). This gives

$$\dot{\Theta}_{uP}(r, \alpha) = \frac{\partial \dot{W}}{\partial d_P} = \frac{\omega F_z \sin(\beta_P - \alpha)}{8 B k_B} \left[H_1^{(2)}(k_B r) - j \frac{2}{\pi} K_1(k_B r) \right] \qquad (9.54)$$

where the angles α and β_P are defined in Figure 9.13b.

Consider now the case when a couple of moment M_u with orientation u acts on a small rigid indenter fixed to the plate. The resulting transverse velocity $\dot{W}$ at position (r, α) can be calculated by adding the velocities produced by a pair of harmonic forces with opposite phase acting along a line d a distance $2e$ apart as shown in Figure 9.13c. These forces act along a line orthogonal to u, and in the limiting case $2e \to 0$. The velocity is given by (Gardonio and Elliott 1998)

$$\dot{W}(r, \alpha) = \frac{\omega M_u \sin(\alpha - \beta)}{8 B k_B} \left[H_1^{(2)}(k_B r) - j \frac{2}{\pi} K_1(k_B r) \right] \qquad (9.55)$$

where the angles α and β are defined as in Figure 9.13c. Figure 9.15 shows the real and imaginary parts of the transverse velocity $\dot{W}(r, \alpha)$ for $k_B r = 0\text{–}5$ and $\alpha = 90°\text{–}360°$. This plot shows that the response to

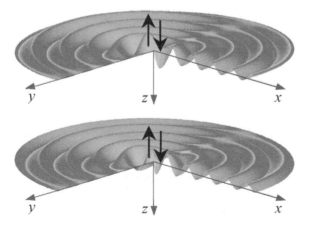

Figure 9.15 Real (top) and Imaginary (bottom) part of the out-of-plane velocity $\dot{W}$ when an infinite plate is excited by a unit moment M_u with u orientation defined by $\beta = 90°$ ($k_B r = 0\text{–}5$).

a moment excitation is not characterised by the strong near-field effect generated by a point force excitation.

The angular velocity $\dot{\Theta}_{uP}$ with orientation u_P (defined by the angle β_P) at a position (r, α), generated by the point moment M_u with orientation u (defined by the angle β) acting on a small rigid indenter fixed to the plate at the origin O of the main system of reference (O, x, y, z), can be derived using the two approaches described above for equations (9.54) and (9.55). In this case, as shown in Figure 9.13d, the gradient along a line d_P, which is orthogonal to the direction u_P, of the transverse velocity field $\dot{W}(r, \alpha)$ is calculated with reference to the vibrating field generated by a pair of harmonic force excitations with opposite phase, so that $F_{z,\mathrm{I}}(\omega) = +F_z$ and $F_{z,\mathrm{II}}(\omega) = -F_z$, acting along a line d which is orthogonal to u, in the limiting case their distance $2e$ goes to zero so that (Gardonio and Elliott 1998)

$$
\dot{\Theta}_{uP}(r, \alpha) = \frac{\omega M_u}{8B} \left\{ \sin(\alpha - \beta_P) \sin(\alpha - \beta) \left[\left(H_0^{(2)}(k_B r) + j\frac{2}{\pi} K_0(k_B r) \right) \right. \right.
$$
$$
\left. - \frac{1}{k_B r} \left(H_1^{(2)}(k_B r) - j\frac{2}{\pi} K_1(k_B r) \right) \right]
$$
$$
\left. + \frac{\cos(\alpha - \beta_P)\cos(\alpha - \beta)}{k_B r} \left(H_1^{(2)}(k_B r) - j\frac{2}{\pi} K_1(k_B r) \right) \right\}
$$

$$(9.56)$$

With equations (9.53)–(9.56), the nine mobility terms for bending motion of an infinite plate given in equation (9.52) can be derived to give

$$
Y_{W,F_z} = \frac{\omega}{8 B k_B^2} \left[H_0^{(2)}(k_B r) - j\frac{2}{\pi} K_0(k_B r) \right] \tag{9.57a}
$$

$$
Y_{W,M_x} = \frac{\omega \sin \alpha_1}{8 B k_B} \left[H_1^{(2)}(k_B r) - j\frac{2}{\pi} K_1(k_B r) \right] \tag{9.57b}
$$

$$
Y_{W,M_y} = -\frac{\omega \cos \alpha_1}{8 B k_B} \left[H_1^{(2)}(k_B r) - j\frac{2}{\pi} K_1(k_B r) \right] \tag{9.57c}
$$

$$
Y_{\Theta_x,F_z} = -\frac{\omega \sin \alpha_1}{8 B k_B} \left[H_1^{(2)}(k_B r) - j\frac{2}{\pi} K_1(k_B r) \right] \tag{9.57d}
$$

$$
Y_{\Theta_x,M_x} = \frac{\omega M_u}{8B} \left\{ \sin^2 \alpha \left[\left(H_0^{(2)}(k_B r) + j\frac{2}{\pi} K_0(k_B r) \right) \right. \right.
$$
$$
\left. - \frac{1}{k_B r} \left(H_1^{(2)}(k_B r) - j\frac{2}{\pi} K_1(k_B r) \right) \right]
$$
$$
\left. + \frac{\cos^2 \alpha}{k_B r} \left(H_1^{(2)}(k_B r) - j\frac{2}{\pi} K_1(k_B r) \right) \right\} \tag{9.57e}
$$

$$Y_{\Theta_x, M_y} = \frac{\omega M_u}{8B} \left\{ -\sin\alpha\cos\alpha \left[\left(H_0^{(2)}(k_B r) + j\frac{2}{\pi} K_0(k_B r) \right) \right. \right.$$
$$\left. -\frac{1}{k_B r} \left(H_1^{(2)}(k_B r) - j\frac{2}{\pi} K_1(k_B r) \right) \right]$$
$$\left. +\frac{\cos\alpha\sin\alpha}{k_B r} \left(H_1^{(2)}(k_B r) - j\frac{2}{\pi} K_1(k_B r) \right) \right\} \tag{9.57f}$$

$$Y_{\Theta_y, F_z} = \frac{\omega\cos\alpha_1}{8Bk_B} \left(H_1^{(2)}(k_B r) - j\frac{2}{\pi} K_1(k_B r) \right) \tag{9.57g}$$

$$Y_{\Theta_y, M_x} = \frac{\omega M_u}{8B} \left\{ -\sin\alpha\cos\alpha \left[\left(H_0^{(2)}(k_B r) + j\frac{2}{\pi} K_0(k_B r) \right) \right. \right.$$
$$\left. -\frac{1}{k_B r} \left(H_1^{(2)}(k_B r) - j\frac{2}{\pi} K_1(k_B r) \right) \right]$$
$$\left. +\frac{\cos\alpha\sin\alpha}{k_B r} \left(H_1^{(2)}(k_B r) - j\frac{2}{\pi} K_1(k_B r) \right) \right\} \tag{9.57h}$$

$$Y_{\Theta_y, M_y} = \frac{\omega M_u}{8B} \left\{ \cos^2\alpha \left[\left(H_0^{(2)}(k_B r) + j\frac{2}{\pi} K_0(k_B r) \right) \right. \right.$$
$$\left. -\frac{1}{k_B r} \left(H_1^{(2)}(k_B r) - j\frac{2}{\pi} K_1(k_B r) \right) \right]$$
$$\left. +\frac{\sin^2\alpha}{k_B r} \left(H_1^{(2)}(k_B r) - j\frac{2}{\pi} K_1(k_B r) \right) \right\} \tag{9.57i}$$

The point mobility terms for an infinite plate can be derived by setting r to zero. The force F_z does not generate an angular velocity $\dot{\Theta}_u$ and the moment M_u does not generate a transverse velocity $\dot{W}$ at the excitation point (Cremer *et al.* 1988). This means that the point mobility matrix is diagonal with the elements given by (Ljunggren 1984; Cremer *et al.* 1988).

$$Y_{W, F_z}(0) = \frac{\omega}{8Bk_B^2} = \frac{1}{8\sqrt{Bm}} \tag{9.58a}$$

$$Y_{\Theta_x, M_x}(0) = Y_{\Theta_y, M_y}(0) = \frac{\omega}{16B} \left\{ 1 - j\frac{4}{\pi}\ln\frac{k_B e}{2} \right\} \tag{9.58b,c}$$

where e is the radius of the indenter. Equation (9.58a) shows that the driving point mobility of a force is purely real and independent of frequency: this shows that it behaves as a damper. The driving point mobility of a moment, however, has both real and imaginary components, showing that it behaves as a damper in parallel with a mass.

9.8.3   In-plane vibration of an infinite plate

Consider the infinite plate shown in Figure 9.16. The relationship between the applied forces and moments at position 1, (x_1, y_1), and the resulting transverse and rotational velocities at position 2, (x_2, y_2), is given by

$$\begin{bmatrix} \dot{U} \\ \dot{V} \\ \dot{\Theta}_z \end{bmatrix}_2 = \begin{bmatrix} Y_{U,F_x} & Y_{U,F_y} & Y_{U,M_z} \\ Y_{V,F_x} & Y_{V,F_y} & Y_{V,M_z} \\ Y_{\Theta_z,F_x} & Y_{\Theta_z,F_y} & Y_{\Theta_z,M_z} \end{bmatrix} \begin{bmatrix} F_x \\ F_y \\ M_z \end{bmatrix}_1 \qquad (9.59)$$

where the mobility terms related to the in-plane longitudinal and shear waves are given by

$$Y_{U,F_x} = \left.\frac{\dot{U}}{F_x}\right|_{F_y,M_z=0} \qquad Y_{U,F_y} = \left.\frac{\dot{U}}{F_y}\right|_{F_x,M_z=0} \qquad Y_{U,M_z} = \left.\frac{\dot{U}}{M_z}\right|_{F_x,F_y=0}$$

$$Y_{V,F_x} = \left.\frac{\dot{V}}{F_x}\right|_{F_y,M_z=0} \qquad Y_{V,F_y} = \left.\frac{\dot{V}}{F_y}\right|_{F_x,M_z=0} \qquad Y_{V,M_z} = \left.\frac{\dot{V}}{M_z}\right|_{F_x,F_y=0}$$

$$Y_{\Theta_z,F_x} = \left.\frac{\dot{\Theta}_z}{F_x}\right|_{F_y,M_z=0} \qquad Y_{\Theta_z,F_y} = \left.\frac{\dot{\Theta}_z}{F_y}\right|_{F_x,M_z=0} \qquad Y_{\Theta_z,M_z} = \left.\frac{\dot{\Theta}_z}{M_z}\right|_{F_x,F_y=0}$$

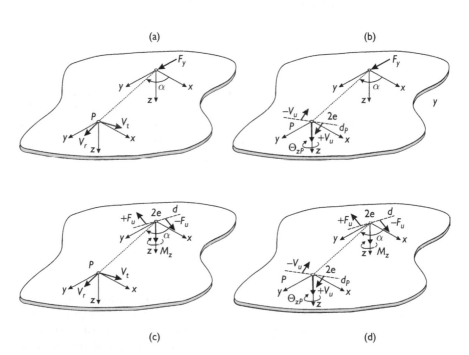

Figure 9.16 Notation for translational-angular velocities and force-moment excitations related to out-of-plane flexural waves in a plate.

Assuming 'classical plate theory' for thin plates, the two coupled differential equations of motion for in-plane vibration of an infinite thin plate suggest to distributed in-plane forces per unit area $p_x(x, y)$ and $p_y(x, y)$ are (Reddy 1984; Cremer *et al.* 1988)

$$\frac{Eh}{1 - \nu^2} \left(\frac{\partial^2 u(x, y)}{\partial x^2} + \frac{1 - \nu}{2} \frac{\partial^2 u(x, y)}{\partial y^2} \right) + \frac{Eh}{2(1 - \nu)} \frac{\partial^2 v(x, y)}{\partial x \partial y}$$

$$- \rho h \frac{\partial^2 u(x, y, t)}{\partial t^2} = -p_x(x, y, t) \tag{9.60a}$$

$$\frac{Eh}{2(1 - \nu)} \frac{\partial^2 u(x, y)}{\partial y \partial x} + \frac{Eh}{1 - \nu^2} \left(\frac{1 - \nu}{2} \frac{\partial^2 v(x, y)}{\partial x^2} + \frac{\partial^2 v(x, y)}{\partial y^2} \right)$$

$$- \rho h \frac{\partial^2 v(x, y, t)}{\partial t^2} = -p_y(x, y, t) \tag{9.60b}$$

The in-plane steady-state radial and tangential linear velocities at position (r, α) are given by Ljunggren (1984) as

$$\dot{V}_r(r, \alpha) = F_y \left[\frac{\omega}{4D} \left(H_0^{(2)}(k_L r) - \frac{1}{k_L r} H_1^{(2)}(k_L r) \right) \right.$$

$$\left. + \frac{\omega}{4S} \left(\frac{1}{k_T r} H_1^{(2)}(k_T r) \right) \right] \sin \alpha \tag{9.61a}$$

$$\dot{V}_t x(r, \alpha) = F_y \left[\frac{\omega}{4D} \left(\frac{1}{k_L r} H_1^{(2)}(k_L r) \right) \right.$$

$$\left. + \frac{\omega}{4S} \left(H_0^{(2)}(k_T r) - \frac{1}{k_T r} H_1^{(2)}(k_T r) \right) \right] \cos \alpha \tag{9.61b}$$

where $D = Eh/(1 - \nu^2)$ and $S = hG = Eh/2(1 + \nu)$ are the longitudinal and shear stiffness respectively, and G is the shear modulus of elasticity. $k_L = \sqrt{(\rho(1 - \nu^2)/E)}\,\omega$ and $k_T = \sqrt{(2\rho(1 + \nu)/E)}\,\omega$ are the wave numbers of the quasi-longitudinal and shear waves in thin plates respectively. $H_i^{(2)}(kr)$ is an ith order Hankel function of the second kind. Figure 9.17 shows the real and imaginary parts of the radial and tangential velocities $\dot{V}_r$ and $\dot{V}_t$ due to a unit harmonic force F_y with $k_L r = 0{-}5$ and $\alpha = 0°{-}360°$. These two plots show that both the radial and tangential components of the velocity are characterised by a dipole field which are rotated by 90° one from the other so that in the direction of the excitation, there is the larger radial velocity and there is no tangential velocity, while in the direction orthogonal to that of the excitation, there is the larger tangential velocity and there is no radial velocity. As for the case of out-of-plane vibration due to a moment excitation, for small values of $k_L r$ both the real and the imaginary parts of the velocity are not characterised by the strong near-field effect generated by a point force excitation.

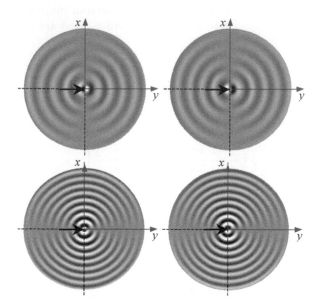

Figure 9.17 Real (left) and imaginary (right) parts of the in-plane radial (top) and tangential (bottom) velocity components generated by an in-plane unit harmonic force F_y acting at the origin of the system of reference ($k_L r = $ 0–5).

The in-plane steady-state angular velocity $\dot{\Theta}_z(r, \alpha)$ with orientation z, generated by an in-plane force F_y is given by,

$$\dot{\Theta}_z(r, \alpha) = \frac{1}{2}\left(\frac{\partial \dot{V}_t(r, \alpha)}{\partial r} - \frac{1}{r}\frac{\partial \dot{V}_r(r, \alpha)}{\partial \alpha}\right)$$

$$= F_y\left[\frac{\omega}{8D}\left(-\frac{1}{r}H_2^{(2)}(k_L r) - \frac{1}{r}H_0^{(2)}(k_L r) + \frac{1}{k_L r^2}H_1^{(2)}(k_L r)\right)\right.$$

$$\left. + \frac{\omega}{8S}\left(-k_T H_1^{(2)}(k_T r) + \frac{1}{r}H_2^{(2)}(k_T r) - \frac{1}{k_T r^2}H_1^{(2)}(k_T r)\right)\right]\cos\alpha$$

$$(9.62)$$

If, instead of a point force, the excitation is a point moment M_z with orientation z, then the in-plane vibration field is characterised only by tangential displacements. The tangential velocity $\dot{V}_t(r, \alpha)$ for a moment excitation M_z is given by Ljunggren (1984) as

$$\dot{V}_t(r, \alpha) = M_z\frac{\omega k_T}{8S}H_1^{(2)}(k_T r)$$

$$(9.63)$$

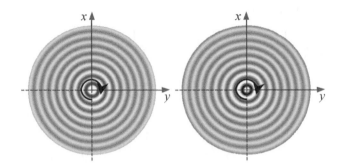

Figure 9.18 Real (left) and imaginary (right) parts of the in-plane tangential velocity $\dot{V}_t$ generated by an in-plane unit harmonic moment M_z acting at the origin of the system of reference ($k_L r = 0$–5).

Figure 9.18 shows the real and imaginary parts of the tangential velocity $\dot{V}_t$ due to a unit harmonic moment M_z with $k_L r = 1$–5 and $\alpha = 0°$ –360°. The in-plane steady-state angular velocity $\dot{\Theta}_z(r, \alpha)$ with orientation z, generated by a moment M_z, is given by

$$\dot{\Theta}_z(r, \alpha) = M_z \frac{\omega k_T^2}{16S} \left(H_0^{(2)}(k_T r) - H_2^{(2)}(k_T r) \right) \tag{9.64}$$

With equations (9.61)–(9.64), the nine mobility terms for in-plane motion of an infinite plate given in equation (9.59) can be derived to give

$$Y_{U,F_x} = \left[\frac{\omega}{4D} T_0(k_L r) + \frac{\omega}{4S} T_1(k_T r) \right] \cos^2 \alpha$$
$$+ \left[\frac{\omega}{4D} T_1(k_L r) + \frac{\omega}{4S} T_0(k_T r) \right] \sin^2 \alpha \tag{9.65a}$$

$$Y_{U,F_y} = \left[\frac{\omega}{4D} T_0(k_L r) + \frac{\omega}{4S} T_1(k_T r) \right] \sin \alpha \cos \alpha$$
$$- \left[\frac{\omega}{4D} T_1(k_L r) + \frac{\omega}{4S} T_0(k_T r) \right] \sin \alpha \cos \alpha \tag{9.65b}$$

$$Y_{U,M_z} = -\frac{\omega k_T}{8S} H_1^{(2)}(k_T r) \sin \alpha \tag{9.65c}$$

$$Y_{V,F_x} = \left[\frac{\omega}{4D} T_0(k_L r) + \frac{\omega}{4S} T_1(k_T r) \right] \sin \alpha \cos \alpha$$
$$- \left[\frac{\omega}{4D} T_1(k_L r) + \frac{\omega}{4S} T_0(k_T r) \right] \sin \alpha \cos \alpha \tag{9.65d}$$

$$Y_{V,F_y} = \left[\frac{\omega}{4D} T_0(k_L r) + \frac{\omega}{4S} T_1(k_T r) \right] \sin^2 \alpha$$
$$+ \left[\frac{\omega}{4D} T_1(k_L r) + \frac{\omega}{4S} T_0(k_T r) \right] \cos^2 \alpha \tag{9.65e}$$

$$Y_{V,M_z} = \frac{\omega k_T}{8S} H_1^{(2)}(k_T r) \cos \alpha \qquad (9.65f)$$

$$Y_{\Theta_z,F_x} = -\left[\frac{\omega}{8D} \left(-\frac{1}{r} H_2^{(2)}(k_L r) - \frac{1}{r} H_0^{(2)}(k_L r) + \frac{1}{k_L r^2} H_1^{(2)}(k_L r) \right) \right.$$
$$\left. + \frac{\omega}{8S} \left(-k_T H_1^{(2)}(k_T r) + \frac{1}{r} H_2^{(2)}(k_T r) - \frac{1}{k_T r^2} H_1^{(2)}(k_T r) \right) \right] \sin \alpha$$
$$(9.65g)$$

$$Y_{\Theta_z,F_y} = \left[\frac{\omega}{8D} \left(-\frac{1}{r} H_2^{(2)}(k_L r) - \frac{1}{r} H_0^{(2)}(k_L r) + \frac{1}{k_L r^2} H_1^{(2)}(k_L r) \right) \right.$$
$$\left. + \frac{\omega}{8S} \left(-k_T H_1^{(2)}(k_T r) + \frac{1}{r} H_2^{(2)}(k_T r) - \frac{1}{k_T r^2} H_1^{(2)}(k_T r) \right) \right] \cos \alpha$$
$$(9.65h)$$

$$Y_{\Theta_z,M_z} = \frac{\omega k_T^2}{16S} \left[H_0^{(2)}(k_T r) - H_2^{(2)}(k_T r) \right] \qquad (9.65i)$$

where

$$T_0(kr) = \left[H_0^{(2)}(kr) - \frac{1}{kr} H_1^{(2)}(kr) \right] \quad \text{and} \quad T_1(kr) = \frac{1}{kr} H_1^{(2)}(kr)$$

and k could be either equal to k_L or k_T. Considering equations (9.61a,b) and (9.62) it can be seen that the in-plane force F_y generates only linear velocity $\dot{U}$ at the excitation point, and F_x generates only linear velocity $\dot{V}$. Moreover, the moment excitation M_z produces only angular velocity Θ_z at the excitation point (Ljunggren 1984). This means that the driving point mobility matrix is diagonal with the terms in the matrix given by (Ljunggren 1984)

$$Y_{U,F_x} = Y_{V,F_y} = \frac{\omega}{8D} \left[1 - j\frac{2}{\pi} \ln \frac{k_L e}{2} \right] + \frac{\omega}{8S} \left[1 - j\frac{2}{\pi} \ln \frac{k_T e}{2} \right] \qquad (9.66a,b)$$

$$Y_{\Theta_z,M_z} = \frac{\omega}{16S} \left(k_T^2 + j\frac{4}{\pi e^2} \right) \qquad (9.66c)$$

where e is the radius of the indenter.

9.8.4 Out-of-plane vibration of rectangular plates

The mobilities of finite plates can be written in terms of a modal summation (Soedel 1993). The case of a rectangular plate of dimensions $l_x \times l_y$ is considered. Figure 9.19 shows the notation used for the force-moment excitation parameters at position 1, (x_1, y_1), and for the linear-angular velocity

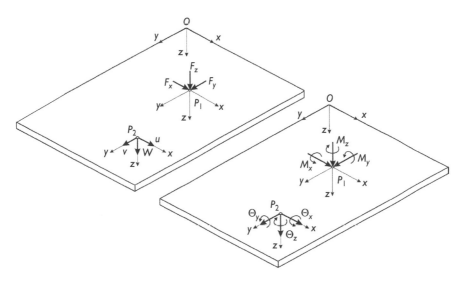

Figure 9.19 Notation for linear displacements and force excitations (left) and for angular displacements and moment excitations (right) acting on a thin rectangular plate.

parameters at position 2, (x_2, y_2). The main Cartesian co-ordinate system of reference (O, x, y, z) is located at a corner of the plate with the z axis orthogonal to the surface of the plate. The steady-state out-of-plane velocity $\dot{W}(x_2, y_2)$ is given by Soedel (1993) as

$$\dot{W}(x_2, y_2) = j\omega \sum_{m=1}^{\infty} \sum_{n=1}^{\infty} \frac{\psi_{mn}(x_2, y_2)\psi_{mn}(x_1, y_1)}{\rho h l_x l_y [\omega_{mn}^2(1 + j\eta) - \omega^2]} F_z(x_1, y_1) \tag{9.67}$$

where $\psi_{mn}(x, y)$ is the (m, n)th bending natural mode and ω_{mn} is the (m, n)th natural frequency and η is the loss factor (Cremer *et al.* 1988). The natural frequencies for rectangular plates with any type of boundary conditions are given by Warburton (1951) as

$$\omega_{mn} = \sqrt{\frac{Eh^2}{12\rho(1 - \nu^2)}} \cdot \left(\frac{\pi}{l_x}\right)^2 q_{mn} \tag{9.68}$$

where
$$q_{mn} = \sqrt{G_x^4(m) + G_y^4(n)\left(l_x/l_y\right)^4 + 2\left(l_x/l_y\right)^2 \left[\nu H_x(m)H_y(n) + (1 - \nu)J_x(m)J_y(n)\right]}.$$

The constants G_x, H_x, J_x and G_y, H_y, J_y are given in Table 9.9 for the six most common combinations of boundary conditions along two opposite edges. The subscripts x and y for the constants G, H and J are either defined with reference to the boundary conditions for $x = 0$ and $x = l_x$ or

Table 9.9 Values for the constants G_x, H_x, J_x (after Warburton 1951). Considering a given pair of boundary conditions along the edges $y=0$ and $y=l_y$, the constants G_y, H_y, J_y are the same as those in the x direction with equivalent boundary conditions along the edges $x=0$ and $x=l_x$

Boundary conditions	m	G_x	H_x	J_x
Pinned–pinned $W(0)=0 \quad W''(0)=0$ $W(l)=0 \quad W''(l)=0$	$1,2,3,\ldots$	m	m^2	m^2
Clamped–clamped $W(0)=0 \quad W'(0)=0$ $W(l)=0 \quad W'(l)=0$	1	1.506	1.248	1.248
	$2,3,4,\ldots$	$m+\dfrac{1}{2}$	$\left(m+\dfrac{1}{2}\right)^2\left[1-\dfrac{4}{(2m+1)\pi}\right]$	$\left(m+\dfrac{1}{2}\right)^2\left[1-\dfrac{4}{(2m+1)\pi}\right]$
Free–free $W''(0)=0 \quad W'''(0)=0$ $W''(l)=0 \quad W'''(l)=0$	Even mode	0	0	0
	Rocking mode	0	0	$12/\pi^2$
	1	1.506	1.248	5.017
	$2,3,4,\ldots$	$m+\dfrac{1}{2}$	$\left(m+\dfrac{1}{2}\right)^2\left[1-\dfrac{4}{(2m+1)\pi}\right]$	$\left(m+\dfrac{1}{2}\right)^2\left[1+\dfrac{12}{(2m+1)\pi}\right]$
Clamped–free $W(0)=0 \quad W'(0)=0$ $W''(l)=0 \quad W'''(l)=0$	1	0.597	-0.0870	0.471
	2	1.494	1.347	3.284
	$3,4,\ldots$	$m+\dfrac{1}{2}$	$\left(m+\dfrac{1}{2}\right)^2\left[1-\dfrac{4}{(2m+1)\pi}\right]$	$\left(m+\dfrac{1}{2}\right)^2\left[1+\dfrac{4}{(2m+1)\pi}\right]$
Clamped–pinned $W(0)=0 \quad W'(0)=0$ $W(l)=0 \quad W''(l)=0$	$1,2,\ldots$	$m+\dfrac{1}{4}$	$\left(m+\dfrac{1}{4}\right)^2\left(1-\dfrac{4}{(4m+1)\pi}\right)$	$\left(m+\dfrac{1}{4}\right)^2\left(1-\dfrac{4}{(4m+1)\pi}\right)$
Free–pinned $W''(0)=0 \quad W'''(0)=0$ $W(l)=0 \quad W''(l)=0$	Rocking mode	0	0	$3/\pi^2$
	$1,2,3,\ldots$	$m+\dfrac{1}{4}$	$\left(m+\dfrac{1}{4}\right)^2\left(1-\dfrac{4}{(4m+1)\pi}\right)$	$\left(m+\dfrac{1}{4}\right)^2\left(1+\dfrac{12}{(4m+1)\pi}\right)$

with reference to the boundary conditions for $y = 0$ and $y = l_y$. The plate natural modes $\psi_{mn}(x, y)$ are given by the product of the 'characteristic beam functions', $\phi_m(x)$ and $\phi_n(x)$, that is $\psi_{mn}(x, y) = \phi_m(x)\phi_n(y)$. The characteristic beam functions for the most common pairs of boundary conditions are given in Table 9.10 with the zeros of the 'gama-functions' in Table 9.11.

Three other expressions are necessary to derive the nine mobility terms of equation (9.52). These three relations can be derived with the same approach as above and are given by

$$\dot{\Theta}_u(x_2, y_2) = j\omega \sum_{m=1}^{\infty} \sum_{n=1}^{\infty} \frac{\psi_{mn}^{u2}(x_2, y_2)\psi_{mn}(x_1, y_1)}{\rho h l_x l_y [\omega_{mn}^2(1 + j\eta) - \omega^2]} F_z(x_1, y_1) \tag{9.69}$$

where

$$\psi_{mn}^{u2}(x, y) = -\sin\beta_2 \frac{\partial \psi(x, y)}{\partial x} + \cos\beta_2 \frac{\partial \psi(x, y)}{\partial y} \tag{9.70}$$

$$\dot{\Theta}_u(x_2, y_2) = j\omega \sum_{m=1}^{\infty} \sum_{n=1}^{\infty} \frac{\psi_{mn}^{u2}(x_2, y_2)\psi_{mn}^{u1}(x_1, y_1)}{\rho h l_x l_y [\omega_{mn}^2(1 + j\eta) - \omega^2]} M_u(x_1, y_1) \tag{9.71}$$

where both $\psi_{mn}^{u1}(x, y)$ and $\psi_{mn}^{u2}(x, y)$ are given by equation (9.70) with $\beta = \beta_1$ and $\beta = \beta_2$ respectively, and

$$\dot{W}(x_2, y_2) = j\omega \sum_{m=1}^{\infty} \sum_{n=1}^{\infty} \frac{\psi_{mn}(x_2, y_2)\psi_{mn}^{u1}(x_1, y_1)}{\rho h l_x l_y [\omega_{mn}^2(1 + j\eta) - \omega^2]} M_u(x_1, y_1) \tag{9.72}$$

The angles β_1 and β_2 are defined as shown in Figure 9.13. Using equations (9.67), (9.69), (9.71) and (9.72), the nine mobility equations can be derived to give

$$Y_{W,F_z} = j\omega \sum_{m=1}^{\infty} \sum_{n=1}^{\infty} \frac{\psi_{mn}(x_2, y_2)\psi_{mn}(x_1, y_1)}{\rho h l_x l_y [\omega_{mn}^2(1 + j\eta) - \omega^2]} \tag{9.73a}$$

$$Y_{W,M_x} = j\omega \sum_{m=1}^{\infty} \sum_{n=1}^{\infty} \frac{\psi_{mn}(x_2, y_2)\psi_{mn}^x(x_1, y_1)}{\rho h l_x l_y [\omega_{mn}^2(1 + j\eta) - \omega^2]} \tag{9.73b}$$

$$Y_{W,M_y} = j\omega \sum_{m=1}^{\infty} \sum_{n=1}^{\infty} \frac{\psi_{mn}(x_2, y_2)\psi_{mn}^y(x_1, y_1)}{\rho h l_x l_y [\omega_{mn}^2(1 + j\eta) - \omega^2]} \tag{9.73c}$$

$$Y_{\Theta_x, F_z} = j\omega \sum_{m=1}^{\infty} \sum_{n=1}^{\infty} \frac{\psi_{mn}^x(x_2, y_2)\psi_{mn}(x_1, y_1)}{\rho h l_x l_y [\omega_{mn}^2(1 + j\eta) - \omega^2]} \tag{9.73d}$$

$$Y_{\Theta_x, M_x} = j\omega \sum_{m=1}^{\infty} \sum_{n=1}^{\infty} \frac{\psi_{mn}^x(x_2, y_2)\psi_{mn}^x(x_1, y_1)}{\rho h l_x l_y [\omega_{mn}^2(1 + j\eta) - \omega^2]} \tag{9.73e}$$

Table 9.10 Characteristic beam functions (with $\int_0^l \psi^2(x)dx = \int_0^l \phi^2(x)\phi^2(y)dx = l$)

Boundary conditions	$\phi_{1,3,5,\ldots}(x)$ with $i = (n+1)/2$	$\phi_{2,4,6,\ldots}(x)$ with $j = n/2$
Pinned–pinned $W(0)=0 \quad W''(0)=0$ $W(l)=0 \quad W''(l)=0$	$\phi_n(x) = \sqrt{2}\sin\left(\dfrac{n\pi x}{l_x}\right)$	
Clamped–clamped $W(0)=0 \quad W'(0)=0$ $W(l)=0 \quad W'(l)=0$	$\phi_n(x) = \sqrt{2}\left\{\cos\gamma_i\left(\dfrac{x}{l_x}-\dfrac{1}{2}\right)+k_n\cosh\gamma_i\left(\dfrac{x}{l_x}-\dfrac{1}{2}\right)\right\}$ $k_n = -\dfrac{\sin\frac{1}{2}\gamma_i}{\sinh\frac{1}{2}\gamma_i}$ with $\tan\frac{1}{2}\gamma_i + \tanh\frac{1}{2}\gamma_i = 0$	$\phi_n(x) = \sqrt{2}\left\{\sin\gamma_j\left(\dfrac{x}{l_x}-\dfrac{1}{2}\right)+k_n\sinh\gamma_j\left(\dfrac{x}{l_x}-\dfrac{1}{2}\right)\right\}$ $k_n = -\dfrac{\sin\frac{1}{2}\gamma_j}{\sinh\frac{1}{2}\gamma_j}$ with $\tan\frac{1}{2}\gamma_j - \tanh\frac{1}{2}\gamma_j = 0$
Free–free $W''(0)=0 \quad W'''(0)=0$ $W''(l)=0 \quad W'''(l)=0$	$\phi_{even}(x) = 1$ $\phi_{roking}(x) = \sqrt{3}\,(1-2x/l)$	
	$\phi_n(x) = \sqrt{2}\left\{\cos\gamma_i\left(\dfrac{x}{l_x}-\dfrac{1}{2}\right)+k_n\cosh\gamma_i\left(\dfrac{x}{l_x}-\dfrac{1}{2}\right)\right\}$ $k_n = -\dfrac{\sin\frac{1}{2}\gamma_i}{\sinh\frac{1}{2}\gamma_i}$ with $\tan\frac{1}{2}\gamma_i + \tanh\frac{1}{2}\gamma_i = 0$	$\phi_n(x) = \sqrt{2}\left\{\sin\gamma_j\left(\dfrac{x}{l_x}-\dfrac{1}{2}\right)+k_n\sinh\gamma_j\left(\dfrac{x}{l_x}-\dfrac{1}{2}\right)\right\}$ $k_n = -\dfrac{\sin\frac{1}{2}\gamma_j}{\sinh\frac{1}{2}\gamma_j}$ with $\tan\frac{1}{2}\gamma_j - \tanh\frac{1}{2}\gamma_j = 0$

Clamped–free

$W(0) = 0 \quad W'(0) = 0$
$W''(l) = 0 \quad W'''(l) = 0$

$$\phi_n(x) = \sqrt{2}\left\{\cos\frac{\gamma_n x}{l_x} - \cosh\frac{\gamma_n x}{l_x} + k_n\left(\sin\frac{\gamma_n x}{l_x} - \sinh\frac{\gamma_n x}{l_x}\right)\right\}$$

$$k_n = \frac{\sin\gamma_n - \sinh\gamma_n}{\cos\gamma_n - \cosh\gamma_n} \quad \text{with } \cos\gamma_n\cosh\gamma_n = -1$$

Clamped–pinned

$W(0) = 0 \quad W'(0) = 0$
$W(l) = 0 \quad W''(l) = 0$

$$\phi_n(x) = \sqrt{2}\left\{\sin\gamma_n\left(\frac{x}{2l_x} - \frac{1}{2}\right) + k_n\sinh\gamma_n\left(\frac{x}{2l_x} - \frac{1}{2}\right)\right\}$$

$$k_n = -\frac{\sin\frac{1}{2}\gamma_n}{\sinh\frac{1}{2}\gamma_n} \quad \text{with } \tan\frac{1}{2}\gamma_n - \tanh\frac{1}{2}\gamma_n = 0$$

Free–pinned

$W'''(0) = 0 \quad W''''(0) = 0$
$W(l) = 0 \quad W''(l) = 0$

$$\phi_{rocking}(x) = \sqrt{3}(l-x)/l$$

$$\phi_n(x) = \sqrt{2}\left\{\sin\gamma_n\left(\frac{x}{2l_x} - \frac{1}{2}\right) + k_n\sinh\gamma_n\left(\frac{x}{2l_x} - \frac{1}{2}\right)\right\}$$

$$k_n = \frac{\sin\frac{1}{2}\gamma_n}{\sinh\frac{1}{2}\gamma_n} \quad \text{with } \tan\frac{1}{2}\gamma_n - \tanh\frac{1}{2}\gamma_n = 0$$

Table 9.11 Zeros of the gamma-functions γ in Table 9.10

	$\tan \frac{1}{2}\gamma_i - \tanh \frac{1}{2}\gamma_i = 0$	$\tan \frac{1}{2}\gamma_j + \tanh \frac{1}{2}\gamma_j = 0$	$\cos \gamma_n \cosh \gamma_n = -1$
1	7.8532	4.73004	1.87510
2	14.13716	10.9956	4.69409
3	20.4204	17.27876	7.85476
4	26.7036	23.5620	10.9955
5	32.9868	29.8452	14.1372
6, 7, 8, . . .	$\dfrac{(4i+1)\pi}{2}$	$\dfrac{(4j-1)\pi}{2}$	$\dfrac{(2n-1)\pi}{2}$

$$Y_{\Theta_x, M_y} = j\omega \sum_{m=1}^{\infty} \sum_{n=1}^{\infty} \frac{\psi_{mn}^x(x_2, y_2)\psi_{mn}^y(x_1, y_1)}{\rho h l_x l_y [\omega_{mn}^2(1+j\eta) - \omega^2]} \tag{9.73f}$$

$$Y_{\Theta_y, F_z} = j\omega \sum_{m=1}^{\infty} \sum_{n=1}^{\infty} \frac{\psi_{mn}^y(x_2, y_2)\psi_{mn}(x_1, y_1)}{\rho h l_x l_y [\omega_{mn}^2(1+j\eta) - \omega^2]} \tag{9.73g}$$

$$Y_{\Theta_y, M_x} = j\omega \sum_{m=1}^{\infty} \sum_{n=1}^{\infty} \frac{\psi_{mn}^y(x_2, y_2)\psi_{mn}^x(x_1, y_1)}{\rho h l_x l_y [\omega_{mn}^2(1+j\eta) - \omega^2]} \tag{9.73h}$$

$$Y_{\Theta_y, M_y} = j\omega \sum_{m=1}^{\infty} \sum_{n=1}^{\infty} \frac{\psi_{mn}^y(x_2, y_2)\psi_{m,n}^y(x_1, y_1)}{\rho h l_x l_y [\omega_{mn}^2(1+j\eta) - \omega^2]} \tag{9.73i}$$

where the two functions $\psi_{mn}^x(x, y)$ and $\psi_{mn}^y(x, y)$ are given by

$$\psi_{mn}^x(x, y) = \phi_m(x)\frac{\partial \phi_n(y)}{\partial y} \qquad \psi_{mn}^y(x, y) = -\frac{\partial \phi_m(x)}{\partial x}\phi_n(y) \tag{9.74}$$

The driving point mobility terms can be determined by evaluating equations (9.73a–i) at (x_1, y_1).

9.8.5 In-plane vibration of a rectangular plates

The nine mobility parameters related to in-plane vibration of a finite plate can be derived in a similar way to the previous section. The in-plane velocities $\dot{U}(x_2, y_2)$ and $\dot{V}(x_2, y_2)$ are due to the in-plane force excitations $F_x(x_1, y_1)$ and $F_y(x_1, y_1)$. For example, the in-plane velocity $\dot{U}(x_2, y_2)$ in the x direction, due to the point force $F_y(x_1, y_1)$ in the y direction is given by

$$\dot{U}(x_2, y_2) = j\omega \sum_{m=0}^{\infty} \sum_{n=0}^{\infty} \frac{\lambda_{mn}^x(x_2, y_2)\lambda_{mn}^y(x_1, y_1)}{\rho h l_x l_y [\omega_{mn}^2(1+j\eta) - \omega^2]} F_y(x_1, y_1) \tag{9.75}$$

where the in-plane natural modes $\lambda_{mn}^x(x, y)$ and $\lambda_{mn}^y(x, y)$ are given for the most common pairs of boundary conditions in Table 9.12. Three other expressions are necessary to derive the nine mobility expressions.

Table 9.12 Natural frequencies and modes for in-plane vibration of rectangular plates

$$\omega_{mn} = \sqrt{\frac{E}{\rho(1-\nu^2)}}q_{mn} \quad \text{and} \quad \omega_{mn} = \sqrt{\frac{G}{\rho}}q_{mn}$$

Boundary conditions			$\lambda^x(x,y)$	$\lambda^y(x,y)$
	for y = 0 and y = l_y	*for x = 0 and x = l_x*		
Cross sliding–cross sliding	$U=0 \;\; \dfrac{\partial V}{\partial y}=0$	$\dfrac{\partial U}{\partial x}=0 \;\; V=0$	$q_{mn} = \sqrt{\left(\dfrac{m\pi}{l_x}\right)^2 + \left(\dfrac{n\pi}{l_y}\right)^2}$ $\quad 2\cos\dfrac{m\pi x}{l_x}\sin\dfrac{n\pi y}{l_y}$	$2\sin\dfrac{m\pi x}{l_x}\cos\dfrac{n\pi y}{l_y}$
Edge sliding–edge sliding	$\dfrac{\partial U}{\partial x}=0 \;\; V=0$	$U=0 \;\; \dfrac{\partial V}{\partial y}=0$	$q_{mn} = \sqrt{\left(\dfrac{m\pi}{l_x}\right)^2 + \left(\dfrac{n\pi}{l_y}\right)^2}$ $\quad 2\sin\dfrac{m\pi x}{l_x}\cos\dfrac{n\pi y}{l_y}$	$2\cos\dfrac{m\pi x}{l_x}\sin\dfrac{n\pi y}{l_y}$

The angular velocity $\dot{\theta}_z(x_2, y_2)$ is given by $\dot{\theta}_z(x_2, y_2) = 1/2\,(\partial\dot{v}(x_2, y_2)/\partial x - \partial\dot{u}(x_2, y_2)/\partial y)$ so that

$$\dot{\Theta}_z(x_2, y_2) = j\omega \sum_{m=0}^{\infty} \sum_{n=0}^{\infty} \frac{\chi_{mn}(x_2, y_2)\lambda_{mn}^x(x_1, y_1)}{\rho h l_x l_y [\omega_{mn}^2\,(1+j\eta) - \omega^2]} F_x(x_1, y_1) \tag{9.76}$$

where

$$\chi_{mn}(x, y) = \frac{1}{2}\left(\frac{\partial\lambda_{mn}^y(x, y)}{\partial x} - \frac{\partial\lambda_{mn}^x(x, y)}{\partial y}\right) \tag{9.77}$$

The in-plane velocities $\dot{U}(x_2, y_2)$ and $\dot{V}(x_2, y_2)$ due to moment excitation $M_z(x_1, y_1)$ are given by

$$\dot{U}(x_2, y_2) = j\omega \sum_{m=0}^{\infty} \sum_{n=0}^{\infty} \frac{\lambda_{mn}^x(x_2, y_2)\chi_{mn}(x_1, y_1)}{\rho h l_x l_y [\omega_{mn}^2(1+j\eta) - \omega^2]} M_z(x_1, y_1) \tag{9.78}$$

$$\dot{V}(x_2, y_2) = j\omega \sum_{m=0}^{\infty} \sum_{n=0}^{\infty} \frac{\lambda_{mn}^y(x_2, y_2)\chi_{mn}(x_1, y_1)}{\rho h l_x l_y [\omega_{mn}^2(1+j\eta) - \omega^2]} M_z(x_1, y_1) \tag{9.79}$$

and the angular velocity $\dot{\Theta}_z(x_2, y_2)$ due to a point moment $M_z(x_1, y_1)$ is given by

$$\dot{\Theta}(x_2, y_2) = j\omega \sum_{m=0}^{\infty} \sum_{n=0}^{\infty} \frac{\chi_{mn}(x_2, y_2)\chi_{mn}(x_1, y_1)}{\rho h l_x l_y [\omega_{mn}^2(1+j\eta) - \omega^2]} M_z(x_1, y_1) \tag{9.80}$$

Using equations (9.75), (9.76), (9.78), (9.79) and (9.80), the nine mobility expressions can be derived to give

$$Y_{U,F_x} = j\omega \sum_{m=0}^{\infty} \sum_{n=0}^{\infty} \frac{\lambda_{mn}^x(x_2, y_2)\lambda_{mn}^x(x_1, y_1)}{\rho h l_x l_y [\omega_{mn}^2(1+j\eta) - \omega^2]} \tag{9.81a}$$

$$Y_{U,F_y} = j\omega \sum_{m=0}^{\infty} \sum_{n=0}^{\infty} \frac{\lambda_{mn}^x(x_2, y_2)\lambda_{mn}^y(x_1, y_1)}{\rho h l_x l_y [\omega_{mn}^2(1+j\eta) - \omega^2]} \tag{9.81b}$$

$$Y_{U,M_z} = j\omega \sum_{m=0}^{\infty} \sum_{n=0}^{\infty} \frac{\lambda_{mn}^x(x_2, y_2)\chi_{mn}(x_1, y_1)}{\rho h l_x l_y [\omega_{mn}^2(1+j\eta) - \omega^2]} \tag{9.81c}$$

$$Y_{V,F_x} = j\omega \sum_{m=0}^{\infty} \sum_{n=0}^{\infty} \frac{\lambda_{mn}^y(x_2, y_2)\lambda_{mn}^x(x_1, y_1)}{\rho h l_x l_y [\omega_{mn}^2(1+j\eta) - \omega^2]} \tag{9.81d}$$

$$Y_{V,F_y} = j\omega \sum_{m=0}^{\infty} \sum_{n=0}^{\infty} \frac{\lambda_{mn}^y(x_2, y_2)\lambda_{mn}^y(x_1, y_1)}{\rho h l_x l_y [\omega_{mn}^2(1+j\eta) - \omega^2]} \tag{9.81e}$$

$$Y_{V,M_z} = j\omega \sum_{m=0}^{\infty} \sum_{n=0}^{\infty} \frac{\lambda_{mn}^y(x_2, y_2)\chi_{mn}(x_1, y_1)}{\rho h l_x l_y [\omega_{mn}^2(1+j\eta) - \omega^2]} \tag{9.81f}$$

$$Y_{\Theta_z, F_x} = j\omega \sum_{m=0}^{\infty} \sum_{n=0}^{\infty} \frac{\chi_{mn}(x_2, y_2)\lambda_{mn}^x(x_1, y_1)}{\rho h l_x l_y [\omega_{mn}^2(1 + j\eta) - \omega^2]} \tag{9.81g}$$

$$Y_{\Theta_z, F_y} = j\omega \sum_{m=0}^{\infty} \sum_{n=0}^{\infty} \frac{\chi_{mn}(x_2, y_2)\lambda_{mn}^y(x_1, y_1)}{\rho h l_x l_y [\omega_{mn}^2(1 + j\eta) - \omega^2]} \tag{9.81h}$$

$$Y_{\Theta_z, M_z} = j\omega \sum_{m=0}^{\infty} \sum_{n=0}^{\infty} \frac{\chi_{mn}(x_2, y_2)\chi_{mn}(x_1, y_1)}{\rho h l_x l_y [\omega_{mn}^2(1 + j\eta) - \omega^2]} \tag{9.81i}$$

The driving point mobility terms can be evaluated by simply evaluating the expressions at (x_1, y_1).

References

Bauer, B.B. (1953) Transformer couplings for equivalent network synthesis. *Journal of the Acoustical Society of America*, **25**(5), 837–840.

Beranek, L.L. (1954) *Acoustics*. Ch. 3, Electro-Mechano-Acoustical Circuits. New York: McGraw-Hill.

Bishop, R.E.D. and Johnson, D.C. (1960) *The Mechanics of Vibration*. Cambridge University Press.

Bloch, A. (1945) Electromechanical analogies and their use for the analysis of mechanical and electromechanical systems. *Journal of the Institution of Electrical Engineers*, **92**, 157–169.

Bonello, P. and Brennan, M.J. (2001) Modelling the dynamic behaviour of a super-critical rotor on a flexible foundation using the mechanical impedance technique. *Journal of Sound and Vibration*, **239**(3), 445–466.

Brennan, M.J. and Dayou, J. (2000). Global control of vibration using a tunable vibration neutraliser. *Journal of Sound and Vibration*, **232**(3), 587–602.

Crandall, S.H. and Mark, W.D. (1963) *Random Vibration in Mechanical Systems*. New York: Academic Press.

Cremer, L., Heckl, M. and Ungar, E.E. (1988) *Structure-Borne Sound*. Berlin, Heidelberg, New York: Springer-Verlag (2nd edn).

Darrieus, M. (1929) Les modéles mécaniques en électrotechnique leur application aux problémes de stabilité. *Bull. Soc. Fren.*, **96**, 794–809.

Ewins, D.J. (2000) *Modal Testing: Theory and Practice*. Letchworth, Hertfordshire, England: Research Studies Press.

Firestone, F.A. (1933) A new analogy between mechanical and electrical systems. *Journal of the Acoustical Society of America*, **4**, 249–267.

Firestone, F.A. (1938) The mobility method of computing the vibration of linear mechanical and acoustical systems: mechanical-electrical analogies. *Journal of Applied Physics*, **9**, 373–387.

Firestone, F.A. (1956) Twixt earth and sky with road and tube; the mobility and classical impedance analogies. *Journal of the Acoustical Society of America*, **28**(6), 1117–1153.

Gardonio, P. and Brennan, M.J. (2002) On the origins and development of mobility and impedance methods in structural dynamics. *Journal of Sound and Vibration*, **249**(3), 557–573.

Gardonio, P. and Elliott, S.J. (1998) Driving point and transfer mobility matrices for thin plates excited in flexure. *ISVR Technical Report*, No. 277.

Gardonio, P. and Elliott, S.J. (1999) Active control of structure-borne and air-borne sound transmission through double panels. *Journal of Aircraft*, 36(6), 1023–1032.

Gardonio, P. and Elliott, S.J. (2000) Passive and active isolation of structural vibration transmission between two plates connected by a set of mounts. *Journal of Sound and Vibration*, 237(3), 483–511.

Gardonio, P., Elliott, S.J. and Pinnington, R.J. (1997) Active isolation of structural vibration on multiple degree of freedom systems. Part I: Dynamics of the system. *Journal of Sound and Vibration*, 207(1), 61–93.

Guillemin, E.A. (1953) *Introductory Circuit Theory*. New York: McGraw-Hill.

Hähnle, W. (1932) Die darstellung elektromechanischer gebilde durch rein elektrische schaltbilder. *Wiss Verloff and Siemens-Kohzerh*, 11(1), 1–23.

Herbert, R.J. (1932) A mechanical analogy for coupled electrical circuits. *Review of Scientific Instruments*, 3, 287.

Hixson, E.L. (1976) Chapter 10 in *Shock and Vibration Handbook*. C.M. Harris and C.E. Crede (eds). New York: McGraw-Hill.

Hurty, W.C. and Rubinstein, M.F. (1964) *Dynamics of Structures*. Englewood Cliffs, NJ Prentice-Hall.

Kim, S.M. and Brennan, M.J. (1999) A compact matrix formulation using the impedance and mobility approach for the analysis of structural-acoustic systems. *Journal of Sound and Vibration*, 223(1), 97–113.

Koenig, H.E. and Blackwell, W.A. (1961) *Electromechanical System Theory*. New York: McGraw-Hill.

le Corbeiller P. (1929) Origine des terms gyroscopiques. *Annales des Postes Télégraphes*, 18, 1–22.

Ljunggren, S. (1983) Generation of waves in an elastic plate by a vertical force and by a moment in the vertical plane. *Journal of Sound and Vibration*, 90(4), 559–584.

Ljunggren, S. (1984) Generation of waves in an elastic plate by a torsional moment and a horizontal force. *Journal of Sound and Vibration*, 93(2), 161–187.

Mason, W.P. (1942) *Electromechanical Transducers and Wave Filters*. London: Van Nostrad Reinhold Company.

Maxfield, J.P. and Harrison, H.C. (1926) Methods of high quality recording and reproducing of music and speech based on telephone research. *Transaction of the American Institute Electrical Engineers*, 45, 334–348 (also in *Bell System Technical Journal*, 5(3), 493–523).

Mead, D.J. (1986) Chapter 9 in *Noise and Vibration*. R.G. White and J.G. Walker (eds). Ellis Horwood Publishers.

Nickle, C.A. (1925) Oscillographic solution of electromechanical systems. *Transactions A.I.E.E.*, 44, 844–856.

O'Hara, G.J. (1966) Mechanical impedance and mobility concepts. *Journal of the Acoustical Society of America*, 41(5), 1180–1184.

Olson, H.F. (1943) *Dynamical Analogies*. Science Park Road, State College, PA 16801: Jostens Printing and Publishing Company.

Pawley, M. (1937) The design of a mechanical analogy for the general linear electrical network with lumped parameters. *Franklin Institute Journal*, 223, 179–198.

Plunket, R. (1958) *Colloquium on Mechanical Impedance Methods for Mechanical Vibrations*. The American Society of Mechanical Engineers. Presented at the ASME Annual Meeting, New York, 2 December.

Reddy, J.N. (1984) *Energy and Variational Methods in Applied Mechanics*. New York: John Wiley & Sons.

Rubin, S. (1967) Mechanical immitance- and transmission matrix concepts. *Journal of the Acoustical Society of America*, **41**(5), 1171–1179.

Skudrzyk, E. (1968) *Simple and Complex Vibratory Systems*. London: The Pennsylvania State University Press.

Soedel, W. (1993) *Vibrations of Shells and Plates*. New York: Marcel Dekker Inc (2nd edn).

Thomson, W.T. (1993) *Theory of Vibration*. Englewood Cliffs: Prentice-Hall.

Ungar, E.E. and Dietrich, C.W. (1966) High-frequency vibration isolation. *Journal of Sound and Vibration*, **4**(2), 224–241.

Warburton, G.B. (1951) The vibration of rectangular plates. *Proceedings of the Institute of Mechanical Engineering*, **168**, 371–384.

Webster, A.G. (1919) Acoustical impedance and theory of horns and of the phonograph. *Proceedings of the National Academy of Sciences* (Washington), **5**, 275–282.

Wegel, R.L. (1921) Theory of magneto-mechanical systems as applied to telephone receivers and similar structures. *Journal American Institute Electrical Engineering*, **40**, 791–802.

Wylie, C.R. and Barrett, L.C. (1982) *Advanced Engineering Mathematics*. New York: McGraw-Hill.

Chapter 10

Finite element techniques for structural vibration

M. Petyt and P. Gardonio

10.1 Introduction

The dynamic response of simple structures, such as uniform beams, plates and cylindrical shells, may be obtained by solving their equations of motion. However, in many practical situations either the geometrical or material properties vary, or the shape of the boundaries cannot be described in terms of known mathematical functions. Also, practical structures consist of an assemblage of components of different types, namely beams, plates, shells and solids. In these situations it is impossible to obtain analytical solutions to the equations of motion. This difficulty is overcome by seeking some form of approximate solution. There are a number of techniques available for determining approximate solutions, the most widely used one being the Rayleigh-Ritz method. An extension of this method, which is known as the 'Finite element displacement method', can be used to analyse any structure, however complex. The salient features of the method are described. Fuller details can be found in a number of references (Ashwell and Gallagher 1976; Weaver and Johnston 1987; Cook 1989; Zienkiewicz and Taylor 1989a,b; Cheung and Tham 1997; Petyt 1998).

10.2 Formulation of the equations of motion

The first step in the analysis of any linear structural vibration problem is the formulation of the equations of motion. There are a number of ways of doing this, but the simplest, since it is less prone to error, is to use Hamilton's Principle. This principle states that among all displacements which satisfy the prescribed (geometric) boundary conditions and the prescribed conditions at times $t = t_1$ and $t = t_2$, the actual solution satisfies the equation

$$\int_{t_1}^{t_2} [\delta (T - U) + \delta W] \, \mathrm{d}t = 0 \tag{10.1}$$

where T denotes the kinetic energy of the system, U the strain energy, δW the virtual work done by the non-conservative forces (e.g. external forces and friction forces) and δ denotes the first variation of a quantity. Equation (10.1) can be applied to both continuous and discrete dynamic systems.

In the case of a multi-degrees-of-freedom system, the deformation of which is described by n independent displacements $q_1, q_2, \ldots q_n$, the condition that Hamilton's integral is satisfied reduces to

$$\frac{d}{dt}\left(\frac{\partial T}{\partial \dot{q}_i}\right) + \left(\frac{\partial D}{\partial \dot{q}_i}\right) + \left(\frac{\partial U}{\partial q_i}\right) = Q_i \qquad i = 1, 2, \ldots n \qquad (10.2)$$

which are known as Lagrange's equations. In these equations, D is a dissipation function which represents the instantaneous rate of energy dissipation caused by the viscous damping forces and Q_i is a generalised force which is equal to the work done by the applied forces when the component q_i undergoes a unit displacement.

10.3 Rayleigh-Ritz method

Consider a one-dimensional structure whose axial coordinate is x and displacement component $u(x, t)$. The Rayleigh-Ritz method approximates the solution by a finite series expansion of the form

$$u(x, t) = \sum_{i=1}^{n} d_i(x) q_i(t) \qquad (10.3)$$

where the $d_i(x)$ are prescribed functions of x and the $q_i(t)$ unknown functions of t.

A continuous deformable body consists of an infinity of points, and therefore it has infinitely many degrees of freedom. By assuming that the motion is given by the expression (10.3), the continuous system has been reduced to an equivalent, stiffer system, with a finite number of degrees of freedom, since according to equation (10.3) $q_{n+1} = q_{n+2} = \cdots = 0$. Substituting (10.3) into the appropriate energy expressions T, D, U and δW reduces the continuous structure to a multi-degrees-of-freedom system with $q_1, q_2, \ldots q_n$ as degrees of freedom. Substituting into equation (10.2) gives the equations of motion as

$$[M]\{\ddot{q}\} + [C]\{\dot{q}\} + [K]\{q\} = \{Q\} \qquad (10.4)$$

where $[M]$ is the inertia matrix, $[C]$ the damping matrix and $[K]$ the stiffness matrix. $\{Q\}$ is a column matrix of generalised forces corresponding to the generalised displacements $\{q\}$.

Equation (10.4) is solved for the unknowns $\{q\}$, which are then substituted into equation (10.3) to give the required approximate solution. The accuracy of the solution obtained can be increased by increasing the number of terms in equation (10.3), provided the prescribed functions $d_i(x)$ satisfy certain criteria. If the integral in equation (10.1) involves derivatives up to order p, then the functions $d_i(x)$ must satisfy the following conditions (Meirovitch 1967):

(i)   be linearly independent;
(ii) be p times differentiable;
(iii) satisfy the geometric boundary conditions (these will involve derivatives up to order $(p-1)$);
(iv) form a complete series.

Polynomials in powers of x, trigonometric functions, Legendre, Tchebycheff and Jacobi or hypergeometric polynomials are all complete series of functions. It should be noted that in using the Rayleigh-Ritz method the equations of motion and natural boundary conditions will only be satisfied approximately.

Example

Calculation of the lower natural frequencies and natural modes of the undamped fixed-free rod shown in Figure 10.1 using the Rayleigh-Ritz method. The natural frequencies and associated natural modes are derived by considering free undamped axial vibrations of the rod in Figure 10.1. In this case, $\delta W = 0$ so that only the kinetic energy and strain energy expressions are required in equation (10.1). The kinetic and strain energy quantities relative to axial vibration of a beam are given by the following expressions (Petyt 1998):

$$T_{\text{rod}} = \frac{1}{2} \int_0^L \rho A \left(\frac{\partial u(x, t)}{\partial t} \right)^2 dx \tag{10.5}$$

$$U_{\text{rod}} = \frac{1}{2} \int_0^L EA \left(\frac{\partial u(x, t)}{\partial x} \right)^2 dx \tag{10.6}$$

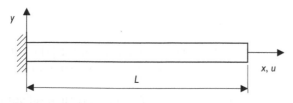

Figure 10.1 Clamped-free rod (axial vibration).

where E is the Young's modulus of elasticity, ρ is the density of the material and A is the cross-sectional area of the rod. According to these two energy expressions, the integral in equation (10.1) involves derivatives up to the order $p = 1$. Therefore, the prescribed functions $d_i(x)$ to be used in the finite series expansion of equation (10.3) must be at least once differentiable. Also, these functions must be linearly independent, form a complete series and satisfy the geometric boundary condition; that is, $u(0, t) = 0$. A set of functions that satisfy all these conditions is, for example, a polynomial of order $i \geq 1$

$$d_i(x) = x^i \tag{10.7}$$

When the displacement of the rod is approximated by a finite expansion, as in equation (10.3), the mathematical formulation of the kinetic energy and strain energy, as given in equations (10.5) and (10.6), is reduced to that of a discrete system with a finite number of degrees of freedom (the $q_i(t)$ unknown functions of t). If the finite series expansion of equation (10.3) is expressed in matrix form as follows,

$$u(x, t) = \lfloor \mathbf{d}(x) \rfloor \{\mathbf{q}(t)\} = \lfloor d_1(x) d_2(x) \cdots d_n(x) \rfloor \begin{bmatrix} q_1(t) \\ q_2(t) \\ \vdots \\ q_n(t) \end{bmatrix} \tag{10.8}$$

then it can be shown that the two energy expressions can be expressed in the following form:

$$T_{\text{rod}} = \frac{1}{2} \{\dot{\mathbf{q}}(t)\}^{\text{T}} [\mathbf{M}] \{\dot{\mathbf{q}}(t)\} \tag{10.9}$$

$$U_{\text{rod}} = \frac{1}{2} \{\mathbf{q}(t)\}^{\text{T}} [\mathbf{K}] \{\mathbf{q}(t)\} \tag{10.10}$$

where $[\mathbf{M}]$ and $[\mathbf{K}]$ are respectively the inertia and the stiffness matrices which, for axial vibration of a rod, are given by the following expressions:

$$[\mathbf{M}] = \rho A \int_0^L \lfloor \mathbf{d}(x) \rfloor^{\text{T}} \lfloor \mathbf{d}(x) \rfloor \, \mathrm{d}x \tag{10.11}$$

with

$$M_{ik} = \int_0^L \rho A d_i(x) d_k(x) \, \mathrm{d}x = \frac{1}{(i+k+1)} \rho A L^{i+k+1} \tag{10.12}$$

and

$$[\mathbf{K}] = EA \int_0^L \left\lfloor \frac{\partial \mathbf{d}(x)}{\partial x} \right\rfloor^{\text{T}} \left\lfloor \frac{\partial \mathbf{d}(x)}{\partial x} \right\rfloor \, \mathrm{d}x \tag{10.13}$$

with

$$K_{ik} = \int_0^L EA d_i'(x) d_k'(x)\, dx = \frac{ik}{(i+k-1)} EA L^{i+k-1} \tag{10.14}$$

Assuming the dissipation function and generalised forces to be zero, equation (10.4) reduces to the following system of n ordinary differential equations

$$[\mathbf{M}]\{\ddot{\mathbf{q}}\} + [\mathbf{K}]\{\mathbf{q}\} = \{0\} \tag{10.15}$$

Assuming harmonic motion in the form $\{\mathbf{q}\} = \{\mathbf{A}\}\sin\omega t$, where ω is the circular frequency and the amplitudes $\{\mathbf{A}\}$ are independent of time, equation (10.15) gives

$$[\mathbf{K} - \omega^2\mathbf{M}]\{\mathbf{A}\} = \{0\} \tag{10.16}$$

which represent a set of n linear homogeneous equations in the unknowns $A_1, A_2, \ldots, A_n$. The condition for these equations to have a non-zero solution is that the determinant of coefficients should vanish $|\mathbf{K} - \omega^2\mathbf{M}| = 0$. This condition leads to a polynomial of degree n in ω^2. The n roots of this polynomial $\omega_1^2, \omega_2^2, \ldots, \omega_n^2$, which are all real and positive (Bishop *et al.* 1965), are indeed the squares of the approximate values of the first n natural frequencies of the system. These approximate natural frequencies will be greater than the true natural frequencies of the system. This is because the expressions for the strain energy and kinetic energy have been calculated for an equivalent *stiffer rod* with only n degrees of freedom (in fact the remaining degrees of freedoms have been set equal to zero).

Corresponding to each natural frequency ω_k, a unique solution exists (to within an arbitrary constant) to equation (10.16) for $\{\mathbf{A}\}_k$ which, when combined with the prescribed functions $d_i(x)$, gives the shapes of the natural modes of vibration in an approximate sense. The approximate shape of the kth natural mode is therefore given by

$$\phi_k(x) = \sum_{i=1}^n d_i(x) A_{i,k} = \lfloor d_1(x) d_2(x) \cdots d_n(x) \rfloor \begin{bmatrix} A_1 \\ A_2 \\ \vdots \\ A_n \end{bmatrix}_k \tag{10.17}$$

Convergence to the true frequencies and mode shapes is obtained as the number of terms in the approximating expression (10.3) is increased.

If only one term is used for the prescribed functions used in the approximate expression of the displacement for the rod of Figure 10.1, i.e. $d_1(x) = x$, equation (10.16) reduces to

$$\left(EAL - \omega_1^2 \frac{\rho A L^3}{3} \right) A_{1,1} = 0 \tag{10.18}$$

whose solution gives $\omega_1 = 1.732(E/\rho L^2)^{1/2}$. If the number of terms in the series (10.3) is increased to two, i.e. $d_1(x) = x$ and $d_2(x) = x^2$, then equation (10.16) reduces to

$$\left[EA \begin{bmatrix} L & L^2 \\ L^2 & \dfrac{4L^3}{3} \end{bmatrix} - \omega^2 \rho A \begin{bmatrix} \dfrac{L^3}{3} & \dfrac{L^4}{4} \\ \dfrac{L^4}{4} & \dfrac{L^5}{5} \end{bmatrix} \right] \begin{bmatrix} A_1 \\ A_2 \end{bmatrix} = \begin{bmatrix} 0 \\ 0 \end{bmatrix} \tag{10.19}$$

Letting $\omega^2 \rho L^2 / E = \lambda$, the above equation simplifies to

$$\begin{bmatrix} \left(1 - \dfrac{\lambda}{3}\right) & \left(1 - \dfrac{\lambda}{4}\right) \\ \left(1 - \dfrac{\lambda}{4}\right) & \left(\dfrac{4}{3} - \dfrac{\lambda}{5}\right) \end{bmatrix} \begin{bmatrix} A_1 \\ A_2 L \end{bmatrix} = \begin{bmatrix} 0 \\ 0 \end{bmatrix} \tag{10.20}$$

This equation has a non-zero solution, provided,

$$\det\left(\begin{bmatrix} \left(1 - \dfrac{\lambda}{3}\right) & \left(1 - \dfrac{\lambda}{4}\right) \\ \left(1 - \dfrac{\lambda}{4}\right) & \left(\dfrac{4}{3} - \dfrac{\lambda}{5}\right) \end{bmatrix} \right) = 0 \tag{10.21}$$

which yields

$$\frac{\lambda^2}{240} - \frac{13\lambda}{90} + \frac{1}{3} = 0 \tag{10.22}$$

The two roots of this equation are $\lambda_1 = 2.486$ and $\lambda_2 = 32.18$, and the first two natural frequencies can be calculated from the relation

$$\omega_k = \lambda_k^{\frac{1}{2}} \left(\frac{E}{\rho L^2} \right)^{\frac{1}{2}} \tag{10.23}$$

where $k = 1, 2$. From the homogeneous equations in equation (10.16) the following relation is derived:

$$A_{2,k} = -\frac{\left(1 - \dfrac{\lambda}{3}\right)}{\left(1 - \dfrac{\lambda}{4}\right) L} A_{1,k} \tag{10.24}$$

with $k = 1, 2$. Therefore, assuming the first term to be equal to one $A_{1,k} = 1$, the two vectors $\{A\}$ are given by:

$$\lambda_1 = 2.486, \qquad \left\{ \begin{array}{c} A_{1,1} \\ A_{2,1} \end{array} \right\} = \left\{ \begin{array}{c} 1 \\ \dfrac{-0.4527}{L} \end{array} \right\} \tag{10.25}$$

$$\lambda_2 = 32.18, \qquad \left\{ \begin{array}{c} A_{1,2} \\ A_{2,2} \end{array} \right\} = \left\{ \begin{array}{c} 1 \\ \dfrac{-1.3806}{L} \end{array} \right\} \tag{10.26}$$

In conclusion, the two natural frequencies and mode shapes of the fixed-free rod are given by the following expressions:

$$\omega_1 = 1.577 \left(\frac{E}{\rho L^2} \right)^{\frac{1}{2}}, \quad \phi_1(x) = L \left\{ \frac{x}{L} - 0.4527 \left(\frac{x}{L} \right)^2 \right\} \tag{10.27a,b}$$

$$\omega_2 = 5.673 \left(\frac{E}{\rho L^2} \right)^{\frac{1}{2}}, \quad \phi_2(x) = L \left\{ \frac{x}{L} - 1.3806 \left(\frac{x}{L} \right)^2 \right\} \tag{10.28a,b}$$

Table 10.1 Comparison of approximate frequencies with exact solution for a rod

Mode	R–R solutions		Exact solution
	I term	2 terms	
1	1.732	1.577	1.571
2	–	5.673	4.712

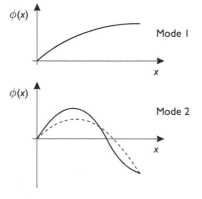

Figure 10.2 Axial modes of vibration of a rod (——— exact; - - - - - approximate (R–R)).

The approximate values of $\lambda_{k=1,2}$ are compared with the exact values in Table 10.1. The natural frequencies are greater than the exact ones and approach the exact ones as the number of terms is increased.

The approximate mode shapes for the two terms solution are compared with the exact mode shapes in Figure 10.2. The difference between the approximate and exact shapes for the first natural mode is too small to show up on the scale used.

10.4   Finite element method

The major drawback to the Rayleigh-Ritz method is the difficulty in constructing a set of prescribed functions, particularly for a built-up structure. This difficulty can be overcome by using the finite element displacement method that provides an automatic procedure for constructing such functions. In the finite element displacement method the prescribed functions are constructed in the following manner:

1. select a set of reference or 'node points' on the structure;
2. associate with each node point a given number of degrees of freedom (displacement, slope, etc.);
3. construct a set of functions such that each one gives a unit value for one degree of freedom and zero values for all the others.

This procedure is illustrated for the axial motion of a rod in Figure 10.3, where five node points have been selected at equal intervals. The portion of the rod between two adjacent nodes is called an 'element'. The highest derivative appearing in the energy expressions for a rod is the first (see equations (10.5) and (10.6)). Therefore, only the prescribed functions themselves need to be continuous, and so the axial displacement $u(x, t)$ is the only degree of freedom required at each node point. Five prescribed functions are illustrated in Figure 10.3. They have been constructed by giving each node point in turn a unit displacement, whilst maintaining zero displacement at all other nodes. The geometric boundary conditions for the particular problem to be analysed can be satisfied by omitting any of the functions constructed which do not satisfy them. For example, if node 1 is fixed, then the function $\phi_1(x)$ is omitted. These functions are referred to as 'element displacement functions' or 'shape functions'. In order to satisfy the convergence criteria of the Rayleigh-Ritz method, the element displacement functions should satisfy the following conditions:

 (i) be linearly independent;
(ii) be a continuous function which is p times differentiable within the element and whose derivatives up to order $(p - 1)$ are continuous across element boundaries. An element which satisfies this condition is referred to as a 'conforming' element;

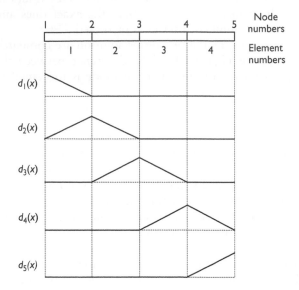

Figure 10.3 Prescribed functions of a rod.

(iii) if the element displacement functions are polynomial functions, then they must be complete polynomials of at least degree p. If any terms of degree greater than p are used, they need not be complete. However, the rate of convergence to the exact solution of the dynamic problem under study is governed by the order of completeness of the polynomial. The element displacement functions need not be polynomials;

(iv) satisfy the geometric boundary conditions.

In the Rayleigh-Ritz method, convergence is obtained as the number of prescribed functions is increased. To increase the number of prescribed functions in the finite element method, the number of node points, and therefore the number of elements, is increased (Oden 1972).

It can be seen in Figure 10.3 that each element deforms in only two deformation patterns and that the deformation patterns for each element are the same. Because of this, the emphasis is on determining deformation patterns for individual elements and not for the whole structure. There are a number of ways of determining the displacement function of an element. The most common of these are as follows:

(i) by inspection;

(ii) by assuming a polynomial function having the appropriate number of terms;

(iii) by solving the equations of static equilibrium to determine the deformation of the element due to prescribed boundary displacement.

In practice, the most appropriate method is used for each type of element.
Petyt (1998) showed that all these approaches result in the axial displacement
variation of a rod being a polynomial function of the first order. Therefore,
the displacement function for a rod element, as shown in Figure 10.4, can be
expressed in terms of the dimensionless coordinate $\xi = x/a$ as follows:

$$u(\xi) = \alpha_1 + \alpha_2 \xi \tag{10.29}$$

Evaluating this function at $\xi = \mp 1$ gives $u_1 = \alpha_1 - \alpha_2$ and $u_2 = \alpha_1 + \alpha_2$,
where u_1 and u_2 are axial displacements of nodes 1 and 2. Solving for α_1
and α_2 gives

$$\alpha_1 = \frac{1}{2}(u_1 + u_2), \qquad \alpha_2 = \frac{1}{2}(u_1 - u_2) \tag{10.30a,b}$$

Therefore the displacement variation for the rod element is given by

$$u(\xi, t) = \frac{1}{2}(1 - \xi) u_1(t) + \frac{1}{2}(1 + \xi) u_2(t) \tag{10.31}$$

This expression can be rewritten in the following matrix form:

$$u(\xi, t) = \lfloor N_1(\xi) N_2(\xi) \rfloor \begin{bmatrix} u_1(t) \\ u_2(t) \end{bmatrix} = \lfloor \mathbf{N}(\xi) \rfloor \{\mathbf{u}(t)\}_e \tag{10.32}$$

where

$$N_1(\xi) = \frac{1}{2}(1 - \xi) \qquad N_2(\xi) = \frac{1}{2}(1 + \xi) \tag{10.33a,b}$$

Once the matrix relation in equation (10.32) has been found for each
element, it is relatively simple to evaluate the energy expressions (10.5) and

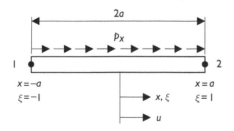

Figure 10.4 Geometry of a single axial element.

(10.6) by calculating the kinetic and strain energies for each element and then adding the contributions from the elements together. When approached in this way, there is no reason why the elements should be identical. For example, in Figure 10.3, the elements could have different cross-sectional areas and densities as well as different lengths. The energy expressions and virtual work for a single element are

$$T_e = \frac{1}{2} \int_{-1}^{+1} \rho A \dot{u}^2 a \, d\xi \tag{10.34}$$

$$U_e = \frac{1}{2} \int_{-1}^{+1} EA \frac{1}{a^2} \left(\frac{\partial u}{\partial \xi} \right)^2 a \, d\xi \tag{10.35}$$

$$\delta W_e = \frac{1}{2} \int_{-1}^{+1} p_x \delta u a \, d\xi \tag{10.36}$$

where A is the cross-sectional area, E is Young's modulus, ρ is the density and p_x is the distributed axial load per unit length. Substituting equation (10.32) into equations (10.34), (10.35) and (10.36) gives

$$T_e = \frac{1}{2} \{\dot{u}\}_e^T [m]_e \{\dot{u}\}_e \tag{10.37a}$$

$$U_e = \frac{1}{2} \{u\}_e^T [k]_e \{u\}_e \tag{10.37b}$$

$$\delta W_e = \frac{1}{2} \{\delta u\}_e^T \{f\}_e \tag{10.37c}$$

where

$$[m]_e = \rho A a \int_{-1}^{+1} \lfloor N(\xi) \rfloor^T \lfloor N(\xi) \rfloor \, d\xi = \rho A a \begin{bmatrix} \frac{2}{3} & \frac{1}{3} \\ \frac{1}{3} & \frac{2}{3} \end{bmatrix} \tag{10.38}$$

is the element inertia matrix,

$$[k]_e = \frac{EA}{a} \int_{-1}^{+1} \lfloor N'(\xi) \rfloor^T \lfloor N'(\xi) \rfloor \, d\xi = \frac{EA}{a} \begin{bmatrix} \frac{1}{2} & -\frac{1}{2} \\ -\frac{1}{2} & \frac{1}{2} \end{bmatrix} \tag{10.39}$$

is the element stiffness matrix and

$$\{f\}_e = a \int_{-1}^{+1} p_x \lfloor N(\xi) \rfloor^T \, d\xi = a p_x^e \begin{bmatrix} 1 \\ 1 \end{bmatrix} \tag{10.40}$$

is the element load matrix, for $p_x = p_x^e$ (a constant) over the element.

The energy expressions for the complete rod are obtained by adding together the energies for all the individual elements. Before carrying this out it is necessary to relate the degrees of freedom of a single element $\{u\}_e$ to the set of degrees of freedom for the complete rod $\{u\}$. For the rod shown in Figure 10.3 this is

$$\{u\}^T = \lfloor u_1 \quad u_2 \quad u_3 \quad u_4 \quad u_5 \rfloor \tag{10.41}$$

For element e the relationship is

$$\{u\}_e = [a]_e \{u\} \tag{10.42}$$

where, for the four elements, $[a]_e$ takes the form

$$[a]_1 = \begin{bmatrix} 1 & 0 & 0 & 0 & 0 \\ 0 & 1 & 0 & 0 & 0 \end{bmatrix} \qquad [a]_2 = \begin{bmatrix} 0 & 1 & 0 & 0 & 0 \\ 0 & 0 & 1 & 0 & 0 \end{bmatrix}$$

$$[a]_3 = \begin{bmatrix} 0 & 0 & 1 & 0 & 0 \\ 0 & 0 & 0 & 1 & 0 \end{bmatrix} \qquad [a]_4 = \begin{bmatrix} 0 & 0 & 0 & 1 & 0 \\ 0 & 0 & 0 & 0 & 1 \end{bmatrix} \tag{10.43}$$

For the kinetic energy, substituting equation (10.42) into equation (10.37a) and summing over all the elements gives

$$T - \frac{1}{2}\{\dot{u}\}^T [M] \{\dot{u}\} \tag{10.44}$$

where

$$[M] = \sum_{e=1}^{4} [a]_e^T [m]_e [a]_e \tag{10.45}$$

Using equations (10.38) and (10.43) in equation (10.45) gives

$$[M] = \frac{\rho A a}{3} \begin{bmatrix} 2 & 1 & 0 & 0 & 0 \\ 1 & 4 & 1 & 0 & 0 \\ 0 & 1 & 4 & 1 & 0 \\ 0 & 0 & 1 & 4 & 1 \\ 0 & 0 & 0 & 1 & 2 \end{bmatrix} \tag{10.46}$$

For the strain energy, substituting equation (10.42) into equation (10.37b) and summing over all the elements gives

$$U = \frac{1}{2}\{u\}^T [K] \{u\} \tag{10.47}$$

where

$$[K] = \sum_{e=1}^{4} [a]_e^T [k]_e [a]_e \tag{10.48}$$

Using equations (10.39) and (10.43) in equation (10.48) gives

$$[\mathbf{K}] = \frac{EA}{2a} \begin{bmatrix} 1 & -1 & 0 & 0 & 0 \\ -1 & 2 & -1 & 0 & 0 \\ 0 & -1 & 2 & -1 & 0 \\ 0 & 0 & -1 & 2 & -1 \\ 0 & 0 & 0 & -1 & 1 \end{bmatrix} \tag{10.49}$$

Finally, the matrix expression of the virtual work done by the external axial forces, as a function of the nodal displacements, can be derived by substituting equation (10.42) into equation (10.37c) and summing over all elements giving

$$\delta W = \{\delta \mathbf{u}\}^{\mathrm{T}} \{\mathbf{f}\} \tag{10.50}$$

where

$$\{\mathbf{f}\} = \sum_{e=1}^{4} [\mathbf{a}]_e^{\mathrm{T}} \{\mathbf{f}\}_e \tag{10.51}$$

is a column matrix of equivalent nodal forces. Using equations (10.40) and (10.43) in equation (10.51) gives

$$\{\mathbf{f}\} = a \begin{bmatrix} p_x^1 \\ p_x^1 + p_x^2 \\ p_x^2 + p_x^3 \\ p_x^3 + p_x^4 \\ p_x^4 \end{bmatrix} \tag{10.52}$$

The matrix products $[\mathbf{a}]_e^{\mathrm{T}} [\mathbf{m}]_e [\mathbf{a}]_e$, in equation (10.45), and $[\mathbf{a}]_e^{\mathrm{T}} [\mathbf{k}]_e [\mathbf{a}]_e$, in equation (10.48), effectively locate the positions in $[\mathbf{M}]$ and $[\mathbf{K}]$ to which the elements of $[\mathbf{m}]_e$ and $[\mathbf{k}]_e$, respectively, have to be added. However, in practice, it is unnecessary to form the four matrices $[\mathbf{a}]_e$ and carry out the matrix multiplication. The information required to form the matrices $[\mathbf{M}]$ and $[\mathbf{K}]$ can in fact be obtained from the indices given to the nodal points. Nodes e and $(e+1)$ correspond to the element e. Therefore, the two rows and columns of the element inertia or stiffness matrices (equations (10.38) and (10.39)) are added into the rows and the columns e and $(e+1)$ of the inertia and stiffness matrices respectively for the complete rod. This procedure is known as the 'assembly process'.

Also, note that if the elements and nodes are numbered progressively, as in Figure 10.3, then the matrices $[\mathbf{M}]$ and $[\mathbf{K}]$ are symmetrical and banded. These properties can be exploited when writing computer programs. All the information displayed in equations (10.46) and (10.49) can be stored in nine locations, whereas 25 locations would be required to store the complete matrix $[\mathbf{M}]$ or $[\mathbf{K}]$.

Once the inertia and stiffness matrices and the vector of external excitations (equations (10.46), (10.49) and (10.52)) are derived, so that the kinetic and strain energies and the virtual work done by the external excitations are algebraically formulated, the geometric boundary conditions have to be satisfied. For example, if the finite element model of Figure 10.3 is used to analyse the dynamic response of a fixed-free rod, as in Figure 10.1, the nodal displacement u_1 is zero. This condition is simply introduced by omitting u_1 from the set of degrees of freedom for the complete rod (equation (10.41)) and at the same time omitting the first row and column from the inertia and stiffness matrices for the complete rod given by equations (10.46) and (10.49). Finally the first row of the column vector of excitations, equation (10.52), has to be removed.

As in the Rayleigh-Ritz method, the total energy expressions, equations (10.44), (10.47) and (10.50), are substituted into Lagrange's equation (10.2) to give the equations of motion in the form of equation (10.4), namely

$$[M]\{\ddot{u}\} + [C]\{\dot{u}\} + [K]\{u\} = \{f\} \tag{10.53}$$

where $\{u\}$ represents a column matrix of nodal degrees of freedom (axial displacements in this example, as shown in equation (10.41)).

It is not usual to derive the dissipation function for each element in the same manner as the kinetic and strain energies. This is because damping is not necessarily an inherent property of the vibrating structure. Damping forces depend not only on the structure itself, but also on the surrounding medium. Structural damping is caused by internal friction within the material and at joints between components. Viscous and sound radiation damping occurs when a structure is moving in air or other fluids. It is usual to use simplified damping models for the complete structure, rather than for individual elements. Suitable forms for the matrix [C] will be introduced in Sections 10.6.4 and 10.6.5 when discussing the solution of equation (10.53).

In the Rayleigh-Ritz method the accuracy of the solution is increased by increasing the number of prescribed functions in the assumed series. To increase the number of prescribed functions in the finite element method, the number of node points, and therefore the number of elements, is increased. The displacement function for each element should satisfy the Rayleigh-Ritz convergence criteria. In particular, the functions and their derivatives up to order $(p-1)$ should be continuous across element boundaries. This version of the finite element method is often referred to as the *h-method*, where *h* refers to the size of the element (Zienkiewicz and Taylor 1989a,b).

Example

Calculation of the lower natural frequencies and natural modes of the clamped-free rod shown in Figure 10.1 using the finite element method. As in the previous example, where the Rayleigh-Ritz approximate method was used, the natural frequencies and associated natural modes are derived by considering free undamped axial vibrations of the rod in Figure 10.1. Therefore, also in this analysis, the virtual work done by the dissipative and external forces is neglected, $\delta W = 0$, so that only the kinetic energy and strain energy expressions are derived and used in the Lagrange's formulation to derive the free axial vibration of the rod (see equation (10.15)).

Following the procedure already described and shown in Figure 10.3, the rod is divided into a set of n elements and only one degree of freedom, the axial displacement $u(x, t)$, is associated with the $(n+1)$ node points, as shown in Figure 10.4. The two displacement functions are taken to be polynomial functions so that, defining the dimensionless coordinate $\xi = x/a$, the displacement $u(\xi, t)$ within an element is related to the displacements at the two nodes u_1 and u_2, as in equation (10.32). Therefore the inertia and stiffness matrices associated with each element are given by equations (10.38) and (10.39). If only one element is used, the total kinetic and strain energy terms are given by the following relations:

$$T = \frac{1}{2} \lfloor \dot{u}_1 \ \dot{u}_2 \rfloor \frac{\rho A L}{6} \begin{bmatrix} 2 & 1 \\ 1 & 2 \end{bmatrix} \begin{bmatrix} \dot{u}_1 \\ \dot{u}_2 \end{bmatrix}, \qquad U = \frac{1}{2} \lfloor u_1 \ u_2 \rfloor \frac{EA}{L} \begin{bmatrix} 1 & -1 \\ -1 & 1 \end{bmatrix} \begin{bmatrix} u_1 \\ u_2 \end{bmatrix}$$

$$(10.54a,b)$$

where E is the Young's modulus of elasticity, ρ is the density of the material and L is the length of the beam. Imposing the geometric boundary condition at node one, so that $u_1 = 0$, substituting the two energy expressions into Lagrange's equations and, assuming harmonic motion of the form $u_2 = A_2 \sin \omega t$, where ω is the circular frequency, the following equation for free axial vibration is obtained:

$$\left[\frac{EA}{L} \cdot 1 - \omega^2 \frac{\rho A L}{6} \cdot 2 \right] A_2 = 0 \tag{10.55}$$

Thus, the first natural frequency of axial vibration of the rod in Figure 10.1 is given by

$$\omega_1 = 1.732 \left(\frac{E}{\rho L^2} \right)^{\frac{1}{2}} \tag{10.56}$$

If the rod is now divided into two elements of length $L/2$, the equation for free axial vibration becomes

$$\left[\frac{2EA}{L} \begin{bmatrix} 2 & -1 \\ -1 & 1 \end{bmatrix} - \omega^2 \begin{bmatrix} 4 & 1 \\ 1 & 2 \end{bmatrix} \right] \begin{bmatrix} A_2 \\ A_3 \end{bmatrix} = \begin{bmatrix} 0 \\ 0 \end{bmatrix} \tag{10.57}$$

Letting $(\omega^2 \rho L^2/24E) = \lambda$, the above equation simplifies to

$$\begin{bmatrix} (2-4\lambda) & -(1+\lambda) \\ -(1+\lambda) & (1-2\lambda) \end{bmatrix} \begin{bmatrix} A_2 \\ A_3 \end{bmatrix} = \begin{bmatrix} 0 \\ 0 \end{bmatrix} \tag{10.58}$$

This equation has a non-zero solution, provided that,

$$\det\left(\begin{bmatrix} (2-4\lambda) & -(1+\lambda) \\ -(1+\lambda) & (1-2\lambda) \end{bmatrix} \right) = 0 \tag{10.59}$$

which yields

$$7\lambda^2 - 10\lambda + 1 = 0 \tag{10.60}$$

The two roots of this equation are $\lambda_1 = 0.108$ and $\lambda_2 = 1.320$. Thus the two natural frequencies of axial vibration of the fixed-free rod are

$$\omega_1 = (24\lambda_1)^{\frac{1}{2}} \left(\frac{E}{\rho L^2} \right)^{\frac{1}{2}} = 1.610 \left(\frac{E}{\rho L^2} \right)^{\frac{1}{2}} \tag{10.61}$$

$$\omega_2 = (24\lambda_2)^{\frac{1}{2}} \left(\frac{E}{\rho L^2} \right)^{\frac{1}{2}} = 5.628 \left(\frac{E}{\rho L^2} \right)^{\frac{1}{2}} \tag{10.62}$$

From the homogeneous equation (10.58) the values of the eigenvectors are found to be

$$A_{2,k} = \frac{(1+\lambda)}{(2-4\lambda)L} A_{3,k} \tag{10.63}$$

with $k = 1, 2$. Therefore, assuming the third term equal to one $A_{3,k} = 1$, the two eigenvectors are given by

$$\lambda_1 = 0.108 \qquad \begin{bmatrix} A_{1,1} \\ A_{2,1} \\ A_{3,1} \end{bmatrix} = \begin{bmatrix} 0 \\ 0.707 \\ 1.0 \end{bmatrix} \tag{10.64}$$

$$\lambda_2 = 1.320 \qquad \begin{bmatrix} A_{1,2} \\ A_{2,2} \\ A_{3,2} \end{bmatrix} = \begin{bmatrix} 0 \\ -0.707 \\ 1.0 \end{bmatrix} \tag{10.65}$$

The approximate values of $\lambda_{k=1,2}$ are compared with the exact values in Table 10.2. The natural frequencies are greater than the exact ones since, as discussed for the Rayleigh-Ritz formulation, the approximate values of the total kinetic and strain energies refer to a stiffer rod with only one $(n = 1)$

Table 10.2 Comparison of approximate frequencies
with exact solution for a rod

Mode	FEM solution		Exact solution
	1 element	2 elements	
1	1.732	1.610	1.571
2	–	5.628	4.712

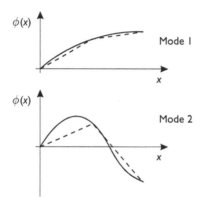

Figure 10.5 Axial modes of vibration of a rod (—— exact; - - - - approximate (FEM)).

or two ($n = 2$) degrees of freedom set to be free in the two cases, and all the other set to be equal to zero. Moreover, this example highlights the fact that the approximate prediction gets more and more accurate as the number of elements increases. For example, if four elements were used in the model the error of the approximate first natural frequency would be less than 1 per cent. Figure 10.5 shows the approximate shapes of the first two natural modes of vibration of the fixed-free rod derived with the finite element approach. According to equation (10.31), the displacement function within each element is linear; thus the mode shapes have been derived with straight lines connecting the amplitude of the displacements A_1, A_2 and A_3 at the three nodal points.

10.5   Finite element models

The inertia, stiffness and consistent load matrices can be derived for any finite element model once the energy and virtual work expressions, T, U, δW and the displacement functions $\lfloor \mathbf{N} \rfloor$ are known. In this section a number of common finite element models are reviewed.

10.5.1 Beams

Although the beam is one of the simplest of structural elements, it is difficult to model mathematically owing to the large number of different forms it can take. It is used by engineers to represent the behaviour of everything from a uniform, rectangular section beam to the hull of a ship.

The simplest form is a uniform, slender beam vibrating in one of its principal planes. The energy expressions for a single element of length $2a$ are

$$T_e = \frac{1}{2} \int_{-a}^{+a} \rho A \dot{v}^2 \, dx \tag{10.66}$$

$$U_e = \frac{1}{2} \int_{-a}^{+a} EI_z \left(\frac{\partial^2 v}{\partial x^2} \right)^2 \, dx \tag{10.67}$$

$$\delta W_e = \frac{1}{2} \int_{-a}^{+a} p_y \delta v \, dx \tag{10.68}$$

where A is the cross-sectional area, I_z is the second moment of area of the cross-section about the principal axis which is perpendicular to the plane of vibration, E is Young's modulus, ρ is the density, p_y is the distributed lateral load per unit length and v is the lateral displacement of the centroid. The highest derivative appearing in the energy expressions is the second. Hence, it is necessary to take v and $\theta_z = (\partial v/\partial x)$ as degrees of freedom at each node to ensure that v and its first derivative are continuous between elements. With four degrees of freedom (two at two nodes) it is possible to represent the displacement v by a cubic polynomial. Details of the element are given by Leckie and Lindberg (1963).

When analysing three-dimensional frameworks, each member is capable of axial deformation, bending in two principal planes and torsion about its axis. The axial and bending deformations are treated as described previously. The torsional deformation can be treated in a similar way to the axial deformation. After deriving the energy expressions for a single element with respect to local axes, the nodal components of displacement and rotation are transformed to global axes. The energies of each element are then added together as before.

When analysing deep beams at low frequencies and slender beams at high frequencies it is necessary to include the effect of shear deformation and rotary inertia in the energy expressions. A survey of methods of analysing such beams is given in Thomas et al. (1973). Other factors to be considered are the offset of the centroid from the node point and the offset of the shear centre from the centroid. In the case of thin-walled, open-section beams, the effect of warping of the cross-section must also be included in the energy expressions (Petyt 1977). Beams with variable cross-sections can also be analysed by treating A and I_z as functions of x in equations (10.66) and (10.67) (see, for example, Lindberg 1963). Tapered and twisted beam elements have also been developed for analysing turbine blades (Gupta and Rao 1978). Curved beam elements have been developed for analysing in-plane vibrations (Petyt and Fleischer 1971; Davis *et al.* 1972a) and out-of-plane vibrations (Davis *et al.* 1972b).

10.5.2 Membranes

Membrane elements are used to analyse the low-frequency vibrations of complex shell-type structures such as aircraft and ships. The highest derivative in the energy expressions for such elements is the first, and so only components of in-plane displacements are used as nodal degrees of freedom. Some typical shapes are shown in Figure 10.6, details of which are given in Petyt (1998). The accuracy of the two elements with nodes at the vertices only is not good, as the displacements vary linearly. The introduction of additional nodes at the mid-points of the sides allows parabolic variation of displacement. The displacement functions of these latter two elements can be used to transform the elements into curvilinear ones, as shown in Figure 10.7. These elements are known as isoparametric elements (Zienkiewicz and Taylor 1989a). The inertia, stiffness and load matrices of this type of element are obtained using numerical integration (Zienkiewicz and Taylor 1989a).

10.5.3 Plates

The highest derivative in the energy expressions for a thin, flat plate-bending element is the second. Therefore, the displacement function for the lateral displacement w must ensure continuity of w and its first derivatives between elements. Continuity of w and its tangential derivative $\partial w/\partial s$ can easily be

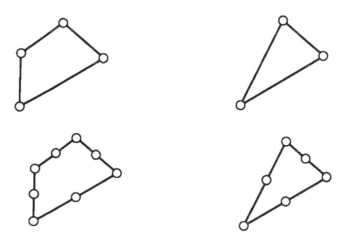

Figure 10.6 Typical shapes for two-dimensional elements (from Ewins and Imregun 1988).

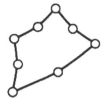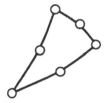

Figure 10.7 Two-dimensional isoparametric elements.

achieved by ensuring that w varies cubically along an edge and taking w and $\partial w/\partial s$ as nodal degrees of freedom. Continuity of the normal derivative $\partial w/\partial n$ has proved a little more difficult to achieve.

The first way of achieving continuity of normal slope between elements is to ensure that the normal slope varies linearly between node points. Deak and Pian (1967) use this method for a rectangular element. The element has nodes at the four corners only, and the degrees of freedom at each node are w, $\partial w/\partial x$ and $\partial w/\partial y$, x and y being parallel to the sides of the element. The displacement function is defined by dividing the rectangle into four triangles by means of two diagonals.

Another way of ensuring continuity of the normal slope is to introduce an additional node point at the mid-point of each edge. If w, $\partial w/\partial x$ and $\partial w/\partial y$ are taken as degrees of freedom at the corner nodes and $\partial w/\partial n$ as a degree of freedom at the mid-point nodes, then $\partial w/\partial n$ will vary quadratically along each edge. Fraeijs de Veubeke (1968) and Orris and Petyt (1973) give two alternative derivations of the same quadrilateral element which uses this technique. Again, the quadrilateral is divided into four triangles by means of the two diagonals. The quadrilateral element is more useful than the rectangle because of its ability to model complex polygonal shapes. However, in some situations, it is necessary to use triangular elements. These should be capable of maintaining the necessary continuity when used in conjunction with either rectangular or quadrilateral elements.

A triangular element with a linear variation of normal slope along its edges can be obtained by dividing it into three sub-triangles, by connecting the centroid to the three vertices (Clough and Tocher 1965; Dickinson and Henshell 1969). Within each sub-triangle, w is represented by an incomplete cubic to ensure that $\partial w/\partial n$ varies linearly over the external edge. The degrees of freedom are w, $\partial w/\partial x$ and $\partial w/\partial y$ at the three vertices of the main triangle. A triangular element with quadratic variation of normal slope can be obtained by using a complete cubic polynomial within each sub-triangle (Clough and Felippa 1968). As in the case of the quadrilateral, $\partial w/\partial n$ is introduced as an additional degree of freedom at the mid-points of the edges.

For thick plates, the effect of shear deformation and rotary inertia should be included in the energy expressions. This results in expressions which involve the lateral displacement w, the cross-sectional rotations θ_x and θ_y and their first derivatives. If w, θ_x and θ_y are represented by independent functions, it is only necessary to satisfy continuity of these quantities between elements. In this case the displacement functions developed for membrane elements can be used (Rock and Hinton 1974, 1976), and typical element shapes are as shown in Figures 10.6 and 10.7. It can be shown that these elements can also be used to model the behaviour of thin plates, provided reduced integration techniques are used.

10.5.4 Solids

Three-dimensional solid elements should be used for very thick structures. The energy expressions contain the first derivatives of the displacement components u, v and w; therefore, only continuity of displacements is required. Suitable displacement functions can be obtained by generalising the membrane functions into three dimensions (Petyt 1998). Typical element shapes are shown in Figure 10.8.

10.5.5 Shells

There are many theories available for analysing thin shells. A critical review of many of these can be found in Kraus (1967). One which is commonly used for doubly curved shells is given by Novozhilov (1964).

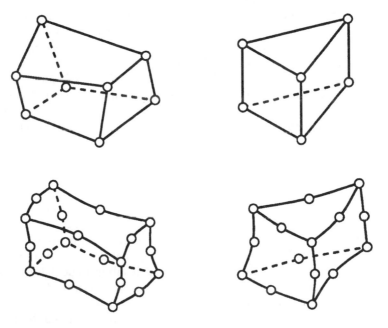

Figure 10.8 Three-dimensional elements.

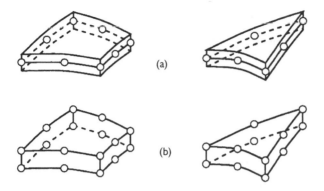

Figure 10.9 Shell elements: (a) middle surface type elements; (b) solid type elements.

The experience gained in constructing flat plate displacement functions can be used in developing suitable displacement functions for doubly curved shells. If the element is assumed shallow, then everything can be referred to the base plane. In this case, any of the shape functions described for thin, flat plate elements in Section 10.5.3 can be used to represent the shell normal displacement. All that remains is to develop suitable displacement functions for the components of displacement which are tangential to the middle surface. Experience has shown that it is advantageous to represent all three components by polynomials of equal degree. Typical examples of such elements are given in Cowper *et al.* (1970) and Petyt and Fleischer (1972). A survey of user experience with various shell elements is given in Ashwell and Gallagher (1976).

Thick shells can be analysed by using middle surface elements which can be constructed by methods which are similar to the ones used for thick plate elements (Zienkiewicz and Taylor 1989b). Again, they can be used for thin shells, provided reduced integration techniques are used. Solid-type shell elements (Nicolas and Citipitioglu 1977) can be constructed by modifying the solid elements described in Section 10.5.4. Both types of elements are shown in Figure 10.9.

10.5.6 Axisymmetric elements

Many structures, such as cooling towers, pressure vessels and parts of aerospace and marine vehicles, are axisymmetric. In this case they may be idealised by an assembly of ring elements of the type shown in Figure 10.10. The variation of displacement in the angular direction can be represented exactly by means of trigonometric functions. The problem is then reduced to a two-dimensional one, and the necessary displacement functions can be

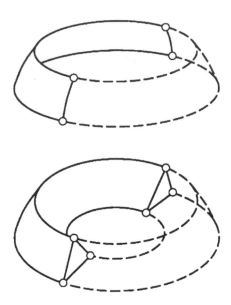

Figure 10.10 Axisymmetric ring elements.

constructed in a similar way to membrane elements (Zienkiewicz and Taylor 1989a; Petyt 1998).

10.5.7 Finite strip elements

In many problems the geometry and material properties do not vary in one coordinate direction. The analysis can be simplified by using orthogonal functions to represent the variation of displacement in this direction. Details of the method are given in Cheung and Tham (1997).

10.5.8 Hierarchical elements

Another way of increasing the accuracy of the solution is to keep the number of elements fixed and to increase the number of assumed functions within each element, which implies increasing the degree p of the polynomials. This approach is known as the p-version of the finite element method (Meirovitch 1986). The additional polynomials are chosen so that they have zero value in each degree of freedom at all the nodes of the element. Thus, the additional degrees of freedom can be considered to be internal degrees of freedom. It is possible to choose from a variety of sets of polynomials, but the most common ones used are the so-called hierarchical ones. These have the property that the set of functions corresponding to a polynomial approximation of order

p constitutes a subset of the set of functions corresponding to the approximation of order $(p+1)$. Experience has shown that the rate of convergence of the p-method can, and usually is, higher than the h-version. However, there is a limit to the number of polynomial functions which can be used (West *et al.* 1992). This can be overcome by combining both the h- and p-methods. The result is the h–p finite element method.

10.5.9   Other types of elements

In the preceding sections the emphasis has been on displacement type elements which satisfy all the necessary convergence criteria. Not all elements to be found in the literature satisfy all of these criteria. In this case, the convergence of the resulting solutions cannot be guaranteed. It has been suggested (Davies *et al.* 1993) that all elements should be tested by a recognised set of tests in order to indicate their accuracy. Other types of models exist such as hybrid elements and mixed variable elements. These will not be discussed here.

10.6   Methods of solution

Methods of solving the equation of motion (10.53) will be reviewed in this section. Both free and forced vibrations will be considered.

10.6.1   Free vibration

The free motion of an undamped structure is given by the solution of the equation

$$[M]\{\ddot{u}\} + [K]\{u\} = \{0\} \tag{10.69}$$

Since the free motion is harmonic, the solution of equation (10.69) can be written in the following complex form:

$$\{u\} = \{\phi\}\, e^{j\omega t} \tag{10.70}$$

where ω is the frequency of vibration and $j = \sqrt{-1}$. Substituting equation (10.70) into equation (10.69) gives

$$[[K] - \omega^2 [M]]\{\phi\} = \{0\} \tag{10.71}$$

This is a linear eigenvalue problem which can be solved by a number of methods. If $[K]$ and $[M]$ are square matrices of order n, then equation (10.71) represents a set of n linear homogeneous equations. These equations have a non-zero solution when the determinant of the coefficients vanish, so that, assuming $\lambda = \omega^2$,

$$\det([[K] - \lambda [M]]) = |K - \lambda M| = 0 \tag{10.72}$$

This equation can be expanded to give a polynomial of degree n in λ. Therefore, this polynomial equation has n roots, $\lambda_1, \lambda_2, \ldots, \lambda_n$, which are called 'eigenvalues'. Because the kinetic energy of a mechanical system cannot be negative and is never zero unless $\{\dot{u}\}$ is a null matrix the inertia matrix $[M]$ is positive definite. In contrast, the strain energy of a mechanical system can be greater than or equal to zero, so that $U = \frac{1}{2}\{u\}^T [K]\{u\} \geq 0$, making the stiffness matrix $[K]$ either positive definite or positive semi-definite. Because of these properties of the inertia and stiffness matrices, the eigenvalues are all real and either positive or zero (Bishop *et al.* 1965). However, they are not necessarily all different from one another. Indeed it is possible to have an eigenvalue which occurs m times in which case the eigenproblem is said to have multiple eigenvalues of multiplicity m.

When all eigenvalues are different, then corresponding to each one there exists a non-trivial solution to equation (10.71) for $\{\phi\}_r$ which are called 'eigenvectors'. An eigenvector is defined apart from an arbitrary constant such that a scalar multiple of it is also a solution of equation (10.71). In general, it is convenient to choose this multiplier in such a way that each eigenvector $\{\phi\}_r$ is normalised. The most common normalisation procedures are either to scale the eigenvectors $\{\phi\}_r$ such that their largest component is unity or to scale the eigenvectors so that

$$\{\phi\}_r^T [M]\{\phi\}_r = 1 \tag{10.73}$$

If the rth eigenvalue λ_r has multiplicity m, then there are exactly m eigenvectors corresponding to it. All the eigenvectors, including the multiple ones, satisfy the conditions of orthogonality which for $r \neq s$ state:

$$\{\phi\}_r^T [M]\{\phi\}_s = 0, \qquad \{\phi\}_r^T [K]\{\phi\}_s = 0 \tag{10.74a,b}$$

The eigenvalues λ_r, which are positive, represent the square of the natural frequencies ω_r^2, and the corresponding eigenvectors $\{\phi\}_r$ give the shape of the modes of vibration.

Methods of solving the eigenvalue problem (equation (10.71)) can be divided into four basic groups: vector iteration, transformation, polynomial iteration and Sturm sequence methods. In fact, some methods combine some of these techniques. The choice of method is mostly influenced by the order of the matrices $[K]$ and $[M]$ and the number of eigenvalues required (that is, all or a few). In finite element analysis the order of $[K]$ and $[M]$ is usually quite large (several hundred to a few thousand), and only a relatively few eigenvalues are required. In this case, methods based upon sub-space iteration, simultaneous iteration and Sturm sequences are to be recommended. Details of these methods can be found in Sehmi (1989), Jennings and McKeown (1991), Bathe (1996) and Petyt (1998).

10.6.2 Reduction of degrees of freedom

The number of degrees of freedom in any vibration problem can be reduced in three ways: making use of symmetry, employing a direct reduction technique and using component mode synthesis.

If a structure has a plane of symmetry, then a single eigenvalue problem for the complete structure can be replaced by two eigenvalue problems for the half-structure. These two eigenvalue problems will give the symmetric and anti-symmetric modes separately if the appropriate boundary conditions are applied on the plane of symmetry. This method greatly reduces both computer storage and computing time.

In the case of cyclic structures, such as bladed disc assemblies and cooling towers on column supports, the theory of periodic structures can be used (Petyt 1998). In this method, only one of the repeating sections is modelled by an assemblage of finite elements. The natural frequencies of the complete structure can be found by solving an eigenvalue problem associate with this single section, once the wave numbers corresponding to standing waves in the complete structure have been determined.

For more general structures it is necessary to employ a technique for reducing the number of degrees of freedom directly (Guyan 1965; Irons 1965). In this method only a small proportion of the total number of degrees of freedom, referred to as 'masters', are retained. The remaining 'slave' degrees of freedom take the values giving least strain energy, regardless of what this does to the kinetic energy. In practice, large reductions in the numbers of degrees of freedom can be introduced without introducing significant errors in the lower natural frequencies. An automatic procedure for selecting the master degrees of freedom is available (Henshell and Ong 1975).

For very large structures, using the reduction technique is not an efficient method of analysis. In this case a method known as 'component mode synthesis' should be used. This method is also known as 'substructure analysis'. The first step in the analysis is to divide the complete structure into a number of substructures. Each substructure is then represented by a finite element model. The next step involves reducing the number of degrees of freedom for each substructure by modal substitution. Finally, the substructures are coupled together and the complete structure is analysed. There are various ways of carrying out this analysis. They can be classified as fixed interface methods, modal substitution, free interface and hybrid methods. A survey of these techniques is given by Stavrinidis (1978).

10.6.3 Forced vibration

The equation of forced vibration of a structure takes the form

$$[M]\{\ddot{u}\} + [C]\{\dot{u}\} + [K]\{u\} = \{f\} \tag{10.75}$$

The method of solving this equation depends upon whether the exciting forces are harmonic, periodic, transient or random. Whatever the nature of the exciting force, equation (10.75) can be solved either directly or after reduction by modal substitution. In finite element analysis the number of degrees of freedom is usually quite large, and so it is usual to use modal substitution. In this method, use is made of the transformation

$$\{\mathbf{u}(t)\} = [\mathbf{\Phi}]\{\mathbf{q}(t)\} \tag{10.76}$$

where $[\mathbf{\Phi}]$ is a matrix whose columns represent the natural modes of vibration of the structure

$$[\mathbf{\Phi}] = [\{\mathbf{\phi}\}_1 \{\mathbf{\phi}\}_2 \cdots \{\mathbf{\phi}\}_n] \tag{10.77}$$

When only the lower natural modes of vibration contribute significantly to the response, the number of elements in $\{\mathbf{q}\}$ is far fewer than the number in $\{\mathbf{u}\}$. Substituting equation (10.76) into the energy expressions

$$T = \frac{1}{2} \{\dot{\mathbf{u}}\}^{\mathrm{T}} [\mathbf{M}] \{\dot{\mathbf{u}}\} \qquad D = \frac{1}{2} \{\dot{\mathbf{u}}\}^{\mathrm{T}} [\mathbf{C}] \{\dot{\mathbf{u}}\}$$

$$\tag{10.78}$$

$$U = \frac{1}{2} \{\mathbf{u}\}^{\mathrm{T}} [\mathbf{K}] \{\mathbf{u}\} \qquad \delta W = \delta \{\mathbf{u}\}^{\mathrm{T}} \{\mathbf{f}\}$$

and using Lagrange's equations the following inhomogeneous set of equations is obtained

$$\left[\overline{\mathbf{M}}\right] \{\ddot{\mathbf{q}}\} + \left[\overline{\mathbf{C}}\right] \{\dot{\mathbf{q}}\} + \left[\overline{\mathbf{K}}\right] \{\mathbf{q}\} = \{\mathbf{Q}\} \tag{10.79}$$

where

$$\left[\overline{\mathbf{M}}\right] = [\mathbf{\Phi}]^{\mathrm{T}} [\mathbf{M}][\mathbf{\Phi}] \qquad \left[\overline{\mathbf{C}}\right] = [\mathbf{\Phi}]^{\mathrm{T}} [\mathbf{C}][\mathbf{\Phi}]$$

$$\tag{10.80}$$

$$\left[\overline{\mathbf{K}}\right] = [\mathbf{\Phi}]^{\mathrm{T}} [\mathbf{K}][\mathbf{\Phi}] \qquad \{\mathbf{Q}\} = [\mathbf{\Phi}]^{\mathrm{T}} \{\mathbf{f}\}$$

According to the orthogonality conditions given in equations (10.74a,b), the inertia and stiffness matrices are diagonal whereas, in general, the damping matrix will not be diagonal. Also, if the eigenvectors are normalised according to equation (10.73), then equation (10.75) reduces to the following matrix expression

$$\{\ddot{\mathbf{q}}\} + [\overline{\mathbf{C}}]\{\dot{\mathbf{q}}\} + [\mathbf{\Lambda}]\{\mathbf{q}\} = \{\mathbf{Q}\} \tag{10.81}$$

where $[\Lambda]$ is a diagonal matrix of eigenvalues (solutions of equation (10.71) with $\Lambda_r = \omega_r^2$, $\omega_r = $ frequency of mode r)

$$[\Lambda] = \begin{bmatrix} \lambda_1 & & & \\ & \lambda_2 & & \\ & & \ddots & \\ & & & \lambda_n \end{bmatrix} = \begin{bmatrix} \omega_1^2 & & & \\ & \omega_2^2 & & \\ & & \ddots & \\ & & & \omega_n^2 \end{bmatrix} \tag{10.82}$$

As mentioned in Section 10.4, the dissipation function D is not evaluated in the same manner as the kinetic and strain energies. Dissipation is not necessarily an inherent property of the vibrating structure; thus it is difficult to formulate explicit expressions for the dissipating forces in a structure. Structural damping is caused by internal friction within the material and at joints between components. Viscous damping occurs when a structure is moving in air or other fluids. Therefore, simplified models, based more on mathematical convenience than physical representation, are used.

Inspection of equation (10.81) reveals that the equations of motion for the individual natural modes are coupled via the generalised damping matrix $[\overline{C}]$ only. These equations would be very much easier to solve if they were completely uncoupled. In Sections 10.6.4 and 10.6.5 two different models of dissipative phenomena are examined and a formulation that produces a set of uncoupled equations of motion in the two cases is presented.

10.6.4 Structural damping

Experimental investigations have indicated that for most structural metals, the energy dissipated per cycle is independent of frequency, over a wide frequency range, and proportional to the square of the amplitude of vibration. Thus, structural damping can be modelled in terms of a frequency-dependent damping coefficient $c(\omega) = h/\omega$. In this case, the structural damping force is given by $f_d = -(h/\omega)\dot{u} = -jhu$ for harmonic motion, where u represents displacement. The damping force due to structural dissipation effects is thus in anti-phase to velocity and proportional to displacement. The equation of harmonic motion of a one-degree-of-freedom system consisting of a mass m, a dissipation element with structural damping coefficient $c(\omega) = h/\omega$ and a spring of stiffness k connected in parallel, is given by $m\ddot{u} + jhu + ku = f$, where f is the harmonic excitation force $f = f_o \exp(j\omega t)$. The quantity $(k + jh)$ is called a 'complex stiffness' and it is usually written in the form $k(1 + j\eta)$ where η is the 'loss factor' of the system. Generalising this formulation of structural damping to a multiple-degrees-of-freedom system, equation (10.75) is written in the following form:

$$[M]\{\ddot{u}\} + [K + jH]\{u\} = \{f\} \tag{10.83}$$

The complex matrix $[K+jH]$ is derived by replacing Young's modulus E by the complex form $E(1+j\eta)$, where η is the material loss factor, when deriving the element stiffness matrices. Typical values of the structural loss factor range from 1×10^{-5} for pure aluminium or steel up to 1.0 for rubber materials.

Equation (10.83) can be simplified using the modal analysis already described so that

$$\{\ddot{q}\} + \left[\Lambda + j\overline{H} \right]\{q\} = \{Q\} \tag{10.84}$$

where $[\Lambda]$ and $[Q]$ are defined by equations (10.82) and (10.80) respectively. In general, $[\overline{H}]$ will be a fully populated matrix, unless every element has the same loss factor. In this case,

$$[H] = \eta[K], \qquad \left[\overline{H}\right] = \eta[\Lambda] \tag{10.85a,b}$$

and equation (10.83) now becomes

$$\{\ddot{q}\} + (1+j\eta)[\Lambda]\{q\} = \{Q\} \tag{10.86}$$

All the equations in the matrix expression (10.86) are uncoupled and each one assumes the same form as that for a single-degree-of-freedom system.

10.6.5   Viscous damping

The most common form of viscous damping is the so-called Rayleigh-type damping given by

$$[C] = a_1[M] + a_2[K] \tag{10.87}$$

This type of damping is characterised by two terms, mass-proportional damping and stiffness-proportional damping. When the damping matrix is defined with equation (10.87), the matrix $[\overline{C}]$ is diagonal since

$$\left[\overline{C}\right] = a_1[I] + a_2[\Lambda] \tag{10.88}$$

Therefore, with the Rayleigh-type damping the set of equations in equation (10.81) is uncoupled and each one of these equations assumes the form

$$\ddot{q}_r + 2\gamma_r \omega_r \dot{q}_r + \omega_r^2 q_r = Q_r \tag{10.89}$$

where γ_r is the modal damping ratio and

$$2\gamma_r \omega_r = a_1 + \omega_r^2 a_2 \tag{10.90}$$

Thus, the two coefficients a_1 and a_2 can be calculated from a knowledge (from test or experience) of the damping ratio in two modes. If, for example, the damping ratios γ_1 and γ_2 and natural frequencies ω_1 and ω_2 of the first two modes are known, then,

$$a_1 = 2\omega_1\omega_2 \frac{(\omega_2\gamma_1 - \omega_1\gamma_2)}{(\omega_2^2 - \omega_1^2)} \qquad (10.91a)$$

$$a_2 = 2\frac{(\omega_2\gamma_2 - \omega_1\gamma_1)}{(\omega_2^2 - \omega_1^2)} \qquad (10.91b)$$

The damping ratios in the other modes are then given by equation (10.90) as

$$\gamma_r = \frac{a_1}{2\omega_r} + \frac{a_2\omega_r}{2} \qquad (10.92)$$

The values of a_1 and a_2 given by equation (10.91) can also be used in equation (10.87) to calculate the matrix $[C]$ so that the equations of motion in the form of equation (10.75) are then used.

Mass-proportional and stiffness-proportional damping can be used separately. For mass-proportional damping, $a_2 = 0$, so that from equation (10.90) $a_1 = 2\gamma_1\omega_1$ and therefore $\gamma_r = \gamma_1\omega_1/\omega_r$. Thus, mass-proportional damping ratios decrease with increase of mode frequency. Alternatively, for stiffness-proportional damping, $a_1 = 0$, so that from equation (10.90) $a_2 = 2\gamma_1/\omega_1$ and $\gamma_r = \gamma_1\omega_r/\omega_1$. In this case the stiffness-proportional damping ratios increase with increase of mode frequency. In practice, it has been found that mass-proportional damping can represent friction damping, whilst stiffness-proportional damping can represent internal material damping. Values of damping ratios for typical forms of construction vary from 0.01 for small diameter piping systems, to 0.07 for bolted joint and reinforced concrete structures.

This procedure for uncoupling the equations of motion can only be carried out if the damping is light. Fortunately, for most structural configurations, this is the case. If the damping is heavy (for example, if the structure is coated with a significant amount of viscoelastic material or submerged in a fluid and there is significant sound radiation) then the equations of motion can only be uncoupled by using certain modes of the damped system which are complex.

10.6.6 Response to harmonic excitation

If the excitations are harmonic, all with the same frequency ω, but with different amplitude and phase, then the excitation vector $\{f(t)\}$ is defined as follows

$$\{f(t)\} = \{f\}\exp(j\omega t) \qquad (10.93)$$

where the elements of the vector $\{f\}$ are complex and of the form $f_k = |f_k| \exp(j\theta_k)$, where θ_k is the phase of force f_k relative to time $t = 0$. Two approaches can be used to derive the solution of the system of equation (10.75) when the excitation vector $\{f(t)\}$ assumes the form in equation (10.93); first, the modal approach and second, the direct approach.

(i)   Modal analysis
For harmonic excitations, the generalised force vector in equation (10.81) becomes

$$\{Q(t)\} = [\Phi]^T \{f\} \exp(j\omega t) = \{Q\} \exp(j\omega t) \tag{10.94}$$

Thus, the steady-state harmonic response can be derived directly from equation (10.81) in terms of the generalised displacements $\{q(t)\} = \{q\} \exp(j\omega t)$

$$\{q(t)\} = \left[\Lambda - \omega^2 I + j\omega \overline{C}\right]^{-1} \{Q\} \exp(j\omega t) \tag{10.95}$$

If either structural or proportional damping is used, the matrix to be inverted is diagonal. Its inverse is also a diagonal matrix with the following diagonal elements:
structural damping

$$\left(\omega_r^2 - \omega^2 + j\eta\omega_r^2\right)^{-1} \tag{10.96a}$$

proportional damping

$$\left(\omega_r^2 - \omega^2 + j2\gamma_r\omega_r\omega\right)^{-1} \tag{10.96b}$$

Substituting equations (10.95) and (10.94) into equation (10.76) gives

$$\{u\} = [\Phi]\left[\Lambda - \omega^2 I + j\omega\overline{C}\right]^{-1} [\Phi]^T \{f\} \exp(j\omega t) \tag{10.97}$$

Equation (10.97) is often written in the form $\{u\} = [\alpha]\{f\} \exp(j\omega t)$, where the matrix $[\alpha] = [\Phi][\Lambda - \omega^2 I + j\omega\overline{C}]^{-1} [\Phi]^T$ is a matrix of receptances. $\alpha_{rk}(\omega)$ is a transfer receptance which represents the response in degree of freedom r due to a harmonic force of unit magnitude and frequency ω applied in degree of freedom k. Likewise, $\alpha_{rr}(\omega)$ represents a point receptance. The receptance terms take the following forms in the two cases of structural and proportional damping:
structural damping

$$\alpha_{rk}(\omega) = \sum_{i=1}^{n} \frac{\phi_{ri}\phi_{ki}}{\left(\omega_i^2 - \omega^2 + j\eta\omega_i^2\right)} \tag{10.98a}$$

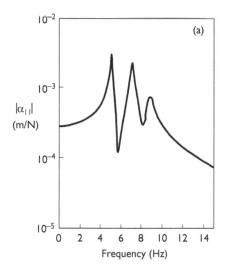
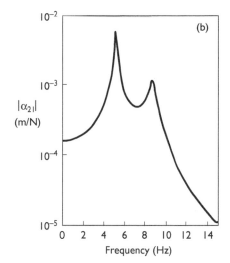

Figure 10.11 Point α_{11} (a) and transfer α_{21} (b) receptances of a three-degrees-of-freedom system composed of three masses linked by two springs.

proportional damping

$$\alpha_{rk}(\omega) = \sum_{i=1}^{n} \frac{\phi_{ri}\phi_{ki}}{\left(\omega_i^2 - \omega^2 + j2\gamma_i\omega_i\omega\right)} \tag{10.98b}$$

Figure 10.11 shows the point α_{11} (a) and transfer α_{21} (b) receptances of a three-degrees-of-freedom system composed of three masses supported by four springs. The damping matrix is assumed to be proportional to the stiffness matrix.

(ii)   *Direct analysis*
The steady state response to harmonic excitation can also be obtained by direct solution of equation (10.75) so that the natural frequencies and natural modes of the system are not required. The solution of equation (10.75) is given by

$$\{\mathbf{u}\} = \left[\mathbf{K} - \omega^2\mathbf{M} + j\omega\mathbf{C}\right]^{-1}\{\mathbf{f}\}\exp(j\omega t) \tag{10.99}$$

which can be evaluated in various ways (Petyt 1998).

10.6.7   Response to periodic excitation

Periodic excitations are typical of rotating machinery. This type of excitation can be represented by means of a Fourier series. Thus, when the nodal forces acting on the system are periodic, then the load vector $\{f(t)\}$ can be expressed in the following matrix form

$$\{f(t)\} = \sum_{i=-\infty}^{+\infty} \{c\}_i \exp(j\omega_i t) \tag{10.100}$$

where $\omega_i = i(2\pi/T)$ and

$$\{c\}_i = \frac{1}{T} \int_0^T \{f(t)\} \exp(-j\omega_i t)\, dt \tag{10.101}$$

with T being the period of the excitation. When the periodic forces cannot be expressed as a mathematical function, then a discrete Fourier series is derived for each force function f_k from a set of values $f_{k,0}, f_{k,1}, \ldots, f_{k,N}$ taken at times $t_0, t_1, \ldots, t_N = T$, as shown in Figure 10.12. The discrete Fourier series can be derived with equation (10.100) by considering the summation between $-N/2$ and $+N/2$, where N is taken to be even, and

$$\{c\}_i = \frac{1}{N} \sum_{s=1}^{N} \{f\}_s \exp\left\{-j\left(\frac{2\pi}{N}\right)si\right\} \tag{10.102}$$

The highest frequency present in the Fourier series is then $\omega_{N/2} = N\pi/T = \pi/\Delta t\,[\text{rad/s}] = 1/2\Delta t\,[\text{Hz}]$. If there are any frequencies higher than this present in the periodic force, they will contribute to frequencies below this maximum. This is a problem of aliasing, which can be avoided by selecting Δt to be small enough. The efficiency of computing the coefficients in equation (10.102) can be increased by means of a Fast Fourier Transform (FFT) algorithm.

Using the Fourier series, the periodic excitations have been transformed into a series of harmonic excitations so that equation (10.75) becomes

$$[\mathbf{M}]\{\ddot{u}\} + [\mathbf{C}]\{\dot{u}\} + [\mathbf{K}]\{u\} = \sum_{i=-\infty}^{+\infty} \{c\}_i \exp(j\omega_i t) \tag{10.103}$$

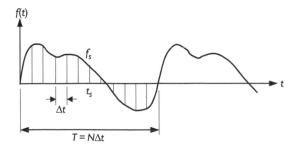

Figure 10.12 Discrete representation of periodic excitation.

This is a linear non-homogeneous equation whose solutions corresponding to each term on the right-hand side can be calculated separately and summed to give the complete solution. Thus

$$\{\mathbf{u}\} = \sum_{i=-\infty}^{+\infty} [\alpha(\omega_i)]\{\mathbf{c}\}_i \exp(j\omega_i t) \qquad (10.104)$$

where $[\alpha(\omega_i)]$ is the matrix of receptances defined in the Section 10.6.6.

10.6.8 Response to transient excitation

When a structure is excited by a non-periodic excitation which is deterministic, the excitation is called 'transient'. To be precise, the term 'transient' should be used for situations where the excitations are applied for a finite interval of time. Subsequent motion of the structure is free vibration, which will decay due to the damping present. However, it is often applied to a continually changing situation for an indefinite period of time. The response to transient excitation can be derived with two formulations: first, modal analysis and second, direct analysis.

(i) *Modal analysis*
Considering the case of viscous damping, the solution at time t of the set of uncoupled equations in the generalised coordinates q_r given in equation (10.81) as

$$\ddot{q}_r + 2\gamma_r \omega_r \dot{q}_r + \omega_r^2 q_r = Q_r(t) \qquad (10.105)$$

is given by the Duhamel integral

$$q_r(t) = \int_0^t Q_r(\tau) h_r(t - \tau) d\tau \qquad (10.106)$$

when the motion starts form rest at $t = 0$; $h_r(t)$ is the impulse response function which is given by

$$h_r(t) = \frac{1}{\omega_{dr}} \exp(-\gamma_r \omega_r t) \sin \omega_{dr} t \qquad (10.107)$$

where $\omega_{dr} = \omega_r \sqrt{1 - \gamma_r^2}$ is the damped natural frequency of mode r. When the forces are applied for a finite interval of time the maximum response will occur during the first few oscillations. If the damping is small, then the free response will last longer. An analytical solution of the Duhamel integral can be derived for time histories of the excitation given in an analytical form. The responses to typical loading functions are tabulated in Przemieniecki (1968).

When the time history of the forces cannot be expressed analytically, then the transient response can be derived numerically with a step-by-step procedure. In this formulation, equation (10.105) is rewritten in the standard form

$$m\ddot{q} + c\dot{q} + kq = f(t) \tag{10.108}$$

where $q = q_r$, $m = 1$, $c = 2\gamma_r \omega_r$, $k = \omega_r^2$ and $f = Q_r$. Knowing the initial displacement $q(0) = q_0$, velocity $\dot{q}(0) = \dot{q}_0$ and force $f(0) = f_0$, the acceleration at $t = 0$ can be calculated from equation (10.108) as follows:

$$\ddot{q}(0) = \ddot{q}_0 = \frac{f_0 - c\dot{q}_0 - kq_0}{m} \tag{10.109}$$

In order to evaluate the response at time t, the time interval $(0, t)$ is divided into N equal time intervals $\Delta t = t/N$, as shown in Figure 10.13. The response $(q, \dot{q}$ and $\ddot{q})$ is then evaluated with an approximate technique at the times $\Delta t, 2\Delta t, 3\Delta t, \ldots, t$.

The most common numerical techniques used are the following: first, the central difference method, second, the Houbolt method (backward difference method) and third, the Newmark method (generalisation of the linear acceleration method). The central difference method is briefly reviewed in this section, while a complete description of all three techniques can be found in Petyt (1998).

In the central difference method, the velocity and acceleration are expressed at time t_i in terms of the displacements at times t_{i-1}, t_i and t_{i+1} using central finite difference formulae. These are obtained by approximating the response curve, shown in Figure 10.14, with a quadratic polynomial within the interval (t_{i-1}, t_{i+1}). In this way the velocity and acceleration at the step i are found to be given by the following expressions:

$$\dot{q}_i = \frac{1}{2\Delta t}(q_{i+1} - q_{i-1}), \qquad \ddot{q}_i = \frac{1}{(\Delta t)^2}(q_{i+1} - 2q_i + q_{i-1}) \tag{10.110a,b}$$

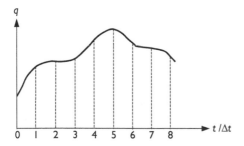

Figure 10.13 Discrete representation of transient response.

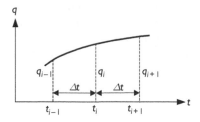

Figure 10.14 Central difference method.

The response at time t_{i+1} is obtained by substituting equation (10.110) into the equation of motion (equation (10.108)) evaluated at time t_i. Methods based upon equation (10.108) are called explicit methods. Performing the substitution and solving for q_{i+1} gives

$$q_{i+1} = \alpha_1^{-1} \left\{ f_i + \left[\frac{2m}{(\Delta t)^2} - k \right] q_i \right\} - \alpha_1^{-1} \alpha_2 q_{i-1} \tag{10.111}$$

where

$$\alpha_1 = \left(\frac{m}{(\Delta t)^2} + \frac{c}{2\Delta t} \right), \qquad \alpha_2 = \left(\frac{m}{(\Delta t)^2} - \frac{c}{2\Delta t} \right) \tag{10.112a,b}$$

Hence, if the displacements q_{i-1} and q_i are known, the displacement q_{i+1} can be calculated. The time history of the response can be obtained by taking $i = 1, 2, \ldots$, provided that q_1 and q_0 are known. q_1 can be determined by evaluating equation (10.110) for $i = 0$ giving

$$\dot{q}_0 = \frac{1}{2\Delta t} (q_1 - q_{-1}), \qquad \ddot{q}_0 = \frac{1}{(\Delta t)^2} (q_1 - 2q_0 + q_{-1}) \tag{10.113a,b}$$

and eliminating q_{-1} to give

$$q_1 = q_0 + \Delta t \dot{q}_0 + \frac{1}{2} (\Delta t)^2 \ddot{q}_0 \tag{10.114}$$

As previously mentioned, q_0 and $\dot{q}_0$ are given and $\ddot{q}_0$ can be calculated using equation (10.109). The velocity and acceleration at each time step can be obtained using equation (10.110).

(ii) *Direct analysis*
The three methods listed above can be used to solve equation (10.75) directly so that the calculation of the natural frequencies and natural modes of the undamped system do not have to be performed prior to the response analysis.

The central difference procedure for a single-degree-of-freedom system can be extended to the matrix expression of equation (10.75) for a multiple-degrees-of-freedom system. Indeed, knowing the vectors of the initial displacement $u(0) = u_0$, velocity $\dot{u}(0) = \dot{u}_0$ and force $f(0) = f_0$, the set of accelerations for $t = 0$ can be derived with similar equation to the expression in equation (10.109); that is

$$\{\ddot{u}(0)\} = \{\ddot{u}\}_0 = [M]^{-1} [f_0 - C\dot{u}_0 - Ku_0] \tag{10.115}$$

The column vector of displacements for $i = 1$ is then derived from the relation

$$\{u\}_1 = \{u\}_0 + \Delta t \{\dot{u}\}_0 + \left\{ \frac{(\Delta t)^2}{2} \right\} \{\ddot{u}\}_0 \tag{10.116}$$

while for $i > 1$, the displacement vector is derived from the following matrix expression

$$[a_1 M + a_2 C] \{u\}_{i+1} = \{f\}_i + [2a_1 M - K] \{u\}_i - [a_2 M - a_2 C] \{u\}_{i-1} \tag{10.117}$$

where

$$a_1 = \frac{1}{(\Delta t)^2} \quad \text{and} \quad a_2 = \frac{1}{2\Delta t} \tag{10.118}$$

It is necessary to solve a set of linear equations in order to determine both $\{\ddot{u}\}_0$ and $\{u\}_{i+1}$. These can be solved using the modified Cholesky symmetric decomposition, as described by Petyt (1998).

10.7   Computer programs

With the aid of the finite element method, the prediction of the dynamic response of structures is reduced to a routine procedure. It is therefore possible to write general computer codes for analysing any structure, the particular structure being specified by means of input data. Details of programming aspects can be found in several references (Hinton and Owen 1979, 1984; Hinton 1988; Cook 1989).

A typical finite element analysis consists of three phases. The first, known as the pre-processing phase, consists of specifying and checking the input data (geometry of the system, type of element, mesh of elements, physical properties, boundary conditions). This is followed by the solution phase in which the analysis is carried out (numerical solution of the free vibration or force vibration problems using the appropriate numerical approach). The final phase, which is known as the post-processing phase, is concerned with the interpretation and presentation of the results of the analysis (list of the natural frequencies, plot of the mode shapes, plot of the frequency response function or time response, which can also be animated).

A new user of finite element analysis is unlikely to start writing a computer program. The reason for this is that there is a large number of general-purpose finite element programs which can be obtained commercially. All are available on a wide range of mainframes, workstations and personal computers. The choice of which commercial program to use depends upon many factors. If several types of analyses are required, then the choice is between the general purpose programs. However, if only one type is required, then a special-purpose program might prove more efficient. The quality of the pre- and post-processors may also affect the choice. Ease of data input, good graphics capability and suitable post-processing facilities are essential.

During the 1980s two surveys were organised in order to assess the reliability of structural dynamic analysis capabilities using finite element methods (Keilb and Leissa 1985; Ewins and Imregun 1988). The first was concerned with the prediction of the natural frequencies of twisted cantilever plates, representative of turbine blades. The second, code-named DYNAS (DYNamic Analysis of Strucures), was more wide-ranging, including free vibration and forced response to harmonic and transient excitations. Technical drawings of the I-beam-cylinder-spring structure shown in Figure 10.15 were sent to a number of organisations who were invited to undertake its dynamic analysis using well-established FEM techniques. Among the several conclusions drawn from the survey it was noted that

> What is seen is that inaccurate predictions can be made by poor analysis methods and this is probably the essence of the conclusions throughout. That being the case, there is clearly some need for

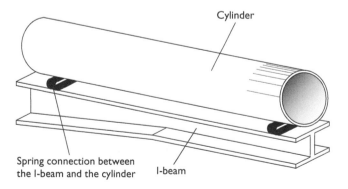

Spring connection between
the I-beam and the cylinder   I-beam

Figure 10.15 Structure studied in DYNAS survey.

procedures which could indicate the quality or reliability of a given prediction so that a good analysis can be identified (as, indeed, can a poor one) (from Ewins and Imregun 1988).

As an example, Figure 10.16 shows the simulation results obtained from the twelve finite element models of the DYNAS survey for the direct receptance at the attachment point of one spring with the I-beam. The plot highlights the variance of the results obtained from the twelve numerical finite element models, even if they had been formulated assuming the same type of geometry and material properties of the system. Among all the possible factors that could influence the numerical analysis with a finite element code, the most important can be listed as follows:

1. choice of the element type;
2. mesh density;
3. solution algorithm;
4. number of modes;
5. choice of master degrees of freedom;
6. time step;
7. damping model and value(s);
8. boundary conditions.

Program developers are increasingly subjecting their programs to stringent quality assurance tests. However, it is very difficult for a developer to envisage

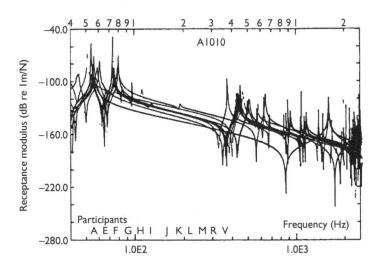

Figure 10.16 Frequency response functions predicted by the 12 members of the DYNAS survey (from Ewins and Imregun 1988).

every possible use a customer might make of the software. The quality assurance tests may, therefore, concentrate more on checking the accuracy of the code. The user can complement this by running a set of benchmarks. These are fully specified standard problems which are used for evaluating the performance of element assemblies. Such sets have been proposed for free and forced vibration analyses (Knowles 1987; Rahman and Petyt 1996), respectively. Possible reasons for different analysts using the same program producing different solutions are different choices of idealisation and/or element types. The accuracy of the computed results is, therefore, a function of the experience of the user as well as the accuracy of the programs. It is essential therefore that users of finite element programs be well-trained in both the finite element method and the use of the program to be used.

Acknowledgement

Some parts of this chapter are inspired by the book *Introduction to Finite Element Vibration Analysis* by M. Petyt (Hardback 1990, Paperback 1998). The kind permission of Cambridge University Press to refer to this work is gratefully acknowledged.

References

Ashwell, D.G. and Gallagher, R.H. (eds) (1976) *Finite Elements for Thin Shells and Curved Members* (London: Wilcy).

Bathe, K.J. (1996) *Finite Element Procedures* (Englewood Cliffs: Prentice-Hall).

Bishop, R.E.D., Gladwell, G.M.L. and Michaelson, S. (1965) *The Matrix Analysis of Vibration* (Cambridge: Cambridge University Press).

Cheung, Y.K. and Tham, L.G. (1997) *Finite Strip Method* (USA: CRC Press).

Clough, R.W. and Felippa, C.A. (1968) A refined quadrilateral element for analysis of plate bending. *Proceedings of the Second Conference on Matrix Methods in Structural Mechanics*, AFF DL-TR-68-150, Wright-Patterson Airforce Base, Ohio.

Clough, R.W. and Tocher, J.L. (1965) Finite element stiffness matrices for analysis of plate bending. *Proceedings of the First Conference on Matrix Methods in Structural Mechanics*, AFF DL-TR-66-80, Wright-Patterson Airforce Base, Ohio.

Cook, R.D. (1989) *Concepts and Applications of Finite Element Analysis* (3rd edn) (New York: Wiley).

Cowper, C.R., Lindberg, G.M. and Olson, M.D. (1970) A shallow shell finite element of triangular shape. *International Journal of Solids and Structures*, 8, 1133–1156.

Davies, G.A.O., Fenner, R.T. and Lewis, R.W. (1993) Background to benchmarks. NAFEMS Report, NAFEMS, Glasgow.

Davis, R., Henshall, R.D. and Warburton, G.B. (1972a) Constant curvature beam finite elements for in-plane vibration. *Journal of Sound and Vibration*, 25, 561–576.

Davis, R., Henshall, R.D. and Warburton, G.B. (1972b) Curved beam finite elements for coupled bending and torsional vibration. *International Journal of Earthquake Engineering and Structural Dynamics*, 1, 165–175.

Deak, A.L. and Pian, T.H.H. (1967) Application of the smooth surface interpolation to the finite element analysis. *AIAA Journal*, **5**, 187–189.

Dickinson, S.M. and Henshell, R.D. (1969) Clough-Tocher triangular plate-bending element in vibration. *AIAA Journal*, **7**, 560–561.

Ewins, D.J. and Imregun, M. (1988) On the reliability of computational dynamic response prediction capabilities (DYNAS). *Journal of the Society of Environmental Engineers*, 3–28.

Fraeijs de Veubeke, B. (1968) A conforming finite element for plate bending. *International Journal of Solids and Structures*, **4**, 95–108.

Gupta, R.S. and Rao, S.S. (1978) Finite element eigenvalue analysis of tapered and twisted Timoshenko beams. *Journal of Sound and Vibration*, **56**, 187–200.

Guyan, R.J. (1965) Reduction of stiffness and mass matrices. *AIAA Journal*, **3**, 380.

Henshell, R.D. and Ong, J.H. (1975) Automatic masters for eigenvalue economization. *Earthquake Engineering and Structural Dynamics*, **3**, 375–383.

Hinton, E. (1988) *Numerical Methods and Software for Dynamic Analysis of Plates and Shells* (Swansea: Pineridge Press).

Hinton, E. and Owen, D.R. (1979) *An Introduction to Finite Element Computations* (Swansea: Pineridge Press).

Hinton, E. and Owen, D.R. (1984) *Finite Element Software for Plates and Shells* (Swansea: Pineridge Press).

Irons, B.M. (1965) Structural eigenvalue problems – elimination of unwanted variables. *AIAA Journal*, **3**, 961–962.

Jennings, A. and McKeown, J.J. (1991) *Matrix Computation for Engineers and Scientists* (2nd edn) (London: Wiley).

Keilb, R.E. and Leissa, W.A. (1985) Vibrations of twisted cantilever plates – a comparison of theoretical results. *International Journal for Numerical Methods in Engineering*, **21**, 1365–1380.

Knowles, N. (1987) Selected benchmarks for natural frequency analysis. NAFEMS Report No. P17. NAFEMS, Glasgow.

Kraus, H. (1967) *Thin Elastic Shells* (New York: Wiley).

Leckie, F.A. and Lindberg, G.M. (1963) The effect of lumped parameters on beam frequencies. *Aeronautical Quarterly*, **14**, 224–240.

Lindberg, G.M. (1963) Vibration of non-uniform beams. *Aeronautical Quarterly*, **14**, 387–395.

Meirovitch, L. (1967) *Analytical Methods in Vibrations* (New York: Macmillan).

Meirovitch, L. (1986) *Elements of Vibration Analysis* (2nd edn) (New York: McGraw-Hill).

Nicolas, V.T. and Citipitioglu, E. (1977) A general isoparametric finite element program SDRC SUPERB. *Computers & Structures*, **7**(2), 303–313.

Novozhilov, V.V. (1964) *The Shell Theory* (Groeningen: Noordhoff).

Oden, J.T. (1972) *Finite Elements of Nonlinear Continua* (New York: McGraw-Hill).

Orris, R.M. and Petyt, M. (1973) A finite element study of the vibration of trapezoidal plates. *Journal of Sound and Vibration*, **27**, 325–334.

Petyt, M. (1977) Finite strip analysis of flat skin-stringer structures. *Journal of Sound and Vibration*, **54**, 533–547.

Petyt, M. (1998) *Introduction to Finite Element Vibration Analysis* (Cambridge: Cambridge University Press).

Petyt, M. and Fleischer, C.C. (1971) Free vibrations of a curved beam. *Journal of Sound and Vibration*, **18**, 17–30.

Petyt, M. and Fleischer, C.C. (1972) Vibrations of curved structures using quadrilateral finite elements. In J.R. Whiteman (ed.), *The Mathematics of Finite Elements and Applications* (London: Academic Press), pp. 367–378.

Przemieniecki, J.S. (1968) *Theory of Matrix Structural Analysis* (New York: McGraw-Hill).

Rahman, A. and Petyt, M. (1996) Fundamental tests for forced vibrations of linear elastic structures. NAFEMS Report No. R0034. NAFEMS, Glasgow.

Rock, T.A. and Hinton, E. (1974) Free vibration and transient response of thick and thin plates using the finite element method. *International Journal of Earthquake Engineering and Structural Dynamics*, **3**, 51–63.

Rock, T.A. and Hinton, E. (1976) A finite element method for the free vibration of plates allowing for transverse shear deformation. *Computers and Structures*, **7**, 37–44.

Sehmi, N.S. (1989) *Large Order Structural Eigenanalysis Techniques* (Chichester: John Wiley & Sons).

Stavrinidis, C. (1978) Theory and practice of modal synthesis techniques. In J. Robinson (ed.), *Finite Element Methods in the Commercial Environment* (Okehampton: Robinson & Associates), pp. 307–331.

Thomas, D.L., Wilson, J.M. and Wilson, R.R. (1973) Timoshenko beam finite elements. *Journal of Sound and Vibration*, **31**, 315–330.

Weaver, W. and Johnston, P.R. (1987) *Structural Dynamics by Finite Elements* (Englewood Ciffs: Prentice-Hall).

West, L.J., Bardell, N.S., Dunsdon, J.M. and Loasby, P.M. (1992) Some limitations associated with the use of k-orthogonal polynomials in hierarchical versions of the finite element method. In N.S. Ferguson, H.F. Wolfe and C. Mei (eds), *Structural Dynamics: Recent Advances* (ISVR, University of Southampton), ISBN 0-85432-6375, pp. 217–231.

Zienkiewicz, O.C. and Taylor, R.L. (1989a) *The Finite Element Method* (4th edn) (London: McGraw-Hill), Vol 1.

Zienkiewicz, O.C. and Taylor, R.L. (1989b) *The Finite Element Method* (4th edn) (London: McGraw-Hill), Vol 2.

Chapter 11

High-frequency structural vibration

R.S. Langley and F.J. Fahy

11.1 Introduction

Many engineering structures are subject to excitation by unsteady forces of which the spectra extend far up the audio-frequency range (20 Hz–20 kHz). Examples include railway trains, cars, aircraft, ships, gas pipelines, buildings, industrial plant and space rockets. The qualification 'high' in the title implies that the frequency range of concern extends to many times the fundamental natural frequency of the system; ultrasonic frequencies are not considered here. High-frequency vibration is of concern to engineers because of the potential for excessive sound transmission and radiation and for fatigue damage. The former aspects are commercial, ergonomic, regulatory, environmental and health concerns; the latter is a safety and commercial matter.

This chapter begins with a qualitative introduction to the physics of high-frequency structural vibration and its distinguishing features. It continues with an explanation of the difficulties that it poses to mathematical modellers and analysts. The parameters relevant to the characterisation of high-frequency structural vibration are then introduced in a qualitative manner as an entrée to Section 11.4, which presents a mathematical description of high-frequency structural vibration. Various forms of deterministic model and numerical solution are described, together with their merits and weaknesses. As an alternative, the probabilistic statistical energy analysis (SEA) model is introduced. The two approaches to estimating SEA subsystem coupling coefficients are explained and the power balance equations are presented. The problem of subsystem selection is illustrated by reference to the modelling of a passenger car. The chapter closes with brief accounts of the various alternatives to SEA that are currently being researched.

11.2 The physics of high-frequency vibration

The term 'high frequency' has a relative connotation – 'high' relative to what? It requires qualification. The phenomenon of vibration of distributed structures involves the generation and propagation of time-dependent disturbances in the form of waves that traverse the extent of the structure. In cases where the waves are not rapidly attenuated by strongly dissipative mechanisms (high damping), they *will traverse and re-traverse a structure many times*, being partially reflected, diffracted and scattered as they repeatedly encounter changes of geometry and/or dynamic properties of the supporting medium. Ultimately, all their vibrational energy is either dissipated into heat by frictional forces or transmitted to connected structures or contiguous fluids. Such structures are said to be 'reverberant'.

Following impulsive excitation of a linear reverberant structure, constructive interference between waves travelling freely in different directions produces spectral response peaks at the discrete characteristic (natural) frequencies of the structure. (The individual peaks may not be evident at high frequencies because of a phenomenon termed 'modal overlap', which is explained later in this section.) Each natural frequency is associated with a different spatial interference pattern (or standing wave), called a 'mode'. The natural frequencies and mode shapes are characteristic of a structure and its boundary conditions, but are independent of the form of excitation. The lowest natural frequency is associated with the 'fundamental' mode. In many practical cases, individual modes of distributed structures behave like individual, single-degree-of-freedom, mass-spring-damper systems. Modal natural frequency increases with the 'order' (or spatial complexity) of the mode.

The strengths of reflection and scattering (redistribution of incident wave energy into many directions) of vibrational waves by structural non-uniformities, inhomogeneities or discontinuities of geometry or material properties, generally increase as the wavelength decreases (or as the frequency increases). Hence, minor variations of geometric, material and connection detail perturb the natural frequencies and, to a lesser extent, the mode shapes of a structure to an increasing degree as mode order increases.

If the modal natural frequencies of a structure are well separated, and it is excited at a frequency very close to any one of them, a strong 'resonant' response occurs. The resulting spatial distribution of response corresponds closely to the associated mode (although it coexists with smaller, non-resonant contributions from other modes). If the frequency is varied over a small range centred on the resonance frequency, the response energy will fall smoothly on either sides. The frequency range over which it exceeds 50 per cent of the peak, resonant energy is termed the 'half-power bandwidth' (Figure 11.1). It is characteristic of many plate and shell structures that the half-power bandwidth increases with frequency, and the average separation of resonance frequencies is either more-or-less independent of, or decreases with, frequency. Consequently, as frequency increases, the resonance peaks increasingly overlap each other,

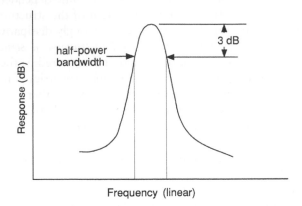

Figure 11.1 Definition of half-power bandwidth.

so that ultimately the response at any one frequency and any position comprises substantial contributions from more than one mode. A typical frequency response curve that exhibits the transition from dominance by individual mode resonances to multi-mode response is presented in Figure 11.2.

Low-order mode resonance frequencies and mode shapes are generally rather insensitive to small variations of structural detail. Minor deviations of a physical system from its abstract representation by a mathematical model generally produce no serious discrepancies between theoretical response estimates and observed behaviour. However, very slight curvature of thin panels can cause considerable discrepancies between the theoretical and observed natural frequencies of some low-order modes. At frequencies far above the fundamental resonance, the resonance frequencies and, to a lesser extent, mode shapes are increasingly sensitive to minor perturbations of the structural detail.

The degree of 'overlap' of adjacent resonant responses is quantified by the 'modal overlap factor', which is the ratio of the average half-power bandwidth of modes having resonance frequencies within some selected frequency bands to the average separation between adjacent resonance frequencies. The latter is the inverse of the 'average modal density' of a system. These parameters are more fully defined in Section 11.5.

Figure 11.3 illustrates what happens to the complex frequency response of a structure if the resonance frequency of one of a few substantially contributing, overlapping modes is altered by one half-power bandwidth; the change can be very large indeed. Consequently, the frequency response at any point on a complex, multi-mode structure is increasingly sensitive to small perturbations of a structure and/or its boundary conditions

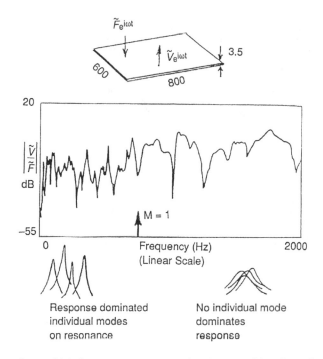

Figure 11.2 Frequency response showing transition from low-frequency, isolated mode, response to high-frequency, overlapped multi-mode response.

as frequency increases. We inevitably lack precision in theoretically modelling structural detail, especially in regard to damping distribution and joint flexibility. Dynamic properties may vary from day to day and with operational conditions. These factors cause *predictive uncertainty*. We can't account in a deterministic manner for small differences of between nominally identical industrial artefacts. Therein lies the origin of the *variance* of frequency response, illustrated by measurements of vibroacoustic transfer functions of mass-produced products such as cars (Figure 11.4).

The increasing sensitivity of the natural frequencies of multi-mode structures to structural perturbation, as mode order increases, is well illustrated by Figure 11.5. It shows the probability distribution of the bending mode natural frequencies of a population of bars having the same uniform bending stiffness and the same total mass, but a random distribution of mass per unit length. The spread of the modal frequencies increases with the order of the mode. The spread is so large for the higher orders of mode that there is a finite probability that the natural frequencies of some orders of modes 'exchange' relative positions. The degree of overlap in this respect

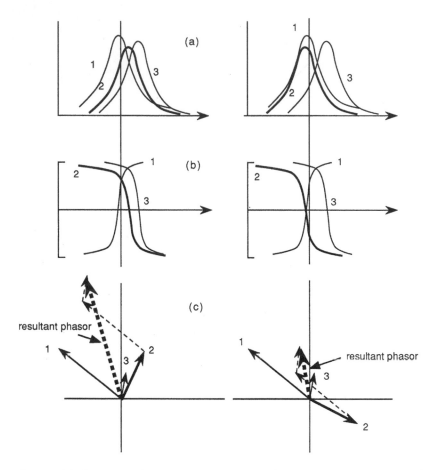

Figure 11.3 Illustration of the sensitivity of multi-mode frequency response to a small shift of individual natural frequency: (a) modal amplitudes; (b) modal phases; (c) resultant complex response.

is quantified by the 'statistical modal overlap factor', which is more fully defined in Section 11.5.

So, we see that the qualification of 'high' in high frequency relates to frequencies many times the fundamental natural frequency of a system. High-frequency vibration involves the resonant response of high-order modes. It has also been defined to be the vibration in which the structural wavelengths are much smaller than the typical system dimension; but this definition is not very satisfactory in cases of strongly non-uniform structures such as rib-stiffened plates.

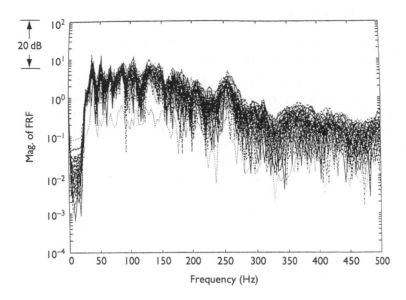

Figure 11.4 Magnitudes of the structure-borne FRFs for the driver microphone of 99 RODEO vehicles. After Kompella and Bernhard (1996).

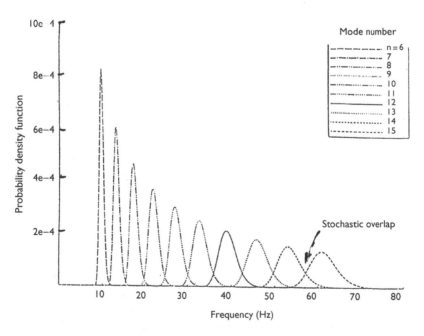

Figure 11.5 Probability density function of the flexural natural frequencies of a set of beams having randomised distributions of mass per unit length (normalised standard deviation = 10 per cent). After Manohar and Keane (1994).

11.3 The modelling problems posed by high-frequency vibration

Most engineering structures that are of concern in respect to high-frequency vibration are complex in terms of geometry, forms of construction, number of distinct components, range of materials and forms of connection. It is, in principle, possible to construct detailed, precise (deterministic) mathematical models of such complex assemblages, and to apply the well-established computational technique, finite element analysis (FEA), to esti-mate the *linear* response to applied forces/displacements of any component at any frequency and at any point. In the application of FEA, the structure is represented by a contiguous array of 'elements', defined by a 'mesh'. These are selected to suit the particular geometric form and composition of the structure, and the required frequency range of analysis.

There are two major impediments to this approach. One is technical in nature; it relates to the size of the theoretical model and the speed of imple-mentation of response calculations. (It will be largely overcome by the future development of faster and more powerful computers.) Part of the problem lies in the requirement that the size of the finite elements used to represent any component must be considerably smaller than the minimum structural wavelength in that component at any frequency. Consequently, the required number of elements (the model size) increases with frequency to a power of between two and three, depending upon whether a thin plate or shell model is adequate, or whether the geometry and structural wavelengths demand a three-dimensional array of elements. The effort required to define an FE model increases with size and with the geometric and material complexity of a model.

This is not the only factor that influences the computational demand. There are two basic methods of estimating the response of a linear structure to applied dynamic forces (see Chapter 10). If it is acceptable to assume that the internal damping is distributed in proportion to the local mass and/or stiffness, the natural modes are 'normal' (mass-weighted orthogonal). In this case, the modes of the undamped structure are real and computed once only. The response at any frequency to any spatial distribution of applied forces may then be expressed as a simple sum of modal responses; this is known as 'modal synthesis'. On the other hand, if the damping is not proportion-ally distributed (as it certainly is not in most passenger vehicles, because of installed damping treatments), the modes are complex, and the forced response problem has to be solved explicitly for each frequency of interest and form of excitation (see Section 11.4.2); this is a very large task indeed.

Even in cases where 'modal synthesis' is adequate, the resulting demand in computational resources necessary to solve 'high frequency' problems exceeds that which is considered by most companies to be acceptable for the task. The number of modes involved can run into millions for large structures such as rocket launchers, buildings, aircraft and ships. Execution time becomes so large that it is impracticable to run a large number of cases in the search for design improvements. This technical impediment will be eventually overcome in the future as the speed and memory capacity of widely affordable computer systems increase.

The second impediment is more profound, and unavoidable, because it is associated with an intrinsic feature of wave dynamics and is caused by the impossibility of possessing *complete* and *precise* knowledge of the dynamic properties of any modelled system. As explained earlier, there exists, in all cases, a degree of uncertainty concerning the dynamic properties of complex structures, particularly in relation to damping distribution, joints and connections between components, and the effects of changes of environmental conditions and operational state. Consequently, there exists *irreducible uncertainty* concerning the high-order natural modes and frequencies. The apparent precision of response prediction based upon a single deterministic model is consequently illusory. In addition, the precise form of excitation forces is rarely known. The reliability of deterministic response estimates correspondingly decreases to the point where it is not sufficient to justify the expenditure of resources necessary to construct the models and implement the computational procedures. In such cases, FEA is not 'wrong'; it is simply inappropriate. Consequently, alternative modelling philosophies and techniques are needed. These are introduced in Sections 11.5 and 11.6.

11.4 Mathematical description of high-frequency vibration

11.4.1 Introduction

In order to develop a mathematical model of the behaviour of a dynamic system it is necessary to identify 'degrees of freedom' that describe the motion of the system. For a discrete system, such as a mass-spring system, this process is straightforward, as the rigid body motions of the various mass elements provide a natural choice. For a continuous elastic system, such as a beam or plate, or a complex engineering structure, the situation is much more complicated; in theory, an infinite number of degrees of freedom are required for a complete description of the displacements of each point on the system. In the finite element method this problem is addressed by specifying the displacement of 'nodal points' distributed throughout the structure, and assuming that the displacement between the nodal points follows a simple pattern described by 'shape functions'. This will provide an accurate description provided there are enough nodal points to capture the physical deformation of the structure – the usual recommendation is at least six nodes per half wavelength of deformation.

Given that the wavelength of vibration decreases with increasing frequency, it is clear that a finite element model will require an increasing number of degrees of freedom as the excitation frequency is increased. As described in the following sections, the problem size can be reduced by a transformation to global mode coordinates or by modal synthesis methods employing subsystem modes. Even so, the required number of degrees of freedom can become unmanageable, and an alternative description of the response is needed. This can be provided by considering the response to be a superposition of elastic waves and/or by focusing interest on average measures of the response rather than the detailed deformation pattern. These issues are also described in the following sections, and this serves in part as an introduction to the SEA described in Section 11.5.

11.4.2 Deterministic modal description – global modes

The equations of motion of a complex structure can be written within the context of the finite element method as

$$\mathbf{M\ddot{y}} + \mathbf{C\dot{y}} + \mathbf{Ky} = \mathbf{F} \tag{11.1}$$

where $\mathbf{y}$ is a vector containing the N nodal degrees of freedom, $\mathbf{M, C}$, and $\mathbf{K}$ are the $N \times N$ mass, damping and stiffness matrices, and $\mathbf{F}$ is the vector of applied forces. The natural frequencies ω_j and mode shapes $\boldsymbol{\phi}_j$ of the system satisfy the equation

$$-\omega_j^2 \mathbf{M} \boldsymbol{\phi}_j = \mathbf{K} \boldsymbol{\phi}_j \tag{11.2}$$

and the number of degrees of freedom used to describe the motion of the system can be reduced by a 'transformation to modal coordinates' in the form

$$\mathbf{y} = \boldsymbol{\Phi}\mathbf{a}, \qquad \boldsymbol{\Phi} = (\boldsymbol{\phi}_1, \boldsymbol{\phi}_2, \ldots, \boldsymbol{\phi}_M) \tag{11.3, 11.4}$$

Physically, equation (11.3) states that the response can be represented in terms of a limited number $(M < N)$ of modes, with the amplitude of the jth mode being written as a_j. The modal amplitudes $\mathbf{a}$ form the new set of degrees of freedom, and combining equations (11.1) and (11.3) leads to the following set of equations of motion

$$\mathbf{\ddot{a}} + \boldsymbol{\Phi}^{\mathrm{T}} \mathbf{C} \boldsymbol{\Phi} \mathbf{\dot{a}} + \boldsymbol{\Lambda}\mathbf{a} = \boldsymbol{\Phi}^{T}\mathbf{F} = \mathbf{G} \tag{11.5}$$

Here $\boldsymbol{\Lambda}$ is a diagonal matrix with $\Lambda_{jj} = \omega_j^2$, and the mode shapes have been scaled to unit generalised mass (i.e. scaled so that the coefficient of each acceleration term is unity). The transformed damping matrix $\boldsymbol{\Phi}^{\mathrm{T}} \mathbf{C} \boldsymbol{\Phi}$ that appears in equation (11.5) is generally coupled, although in the special case of 'Rayleigh damping' (where $\mathbf{C}$ is taken to be a linear combination of $\mathbf{M}$

and **K**), the matrix is diagonal. In this special case, the response of the system to harmonic forcing at frequency ω can be written in the simple form

$$\mathbf{y} = \sum_{j=1}^{M} \frac{\boldsymbol{\phi}_j G_j}{\omega_j^2 - \omega^2 + i\eta_j\omega_j\omega} \tag{11.6}$$

where η_j is the loss factor (arising from **C**) associated with mode j. Equation (11.6) has been derived here within the context of the finite element method, but it can be noted that a continuum analysis leads to a very similar form of result. The response at location **x**, $u(\mathbf{x})$ say, of a structure to a point load F applied at position $\mathbf{x}_0$ can be written as

$$u(\mathbf{x}) = \sum_{j} \frac{F\phi_j(\mathbf{x})\phi_j(\mathbf{x}_0)}{\omega_j^2 - \omega^2 + i\eta_j\omega_j\omega} \tag{11.7}$$

where $\phi_j(\mathbf{x})$ represents the jth mode shape.

Although, in principle, equation (11.6) or (11.7) represents a complete solution of the vibration problem, in practice, there are many practical difficulties in applying the result to the case of high-frequency excitation. The main issues are as follows:

(i) At high frequencies the higher modes of vibration are excited; these tend to have a short wavelength, and the finite element method must contain enough nodal points to capture the detailed structural deformation. Beyond a certain frequency the number of degrees of freedom N required to do this can become impractical, even with modern computational resources.

(ii) The assumption of Rayleigh damping is generally made for the sake of computational convenience, rather than on physical grounds. This means that the transformed matrix $\Phi^{\mathrm{T}}\mathbf{C}\Phi$ will tend to have off-diagonal entries that are ignored by the Rayleigh damping approximation. If the modes of vibration are well separated, in the sense that the modal spacing is large compared to the modal bandwidth, then such off-diagonal terms have a very small effect on the response. In other cases, however, the off-diagonal terms can have a very significant effect, and the Rayleigh damping approximation is not appropriate. The ratio of the modal spacing to the modal bandwidth is measured by the modal overlap factor $M = \omega\eta n(\omega)$, where n is the mean modal density. M generally increases with increasing frequency, and thus the Rayleigh approximation becomes invalid at high frequencies. In principle, the more general case can be dealt with by the direct solution of equation (11.5) and/or the use of complex modes (Newland 1989), although either approach can be cumbersome for systems with many degrees of freedom.

(iii) As explained before, the natural frequency ω_j and mode shape ϕ_j become increasingly sensitive to the structural properties with increasing j. This implies that computed response levels based on nominal design properties may be very different from measured values, and ideally some degree of statistical analysis based on random system properties is required. Within the context of the finite element method, such an analysis would typically consist of a Monte Carlo numerical study in which repeated calculations are performed for randomly generated sets of the system properties; this would require very extensive computer time at high frequencies.

Given these difficulties, it is clear that any variant on the 'direct' finite element approach that leads to a reduced number of degrees of freedom would be welcome. One alternative approach is modal synthesis, in which the degrees of freedom are taken to be the 'local' modes of subsystems of the total structure, rather than the 'global' modes.

11.4.3   Deterministic modal description – local modes

Rather than computing the modes of the whole structure via equation (11.2), an alternative is to divide the structure into regions or subsystems. The modes of each subsystem can then be computed, and the degrees of freedom of the whole system represented as a collection of subsystem modes (Craig 1981). In more detail, a decision must be taken on the boundary conditions to be applied to each subsystem before the modes are computed. The most common approach is to restrain (i.e. clamp) the boundaries of the subsystem and compute the so-called 'blocked' modes. These must then be augmented by 'constraint' modes associated with the boundary degrees of freedom. These are not modes of vibration, but rather static deflection shapes obtained by taking a unit deflection in each boundary degree of freedom in turn. The total degree-of-freedom vector for the system then has the form

$$\mathbf{a}_t^T = (\mathbf{a}_{b1}^T \ \mathbf{a}_{c1}^T \ \mathbf{a}_{b2}^T \ \mathbf{a}_{c2}^T \ldots \mathbf{a}_{bR}^T \ \mathbf{a}_{cR}^T) \tag{11.8}$$

where $\mathbf{a}_{bj}^T$ and $\mathbf{a}_{cj}^T$ are respectively the amplitudes of the blocked and constraint modes in subsystem j, and R is the total number of subsystems. The equations of motion can be formulated in terms of $\mathbf{a}_t$ and then solved to yield the response of the system. The equations of motion will generally be highly coupled due to the presence of the constraint modes, but the total number of degrees of freedom employed will be significantly less than N, as a truncated set of blocked modes will be employed for each subsystem. This approach has computational advantages over the direct approach described in Section 11.4.2, particularly at high frequencies, but the basic difficulties, (i)–(iii), outlined in that section still remain.

Clearly, it is not feasible, or even desirable, to perform a detailed FEA of high-frequency vibration, and some alternative way of modelling the system is required. The modal synthesis approach suggests that the local modes of subsystems of the structure (coupled through constraint modes) might provide a suitable description of the system motion, but it is not reasonable to compute all the local mode shapes deterministically. The question then arises as to whether an alternative modelling approach may be found in which only gross or averaged property of the local modes is required. As described in Section 11.4.4, it is possible to compute important averaged properties of the local modes very easily. It is then shown in Section 11.4.5 that the local modes may be re-interpreted in terms of elastic waves, and this lays the foundation for the SEA prediction technique, described in Section 11.5.

11.4.4 Statistical local modes – modal density, modal overlap and statistical overlap

The modal density of a subsystem is defined as the average number of natural frequencies per unit frequency interval. Thus, the number of natural frequencies to be expected in a narrow frequency band is $n(\omega)d\omega = n(f)df$ (note: $n(f) = 2\pi n(\omega)$). Analytical results are available for the *asymptotic* modal density of many types of components, such as beams, plates, shells or fluid volumes. These results are usually independent of the boundary conditions applied to the component, and they are asymptotic in the sense that they become increasingly accurate with increasing frequency. Some examples of $n(\omega)$ are as follows:

- Beam (flexural modes): $n = (L/2\pi)(m/EI)^{1/4}\omega^{-1/2}$
- Plate (flexural modes): $n = (S_0/4\pi)(\rho h/D)^{1/4}$
- Fluid in rigid-walled rectangular enclosure (acoustic modes): $n = (V\omega^2/2\pi^2 c^3) + (A\omega/8\pi^2 c^2) + (P/16\pi c)$.

L, EI and m are respectively the length, flexural rigidity and mass per unit length of the beam; S_0, D and ρh are respectively the area, flexural rigidity and mass per unit area of the plate; c, V, A and P are respectively the speed of sound, the volume, enclosing surface area and perimeter length of the acoustic enclosure. A full derivation of these results is given by Lyon and DeJong (1995), and more general expressions for one- and two-dimensional systems are derived in Section 11.4.6. The expression for the acoustic modal density of a fluid in a rectangular enclosure having pressure release boundary conditions is at odds with the earlier statement regarding boundary conditions; the second term changes in sign. However, at high frequencies the first term dominates the expression and the asymptotic result does indeed become independent of boundary conditions. It can be seen from these examples that the modal density can either decrease, remain constant or increase with increasing frequency.

As stated previously, the 'modal overlap factor' is defined as the ratio of the modal spacing to the modal bandwidth, and this can be written as $M = \omega\eta n(\omega)$. Given the availability of asymptotic results for n it is possible to estimate M for most common types of subsystem, provided the loss factor η is known. M provides an indication of the smoothness of the response of a system when plotted as a function of frequency. The response given by equation (11.6) or (11.7) will display a peak in the vicinity of each natural frequency, but if the modal overlap is high then these peaks will merge to produce a relatively smooth response function. A quantitative measure of this effect can be obtained by noting that, if the natural frequencies are *evenly spaced* with density n, then the *space-average* point mobility $\overline{Y}$ of an isolated subsystem has the following properties (Langley 1994):

$$\overline{Y}_{max} = \langle\overline{Y}\rangle \coth\left(\pi\frac{M}{2}\right), \qquad \overline{Y}_{min} = \langle\overline{Y}\rangle \tanh\left(\pi\frac{M}{2}\right) \qquad (11.9, 11.10)$$

Here, $\langle\overline{Y}\rangle$ is the frequency-average result, so that the functions $\coth(\pi M/2)$ and $\tanh(\pi M/2)$ form 'envelopes' around the frequency response function. Let us consider two examples: (i) for $M = 0.1$, $\overline{Y}_{max} = 6.4\langle\overline{Y}\rangle$ and (ii) for $M = 1$, $\overline{Y}_{max} = 1.1\langle\overline{Y}\rangle$. The frequency response function can therefore be expected to become relatively smooth as M approaches unity. Another physical interpretation of M is that a harmonic force will produce a significant response in M modes (since any particular frequency lies within the bandwidth of M modes).

Another key parameter in describing the properties of a subsystem is the 'statistical overlap' factor (Manohar and Keane 1994). Random uncertainties in the subsystem properties will produce random uncertainties in the natural frequencies. The statistical overlap factor is defined by $S = 2\sigma n$, where σ is a representative value for the standard deviation of a natural frequency. Experimental results that illustrate the effects of random system properties are shown in Figure 11.6 (Johnson 2001). The structure considered is a beam with twelve attached lumped masses; the various curves represent the transfer function between two fixed points on the beam for different locations of the masses, which have been selected randomly. The response curves form an 'ensemble' of results, and the term 'ensemble statistics' relates to the statistical properties of these curves, either at a fixed frequency or averaged over a frequency band. [It can be noted from the results that the ensemble variance increases with increasing statistical overlap (i.e. increasing frequency), and at high statistical overlap the ensemble mean and variance can be expected to be rather independent of frequency.] Furthermore, it seems to be a reasonable proposition that the response is *ergodic* at high values of statistical overlap, meaning that the

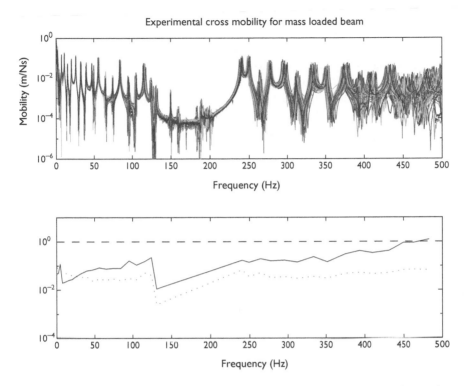

Figure 11.6 Experimental results for a mass-loaded beam. The upper plot shows the transfer function for different locations of the masses. The lower plot shows the statistical overlap factor (solid line) and the modal overlap factor (dotted line).

response statistics across the ensemble at a fixed frequency are the same as those along the frequency axis for a fixed structure. This issue is discussed in more detail in Section 11.4.8.

11.4.5 Wave motion and wave-mode duality

It was shown in Sections 11.4.2 and 11.4.3 that it is not practical to develop a detailed deterministic vibration model based on either the global modes of the whole system or the local modes of the constituent subsystems. However, it was shown in Section 11.4.4 that it is possible to derive a number of important average quantities relating to the subsystem modes. From a modal point of view, the motion of a subsystem can be viewed as a superposition of local modes, each vibrating at some particular phase and amplitude in response to the applied forces; the number of modes responding in a given frequency band can be estimated from the modal density, and the 'smoothness' of the frequency response function can be judged from the modal overlap factor.

An alternative to the modal model is to consider the motion of a subsystem to be made up of elastic plane waves of various amplitudes, headings and phases. To relate this to the modal view, it is helpful to consider initially the special case of a simply supported plate of platform dimensions $L_1 \times L_2$. The natural modes take the form $u(x_1, x_2) = \sin k_{1n}x_1 \sin k_{2m}x_2$, where $k_{1n} = n\pi/L_1$ and $k_{2m} = m\pi/L_2$; the natural frequencies are given by $\omega_{nm} = (D/\rho h)^{1/2}(k_{1n}^2 + k_{2m}^2)$. Each mode can be represented by a point on the k_1–k_2 plane, as shown in Figure 11.7a, with the corresponding natural frequency being proportional to the square of the distance to the origin, k_{nm} say ($k_{1n} = k_{nm} \cos\theta_{nm}$ and $k_{2m} = k_{nm} \sin\theta_{nm}$). Now the response of a typical mode at frequency ω can be written as

$$e^{j\omega t} \sin k_{1n}x_1 \sin k_{2m}x_2 = -\left(\frac{1}{4}\right)\{e^{jk_{1n}x_1+jk_{2m}x_2} - e^{-jk_{1n}x_1+jk_{2m}x_2}$$
$$-e^{jk_{1n}x_1-jk_{2m}x_2} + e^{-jk_{1n}x_1-jk_{2m}x_2}\}e^{j\omega t} \qquad (11.11)$$

where the complex harmonic time dependency $\exp(j\omega t)$ has been adopted. The right-hand side of this expression represents the summation of four plane wave components. These components are indicated in Figure 11.7b, where it can be seen that the physical heading of each wave is given by the polar angle θ subtended on the k_1–k_2 plane. Now consider the case in which the plate is subjected to excitation over the frequency band $\omega_1 \leq \omega \leq \omega_2$. The frequencies ω_1 and ω_2 can each be represented by a circular contour in

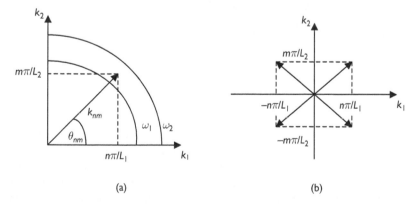

(a)   (b)

Figure 11.7 (a) A mode of vibration represented as a point in k-space; (b) the four wave components that appear in equation (11.11).

the k_1–k_2 plane, and the majority of the response will arise from modes that lie between these contours and are therefore resonated by the excitation. The response will thus have the form

$$\sum_{n,m} A_{nm} e^{j\omega t} \sin k_{1n} x_1 \sin k_{2m} x_2 = \sum_{\pm n, \pm m} B_{nm} e^{-j(x_1 k_{nm} \cos \theta_{nm} + x_2 k_{nm} \sin \theta_{nm}) + j\omega t}$$

$$(11.12)$$

where A_{nm} and B_{nm} are complex amplitudes (whose precise values are not of interest in the present discussion), and the sum on the left-hand side covers the resonant modes. In line with equation (11.11), the sum on the right-hand side will include four wave headings for each mode; thus the response can be viewed as a superposition of plane waves that cover a range of headings. It can be noted that some liberty has been taken with the use of ω in equation (11.11), as an excitation band has been considered and yet a single frequency has been employed in the equation; if the band is narrow, then ω can be identified as the centre frequency, otherwise ω_{nm} might be used for mode nm, assuming that the response is predominantly resonant.

A common assumption in high-frequency vibration analysis is that the right-hand side of equation (11.11) represents a 'diffuse' wavefield, in the sense that equal wave energy flows in all directions. This requires a sufficient number of terms in the sum, together with the condition that the amplitude B_{nm} is of similar magnitude for each term. The latter condition requires that the magnitude of A_{nm} is similar for each term in the modal sum, and this is referred to as 'equipartition of modal energy'. Thus, equipartition of modal energy is the 'modal' equivalent of a diffuse wavefield.

The equivalence between the wave and the modal views of the subsystem response is sometimes referred to as 'wave-mode duality' in the literature – a more detailed discussion of this issue has been given by Lyon and DeJong and Langley in Fahy and Price (1999). As discussed in Section 11.4.7, the wave view can provide a greater physical insight into vibrational energy transfer between subsystems.

11.4.6 Wave properties – phase velocity, group velocity, reverberance

A plane wave propagating in the x-direction has the general complex form

$$u(x, t) = B e^{j(\omega t - kx)} \tag{11.13}$$

where ω is the frequency and k is the wavenumber (note that each wave component in equation (11.11) can be written in this form by making the substitution $x_1 = x \cos \theta_{nm}$ and $x_2 = x \sin \theta_{nm}$). The frequency and wavenumber are connected via the 'dispersion relation' which is derived by substituting equation (11.13) into the differential equation of motion of the system; for example, the dispersion relation of a beam is given by $EIk^4 = m\omega^2$. The speed of propagation of a wave crest is given by the 'phase velocity'

$$c = \frac{\omega}{k} \tag{11.14}$$

Another key parameter is the 'group velocity'

$$c_g = \frac{\partial \omega}{\partial k} \tag{11.15}$$

which is the velocity of the envelope of a group of waves that have near-identical frequencies. Perhaps, more importantly, the group velocity is also the velocity of energy flow in the wave, so that the time-averaged energy flow per unit length of wave crest is given by

$$P = \langle e \rangle c_g \tag{11.16}$$

where $\langle e \rangle$ is the time-averaged energy density (kinetic plus potential). The time-averaged kinetic and potential energies are equal for travelling wave motion, and thus $\langle e \rangle$ can be written as twice the time-averaged kinetic energy density, so that

$$\langle e \rangle = \left(\frac{1}{2}\right) m\omega^2 |B|^2 \tag{11.17}$$

where m is the appropriate mass density.

It is possible to express the modal density of a subsystem in terms of the phase velocity and the group velocity. The general results for one- and two-dimensional subsystems are as follows (Lyon and DeJong 1995):

$$n_{1-D} = \frac{L}{\pi c_g}, \qquad n_{2-D} = \frac{S_0 \omega}{2\pi c c_g} \tag{11.18, 11.19}$$

where L is the length of the one-dimensional system and S_0 is the area of the two-dimensional system. The results quoted in Section 11.4.4 for a beam and a plate are specific applications of these equations. The derivation of equations (11.18) and (11.19) is surprisingly straightforward; for example, the area enclosed under a quarter circle of radius k in Figure 11.7a is $\pi k^2/4$; each modal point on the diagram is associated with an area $\pi^2/L_1 L_2$, so that the number of modes enclosed by the quarter circle is $S_0 k^2/4\pi$; the rate of change of this result with frequency then leads immediately to the modal density equation (11.19).

An important insight into the physical nature of the response of a damped system can be gained by considering the effect of damping on the wave motion. First it can be noted that damping can be introduced into a mathematical model of harmonic vibration by replacing the frequency ω^2 by the complex result $\omega^2(1 - i\eta)$, where η is the loss factor. The change produced in the wavenumber k can be deduced as follows:

$$k \rightarrow k + \frac{\partial k}{\partial \omega} \delta\omega = \frac{k - i\eta\omega}{2c_g} \tag{11.20}$$

Hence equation (11.13) becomes

$$u(x, t) = B e^{\frac{-\eta\omega x}{2c_g}} e^{j(\omega t - kx)} \tag{11.21}$$

If a wave decays significantly over the span of a subsystem then the subsystem is said to be *non-reverberant*; conversely, if the wave shows little reduction in amplitude when reflected many times back and forth across the subsystem, then the subsystem is said to be *reverberant*. Equation (11.21) can thus be used to assess the degree of 'reverberance' of a subsystem. For a *one-dimensional* subsystem of length L, the decay in the wave amplitude across the subsystem is given by $\exp(-\eta\omega L/2c_g) = \exp(-\pi M/2)$, where M is the modal overlap factor. Thus if M is unity or greater, then the wave decays to less than 20 per cent of its original amplitude. This highlights an important physical result; *if the modal overlap factor of a one-dimensional subsystem is greater than unity, then the system is non-reverberant*. For a *two-dimensional* subsystem, the decay over a length L has the form $\exp(-\pi M L/Sk)$. Even for M greater than unity the resulting decay may be relatively small, and thus *a modal overlap factor of unity or greater does NOT imply that a two-dimensional subsystem is non-reverberant*. This very important difference between the behaviour of one- and two- (and higher) dimensional subsystems makes any one-dimensional subsystem a 'special case', and care should be taken in attempting to draw general conclusions from the analysis of a one-dimensional problem.

11.4.7 Vibration transmission from a wave perspective

One advantage of the wave view of high-frequency structural vibration is that it allows a straightforward quantitative analysis of the transmission of vibrational energy from one subsystem (say subsystem 1) to another adjoining subsystem (say subsystem 2). As an example, two two-dimensional subsystems that are coupled along a line junction are shown in Figure 11.8. A wave of amplitude B_i in subsystem 1 that impinges on the boundary will be partly transmitted into B_t and partly reflected into B_r; this can be quantified by calculating the 'reflection and transmission coefficients' of the junction (Graff 1975; Cremer *et al.* 1987). These coefficients are found by considering compatibility and equilibrium between two

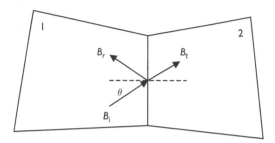

Figure 11.8 Wave transmission across a junction between two subsystems.

semi-infinite subsystems that are connected via an infinitely long line junction. The resulting coefficients are functions of the direction of the incident wave, so that the reflected and transmitted waves are written as

$$B_t = T_{12}(\theta)B_i, \qquad B_r = R_{11}(\theta)B_i \qquad (11.22, 11.23)$$

Given the reflection and transmission coefficients, it is straightforward to derive the 'power transmission and reflection coefficients' $\tau_{11}(\theta)$ and $\rho_{11}(\theta)$ that quantify how much of the power that is incident on the junction is transmitted and reflected. If the incident wavefield is taken to be diffuse, so that equal power is carried in all directions, then it is appropriate to define a 'diffuse field transmission coefficient', which is simply the power transmission coefficient averaged over the range of incident directions

$$\langle \tau_{12} \rangle = \left(\frac{1}{2}\right) \int_{-\frac{\pi}{2}}^{\frac{\pi}{2}} \tau_{12}(\theta) \cos\theta d\theta \qquad (11.24)$$

The term $\cos\theta$ appears in this expression to account for the fact that the power carried by a wave of heading θ must be resolved normal to the junction to calculate the transmitted power. As discussed in Section 11.5, $\langle \tau_{12} \rangle$ plays an important role in SEA.

11.4.8 Spatial-, frequency- and ensemble-averages

In Section 11.4.6 the symbol $\langle e \rangle$ was used to represent the time averaged energy density of a subsystem. If the response is harmonic with frequency ω then the time average simply consists of averaging over a cycle of vibration, and in more detail $\langle e \rangle$ can be written in the form $\langle e(\mathbf{x}, \omega, \mathbf{p}) \rangle$, where $\mathbf{x}$ represents the spatial position and $\mathbf{p}$ is a set of parameters that describe the subsystem properties, such as the geometry and material constants. Clearly $\langle e \rangle$ could be averaged over each of the arguments $\mathbf{x}$, ω and $\mathbf{p}$ to yield respectively a spatial, frequency and ensemble average. The latter average involves random variations in the parameters **p**; as discussed in Section 11.4.4 and shown in Figure 11.6, the effect of these variations will depend upon the statistical overlap factor S. If S is large (say greater than two or three) then it can be postulated that the response is ergodic, in the sense that the frequency and ensemble averages are the same so that

$$E\left[\langle e(\mathbf{x}, \omega, \mathbf{p})\rangle\right] = \frac{1}{2\Delta\omega} \int_{\omega-\Delta\omega}^{\omega+\Delta\omega} \langle e(\mathbf{x}, \omega, \mathbf{p})\rangle\, d\omega \tag{11.25}$$

The expression on the left of this equation is the ensemble average, while the expression on the right is the average over a frequency band of width $2\Delta\omega$. Clearly, the validity of equation (11.25) will depend upon $\Delta\omega$; if the modal overlap factor is small, so that $\langle e \rangle$ fluctuates significantly with frequency, then $\Delta\omega$ may need to cover a reasonable number of resonant modes (say three or more); at high modal overlap, where $\langle e \rangle$ will be a smooth function of ω, $\Delta\omega$ may be very narrow, or even zero.

The concepts of frequency and ensemble averaging are employed in SEA. Clearly, the average is the simplest possible statistical measure, and a more complete description of the response would require knowledge of at least the variance and ideally, the full probability density function. This issue is considered in more detail in Section 11.6.

11.5   Introduction to SEA

11.5.1   The rationale of SEA

Given the intrinsic uncertainty regarding the high-frequency dynamic properties of complex structural assemblages, it is logical to seek to develop theoretical representations that *implicitly incorporate uncertainty*, and to adopt measures of vibrational state that are robust against variations of detail which have no major implications for overall system response, integrity or adequacy. The longest established, and most widely applied, *probabilistic* model for estimating high-frequency response is SEA.

Although the natural frequencies and mode shapes of the high-order modes of complex systems cannot be evaluated theoretically with great precision, it is possible to characterise the high-frequency dynamic properties of structures and associated vibration fields in energetic and statistical terms. In this respect it is helpful to visualise a population of industrial artefacts coming off a production line. If the quality control is good, they will all be very similar. However, it is inevitable that they will differ from each other in detail because of manufacturing and assembly tolerances and variations of material properties. These small differences are sufficient to create substantial differences in the high-frequency response. If an attempt is made to represent the artefact by means of a deterministic mathematical model, the following question arises 'Which member of the population does the model truly represent?' Naturally, we cannot say – maybe none. We may, at best, consider it to represent a virtual item that embodies the average properties of the population of items; these are known as the 'ensemble-mean' properties. Consequently, the vibrational response to any single frequency excitation of any individual member of the population will probably differ from that predicted from the model. The spread of responses of a population of artefacts may be described in terms of a probability distribution.

We actually know very little about the ensemble statistics of the dynamic properties and responses of sets of manufactured products. The principal reason is the prohibitive cost of obtaining the data. It is therefore common in the theory and application of SEA to substitute averages over members of a hypothetical population by averages over the modes of an archetypal system (or over a frequency bandwidth which includes a number of modal natural frequencies). It is known that these averages are equivalent only under particular circumstances, which also happen to be favourable to the reliability of SEA. In this respect, it must be recognised that any individual experimental test specimen, together with any test results, differs to an unknown degree from the ensemble average of the parent population. However, where the target response quantities are *subsystem energies*, rather than local responses, and where the energies are specified in finite bandwidths (such as 1/3 octave bands or dB (A)), the ensemble variance tends to be considerably reduced.

11.5.2 The SEA model

A probabilistic model requires that the parameters which characterise the physical properties of the system, the excitation and the vibrational state (response), are quantified in statistical terms. In SEA modelling

- Analysis is made in finite frequency bands which are broad enough to encompass a suitable large number of modal frequencies but not so large as to obscure the variation of response with frequency.
- A complex system is represented by a network of 'subsystems'. Each subsystem should ideally differ substantially in dynamic properties from its immediate neighbours, so that vibrational/acoustic waves are rather strongly reflected at its mutual interfaces. In technical terms, the wave impedances should be very different (see note 1, p. 527).
- A number of different wave types (e.g. flexural, longitudinal) may coexist within a given structural component. In this case, the component comprises more than one subsystem – one for each wave type.
- Certain subsystems are assumed to be subjected to mutually uncorrelated (statistically independent) time-stationary excitation(s) of which the frequency spectra are normally assumed to be rather smooth and

to encompass many natural frequencies of the system (it is possible to relax this latter condition at the expense of increased uncertainty of predicted response).

- Each subsystem is assumed to support a reverberant vibrational (or acoustic) wave field, thereby excluding any highly damped component which must be modelled so as to account for propagating wave attenuation.

- The vibrational state of a subsystem, denoted by E, is defined in terms of the sum of its time-averaged kinetic and potential energies.

- The 'modal density', denoted by $n(\omega)$ or $n(f)$, quantifies the inverse of the mean separation of adjacent natural frequencies within a range centred on a particular frequency.

- The modal-average 'dissipation loss factor' (DLF) is the arithmetic mean of the dissipation loss factors (twice the damping ratios) of subsystem modes having natural frequencies within a selected range of frequency.

- The modal-average half-power bandwidth equals the product of the centre frequency of the selected band and the modal-average DLF.

- The average number of natural frequencies falling within the average modal half-power bandwidth is expressed by the 'modal overlap factor', denoted by M. When M exceeds unity, more than one mode contributes substantially to the frequency response at any frequency.

- The modal-average 'coupling loss factor' (CLF) relates the time-average energy stored in any one subsystem to the wave energy flux passing across an interface to another subsystem.

In its present state of development, SEA is capable of estimating only the ensemble-mean response in terms of the means of the quantities introduced. However, it is less than satisfactory in statistical terms to make predictions of the ensemble-mean behaviour of a population of similar systems without being able to estimate the associated probability distribution. This lack is not serious in cases where the modal overlap factor exceeds unity and the excitation has a broadband spectrum; but, at frequencies where modal overlap is low, or at any individual frequency, response variance can be of practical concern. Current research is aimed at remedying this lack.

11.5.3 Excitation

External excitation mechanisms are represented in SEA in terms of the time-average mechanical powers that they inject into subsystems. Theoretical and empirical models are available for a range of mechanical and fluid forces. But there exist many physically complex forms of excitation, such as tyre–road interaction, for which fully validated models are not available. For the purpose of design assessment and improvement, it is common to use SEA to predict the subsystem energies normalised by the input power of each source of excitation. The uncertainty of input powers should also be defined.

11.5.4 Stored energy

The principal output of a theoretical SEA analysis is the distribution of time-average vibrational/acoustic energy among the subsystems. The total stored energy of each subsystem is estimated; no spatial distributions are available. This 'global' measure of subsystem response is much less sensitive to small variations of structural detail than the response at any specific point in a subsystem. This is illustrated by Figure 11.9a,b which compare the effects of the attachment of a small mass on the autospectra of local and spatially integrated kinetic energy densities of a flat plate.

In a few ideal cases the energy is simply related to a measurable quantity. For example, the kinetic energy density of a plain panel in bending vibration is simply related to its local mean square normal vibrational velocity. Even the simple modification of attaching stiffeners renders this relation inaccurate because the moving mass is not accurately known. Acoustic energy density is simply related to the local mean square acoustic pressure.

11.5.5 Power flow

SEA deals with the flow of vibrational energy between subsystems. The rate of flow of energy should strictly be termed 'energy flux' to be consistent with other internationally agreed definitions, but the term 'power flow' is now endemic in SEA literature and will be employed herein. It must constantly borne in mind that all the following models and equations relate to a virtual, ensemble-mean system, from which the members of a hypothetical population may deviate to a greater or lesser degree.

11.5.6 The modal model

In 1960, R.H. Lyon approached the general problem of energy transfer between randomly excited structures in terms of a generic model of two linearly coupled, lumped oscillators, each of which was subjected to independent, broadband force excitation. He showed that the time-average power flow between the oscillators is proportional to the difference of their time-average *uncoupled* energies. Later, in collaboration with T.D. Scharton, it was shown that the power flow is also proportional to the difference of the *coupled* energies. Thus,

$$P_{12} = g_{12}(E_1 - E_2) \tag{11.26}$$

It has been shown that a similar relation holds for any system comprising no more than two distributed, multi-mode subsystems, provided that the energies are appropriately defined. But attempts to generalise this fundamental relation to a system represented by three or more arbitrarily coupled

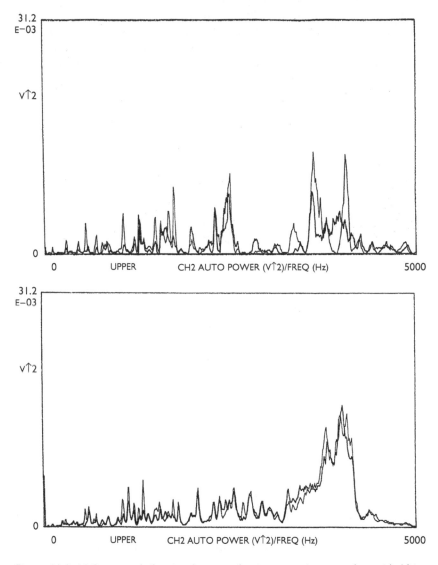

Figure 11.9 (a) Spectra of vibrational energy density at a point on a plate with (thin line) and without (thick line) local 1 per cent mass addition; (b) spectra of total vibrational energy of the plate with and without 1 per cent mass addition.

subsystems have failed. However, it has been shown that the relation holds in the limit of 'weak coupling'. Hence, the power flow between any two weakly coupled subsystems may normally be considered to be independent of the remainder of the system. However, non-zero CLFs may exist between non-contiguous subsystems even in some weakly coupled systems. The condition of weak coupling is further discussed.

The equivalent of equation (11.26) for a pair of weakly coupled subsystems, represented by sets of oscillators (modes) α and β, is obtained by summing the power flows between each pair of oscillators which are coupled through the subsystem interface:

$$P_{\alpha\beta} = \sum_{i}^{N_\alpha}\sum_{j}^{N_\beta} g_{ij}(E_i - E_j) \tag{11.27}$$

On the basis of the fundamental SEA assumption that the energy of each subsystem is *shared equally* between its N modes having natural frequencies in the analysis band

$$P_{\alpha\beta} = \left[\left(\frac{1}{N_\alpha}\right)\sum\sum g_{ij}\right]E_\alpha - \left[\left(\frac{1}{N_\beta}\right)\sum\sum g_{ij}\right]E_\beta \tag{11.28}$$

By analogy with the proportional relation between the power dissipated by a subsystem and its stored energy, which is expressed in terms of a DLF η by $P = \eta\omega E$, the relation between the power flow out of a subsystem and the resident energy may be expressed in terms of a CLF, as follows:

$$P_{\alpha\to\beta} = \eta_{\alpha\beta}\omega E_\alpha \quad \text{and} \quad P_{\beta\to\alpha} = \eta_{\beta\alpha}\omega E_\beta \tag{11.29}$$

in which η denotes CLF. Hence,

$$\eta_{\alpha\beta} = \left(\frac{1}{\omega N_\alpha}\right)\sum\sum g_{ij} \quad \text{and} \quad \eta_{\beta\alpha} = \left(\frac{1}{\omega N_\beta}\right)\sum\sum g_{ij} \tag{11.30}$$

Equation (11.28) may alternatively be written either in terms of *total* subsystem energies as

$$P_{\alpha\beta} = \omega\eta_{\alpha\beta}E_\alpha - \omega\,\eta_{\beta\alpha}E_\beta \tag{11.31}$$

or in terms of a difference of *modal* energies (average energy per mode) as

$$P_{\alpha\beta} = \omega\eta_{\alpha\beta}N_\alpha\left(\frac{E_\alpha}{N_\alpha} - \frac{E_\beta}{N_\beta}\right) \tag{11.32}$$

The number of modes N_α and N_β may be replaced in equation (11.32) by the corresponding modal densities $n_\alpha(\omega)$ and $n_\beta(\omega)$.

Equation (11.32) is more revealing than equation (11.31) because it clearly indicates that it is differences of modal energies (and not of total energies) which 'drives' energy flow between subsystems. Figure 11.10 presents

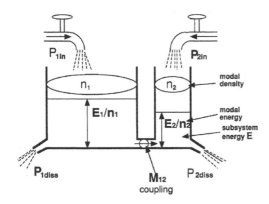

Figure 11.10 Hydraulic analogy of power flow between coupled subsystems.

a rough hydraulic analogy of the expression of equation (11.32). Equation (11.32) is also analogous to the heat conduction equation in which power flow is analogous to heat flow, the term $\omega\eta_{\alpha\beta}N_\alpha$ is analogous to the thermal conductivity and the modal energy is analogous to temperature.

The CLFs satisfy a consistency (sometimes called 'reciprocity') relation

$$n_\alpha\omega\eta_{\alpha\beta} = n_\beta\omega\eta_{\beta\alpha} = M_{\alpha\beta} \tag{11.33}$$

$M_{\alpha\beta}$, which has the same form as the modal overlap factor $M (= \eta\omega n(\omega))$, is not a widely accepted form of representation of the coefficient of proportionality; but it is analytically attractive because it is reciprocal (i.e. $M_{\alpha\beta} = M_{\beta\alpha}$). Unlike the CLF, it is a function of the extent of the interface between subsystems, and not of the overall length, area or volume of the transmitting subsystem. This dependence is more acceptable on physical grounds and in terms of simplicity of notation. Unfortunately, the CLF has become entrenched in SEA lore, and consequently it will be employed hereinafter.

11.5.7   The travelling wave model

The modal SEA model is more useful for the purposes of basic research than for practical applications. Modes are standing waves formed by interference between vibrational/acoustic waves travelling in different directions. If many resonant modes make substantial contributions to the vibration field in two- and three-dimensional subsystems in a given frequency band, many travelling wave directions are present. The ideal model, in which all wave directions are present with equal probability, is called a 'diffuse field'.

Analysis of the relations between energy density and energy flux in diffuse travelling wave fields reveals that the CLF may be written in terms of the diffuse field wave power transmission coefficient $\tau_{\alpha\beta}$ (see Section 11.4.7). This is defined as that proportion of wave energy in subsystem α travelling *towards* an interface with subsystem β that is transmitted through the interface, under conditions where the response of the receiving subsystem does not influence the transmission coefficient. This condition will obtain under one of the following conditions:

(i) The receiving system is highly damped so that very little wave energy is reflected back from its boundaries to the interface (but then it does not qualify as an SEA subsystem).
(ii) The receiving system is infinitely extended so that no wave energy is reflected back to the interface (but then it does not qualify as an SEA subsystem).
(iii) Any waves reflected back to the interface are uncorrelated with the incident waves at the interface. This is one of the essential conditions of *weak coupling*. The condition is promoted by high modal density, irregular shape, broad band excitation and multiple, uncorrelated input forces. It is, of course, violated by an isolated 'global' mode which involves substantial correlated motion in two or more contiguous subsystems.

Diffuse field CLFs for specific forms of interface are calculated by either analytical (see Lyon and DeJong 1995) or numerical models and are available from commercial software.

11.5.8   The effect of non-weak coupling

It has been established by both theoretical and experimental studies that the CLF between two subsystems that are not weakly coupled is less than that based upon the travelling wave (diffuse field) transmission coefficient. One widely accepted indicator of strength of coupling between end-connected waveguides such as beams and rods has been devised independently by Finnveden and Mace (see Fahy and Price 1999). It is symbolised by γ^2 and is given by

$$\gamma^2 = \frac{\tau_{\alpha\beta}}{\pi^2 M_\alpha M_\beta} \tag{11.34}$$

in which M_α and M_β are the modal overlap factors of the subsystems. Webster and Mace (1996) propose the following expression for γ^2 for bending wave coupling between plane *rectangular* plates having bending stiffness B, mass per unit area m, length of coupling interface L and modal overlap factor M:

$$\gamma^2 = \frac{\tau_{\alpha\beta}}{\mu_a \mu_b} \tag{11.35}$$

in which $\mu = (2\pi M/L\omega)^{1/2} (B/m)^{1/4}$. This is a *conservative* estimate for real systems because γ^2 decreases as the *geometrical irregularity* of the coupled subsystems increases. It will be less than the value calculated from equation (11.35) in many practical cases. Note that coupling may be strong if the modal overlap factors are both small even if $\tau_{\alpha\beta} \ll 1$.

If γ^2 is greater than unity, the coupling is not weak. The presence of strong coupling does not invalidate the ensemble- mean power flow-energy difference relation, although some members of the ensemble may not conform. It does, however, imply deviation of the ensemble-mean CLF from the travelling wave field value. The CLF then varies with the value of γ^2, the degree of geometric irregularity of the subsystems and the DLFs of the subsystems. Note that the latter dependence is not recognised by some commercial SEA softwares. Global modes (see Section 11.4.2) are often evident under conditions of strong coupling.

11.5.9 The SEA power balance equations

A schematic of an SEA representation of a system is presented in Figure 11.11. It is implicitly assumed in SEA that the energy flow between any pair of subsystems is not influenced by the presence of any other subsystems. This is a consequence of the condition of weak coupling.

11.5.9.1 The two-subsystem model

Consider just subsystems 1 and 2 in the case that only 1 is excited by external forces.

$$P_1 = \omega E_1(\eta_1 + \eta_{12}) - \eta_{21}\omega E_2 \qquad (11.36)$$

$$O = \omega E_2(\eta_2 + \eta_{21}) - \eta_{12}\omega E_1 \qquad (11.37)$$

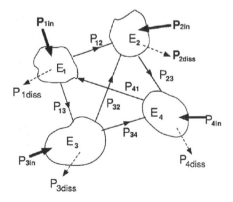

Figure 11.11 SEA subsystem network.

in which η_1 and η_2 are the DLFs of subsystems 1 and 2, and P_1 is the time-average power input to subsystem 1. Solution of Equations (11.36) and (11.37) for the total energies of subsystems 1 and 2, normalised by the input power, are

$$\frac{\omega E_1}{P_1} = \frac{\eta_2 + \eta_{21}}{\eta_{12}\eta_2 + \eta_1\eta_2 + \eta_1\eta_{21}} \tag{11.38}$$

$$\frac{\omega E_2}{P_1} = \frac{\eta_{12}}{\eta_{12}\eta_2 + \eta_1\eta_2 + \eta_1\eta_{21}} \tag{11.39}$$

The implications of these results are that the addition of damping to a subsystem is only effective if the associated DLF substantially exceeds the associated CLF. If the CLF exceeds the DLF, the system tends to equipartition of *modal* energy. Note that this does not mean that the vibration levels in the subsystems are the same, because the modal densities could be quite disparate. It should be noted that the DLFs of untreated, built-up structures commonly exceed the associated CLFs by factors of between three and ten. If they do not, the coupling is almost certainly not weak.

11.5.9.2 Multiple subsystems

The power balance equations for N coupled subsystems may be written in matrix form in terms of the subsystem modal energies as

$$
\omega
\begin{bmatrix}
n_1\left(\eta_{11} + \sum_{i\neq 1}\eta_{1i}\right) & -\eta_{21}n_2 & \cdots & -\eta_{N1}n_N \\
-\eta_{12}n_1 & n_2\left(\eta_{22} + \sum_{i\neq 2}\eta_{2i}\right) & \cdot & \cdot \\
\cdot & & \cdot & \cdot \\
\cdot & & & \cdot \\
\cdot & & & \cdot \\
-\eta_{1N}n_1 & & & n_N\left(\eta_{NN} + \sum_{i\neq N}\eta_{Ni}\right)
\end{bmatrix}
$$

$$
\times
\begin{bmatrix}
\frac{E_1}{n_1} \\
\cdot \\
\cdot \\
\cdot \\
\frac{E_N}{n_N}
\end{bmatrix}
=
\begin{bmatrix}
P_1 \\
\cdot \\
\cdot \\
\cdot \\
P_N
\end{bmatrix}
\tag{11.40}
$$

in which η_{ii} is the DLF of subsystem i. Note that the matrix is symmetric on account of CLF reciprocity and that all off-diagonal terms are negative.

11.5.10 The low-frequency 'limit'

The basic relation between the power flow between two oscillators and their stored energies, expressed by equation (11.26), has no 'low frequency limit'. However, the bandwidth of uniform excitation spectral density must be sufficient to encompass both the coupled natural frequencies of the system, and the power flow and energies are defined as *totals* within this band. Extension of this exact relation to coupled, multi-mode subsystems and the utilisation of diffuse field CLFs implicitly assumes that (i) the subsystems are weakly coupled and (ii) in cases of two- and three-dimensional subsystems, enough modes contribute to the vibration field within the analysis bandwidth for it to constitute a good approximation to a diffuse field. These conditions tend to be violated at low audio frequencies, the problem being exacerbated by analysis in proportional bandwidths, such as one-third octaves, which decrease with frequency. Also, the modal overlap factors of structural components are often low at low frequency, the expressions in equations (11.34) and (11.35) indicating that coupling may not be weak.

The 'penalty' for using SEA at 'low' frequencies is that the probability of large deviation of subsystem response from the SEA estimate becomes unacceptably high because the conditions deviate excessively from those assumed in weak-coupling SEA theory. Consequently, sensitivity analysis on the basis of the SEA parameters then becomes meaningless.

11.5.11 The selection of SEA subsystems

The definition of an SEA model appropriate to the system of concern is a crucial first step because it can have significant repercussions for its validity and therefore the reliability of the prediction therefrom. Unfortunately, very little formal guidance for the inexperienced analyst exists, even in the software packages currently available. The problem is exacerbated by the paucity of reported systematic investigations of the sensitivity of predicted results to the choice of subsystems.

The selection of subsystems should, in principle, be guided by the assumptions upon which the SEA relation between energy exchange and energy storage is based, as set out in Section 11.5.2. However, the reality of the physical features of a system to be modelled demands that the selection is made on a more pragmatic basis. It is general practice to assign subsystem boundaries to junctions or interfaces between distinct structural components, such as those between windscreens and roof panels, doors and pillars, floor panels and sills, on the assumption that they are likely to represent substantial discontinuities of wave impedance, thereby satisfying the weak coupling condition. Of course, a given physical component may support more than one wave type, and therefore more than one subsystem. For example, beam-like components such as pillars and strengthening bars, may support longitudinal, torsional and bending waves, of which the first may not be suitable for representation as an SEA subsystem because of its high group speed and therefore low modal density.

Here, we consider the SEA modelling of automotive vehicles in order to exemplify the modelling problem. The requirement for a subsystem to support a reverberant field, and for the modal overlap factor to be sufficiently large to ensure weak coupling, implies that components of automotive vehicles that are not suitable as candidates to be represented as SEA subsystems include the following:

(i) sound insulation trim and thin cavities containing sound absorbent material (high damping, not reverberant, very low modal density);
(ii) stiff connecting structures (inadequate modal density);
(iii) flexible seals (not modal, sound transmission is diffraction-dominated);
(iv) engine blocks (inadequate modal density, too strongly coupled to subdivide);
(v) gearboxes (as for engine blocks);
(vi) tyres (high damping, very non-uniform distribution of energy density);
(vii) wheel discs (inadequate modal density);
(viii) suspension elements (not modal until high frequency, non-metallic elements non-linear and highly damped);
(ix) seats covers (non-modal);
(x) headliners (non-modal);
(xi) regions of fluid in which there occurs free radiation of sound waves to infinity (no energy storage, non-reverberant).

It is more appropriate to represent (i), (ii), (iii) and (viii) as deterministic transmission elements in terms of transfer mobilities or impedances. The dynamic properties of the last mentioned, which may well be non-linear, must be determined under the appropriate static load condition. Representation of the couplings between such elements and those that are represented as SEA elements will normally assume the modal-average impedances/mobilities of the SEA subsystems at the connections. The trim presents a particular difficulty, because it may vary considerably in composition over rather small distances that may not encompass a complete SEA subsystem, such as a floor panel. The resolution of this problem is not clear.

In the lower portion of the frequency range of interest for noise in passenger cars (30–200 Hz) the subframe and associated structural components to which the panel and window panels are attached are too stiff (and therefore have insufficient modal density) to constitute good candidates for representation by SEA subsystems. Therefore, *hybrid models* are currently being developed, in which these stiff structures are represented by FE models, and the more flexible panel structures are represented as SEA subsystems (see Section 11.6.2.5). This appears to be a logical way to proceed. Little hard evidence of the effectiveness of such hybrid modelling, as applied to complete vehicles, has yet (September 2002) appeared in the open literature.

11.5.12 The insertion loss of vehicle trim

The insertion loss of vehicle sound insulation trim depends upon the form of excitation of the supporting structure. In Figure 11.12, the vibrational (not acoustic) insertion loss of a simple construction comprising a rubber sheet covering a plastic foam sheet mounted on a 1 mm steel panel exhibits a very great difference between the performances under conditions of simulated airborne and actual free-wave structure-borne excitation. The difference will be reduced in cases where the supporting structure is stiffened, beaded or curved. The insertion loss of trim when specifically excited by *structure-borne* sound is not available in some SEA softwares. More research needs to be done on this question.

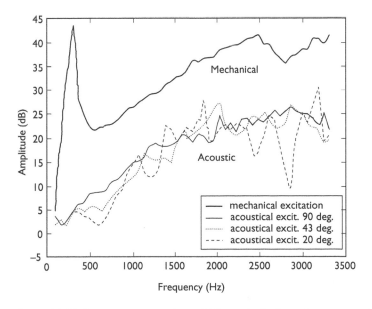

Figure 11.12 Acoustic insertion loss of 10 mm thick plastic foam covered by a 3 mm thick rubber sheet under acoustic and mechanical line-force excitation. After Fahy and Godano (1999).

11.5.13   The merits of SEA

SEA offers two particular advantages over intensive FEA modelling:

(i) The user retains a 'feel' for the physics of the problem. Although this advantage is, to some extent, being eroded by the development of interfaces which are more 'transparent' and 'user-friendly', and which increasingly obscure the physics and mathematics underlying the calculations, the ability to view distributions of energy and power flow on a solid-model offers some compensation.
(ii) The simplicity and speed of the calculation procedure allows a large number of design variants to be evaluated quickly and cheaply.

11.6   Research issues

11.6.1   Research on SEA

The previous section has considered the derivation of the SEA equations and has highlighted sources of uncertainty and inaccuracy in these equations. Current research into these aspects of SEA may be placed in two categories:

(i) that directed at the main SEA parameters, such as modal densities and CLFs;
(ii) that directed at the variance problem and wider statistical issues.

11.6.1.1   SEA parameters

There is much published data on the modal densities of various structural components and the CLFs that are applicable to certain types of structural joint. There are still significant gaps in this knowledge however, particularly for non-isotropic and/or non-homogeneous structural components. Some of these problems can be solved by using standard methods; for example, there are no published results for the CLF between two discretely bolted composite sandwich panels, but this could be calculated by standard wave analysis methods if needed. More difficult problems involve doubly curved structures, such as a segment of a conical shell. In this case, a plane elastic wave description is not appropriate, and the determination of the modal density and any relevant CLF is a less straightforward mathematical problem. Special cases can be studied experimentally or by using the finite element method (Kaminsky 2000).

Another difficult problem is that of a doubly stiffened panel, with or without attached trim, and the radiation efficiencies (or CLF) associated with coupling to an acoustic volume. This has application to both aerospace and marine structures. In this case significant progress has been made, and two approaches to the problem have been described by Bremner (1994) and Paolozzi and Peroni (1996). Other problematic areas include the computation of input power when a stiffened plate structure is forced at a 'hard point', modelling in-situ beams within the context of SEA, the treatment of dissipative couplings via loss factors and CLFs and the definition of subsystems for acoustic volumes that have a non-uniform sound pressure level (for example a vehicle cabin). Some of these problems are conceivably solvable within the context of SEA, while others may require alternative analysis approaches, such as those described in Section 11.6.2.

11.6.1.2   Variance and response statistics

The energy in a subsystem may deviate from the mean value predicted by SEA either in terms of the spatial distribution, the dependency on frequency or the variability across the ensemble. In terms of spatial variability, Norton (1989) has reviewed techniques for estimating local stress levels from SEA subsystem energy levels. Also, the space industry in the USA and Europe has developed methods for estimating the local response of an item of equipment mounted on a satellite panel.

The problem of the frequency or ensemble variability of the response is an extremely difficult one, and at present there is no general robust method for predicting the response variance in a built-up structure. A comprehensive analysis of a single subsystem has been given by Lyon (1969), and simple bounds that might be applied to the response have been presented by Langley (1994). A tentative approach to the prediction of variance in a built-up system has been suggested by Lyon and DeJong (1995), although the method can give poor predictions in some cases. Thus, there is a very pressing need for further research on the issue of variance. On this point it should be noted that variance lies 'outside' the SEA equations, in the sense that variance arises in the main from a breakdown of the standard SEA energy flow equations; the ensemble averaged energy may satisfy the SEA equations, but the response of one particular sample does not, and hence the problem cannot realistically be assessed by studying, for example, the statistics of the CLFs. The study of variance must address the deviation of the individual samples from the mean, and this cannot be done wholly within the context of SEA.

11.6.2   Alternative methods

The fundamental assumptions that underlie the SEA equations are not met in all high-frequency dynamics problems, and for this reason alternative analysis methods have been developed that seek to overcome one or more of the shortcomings of SEA. A number of these methods are outlined in Sections 11.6.2.1–11.6.2.5.

11.6.2.1   Vibrational conductivity

This is a 'continuum' form of SEA, in which the assumption that the energy flow is proportional to the difference in subsystem modal energies is taken to apply on a microscopic level; the energy flow at any point *within* the subsystem is taken to be proportional to the *gradient of the energy density*. Thus the *vector* of energy flow (the Poynting vector) $\mathbf{I}$ is written as

$$\mathbf{I} = -\left(\frac{c_g^2}{\omega\eta}\right)\nabla e \tag{11.41}$$

where e is the energy density. Energy conservation then gives

$$\omega\eta e + \nabla\cdot\mathbf{I} = P \quad \Rightarrow \quad \omega\eta e - \left(\frac{c_g^2}{\omega\eta}\right)\nabla^2 e = P \tag{11.42}$$

where P is the distributed input power. In the vibrational conductivity approach, equation (11.42) is applied within each subsystem, and appropriate energy flow conditions are imposed at the boundaries between subsystems. The fact that equation (11.42) is directly analogous to the heat conduction equation implies that standard finite element numerical procedures can be employed to yield the energy density throughout the system. The method does nothing to capture the 'phase effects' that are omitted from SEA, but unlike SEA it does capture energy gradients within a subsystem – thus it may have most applications to highly damped systems.

There has been much debate regarding the validity of equation (11.42). It is clear that the equation has a firm theoretical basis for one-dimensional systems (Nefske and Sung 1989) but difficulties have been identified for two-dimensional systems (Carcaterra and Sestieri 1995; Langley 1995). Nonetheless, the vibrational conductivity approach has been put forward by proponents as a design tool that might be applied to very general structures (for example, Bernhard and Huff 1999).

11.6.2.2   Wave intensity analysis (WIA)

In this method the 'diffuse wave field' assumption of SEA is relaxed, so that a subsystem may carry different amounts of energy in different wave propagation directions (Langley and Bercin 1994). The motivation for this approach arises from the fact that the transmission coefficient of a structural joint can be highly dependent on the direction of the incident wave; hence, for certain structural geometries, the transmitted wavefield can be highly directional, even though the incident wavefield may be diffuse. In WIA, the distribution of subsystem energy with wave heading is represented by a Fourier series, and the method aims to find the amplitudes of the Fourier terms by considering energy balance. The first term in the series represents a diffuse wave field, and thus SEA is recovered by taking an one-term WIA solution. The effects of phase and coherence are not addressed by WIA, nor does the method allow for significant spatial decay within a subsystem (i.e. a non-reverberant subsystem).

11.6.2.3 Randomised modal synthesis

Mace and Shorter (2000) have presented a randomised version of modal synthesis (see Section 11.3) in which the natural frequencies of the subsystem-blocked modes are taken to be random variables. This allows efficient Monte Carlo simulations to be performed, and furthermore the finite element results can be post-processed to yield the statistics of energy flows and subsystem energy levels. The method neglects the effect of system randomness on the 'constraint' modes, and by necessity imposes a pre-selected probability density function on the subsystem natural frequencies, but nonetheless it offers a promising approach to 'mid-frequency' vibration problems, where neither conventional finite element methods nor SEA are applicable.

11.6.2.4 Fuzzy structure theory

With this approach the system is considered to consist of a master structure, which is modelled deterministically, and a fuzzy system whose effects on the master structure are evaluated statistically. For example, the hull of a submarine may be considered to be a master structure, while the internally mounted equipment might be a fuzzy system. The method was developed by Soize (1993) (see also Ohayon and Soize 1998) and example applications have been given by Strasberg and Feit (1996) and Pierce (1997). In physical terms, the fuzzy system can be viewed as a set of random oscillators attached to the master structure. A key effect of the oscillators is to add damping to the master structure, and fuzzy structure theory leads to the key result that the level of damping is (except in extreme cases) independent of the oscillator internal damping and proportional to the modal density of the oscillators. These results relate to the steady-state response of the system; other studies have considered the effects of the fuzzy system on the transient response of the master structure (for example Weaver 1997).

11.6.2.5 Hybrid FEA/SEA method

In this approach (Langley and Bremner 1999) the total response of a structure is considered to be composed of a 'long' wavelength component and a 'short' wavelength component, so that the equations of motion are written as

$$(-\omega^2 \mathbf{M} + j\omega \mathbf{C} + \mathbf{K})\mathbf{q} = \mathbf{Dq} = \mathbf{F} \tag{11.43}$$

and then partitioned in the form

$$\begin{pmatrix} \mathbf{D}_{gg} & \mathbf{D}_{gl} \\ \mathbf{D}_{gl}^T & \mathbf{D}_{ll} \end{pmatrix} \begin{pmatrix} \mathbf{q}^g \\ \mathbf{q}^l \end{pmatrix} = \begin{pmatrix} \mathbf{F}^g \\ \mathbf{F}^l \end{pmatrix} \tag{11.44}$$

Here $\mathbf{D}$ is termed the dynamic stiffness matrix and $\mathbf{q}$ is the vector of generalised coordinates describing the motion of the system. In equation (11.44) the index g refers to a long wavelength (or 'global') component while l refers to a short wavelength (or 'local') component. Equation (11.44) can be rearranged to yield

$$(\mathbf{D}_{gg} - \mathbf{D}_{gl}\mathbf{D}_{ll}^{-1}\mathbf{D}_{gl}^T)\, \mathbf{q}^g = \mathbf{F}^g - \mathbf{D}_{gl}\mathbf{D}_{ll}^{-1}\mathbf{F}^l \tag{11.45}$$

$$\mathbf{D}_{ll}\mathbf{q}_l = \mathbf{F}^l - \mathbf{D}_{gl}^T\mathbf{q}^g \tag{11.46}$$

In equation (11.45), $\mathbf{D}_{gg}$ and $\mathbf{F}^g$ are evaluated by using the finite element method with a relatively coarse mesh density to capture only the long wavelength deformation. In contrast, the local components of the response are modelled statistically, so that explicit expressions for $\mathbf{D}_{gl}$, $\mathbf{D}_{ll}$ and $\mathbf{F}^l$ are not available; rather the effect of these terms on the equation of motion is estimated by using a fuzzy structure approach. For example, the second matrix on the left of equation (11.45) leads to a fuzzy contribution of the form

$$(\mathbf{D}_{gl}\mathbf{D}_{ll}^{-1}\mathbf{D}_{gl}^T)_{nm} \approx -j\left(\frac{\pi\omega^3}{2}\right)\sum_r j_{mm}^2 \nu_r - \omega^2 \sum_r \alpha_{mm} \tag{11.47}$$

where the sum over r covers the number of subsystems employed in the statistical (SEA) description of the local response, ν_r is the modal density of the rth subsystem and j_{mm}^2 and α_{mm} are coefficients detailed by Langley and Bremner (1999). The second force term that appears on the right of equation (11.45) can be approximated in a similar fashion. Thus, equation (11.45) can be viewed as a coarse mesh finite element model (having relatively few degrees of freedom), whose dynamic stiffness matrix and force vector are augmented by terms in the form of equation (11.47) to account for the presence of short wavelength motion. The equation can then be solved to yield the response $\mathbf{q}^g$. Finally, equation (11.46) is replaced by a set of SEA equations, with an additional (known) power input arising from the second term on the right. In this way equations (11.45) and (11.46) are solved in a non-iterative way to yield both the long and short wavelength components of the response. It can be noted that equation (11.47) includes a fuzzy damping term that is proportional to the modal density of the statistical part of the response; this is consistent with the comments regarding the conclusions from fuzzy structure theory.

The hybrid method has thus far been applied to relatively few example structures, and further development and validation of the method is ongoing. It can be noted that a different approach to coupling high-frequency dynamics to a coarse mesh finite element model has been suggested by Grice and Pinnington (2000); this consists of attaching representative impedances to the nodes of the finite element model.

11.7 Concluding remarks

The methods outlined in Section 11.6 are very much at the research stage in the sense that they cannot as yet be applied to general systems using commercial (or other) software. SEA on the other hand is very well established, and several commercial codes are available, although the accuracy of the method depends to a significant degree on the physics of the structure under consideration and the expertise of the user. Looking towards the future, it is certain that computing power will continue to grow at significant rate, and thus finite element models that are considered large by today's standards will seem routine in years to come. The question of when, and whether, large-scale Monte-Carlo simulations of very large finite element models will be feasible is, however, open. Other questions are whether such a direct numerical approach is actually desirable, or compatible with the design process. It is conceivable that even with such computer power it would be advantageous to use SEA, or one of the variants outlined in Section 11.6, at the preliminary design stage, leaving the intense computational work to the final design stage. The fact that it is not certain what design tools will be employed to analyse high-frequency vibration in 10 or 20 years time makes this an interesting and important area of vibroacoustic research.

Note

1 This condition, which is one of the factors contributing to a state of 'weak coupling' between subsystems, promotes modes of vibration of a complete system which closely resemble in resonance frequency and mode shape those of isolated subsytems, in which the energy of modal vibration tends to be concentrated. These are known as 'local' modes, as opposed to modes in which the vibrational energy is distributed widely over the whole structure, which are known as 'global' modes. Weak coupling also promotes a form of vibration distribution in which subsytem energy decreases substantially as the distance from regions of external excitation increases. This decrease is not due to energy dissipation by damping, but to wave scattering at the impedance discontinuities.

References

Bernhard, R.J. and Huff, J.E. (1999) Structural-acoustic design at high frequency using the energy finite element method. *Journal of Vibration and Acoustics – Transactions of the ASME*, **111**, 295–301.

Bremner, P.G. (1994) Vibro-acoustics of ribbed structures – a compact modal formulation for SEA models. *Proceedings of Noise-Con 94, Fort Lauderdale, Florida, May 1–4* (Institute of Noise Control Engineering).

Carcaterra, A. and Sestieri, A. (1995) Energy density equations and power flow in structures. *Journal of Sound and Vibration*, **188**, 269–282.

Craig, R.R. (1981) *Structural Dynamics: An Introduction to Computer Methods* (New York: John Wiley & Sons).

Cremer L., Heckl, M. and Ungar, E.E. (1987) *Structure-borne Sound* (2nd edn) (Berlin: Springer-Verlag).

Fahy, F.J. and Godano, Ph. (1999) Enabling technology for sound package optimisation. ISVR Contract Report No. 99/13.

Graff, K.F. (1975) *Wave Motion in Elastic Solids* (New York: Dover Publications).

Grice, R.M. and Pinnington, R.J. (2000) Vibration analysis of a thin-plate box using a finite element model which accommodates only in-plane motion. *Journal of Sound and Vibration*, **232**, 449–471.

Johnson, D.F. (2001) *Probabilistic and Possibilistic Models of Uncertainty for Static and Dynamic Structures*. PhD Thesis, University of Cambridge.

Kaminsky, C. (2000) The use of FEM, analytical and experimental SEA in a single model. Case study: the Ariane 5 Vulcain engine. *Spacecraft and Launch Vehicle Dynamics Workshop*.

Kompella, M.S. and Bernhard, R.J. (1996) Variation of structural acoustic characterisitics of automotive structures. *Noise Control Engineering Journal*, **44**(2), 93–99.

Langley, R.S. (1994) Spatially averaged response envelopes for one- and two-dimensional structural components. *Journal of Sound and Vibration*, **178**, 483–500.

Langley, R.S. (1995) On the vibrational conductivity approach to high frequency dynamics for two-dimensional structural components. *Journal of Sound and Vibration*, **182**, 637–657.

Langley, R.S. and Bercin, A.N. (1994) Wave intensity analysis of high frequency vibrations. *Philosophical Transactions of the Royal Society, Series A*, **346**, 489–499.

Langley, R.S. and Bremner, P.G (1999) A hybrid method for the vibration analysis of complex structural-acoustic systems. *Journal of the Acoustical Society of America*, **105**, 1657–1671.

Lyon, R.H. (1969) Statistical analysis of power injection and response in structures and rooms. *Journal of the Acoustical Society of America*, **45**, 545–565.

Mace, B.R. and Shorter, P.J. (2000) Energy flow models from finite element analysis. *Journal of Sound and Vibration*, **233**, 369–389.

Manohar, C.S. and Keane, A.J. (1994) Statistics of energy flows in spring-coupled one-dimensional systems. In, A.J. Keane and W.G. Price (eds) *Statistical Energy Analysis* (Cambridge, UK: Cambridge University Press), pp. 95–112.

Nefske, D.J. and Sung, S.H. (1989) Power flow finite element analysis of dynamic systems: Basic theory and application to beams. *Journal of Vibration, Stress, and Reliability in Design*, **111**, 94–100.

Newland, D.E. (1989) *Mechanical Vibration Analysis and Computation* (Harlow, UK: Longman Scientific and Technical).

Norton, M.P. (1989) *Fundamentals of Noise and Vibration Analysis for Engineers* (Cambridge: Cambridge University Press).

Ohayon, R. and Soize, C. (1998) *Structural Acoustics and Vibration* (San Diego: Academic Press).

Paolozzi, A. and Peroni, I. (1996) Response of aerospace sandwich panels to launch acoustic environment. *Journal of Sound and Vibration*, **196**, 1–18.

Pierce, A.D. (1997) Resonant-frequency-distribution of internal mass inferred from mechanical impedance matrices, with application to fuzzy structure theory. *Journal of Vibration and Acoustics – Transactions of the ASME*, **119**, 324–333.

Soize, C. (1993) A model and numerical method in the medium frequency range for vibroacoustic predictions using the theory of structural fuzzy. *Journal of the Acoustical Society of America*, **94**, 849–865.

Strasberg, M. and Feit, D. (1996) Vibration damping of large structures induced by small resonant structures. *Journal of the Acoustical Society of America*, **99**, 335–344.

Weaver, R.L. (1997) Mean and mean-square responses of a prototypical master/fuzzy structure. *Journal of the Acoustical Society of America*, **101**, 1441–1449.

Webster, E.C.N. and Mace, B.R. (1996) Statistical energy analysis of two edge-coupled rectangular plates: Ensemble averages. *Journal of Sound and Vibration*, **193**, 793–822.

Bibliography

Craik, R.J.M. (1996) *Sound Transmission through Buildings using Statistical Energy Analysis* (Aldershot, UK: Gower Publishers).

Fahy, F.J. and Price, W.G. (eds) (1999) *IUTAM Symposium on Statistical Energy Analysis* (Dordrecht, The Netherlands: Kluwer Academic Publishers).

Keane, A.J. and Price, W.G. (1997) *Statistical Energy Analysis: An Overview with Applications in Structural Dynamics* (Cambridge, UK: Cambridge University Press).

Lyon, R.H. and DeJong, R.G. (1995) *Theory and Application of Statistical Energy Analysis* (2nd edn) (Newton, MA, USA: Butterworth-Heinemann).

Chapter 12

Vibration control

M.J. Brennan and N.S. Ferguson

12.1 Introduction

The control of vibration is becoming more and more important for many industries as their products have to be sold in an increasingly competitive world. This generally has to be achieved without additional cost, and thus detailed knowledge of structural dynamics is required together with familiarity of standard vibration control techniques. There are many textbooks on mechanical vibrations, but there are only a few on vibration control (for example, Nashif *et al.* 1985; Mead 1998), and there are some book-chapters that deal with various control aspects (for example, Mead 1982; Ungar 1992a,b; Crede and Ruzicka 1995; Everett Reed 1995; Newton 1995). Although vibration control is predominantly carried out using passive means, more recently active vibration control has been applied, and there are several textbooks describing this approach (for example, Fuller *et al.* 1997; Preumont 1997). Because of the restrictions of space, this chapter will be confined to passive vibration control.

A generic vibration control problem can be separated into three components: the source, the transmission path and the receiver, as shown in Figure 12.1. The boundaries of the three components are defined by the person solving the problem, and it is probable that different people will define different boundaries. For example, if a gearbox is the vibration source the gears could be considered to be the vibration source, and the shaft and casing, etc. could be the transmission path. Alternatively, the whole gearbox could be considered to be the vibration source. Where the boundaries are drawn is largely unimportant; what is important is that the person who is solving the problem clearly defines the source, the transmission path and the receiver. Once this has been done the vibration control problem reduces to one of changing the size and distribution of mass, stiffness and damping to minimise the vibration at the receiver. Whilst this is very easy to write down, it can be very difficult to do in reality because of the difficulty in determining which

Figure 12.1 Schematic of a general vibration control problem.

parameter to change. In general, vibration control solutions can be classified as follows:

- source modification;
- vibration isolation;
- vibration damping;
- the addition of vibration absorbers/neutralisers;
- structural modification;
- material selection.

Each of these topics is discussed in separate sections in this chapter, with references to further reading.

12.2   Vibration sources

There are many different sources of vibration, and the aim of this section is to describe some of the principal categories and to classify some of the mechanisms. Often the 'black box' approach to source description is adopted and this is briefly mentioned at the end of this section. The sources are grouped as follows: impacts, rotating machinery, acoustic excitation, self-excited systems and flow-induced vibrations. This list is not exhaustive, but indicates the types of vibration source commonly encountered in practice.

12.2.1   Impacts

Impacts occur frequently in machinery, and often tend to be periodic at the operating rate of the machine. To study a single impact, a simple example of an elastic body impacting on a rigid surface is considered (Lyon 1987). Reference to Figure 12.2 shows that when an elastic body of mass m impacts on a rigid surface, there is a local strain because of the contact stiffness k, and the body can be modelled as a mass on a spring. Whilst in contact with the surface, this system will have a natural frequency of $f_n = (1/2\pi)\sqrt{k/m}$ and hence a natural period of $T_n = 1/f_n$. The time that the body will be in contact will be half of the natural period. Thus, contact time

$$T_c = \frac{T_n}{2} = \pi\sqrt{\frac{m}{k}} \qquad (12.1)$$

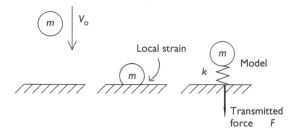

Figure 12.2 Impact of a sphere on a rigid surface.

and the maximum force is given by

$$F_{\max} = V_o \sqrt{km} \tag{12.2}$$

where V_o is the velocity of the impacting body. A half-sine pulse, which is representative of the impact force on the rigid surface is shown in Figure 12.3a. The energy spectrum of the pulse is the Fourier transform of the force time history multiplied by its complex conjugate and divided by 2π to give

$$\text{energy spectrum} = \frac{1}{2\pi} \left(\frac{2mV_o}{1 - \Omega^2} \cos \frac{\pi\Omega}{2} \right)^2 \tag{12.3}$$

where $\Omega = (f/f_n)$ is normalised frequency. The energy spectrum is shown in Figure 12.3b. It is evident from Figure 12.3b that the spectrum of the transmitted force is dependent upon the contact time, which in turn is dependent upon the mass of the object and the contact stiffness. To reduce the high-frequency content of the force either the mass has to increase or the stiffness decrease. In most practical situations it is the latter that is the preferred solution. Note, that in Figure 12.3b, the high-frequency energy spectrum decays at a rate of $1/\omega^4$ or $-12\,\text{dB}$ per octave. If the force were applied in a more abrupt manner, as for example in a square pulse, then the spectrum would decay at $1/\omega^2$ or $-6\,\text{dB}$ per octave. The energy at low frequencies can be found by setting Ω to zero, giving the energy spectrum $= (2mV_o)^2/2\pi$, which shows that this is governed by the change in momentum of the impacting object.

Another case of interest is that of a sphere impacting on an infinite plate. This would represent a situation, where there is an impact on a non-resonant structure, in which energy propagates away from the source, as depicted in Figure 12.4. An infinite plate has the same characteristics as a damper with an equivalent damping coefficient of $c_{eq} = 2.3\rho_m h^2 c_l$ (Cremer *et al.* 1988), where ρ_m is the mass per unit area, h is the thickness and c_l is the

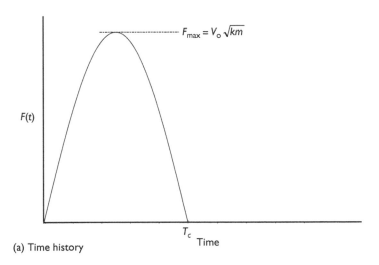

(a) Time history

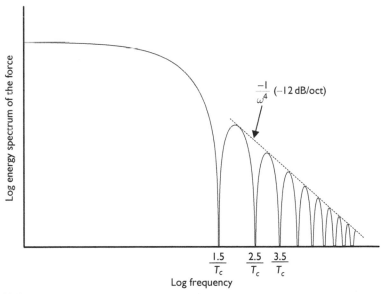

(b) Energy spectrum

Figure 12.3 Time history and energy spectrum of the force generated by an impacting body on a rigid surface.

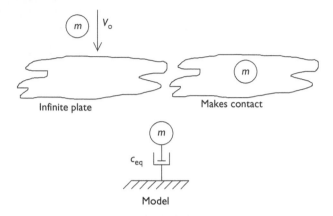

Figure 12.4 Impact of a sphere on an infinite plate.

longitudinal wave speed, which is about $5000\,\text{ms}^{-1}$ for steel or aluminium. The force transmitted to an infinite plate by an object of mass m and velocity of V_o is given by Lyon (1987) as

$$F(t) = c_{eq}V_o e^{\frac{-c_{eq}t}{m}} \tag{12.4}$$

which is plotted in Figure 12.5a. The corresponding energy spectrum can be calculated in a similar way to the example given above, to give

$$\text{energy spectrum} = \frac{1}{2\pi}\frac{\left(c_{eq}V_o\right)^2}{(2\pi f)^2 + \left(\frac{c_{eq}}{m}\right)^2} \tag{12.5}$$

This is plotted in Figure 12.5b, where it can be seen that, after a cut-off frequency of $f_{\text{cut-off}} = (1/2\pi)(c_{eq}/m)$, the spectrum decays at a rate of $1/\omega^2$ or $-6\,\text{dB}$ per octave. The low-frequency content is half that of the case when a body impacts on a rigid surface. This is because the change in momentum is only half of that in the elastic case, because the body does not rebound, remaining in contact with the structure.

12.2.2   Rotating machinery

12.2.2.1   Gears

Meshing gears are common parts found in many types of machinery. They generate vibration at the tooth meshing frequency (product of rotational speed and number of teeth) and its harmonics. In many cases the vibration has high-frequency components ($\approx 1\,\text{kHz}$) and generates subjectively

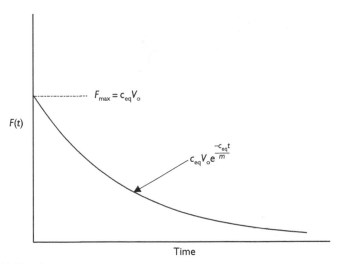

$F_{max} = c_{eq}V_o$

$F(t)$

$c_{eq}V_o e^{\frac{-c_{eq}t}{m}}$

Time

(a) Time history

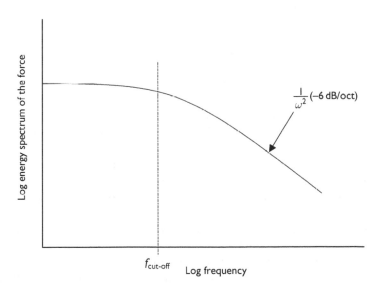

Log energy spectrum of the force

$\frac{1}{\omega^2}$ (–6 dB/oct)

$f_{cut-off}$ Log frequency

(b) Energy spectrum

Figure 12.5 Time history and energy spectrum of the generated force of an impacting body on an infinite plate.

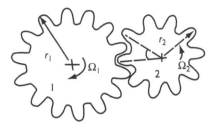

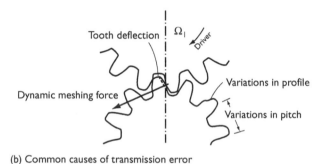

(a) The driving (1) and the driven (2) gears

(b) Common causes of transmission error

Figure 12.6 The meshing of two gears.

annoying noise. It is particularly hard to suppress this vibration at source. To illustrate the source of gear vibration, consider the meshing of two gears of radii r_1 and r_2, as shown in Figure 12.6. The driving wheel is assumed to rotate at constant angular velocity Ω_1 and the driven wheel has an *average* angular displacement of $\theta_2 = \Omega_2 t + \Delta\theta$, where $\Omega_2 = \Omega_1(r_1/r_2)$ and $\Delta\theta$ is the transmission error. Transmission error is defined as the difference between the actual position of the output gear and the position it would take if the gear drive were perfect. Because of a variety of factors, such as tooth deflection under load, Hertzian contact stiffnesses and manufacturing errors, such as variations in tooth profile and variations in pitch, the transmission error is never zero in practice. It will generate a dynamic meshing force that will act on both the driving and the driven gears. A review on the modelling of meshing gears has been presented by Nevzat Özgüven and Houser (1988).

12.2.2.2   Rotating unbalance

Unbalance is perhaps the most common source of vibration in rotating machinery. It is characterised by a forcing function that has the same frequency as the rotational speed of the offending shaft or inertia. To study

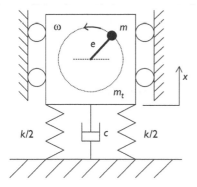

Figure 12.7 Rotating unbalance of a single-degree-of-freedom system.

the fundamental characteristics of rotating unbalance, the single-degree-of-freedom system shown in Figure 12.7 is first considered. The unbalance is characterised by me, where m is an equivalent point mass with an eccentricity e. The equation of motion for this system is given by

$$m_t \ddot{x} + c\dot{x} + kx = me\omega^2 \sin \omega t \tag{12.6}$$

where m_t is the total mass of the machine. Because the system is linear, the response will be harmonic at frequency ω, that is $x = Xe^{j\omega t}$. Noting that $\omega_n = \sqrt{k/m_t}$, $\Omega = \omega/\omega_n$ and $\zeta = c/2\sqrt{m_t k}$, equation (12.6) becomes

$$\frac{me\omega^2}{k} \left(\frac{1}{1 - \Omega^2 + j2\zeta\Omega} \right) \tag{12.7}$$

Multiplying equation (12.7) by m_t, taking the modulus and rearranging gives the non-dimensional amplitude response of the system as a function of non-dimensional frequency (speed)

$$\frac{|X| m_t}{me} = \frac{\Omega^2}{\left\{ (1 - \Omega^2)^2 + (2\zeta\Omega)^2 \right\}^{\frac{1}{2}}} \tag{12.8}$$

This is plotted in Figure 12.8. The following observations can be made. At low speeds $\Omega \ll 1$ and the amplitude of vibration is small. This is because the amplitude of the excitation force is small as it is proportional to the square of the speed. At the resonance frequency when $\Omega = 1$, then $|X| m_t/me = 1/2\zeta$ and the vibration is controlled by damping. At high speeds, when $\Omega \gg 1$, then $|X| m_t/me = 1$, which shows that the vibration amplitude is constant and is independent of frequency and damping in the system. This occurs because in this frequency range the system response attenuates with the square of the rotational speed and the force amplitude

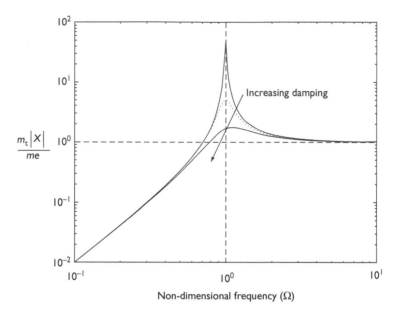

Figure 12.8 Response of a single-degree-of-freedom system to an unbalance.

increases with the square of rotational speed. Because of this, conventional vibration isolation measures discussed in Section 12.3 will not be effective. The most effective way to reduce vibration due to unbalance is simply to balance the rotor.

12.2.2.3   Reciprocating unbalance

If the excitation is from a reciprocating mechanism, such as that shown in Figure 12.9, rather than rotating unbalance, the excitation force contains a fundamental frequency corresponding to the rotational speed together with higher harmonics. It is given approximately by the equation

$$\text{Excitation force} \approx me\omega^2 \left(\sin \omega t + \frac{e}{L} \sin 2\omega t \right) \tag{12.9}$$

from which it can be seen that the magnitude of the second harmonic is dependent upon the dimensions of the mechanism, that is the ratio e/L.

12.2.2.4   Critical speed of a rotating shaft

To study the behaviour of a rotating shaft as a function of speed, the shaft can be represented by the system shown in Figure 12.10a. A massless shaft is

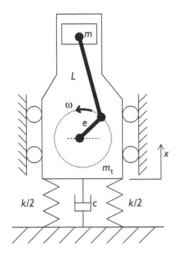

Figure 12.9 Reciprocating unbalance.

supported between two simple supports, which represent the bearings, and a disc carrying an unbalance mass is attached to the shaft at its mid-span position. The shaft will vibrate harmonically in both the x and y directions, and the resulting motion will be a circular orbit about the rotational centre 0. A section through the disc is shown in Figure 12.10b, where S and G are the geometric and mass-centre of the disc, respectively. In addition, the lateral stiffness of the shaft is taken to be k, and the damping coefficient c is proportional to the linear speed of the geometric centre. The resulting

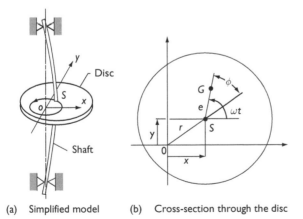

(a)   Simplified model   (b)   Cross-section through the disc

Figure 12.10 Simplified model of a rotating shaft.

equations of motion in the x and y directions are

$$m\ddot{x} + c\dot{x} + kx = me\omega^2 \cos \omega t \qquad (12.10a)$$

$$m\ddot{y} + c\dot{y} + ky = me\omega^2 \sin \omega t \qquad (12.10b)$$

Assuming a harmonic response of the form $x = Xe^{j\omega t}$, $y = Ye^{j\omega t}$, and substituting for $\omega_n = \sqrt{k/m_t}$, $\Omega = \omega/\omega_n$ and $\zeta = c/2\sqrt{m_t k}$, equations (12.10a,b) become

$$\frac{X}{e} = \frac{\Omega^2}{1 - \Omega^2 + j2\zeta\Omega}, \quad Y = Xe^{j\frac{\pi}{2}} \qquad (12.11a,b)$$

Since the two harmonic motions x and y have the same amplitude and a phase difference of 90°, their sum is a circle. If the more general case is considered where the bearings have finite stiffness, which will generally be different in the x and y directions, then the effective stiffness in equations (12.10a,b) will also be different in these directions. This is because the stiffnesses of the shaft and the bearings are in series. The two harmonic motions x and y will then have different amplitudes, but will still have a phase difference of 90°, and the resulting motion will be an ellipse, as shown in Figure 12.11. Another interesting phenomenon can be observed in this figure. When the angular velocity of the shaft rotation is less than the angular natural frequencies in the x and y directions, ω_{nx} and ω_{ny} respectively, the shaft whirls in the same direction as the shaft rotation, but when

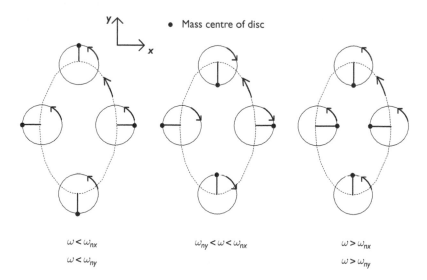

Figure 12.11 Elliptical orbits of a shaft due to rotating unbalance.

the angular velocity of the shaft is between ω_{nx} and ω_{ny}, the shaft whirls in the opposite direction.

12.2.3 Acoustic excitation

When a structure is excited acoustically, large amplitude vibration may occur due to two different phenomena, namely resonance and coincidence. Both of these are illustrated by considering a harmonic plane wave incident upon a uniform beam at angle θ, as shown in Figure 12.12. The sound field of wavelength λ impinges on the beam and propagates along the beam with a trace wavespeed of c_t, which is related to the speed of sound in air by $c_t = c_0 / \sin \theta$. The beam is subject to a pressure distribution given by

$$p(x, t) = Pe^{j(\omega t - k_t x)} \tag{12.12}$$

where $k_t = 2\pi / \lambda_t$ is the trace wavenumber, λ_t is the trace wavelength and ω is circular frequency.

The equation of motion of the beam, neglecting shear deformation and rotary inertia (Euler–Bernoulli beam theory) is

$$EI\frac{\partial^4 w}{\partial x^4} + \rho A \frac{\partial^2 w}{\partial t^2} = p(x, t) \tag{12.13}$$

where w is the transverse displacement and E, I, ρ and A are the Young's modulus, second moment of area, density and cross-sectional area of the beam, respectively. The solution is the sum of the complementary function (CF) and particular integral (PI). The CF is the solution to the free wave equation

$$EI\frac{\partial^4 w}{\partial x^4} + \rho A \frac{\partial^2 w}{\partial t^2} = 0 \tag{12.14}$$

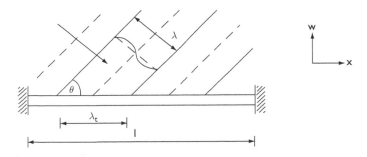

Figure 12.12 Acoustic excitation of a structure.

which is

$$w_{cF} = \left(A_1 e^{k_b x} + A_2 e^{-k_b x} + A_3 e^{jk_b x} + A_4 e^{-jk_b x}\right) e^{j\omega t} \tag{12.15}$$

where k_b is the flexural wavenumber of the beam. To find the PI, a solution of the form $w_{PI} = A_f e^{j(\omega t - k_t x)}$ is assumed, which when substituted into equation (12.13), gives

$$w_{PI} = \frac{P e^{j(\omega t - k_t x)}}{EI k_t^4 - \rho A \omega^2} \tag{12.16}$$

The total solution is thus given by

$$w_{total} = \frac{P e^{j(\omega t - k_t x)}}{EI k_t^4 - \rho A \omega^2} + \left(A_1 e^{k_b x} + A_2 e^{-k_b x} + A_3 e^{jk_b x} + A_4 e^{-jk_b x}\right) \tag{12.17}$$

This must satisfy the boundary conditions of the beam. If the beam is fully fixed at each end, then the displacement $w = 0$ and the slope $w' = 0$ at $x = 0$ and l. This leads to four equations, which can be arranged in matrix form as

$$\begin{bmatrix} 1 & 1 & 1 & 1 \\ k_b & -k_b & jk_b & -jk_b \\ e^{k_b l} & e^{-k_b l} & e^{jk_b l} & e^{-jk_b l} \\ k_b e^{k_b l} & -k_b e^{-k_b l} & jk_b e^{jk_b l} & -jk_b e^{-jk_b l} \end{bmatrix} \begin{bmatrix} A_1 \\ A_2 \\ A_3 \\ A_4 \end{bmatrix}$$

$$= \begin{bmatrix} 1 \\ jk_t \\ -e^{jk_t} \\ jke^{-jk_t} \end{bmatrix} \frac{P}{EI k_t^4 - \rho A \omega^2} \tag{12.18}$$

The wave amplitudes will be large under two different conditions:

(i) when $EI k_t^4 - \rho A \omega^2 = 0$, i.e. when $k_t^4 = \rho A \omega^2 / EI = k_b^4$, which is when the free flexural wavelength in the beam is equal to the trace wavelength. This is the *coincidence* condition, and is independent of the boundary conditions.

(ii) when the determinant of the matrix on the left-hand side of equation (12.18) is zero. This occurs at the natural frequencies of the finite beam, which are dependent on the boundary conditions.

The frequency at which coincidence occurs is dependent upon the angle of the incident acoustic wave, as discussed by Fahy (1987). The lowest frequency ω_c at which it can occur is when $\theta = 90°$. This is called the critical frequency $\omega_c = (\rho A/EI)^{\frac{1}{2}} c_0^2$. The coincidence frequency ω_{co} at other angles of incidence θ is related to this by

$$\omega_{co} = \omega_c \sin^2 \theta \tag{12.19}$$

Thus, at frequencies greater than the critical frequency there is always an angle of incidence when coincidence occurs. Although the above example has illustrated acoustic excitation of a beam, all structures which have distributed mass and stiffness behave in a similar way.

12.2.4 Self-excited vibrations

Forces acting on a vibrating system are rarely independent of the motion of the system, but in many cases the influence of the motion is small. However, in a self-excited system the excitation force is crucially dependent upon the motion of the system. Examples of such systems where this phenomenon occurs are instabilities in rotating shafts, brakes, the flutter of turbine blades, the flow-induced vibration of pipes and the aerodynamically induced motion of bridges (Ehrich 1995).

To examine one of the circumstances that leads to instability, consider a single-degree-of-freedom mass-spring-damper system excited by a force $A\dot{x}$. The equation of motion of the system is

$$m\ddot{x} + c\dot{x} + kx = A\dot{x} \tag{12.20}$$

This can be rearranged to give

$$m\ddot{x} + (c - A)\dot{x} + kx = 0 \tag{12.21}$$

It can be the seen that the effect of the excitation force is to remove damping from the system. If $c > A$ there is residual damping in the system making it stable. If $c < A$ then the damping is negative and the system is unstable (Rao 1995). Any disturbance to this system would result in very large amplitude vibrations. The system would either be destroyed or the vibrations would be limited by non-linearities in the system.

12.2.5 Black box description of vibration sources

A convenient way to describe a vibration source is to use network theory. It can be described as either a 'Thévenin equivalent' or a 'Norton equivalent' source. These are shown in Figure 12.13a,b, respectively. The Thévenin equivalent source consists of a blocked force F_b in parallel with the internal impedance of the source Z_i, and the Norton equivalent source is described

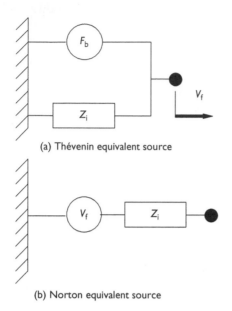

(a) Thévenin equivalent source

(b) Norton equivalent source

Figure 12.13 Idealised models of vibration sources.

by a free velocity V_f in series with the source impedance. These sources are equivalent and are related through the relationship

$$F_b = Z_i V_f \tag{12.22}$$

The concept of mechanical impedance is discussed in Chapter 9. The vibration of the receiver can be calculated from knowledge of the impedances of the transmission path and the receiver. Whilst the description above is restricted to a point-connected system, it can be extended to multi-point-connected systems, where the forces and velocities are grouped into vectors and the source impedance is described by a matrix of impedances.

12.3   Vibration isolation

Vibration isolation is the dynamic decoupling of two connecting systems and is usually achieved by placing a resilient element (isolator) in the transmission path. In practice, two common situations occur:

(a) isolation of a vibrating machine from its surroundings;
(b) isolation of a delicate piece of equipment from a vibrating host structure.

At relatively low frequencies, when flexural wavelengths in the source and receiving structures are long compared to their dimensions, vibration isolation can be considered using simple lumped parameter models, and this is now reviewed. This elementary analysis has some merits, but it also has various deficiencies when considering high-frequency vibration isolation. In this context the term 'high frequency' is used when the vibrating source or the isolator or the receiving structure exhibits resonant behaviour. An alternative modelling approach, which enables the high-frequency dynamics of the system to be accounted for, is also described in this section. Finally, some issues, such as moment excitation and isolator placement are briefly mentioned with references to further reading on these subjects.

12.3.1 Simple approach

The simple approach to the problem of vibration isolation is to view the system as a single-degree-of-freedom system and to examine its frequency response. Cases (a) and (b) in Section 12.3 are depicted in Figure 12.14 where F_e, F_t, X_e, X_b the complex amplitudes. The machine and the equipment are assumed to be point masses and the isolators are modelled as massless springs with hysteretic damping. The frequency-domain ratio of the force transmitted to a rigid foundation to the excitation force, and the ratio of the equipment displacement to the displacement of the host structure turn out to be the same and are called either the force transmissibility (T_f) or the displacement transmissibility (T_x), given by Mead (1998) as

$$T = T_f = T_x = \frac{1 + j\eta}{1 - \Omega^2 + j\eta} \tag{12.23}$$

where $\Omega = \omega/\omega_n$ is the forcing frequency normalised by the undamped natural frequency $\omega_n = \sqrt{k/m}$ of the machine or equipment on the isolator

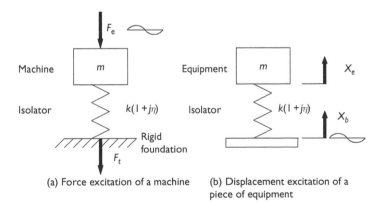

(a) Force excitation of a machine (b) Displacement excitation of a piece of equipment

Figure 12.14 One-dimensional idealised model of a vibration isolation system.

and η is the loss factor of the isolator. It can be seen from equation (12.23) that only two factors affect the transmissibility, namely the non-dimensional frequency Ω and the damping in the isolator η. A non-dimensional graph depicting the modulus of the transmissibility is shown in Figure 12.15. Generally, the phase is of little interest in vibration isolation. The natural frequency of the system can also be calculated from the expression

$$\omega_n = \sqrt{\frac{g}{\delta_s}} \tag{12.24}$$

where g is the acceleration due to gravity and δ_s is the static displacement of the machine or equipment on a linear isolator. Some useful insights can be gained by examining equation (12.23) and Figure 12.15:

- when $\Omega < 1$, $|T| \approx 1$ and the isolator is ineffective;
- when $\Omega = 1$, $|T| \approx \frac{1}{\eta}$ (for light damping) and isolator amplifies the disturbance;
- when $\Omega = \sqrt{2}$, $|T| = 1$. Above this frequency there is isolation;
- when $\Omega \gg \sqrt{2}$, $|T| \approx \frac{1}{\Omega^2}$ (for light damping). The transmitted vibration decreases at 40 dB/decade.

This very clearly shows the problems faced by designers of vibration isolation systems. High-frequency vibration isolation can be achieved at

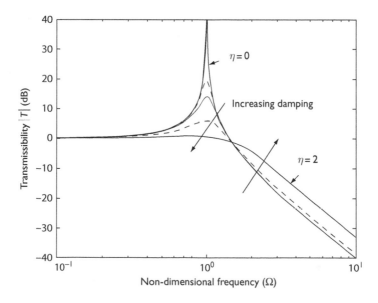

Figure 12.15 Force and displacement transmissibility of the one-dimensional idealised model of a vibration isolation system.

the expense of introducing a resonance, which causes amplification around this frequency. In many applications it is desirable to have a very low resonance frequency, but this is limited by other considerations such as the maximum static displacement allowable or the lateral stability of the machine or equipment.

12.3.2 Two-stage isolation system

Transmitted vibration can be further reduced by inserting an additional mass into the isolation system, as shown in Figure 12.16 where F_e, F_t, X_1, X_2 are complex amplitudes at each frequency. The frequency domain equations of motion for this system are

$$(k_1(1+j\eta) - \omega^2 m_1)X_1 - k_1(1+j\eta)X_2 = F_e \tag{12.25a}$$

$$k_1(1+j\eta)X_1 - ((k_1+k_2)(1+j\eta) - \omega^2 m_2)X_2 = 0 \tag{12.25b}$$

$$k_2(1+j\eta)X_2 = F_t \tag{12.25c}$$

These equations can be combined to give the force transmissibility as

$$T = \frac{(1+j\eta)^2}{s(1-s)\mu\Omega^4 - (1+j\eta)(1+(1-s)\mu)\Omega^2 + (1+j\eta)^2} \tag{12.26}$$

where $s = 1/(1+k_1/k_2)$, which is the combined stiffness of k_1 and k_2 in series divided by k_1, and $\mu = m_2/m_1$ is the ratio of the additional mass to the machine mass. As before, Ω is the forcing frequency normalised by the

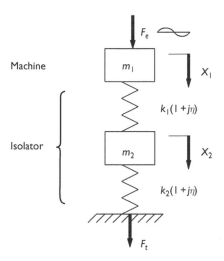

Figure 12.16 Two-stage vibration isolation system.

resonant frequency of the system (with m_2 set to zero). The modulus of the transmissibility is plotted in Figure 12.17, from which the advantages and disadvantages of the two-stage system can be seen. The main advantage of the two-stage system is at high frequencies, where the transmissibility decreases at 80 dB/decade (twice the rate for the single-stage system). This can also be seen by evaluating equation (12.26) at high frequencies, which, provided damping is light, becomes

$$T \approx \frac{1}{s(1-s)\mu\Omega^4} \tag{12.27}$$

This has a maximum value when $s = 0.5$ which means that the optimum stiffness configuration is when $k_1 = k_2$. Moreover, it can be seen that the isolator system increases in effectiveness as μ increases. The disadvantage of the two-stage isolation system is also clear. A second resonance is introduced causing amplification around this frequency, which can be attenuated by increasing the damping in the isolator. Lastly, most vibration isolation systems require the mounting resonance frequencies to be as low as possible compared with the dominant forcing frequencies, and the stiffness and the additional mass may have to be tuned for this, rather than the optimum transmissibility at high frequencies.

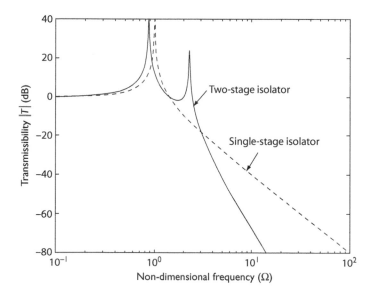

Figure 12.17 Transmissibility of a single- and a two-stage vibration isolation system, $\mu = 1$, $s = 0.5$ and $\eta = 0.01$.

12.3.3 Mobility approach

At high frequencies the source and/or receiver can no longer be considered to be rigid, and the simple model will not give an accurate prediction of the effectiveness of the isolator. An alternative method is required and the mobility approach is appropriate, which is described in Chapter 9. In this section, a relatively simple system is described where a vibrating source is connected to a receiver by a single isolator. Realistic systems will generally have more isolators, but the methodology can be extended relatively easily, and is discussed by Mead (1998). A vibration source is depicted in Figure 12.18a. The source can be described by its free velocity V_f, which is the velocity at point A before anything is attached to it, and its mobility at this position. This can be measured by switching off the vibrating source, and applying a dynamic force F at point A and measuring

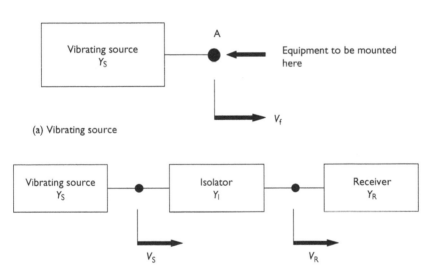

(a) Vibrating source

(b) Receiving structure connected to a vibrating source via an isolator

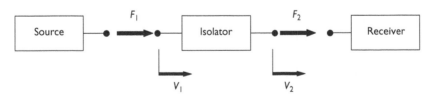

(c) Modelling of an isolator including mass-effects

Figure 12.18 Mobility approach to vibration isolation. All variables are complex quantities.

the velocity response V at the same position. The mobility of the source is given by $Y_s = V/F$. Figure 12.18b depicts the source connected to the receiver via an isolator. If the isolator is assumed to be massless the velocity of the receiver is related to the free velocity of the source by (Ungar 1992a)

$$V_R = V_f \frac{Y_R}{Y_R + Y_S + Y_I} \qquad (12.28a)$$

where Y_R and Y_I are the mobilities of the receiver and the isolator respectively. If the receiver is rigidly connected to the source then the isolator mobility is zero and equation (12.28a) becomes

$$V_R = V_f \frac{Y_R}{Y_S + Y_R} \qquad (12.28b)$$

The effectiveness of the isolator can be defined as

$$E_I = \left| \frac{\text{Velocity of receiver when it is rigidly connected to the source}}{\text{Velocity of the receiver when it is connected via an isloator}} \right|$$

$$(12.29a)$$

Equation (12.29a) can be evaluated by substituting equations (12.28a,b) to give

$$E_I = \left| 1 + \frac{Y_I}{Y_S + Y_R} \right| \qquad (12.29b)$$

By examining equation (12.29b) it can be seen that

- E_I should be as large as possible for good isolation;
- E_I depends on the source, isolator and receiver mobilities.

For good isolation, $|Y_I| \gg |Y_S + Y_R|$, thus high isolator mobility is required. At some frequencies it is possible for:

- Y_S and/or Y_R to become large because of resonances in the source or the receiver.
- Y_I to become very small because of wave effects in the isolator (the above theory is strictly not applicable because of the inertia forces in the isolator). This is discussed in Section 12.3.4.

12.3.4 Wave effects in the isolator

The mass in the isolator cannot be neglected at high frequencies because it will have internal resonances that will reduce its effectiveness. The isolator therefore needs to be modelled as a distributed system and thus a single mobility matrix is replaced by a matrix of point and transfer mobilities. Referring to Figure 12.18c, the matrix equation describing the isolator is given by

$$\begin{bmatrix} V_1 \\ V_2 \end{bmatrix} = \begin{bmatrix} Y_{11} & Y_{12} \\ Y_{21} & Y_{22} \end{bmatrix} \begin{bmatrix} F_1 \\ F_2 \end{bmatrix} \tag{12.30}$$

where Y_{11} and Y_{22} are point mobilities and Y_{12} and Y_{21} are transfer mobilities. Equation (12.28a) then becomes

$$V_R = V_f \frac{Y_{E1}}{Y_S + Y_{E2}} \tag{12.31}$$

where Y_{E1} and Y_{E2} are given by

$$Y_{E1} = \frac{Y_{21} Y_R}{Y_R + Y_{22}} \quad \text{and} \quad Y_{E2} = Y_{11} \frac{-Y_{12} Y_{21}}{Y_R + Y_{22}} \tag{12.32a,b}$$

Dividing equation (12.28b) by equation (12.31) gives the isolator effectiveness

$$E_I = \left| \frac{Y_R (Y_S + Y_{E2})}{Y_{E1} (Y_S + Y_R)} \right| \tag{12.33}$$

At low frequencies, the isolator will behave as a massless spring, and $Y_{E1} \approx Y_R$ and $Y_{E2} \approx Y_R + Y_I$. Equation (12.33) then simplifies to equation (12.29b). Equation (12.33) is plotted in Figure 12.19 for a system where both the source and the receiver have a single mode, and the isolator has distributed mass and stiffness. The fundamental well-damped resonance where the source and receiver bounce on the isolator (which generally has more damping than the source and the receiver) is evident at a low frequency. At higher frequencies the isolator effectiveness increases, except at the resonance frequencies of the source and the receiver, and the resonance frequencies of the isolator. At these resonance frequencies, the damping in the source, the receiver and the isolator govern the effectiveness of the isolator. The isolator effectiveness for a mass-like source and receiver and massless isolator is shown for comparison.

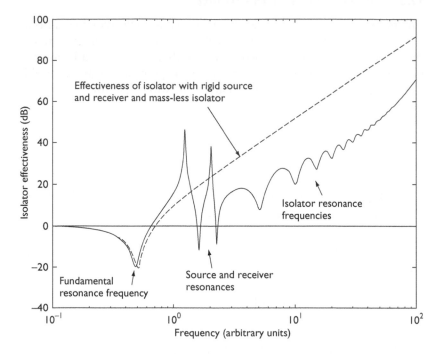

Figure 12.19 Isolator effectiveness of a system where both source and receiver each have a single resonance frequency, and are connected by an isolator that has distributed mass and stiffness.

12.3.5 Power flow

Attempts have been made by researchers to characterise vibration sources and to estimate the effectiveness of vibration isolation systems by estimating the vibrational power transmitted to the receiving structure from the source (by, for example, Pinnington and White 1981; Pinnington 1987; Petersson and Gibbs 1993, 2000). This is problematic when sources have multiple connections to the receiver and it is difficult to measure because of rotational degrees of freedom. To illustrate the principle, a source and receiver connected by a single point is considered. The reader is referred to the references for systems that are connected by several points. This is very much a topic of current research and is not used routinely in practice.

Referring to Figure 12.18, the velocity of the receiver in terms of the free velocity of the source when the isolator is *rigid* is given by equation (12.28b). The force applied to the receiver is given by

$$F_R = \frac{V_f}{Y_S + Y_R} \tag{12.34}$$

Now the power supplied by the source is given by

$$Q = \frac{1}{2} F_R^* V_R \tag{12.35a}$$

where $*$ denotes the complex conjugate. The power is complex, and the real part gives the mean power supplied by the source, while the imaginary part is related to the reactive power that cycles between the source and the receiver. Thus, the real part of the power is used to quantify the vibrational power transmitted to the receiver. Substituting equations (12.28) and (12.34) into equation (12.35a) gives

$$Q = \frac{1}{2} |V_f|^2 \frac{Y_R}{|Y_S + Y_R|^2} \tag{12.35b}$$

which shows that the power supplied by the source to the receiver is dependent upon both the mobility of the source Y_S and the mobility of the receiver Y_R. This makes it impossible to calculate the power that will be supplied by a source, without having knowledge of the structure to which it will be attached. There are two extreme cases that may occur in practice at particular frequencies. The first is where the modulus of the source mobility is small compared to the modulus of the receiver mobility. In this case the velocity of the receiver tends towards that of the free velocity of the source, and equation (12.35b) becomes

$$Q = \frac{1}{2} |V_f|^2 \frac{1}{Y_R^*} \tag{12.36}$$

It can be seen from equation (12.36) that to reduce the power transmitted to the receiver in this case, the receiver mobility should be increased. If the modulus of the source mobility is much larger than the modulus of the receiver mobility the force applied to the receiver tends to that of the blocked force of the source F_b, and equation (12.36) becomes

$$Q = \frac{1}{2} |F_b|^2 Y_R \tag{12.37}$$

where $F_b = V_f / Y_S$. It is clear from equation (12.37) that to reduce the power transmitted to the receiver, in this case, the receiver mobility should be decreased. From this example it is evident that it is vitally important to know the relative magnitudes of the source and the receiver mobilities if the appropriate vibration control measure is to be implemented.

Referring again to Figure 12.18, it can be seen that if an isolator is fitted, then it is in series with the receiver. So, provided the isolator is considered to be massless, then its mobility is simply added to that of the receiver.

Equation (12.35b) can then be modified to give the power supplied by the source as

$$Q = \frac{1}{2} |V_f|^2 \frac{Y_R + Y_I}{|Y_S + Y_R + Y_I|^2} \tag{12.38}$$

Thus, the greater the isolator mobility, the smaller the power supplied by the source.

12.3.6 Other considerations

When choosing and placing vibration isolators there are some other aspects that need consideration. If isolators are placed arbitrarily on a machine, it will generally exhibit coupled translational and rotational rigid body modes. There may be some merit in uncoupling these modes so that there is purely translational motion in one mode and pure rotation in the other. In this case, if the excitation force is translational, then one only needs to consider one of the modes of vibration. Uncoupling of modes can be achieved by choosing appropriate stiffnesses and positions of the isolators; this is discussed in detail by Ungar (1992a) and Crede and Ruzicka (1995).

Other factors that need to be taken into account are the non-linearity of the isolator (many isolators act as stiffening springs) and shock isolation, which is described by Newton (1995).

12.4 Vibration damping

Of all the aspects of vibration engineering, discussions about damping are always controversial. Damping mechanisms in complex structures take various forms, which present considerable difficulties for mathematical representation and are the subject of considerable contention. Consequently there are many theoretical models of damping, and some of those that are commonly used are only valid for harmonic excitation. It is generally agreed, however, that the only way to gain a true estimate of the damping in a built-up structure is by measurement, rather than prediction. In this section, commonly adopted damping mechanisms are reviewed, and two methods of damping treatment are described.

12.4.1 Viscous damping

Most basic texts on vibration cover viscous damping, including the companion volume edited by Fahy and Walker (1998), and thus it is only briefly described here. The equation of motion for *free* vibration of a viscously damped single-degree-of-freedom system is given by

$$m\ddot{x} + c\dot{x} + kx = 0 \tag{12.39}$$

where m, c and k are the mass, the damping coefficient and the stiffness of the system, respectively. From this equation it can be seen that the damping is proportional to the velocity of the system, whereas the inertia force is proportional to acceleration and the stiffness force is proportional to the displacement. Substituting for $\zeta = c/2\sqrt{mk}$ and $\omega_n = \sqrt{k/m}$, where ζ is the damping ratio and ω_n is the undamped natural frequency, gives

$$\ddot{x} + 2\zeta\omega_n\dot{x} + \omega_n^2 x = 0 \tag{12.40}$$

The solution to this equation describes the free vibration of the system, which falls into three categories: (1) underdamped when $\zeta < 1$, (2) critically damped when $\zeta = 1$ and (3) overdamped when $\zeta > 1$. Normally it is only case (1) which is of interest in practical vibration problems. The free vibration of the system is then given by

$$x = Ae^{-\zeta\omega_n t}\sin(\omega_d t + \phi) \tag{12.41}$$

where A and ϕ are constants that depend upon the initial conditions, and $\omega_d = \omega_n\sqrt{1 - \zeta^2}$ is the damped natural frequency. Equation (12.41) is plotted in Figure 12.20, where it can be seen that damping has two effects: (1) it controls the decay of free vibration and (2) it changes the frequency of vibration (although this effect is generally small in practice).

For harmonic excitation, equation (12.39) becomes

$$m\ddot{x} + c\dot{x} + kx = Fe^{j\omega t} \tag{12.42}$$

Assuming a harmonic response of the form $x = Xe^{j\omega t}$ leads to the receptance of the system

$$\frac{X}{F} = \frac{1}{k - \omega^2 m + j\omega c} \tag{12.43a}$$

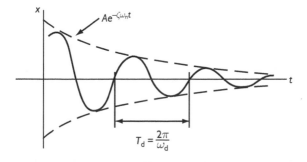

Figure 12.20 Free vibration of a mass-spring-damper system.

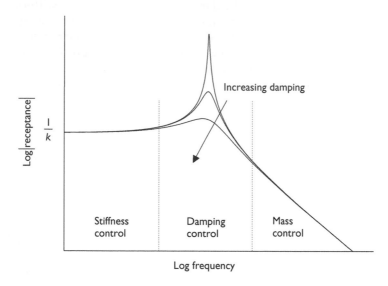

Figure 12.21 Receptance of a mass-spring-damper system.

which can be written as

$$\frac{X}{F} = \frac{1}{k} \left(\frac{1}{1 - \Omega^2 + j2\zeta\Omega} \right) \tag{12.43b}$$

where $\Omega = \omega/\omega_n$ is frequency normalised to the undamped natural frequency. By examining certain frequency regimes in equations (12.43a,b) the effect of damping on the system response can be clearly identified.

When $\omega \to 0 \qquad \dfrac{X}{F} = \dfrac{1}{k}$ (stiffness controlled)

$\omega = \omega_n \qquad \dfrac{X}{F} = \dfrac{1}{j\omega c} = \dfrac{1}{jk2\zeta}$ (damping controlled)

$\omega \gg \omega_n \qquad \dfrac{X}{F} = \dfrac{-1}{\omega^2 m}$ (mass controlled)

This is illustrated in Figure 12.21, where it can be seen that damping only has an appreciable effect around the resonance frequency. This is also true for multi-degrees-of-freedom systems.

12.4.2 Hysteretic damping

Whilst viscous damping receives the most attention in basic vibration texts, it is not the most common form of damping. A model of structural (hysteretic) damping that is widely used stems from a complex elastic or shear modulus, $E(1+j\eta)$ and $G(1+j\beta)$, respectively, where E, G, η and β are the elastic modulus, shear modulus, extensional loss factor and shear loss factor, respectively. This leads to the concept of a complex stiffness $k(1+j\eta)$, and the equation of motion of a harmonically excited single-degree-of-freedom system incorporating a complex spring is given by

$$m\ddot{x} + k(1+j\eta)x = Fe^{j\omega t} \qquad (12.44)$$

Assuming a harmonic response, this leads to

$$\frac{X}{F} = \frac{1}{k}\left(\frac{1}{1-\Omega^2+j\eta\Omega}\right) \qquad (12.45)$$

Setting $\Omega = 1$ in equations (12.43b) and (12.45), and setting the amplitudes of these equations to be equal, results in the following relationship between the loss factor and the damping ratio

$$\eta = 2\zeta \qquad (12.46)$$

The loss factor is widely used by practitioners to incorporate damping into their models. This is because it better represents structural damping, as the damping force is proportional to displacement but in phase with the velocity. However, it cannot be used in time-domain models for transient excitation because it predicts an acausal response of the system, that is, the system responds before it is excited. This problem can be overcome, however, by making the loss factor frequency-dependent, and this is discussed later in this section for viscoelastic materials.

12.4.3 Coulomb damping

In many built-up structures vibration energy is dissipated by sliding friction, and the damping mechanism in this case is called 'Coulomb damping'. With this damping mechanism, the damping force is independent of the velocity once the motion is initiated, but the sign of the damping force is always opposite to that of the velocity. No simple expression exists for this type of damping, but the energy dissipated per cycle ΔE can be equated to that of a viscous damper of damping coefficient c given by Nashif et al. (1985):

$$\Delta E = \pi c\omega|X|^2 \qquad (12.47)$$

where $|X|$ and ω are the amplitude of the response and the frequency of excitation, respectively. Since the damping force for Coulomb damping is given by $4\mu F_N$, where μ is the dynamic coefficient of friction for the two sliding components and F_N is the normal force, the work done in one cycle

Table 12.1 Equivalent viscous damping forces and coefficients

Damping mechanism	Damping force	Equivalent viscous damping coefficient
Viscous	$c\dot{x}$	c
Hysteretic	$k\eta x$	$\dfrac{k\eta}{\omega}$
Coulomb	$\pm\mu F_N$	$\dfrac{4\mu F_N}{\pi\omega X}$
Velocity squared	$\pm a\dot{x}^2$	$\dfrac{8}{3\pi}a\omega X$

is $\Delta E = 4\mu F_N |X|$. Setting this equal to equation (12.47) gives the equivalent damping coefficient:

$$c_{eq} = \frac{4\mu F_N}{\pi\omega|X|} \tag{12.48}$$

It can be seen that this damping is non-linear; higher response amplitudes result in lower damping.

The damping characteristics discussed, together with velocity-squared damping (Tse $et\,al.$ 1978) are summarised in Table 12.1.

12.4.4 Viscoelastic damping

Virtually all structural materials possess both damping and stiffness and are hence viscoelastic. Both damping and stiffness are dependent upon frequency and temperature, and a simple model to describe the frequency dependency is the Maxwell-Voigt model (Mead 1998). This consists of a stiffness in series with a viscous damper which are both in parallel with another stiffness, as shown in Figure 12.22. The dynamic stiffness (see Chapter 9) of this model is given by

$$\frac{F}{X} = \frac{\dfrac{k_2}{c^2} + \dfrac{\omega^2}{k_1}\left(\dfrac{k_1+k_2}{k_1}\right) + j\dfrac{\omega}{c}}{\left(\dfrac{1}{c}\right)^2 + \left(\dfrac{\omega}{k_1}\right)^2} \tag{12.49}$$

The real part of the dynamic stiffness gives the frequency-dependent stiffness of the material, and the ratio of the imaginary part to the real part gives the frequency-dependent loss factor. Hence the stiffness is given by

$$k(\omega) = \mathrm{Re}\left\{\frac{F}{X}\right\} = \frac{\dfrac{k_2}{c^2} + \dfrac{\omega^2}{k_1}\left(\dfrac{k_1+k_2}{k_1}\right)}{\left(\dfrac{1}{c}\right)^2 + \left(\dfrac{\omega}{k_1}\right)^2} \tag{12.50}$$

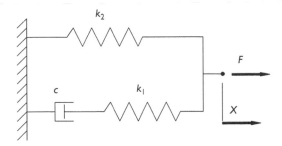

Figure 12.22 Simple model of a viscoeloelastic damping material.

which, when $\omega \to 0$ becomes $k(\omega) = k_2$, and when $\omega \to \infty$ $k(\omega) = k_1 + k_2$. Thus the material stiffness increases with frequency. The loss factor is given by

$$\eta(\omega) = \frac{\mathrm{Im}\left\{\dfrac{F}{X}\right\}}{\mathrm{Re}\left\{\dfrac{F}{X}\right\}} = \frac{\dfrac{\omega}{c}}{\dfrac{k_2}{c^2} + \dfrac{\omega^2}{k_1}\left(\dfrac{k_1 + k_2}{k_1}\right)} \qquad (12.51)$$

When $\omega \to 0$, $\eta(\omega) = \omega c/k_2$, and when $\omega \to \infty$, $\eta(\omega) = 1/(\omega c/k_1)[(k_1 + k_2)/k_1]$. The loss factor is a maximum when $\omega = (k_1/c)[k_2/(k_1 + k_2)]$. The stiffness and loss factor are plotted in Figure 12.23, for arbitrary values of k and c. Three regions are marked on the graph: the rubbery region, the transition region and the glassy region. As the names suggest, in the rubbery region the material is soft and in the glassy region it is stiff. It can be seen that the loss factor is a maximum in the transition region.

Because viscoelastic materials change their behaviour with temperature, the stiffness and loss factors have to be calculated from measured data for different temperatures. These measurements can be conducted on commercially available material, for example by TA Instruments (1997). The measured parameters as a function of frequency and temperature T are shown in Figure 12.24a,b for the elastic or shear modulus and the loss factor, respectively (Nashif *et al.* 1985). To combine the data into a single curve for the stiffness and a single curve for the loss factor, the measured data are shifted left or right by an appropriate amount. This amount is called the shift factor α_T. It is now possible to plot the data against the so-called reduced frequency parameter $f\alpha_T$ on a nomogram, as shown in Figure 12.24c. To use the nomogram for each specific f and T_i one moves along the T_i line until it crosses the horizontal f line. The intersection point X corresponds to the appropriate value of $f\alpha_T$. One then moves vertically up

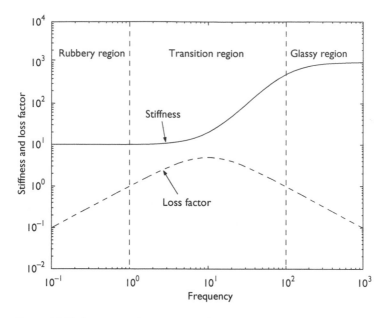

Figure 12.23 Frequency-dependent stiffness and loss factor calculated from equations (12.50) and (12.51) respectively for arbitrary values of k_1, k_2 and c.

to read the value of η and E at points A and B respectively. When selecting a polymer for damping treatment (which is discussed in Section 12.4.5) it is important to consult the nomograms for candidate damping materials to ensure that a material with an appropriate stiffness and damping for a given frequency and temperature is chosen.

12.4.5 Damping treatments

There are two common types of surface damping treatment, namely unconstrained and constrained layer damping. These are illustrated in Figure 12.25a,b, respectively. As the name suggests, the unconstrained layer treatment has a free surface and the energy dissipation in this system is due to extensional strain in the damping layer. Assuming that the damping in the composite structure is due mainly to the damping treatment, then for a uniform beam in flexure the ratio of the loss factor with unconstrained layer damping fitted to the whole of one side, to the loss factor of the damping layer is given approximately by Mead (1998) as

$$\frac{\eta_c}{\eta} = \frac{1}{1 + \dfrac{E_1 I_1}{E_2 I_2}} \tag{12.52}$$

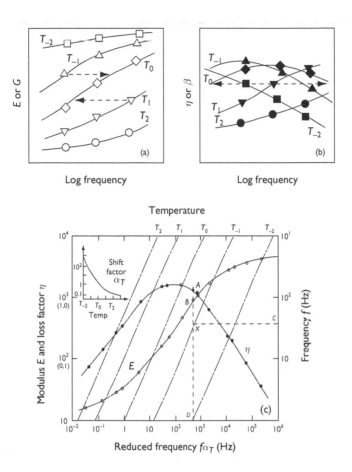

Figure 12.24 Frequency-dependent stiffness and loss factor for a viscoelastic material: (a) elastic or shear modulus for different temperatures *T*; (b) extension or shear loss factor; (c) nomogram showing the elastic modulus and loss factor as a function of temperature and frequency (after Nashif *et al.* 1985).

where the second moments of area (the *I*s) are about the composite neutral axis. By substituting for these, equation (12.52) can be written in terms of h_1 and h_2 and is plotted in Figure 12.26 for different values of E_1/E_2. This is a very useful graph as it can be used to determine rapidly whether a reasonable thickness of unconstrained layer damping treatment will give a suitable composite loss factor. When only one mode is to be damped, it may be preferable to attach the damping layer only in regions of high curvature.

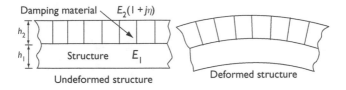

(a) Unconstrained layer damping treatment

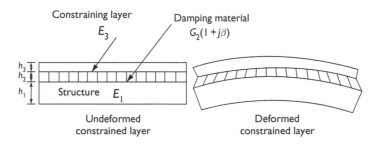

(b) Constrained layer damping treatment

Figure 12.25 Surface damping treatments.

With constrained layer damping, the loss mechanism is through shearing of the sandwiched damping layer, and consequently very thin damping layers can generate high composite loss factors. The two important factors that govern the behaviour are the shear parameter given by

$$g = \frac{G_2}{h_2 k_b^2} \left(\frac{1}{E_1 h_1} + \frac{1}{E_3 h_3} \right) \tag{12.53}$$

and the geometric parameter given by

$$L = \frac{(E_1 h_1)(E_3 h_3)d^2}{(E_1 h_1 + E_3 h_3)(E_1 I_1 + E_3 I_3)} \tag{12.54}$$

where k_b is the flexural wavenumber, d is the distance between the centroids of beams 1 and 3, and the second moments of area are about their own, rather than the composite, neutral axis. For a loss factor of the damping layer of less than 0.5, the ratio of the composite loss factor to that of the damping layer is given approximately by

$$\frac{\eta_c}{\beta} \approx \frac{gL}{1 + g(2 + L) + g^2(1 + L)} \tag{12.55}$$

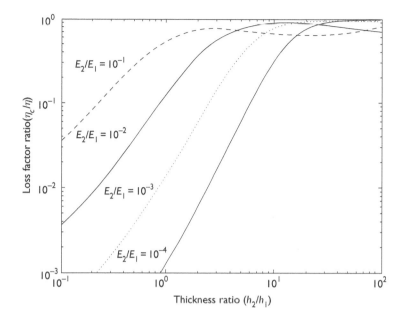

Figure 12.26 Ratio of the composite loss factor to the loss factor of the damping layer for unconstrained layer damping on a beam.

This is plotted in Figure 12.27 as a function of the shear parameter. The shear parameter is inversely proportional to frequency, so it can be seen that the damping in the composite structure is frequency-dependent. The optimal shear parameter for the maximum composite loss factor is obtained in practice by the appropriate choice of thickness and material of the damping layer. The corresponding approximate maximum loss factor ratio is given by

$$\left(\frac{\eta_c}{\beta}\right)_{max} = \frac{L}{2 + L + 2(1 + L)^{\frac{1}{2}}} \qquad (12.56)$$

Because the composite loss factor is dependent on frequency, it is not possible to optimise constrained layer damping for each mode of vibration. It should be noted that as well as increasing damping, both damping treatments also increase the stiffness of the host structure, and this is discussed in detail by Mead (1998).

Another effective damping treatment is squeeze-film damping described by Chow and Pinnington (1987). This is similar to constrained layer damping, but the damping layer is simply a microscopic layer of air. Provided that the constraining layer is thinner than the host structure, the flexural waves have different wavelengths at any one frequency and thus

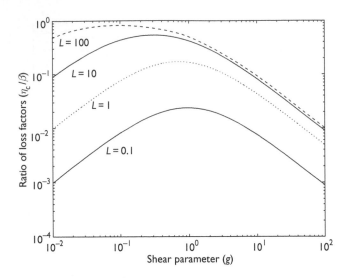

Figure 12.27 Ratio of the composite loss factor to the loss factor of the damping layer for constrained layer damping on a simply supported beam.

the sandwiched air is 'pumped' around dissipating energy in the form of heat. Other forms of squeeze-film dampers exist that use fluids other than air. One such damper is fitted to aero-engines and uses the engine oil as the fluid, as described by Holmes and Box (1992).

12.5   Vibration absorbers and neutralisers

For more than 100 years the vibration absorber has been used as a vibration control device. It is used in one of the two distinct ways; either it is tuned to a troublesome resonance frequency, where it is configured so that it adds damping to the host structure, or it is tuned to an offending forcing frequency, to provide a high impedance so that the response of the host structure is minimised at this frequency. In the former case it is called a 'vibration absorber' and in the latter case it is called a 'vibration neutraliser'. Although the device has many different forms, some of which are described by Den Hartog (1984) and Everett Reed (1995), it can be analysed as a base-excited mass-spring-damper-system. In this section both the vibration absorber and the vibration neutraliser are described, and the important characteristics of each device are highlighted.

A general framework for modelling a vibration absorber or a vibration neutraliser attached to any general vibrating system is shown in Figure 12.28. It is based on the impedance/mobility method described in Chapter 9. The system is excited at point 1, the absorber is fitted at point 2,

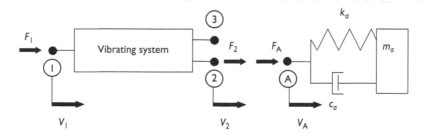

Figure 12.28 A vibration absorber connected to a general vibrating system.

and point 3 is any other point on the vibrating system. Before the absorber is connected to the vibrating system, the equations describing the motion are

$$V_1 = Y_{11}F_1 + Y_{12}F_2 \qquad (12.57a)$$

$$V_2 = Y_{21}F_1 + Y_{22}F_2 \qquad (12.57b)$$

$$V_3 = Y_{31}F_1 + Y_{32}F_2 \qquad (12.57c)$$

and the equation describing the absorber is

$$V_A = Y_A F_A \qquad (12.57d)$$

where the Y_{ii}s are the point mobilities, the Y_{ij}s are the transfer mobilities of the vibrating system and Y_A is the mobility of the absorber, which is the reciprocal of the impedance given by Brennan (1997).

$$Z_A = \frac{1}{Y_A} = \frac{j\omega m_a(1 + j2\zeta\Omega_a)}{1 - \Omega_a^2 + j2\zeta\Omega_a} \qquad (12.58)$$

where $\Omega_a = \omega/\omega_a$ is the forcing frequency normalised by the undamped natural frequency $\omega_a = \sqrt{k_a/m_a}$ and $\zeta = c/2\sqrt{m_a k_a}$ is the damping ratio. When the absorber is connected to the vibrating system, the equations of force equilibrium and continuity of motion are $F_2 = -F_A$ and $V_2 = V_A$, and equations (12.57a–d) combine to give

$$\frac{V_2}{F_1} = \frac{Y_{21}}{1 + \dfrac{Y_{22}}{Y_A}} \qquad (12.59)$$

and

$$\frac{V_3}{F_1} = Y_{31} - \frac{Y_{21}Y_{31}}{Y_{22} + Y_A} \qquad (12.60)$$

Examining these two equations it can be seen that when the mobility of the absorber is zero, that is, its impedance is infinite, the velocity at the attachment point $V_2 \to 0$, but the velocity at point 3 (anywhere else on the vibrating system) does not tend to zero. If the absorber is placed at point 1, where the system is excited, then point $1 \equiv$ point 2, and equation (12.60) becomes

$$\frac{V_3}{F_1} = \frac{Y_{31}}{1 + \dfrac{Y_{11}}{Y_A}} \tag{12.61}$$

It can be seen that when the absorber is placed at the source, then the vibration at *all* points on the vibrating system can be suppressed, provided the impedance of the absorber is infinite. In practical systems it may not be possible to attach the absorber at the position of the source, but it is important to place a vibration absorber as close to it as possible.

12.5.1 Vibration absorber

Because the purpose of a vibration absorber is to control a troublesome resonance frequency it is convenient to examine its effect on the dynamics of an undamped single-degree-of-freedom system, which represents a mode of vibration. This situation is depicted in Figure 12.29. The response of the structure is given by

$$\frac{V}{F} = \frac{1}{Z_S + Z_A} \tag{12.62}$$

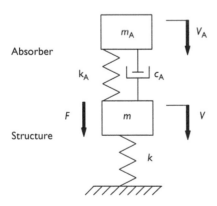

Figure 12.29 A vibration absorber attached to a single-degree-of-freedom system representing a single mode of vibration.

where Z_S is the impedance of the structure given by

$$Z_S = \frac{k}{j\omega} + j\omega m \qquad (12.63)$$

and the impedance of the absorber is given by equation (12.58), which can be written as

$$Z_A = \frac{1}{\dfrac{1}{\dfrac{k_A}{j\omega} + c_A} + \dfrac{1}{j\omega m_A}} \qquad (12.64)$$

Substituting equations (12.63) and (12.64) into equation (12.62), and rearranging to give the displacement of the structure $X = V/j\omega$, normalised to the static displacement X_o, yields (Everett Reed 1995)

$$\frac{X}{X_o} = \left\{ \frac{\left(1 - \Omega_A^2\right)^2 + \left(2\zeta\Omega_A\right)^2}{\left[\left(1 - \Omega_A^2\right)\left(1 - \Omega^2\right) - \Omega^2\mu\right]^2 + \left(2\zeta\Omega_A\right)^2 \left[1 - \Omega^2 - \Omega^2\mu\right]^2} \right\}^{\frac{1}{2}}$$
$$(12.65)$$

where $\Omega_A = \omega/\omega_A$ is the forcing frequency normalised by the undamped natural frequency of the absorber, $\omega_A = \sqrt{k_A/m_A}$, $\zeta = c_A/2\sqrt{m_A k_A}$ is the damping ratio of the absorber, $\mu = m_A/m$ is the ratio of the mass of the absorber to the mass of the structure and $\Omega = \omega/\omega_n$ is the forcing frequency normalised by the undamped natural frequency of the structure. The factors that determine the effectiveness of the vibration absorber are the mass ratio μ, the natural frequency of the absorber compared with the natural frequency of the host structure and the absorber damping ratio. Various optimum parameters have been proposed by Den Hartog (1984), but the following parameters result in an effective practical absorber:

$$\mu = 0.05, \quad \frac{\omega_A}{\omega_n} = \frac{1}{1 + \mu}, \quad \zeta_{opt} = \sqrt{\frac{3\mu}{8\left(1 + \mu\right)^3}} \qquad (12.66a,b,c)$$

Equation 12.65 is plotted in Figure 12.30 for $\mu = 0.05$ and $\omega_A/\omega_n = 1/(1 + \mu)$ for various values of damping. By setting the natural frequency of the absorber as in equation (12.66b), the amplitude of vibration at points A and B in Figure 12.30 are the same. It can be seen that if the damping is too small then although the vibration at the natural frequency of the absorber is small, there is a large response around the two resonance frequencies of the composite system. If the damping is too large ($\zeta > 0.4$)) then the relative motion of the absorber mass compared with that of the host structure is small and it is ineffective. The optimum damping ensures that the maximum amplitude of vibration is roughly the same as that at points A and B.

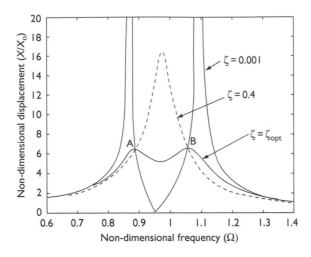

Figure 12.30 Non-dimensional displacement of the vibration of a single-degree-of-freedom system with a vibration absorber attached and tuned so that $\omega_a/\omega_n = 1/(1 + \mu)$ where $\mu = 0.05$ is the mass ratio.

12.5.2   Vibration neutraliser

Consider the vibrating system in Figure 12.28. Before the vibration neutraliser is attached, the velocity at point 2 is given by

$$\frac{V_{2(\text{free})}}{F_1} = Y_{21} \tag{12.67}$$

where the subscript (free) denotes that the neutraliser is not fitted. Dividing equation (12.59) by equation (12.67) gives a simple expression that can be used to examine the parameters which govern the effectiveness of the neutraliser:

$$\frac{V_2}{V_{2(\text{free})}} = \frac{1}{1 + \dfrac{Y_{22}}{Y_N}} \tag{12.68}$$

where the subscript N rather than A is used because the mass-spring-damper system is being used as a neutraliser rather than an absorber. Equation (12.68) shows that the effectiveness of a neutraliser is dependent upon the ratio of the mobility of the host structure, at the point where the neutraliser is to be attached, to the mobility of the neutraliser. These quantities can either be measured or predicted from dynamic models of the vibrating system and the neutraliser. The neutraliser is effective if $|Y_{22}| \gg |Y_N|$, which means that the neutraliser impedance must be much greater than the impedance of the host structure. The impedance of a neutraliser is given by equation (12.63), with the subscript A replaced by N. It can be seen that this is a maximum when the natural frequency is adjusted so that it is coincident with the excitation frequency. If it is assumed that the damping in the neutraliser is small, i.e. $\zeta \ll 1$, then

$$Z_N \approx \frac{\omega_n m_N}{2\zeta} \tag{12.69}$$

which shows that the impedance of a neutraliser increases with frequency, increases with neutraliser mass and decreases with neutraliser damping.

It is convenient to study the vibration neutraliser when it is attached to a single-degree-of-freedom system and the excitation frequency is set to be much greater than the natural frequency of the system. In this case, when the natural frequency of the neutraliser is the same as the excitation frequency and $\zeta \ll 1$, equation (12.68) becomes (Brennan 1997)

$$\left| \frac{V_2}{V_{2(\text{free})}} \right| \approx \frac{2\zeta}{\mu} \tag{12.70}$$

The modulus of equation (12.68) is plotted in Figure 12.31 for a neutraliser fitted to a mass-like host structure. It can be seen that effect of the neutraliser is to put a 'notch' in the frequency response, and the depth of this notch is governed by the damping of the neutraliser and the mass ratio. The mass

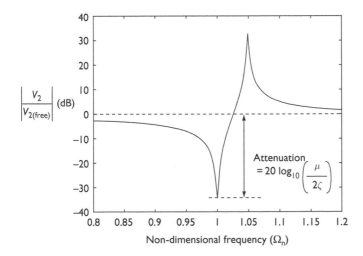

Figure 12.31 Normalised response of a mass-like host structure with a neutraliser attached.

ratio also has another effect in that it governs the proximity of the adjacent resonance ω_r to the notch frequency ω_n. The relationship between these two frequencies is given by

$$\omega_r = \omega_n\sqrt{1+\mu} \tag{12.71}$$

The other parameter of interest is the 3 dB neutraliser bandwidth B, and this is given by

$$B = 2\zeta \tag{12.72}$$

Thus it can be seen that, whilst it might be desirable to improve the effectiveness of the neutraliser by reducing damping, rather than increasing mass, this will mean that the neutraliser will have a narrow bandwidth, which is not desirable in practice. Moreover, a small neutraliser mass will mean that the adjacent resonance frequency will be very close to the notch frequency. This is also undesirable because any mis-tuning could result in amplification rather than attenuation of the vibration of the host structure at the forcing frequency.

12.6   Structural modification

If a potential vibration problem is either identified at the design stage or encountered later during dynamic excitation of the structure, then the options are either to modify or redesign the structure to avoid the problem. An overview of the possible alternatives is presented by Mead (1998), and this section aims to expand on some of these ideas and to indicate which particular approaches may be useful. In modifying a structure, one is attempting to do one or both of the following:

(i) Change the natural frequencies of a structure to avoid one or more of the following:

   (a) exciting a resonance due to discrete frequency excitation components;
   (b) exciting a number of modes in the case of broadband excitation;
   (c) similar natural frequencies of the individual connected components assembled together in the structure, so as to reduce the likelihood of strong coupling.

(ii) Alter the structure such that the spatial variation of the excitation does not couple well with a particular mode, or modes, of the structure.

The general techniques involved in changing the natural frequencies are

(a) detuning, that is separating the natural frequencies from the discrete excitation frequency components;
(b) reducing the number of responding modes.

For changing the spatial coupling between components one can consider

(a) nodalising, that is, designing a structure to exhibit nodal points at the required locations;
(b) decoupling modes having similar mode shapes at the interface;
(c) optimisation of surface geometry or orientation, for example to minimise rms vibration, stress or narrowband response.

These techniques are briefly discussed with appropriate references to further reading.

12.6.1 Natural frequency modification

When a particular excitation frequency, such as that due to a fixed speed rotating unbalanced machine, coincides with a natural frequency then one can try to shift the natural frequency to a frequency either above or below the source input. This is illustrated in Figure 12.32. This can be considered in the most simplistic way as either increasing mass, which normally reduces the natural frequencies if no additional stiffness is provided, or by increasing stiffness, which results in increasing the natural frequencies if no additional mass is provided. A simple structure to consider, that can be analysed exactly, is a simply supported uniform beam to which point masses or stiffnesses can be added. For a more general case, the finite element method can be used for the analysis. Simple analytical formulae based on a modal expansion for the beam can be considered where discrete masses m_i and discrete stiffnesses k_j are attached at x_i and x_j respectively along the length of a uniform simply supported beam. The modes of the modified structure can be expanded in terms of the mode shapes ϕ_s of the original structure. (This is demonstrated later for a beam-stiffened plate.) The additional stiffnesses and masses will introduce potential (PE) and kinetic (KE) energy terms, respectively. The additional PE is given by $\dfrac{1}{2} \sum\limits_{i=1}^{I} k_i \left(\sum\limits_{s=1}^{\infty} q_s \phi_s(x_i) \right)^2$ and the additional KE is $\dfrac{1}{2} \sum\limits_{j=1}^{J} m_j \left(\sum\limits_{s=1}^{\infty} \dot{q}_s \phi_s(x_j) \right)^2$, where the qs are the modal participation factors or generalised coordinates. Both depend on the mode shapes and the location of the attachments, and one extreme case is when all the attachments are placed at nodal points of a particular original mode and then that mode is unaltered in frequency and shape by the modification.

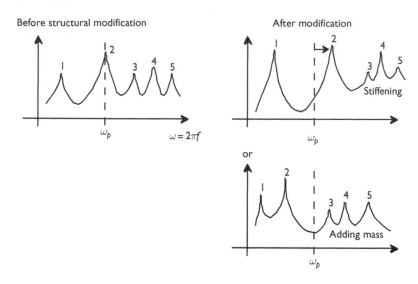

Figure 12.32 Frequency response functions showing the general effect of adding stiffness or mass to remove a resonance from the frequency of excitation ω_p.

The other extreme is when the attachments are at anti-nodes of a particular mode, as shown for point masses in Figure 12.33, and there is maximum effect. If the additions are small and the mode is not substantially altered, it is possible to consider solely the PE and KE in one original mode of the beam; that is, to add to the modal PE and KE the contributions from the stiffnesses and masses, respectively. Assuming the estimate of the 'new' natural frequency of the mode is given by the ratio of the maximum total PE divided by the reference KE (maximum KE divided by ω^2), the new natural frequency estimate ω_{new} is given, for a number of springs of stiffnesses k_i and masses m_j at positions x_i and x_j, respectively, by

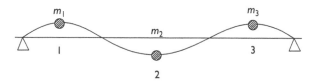

Figure 12.33 Illustration of the optimum positions for the addition of point masses to the antinodes of the third mode of a simply supported beam.

$$\omega_{\text{new}}^2 = \frac{\int_0^l EI \left(\frac{\mathrm{d}^2 \phi_{\mathrm{n}}(x)}{\mathrm{d}x^2} \right) \mathrm{d}x + \sum_{i=1}^I k_i \phi_{\mathrm{n}}(x_i)^2}{\int_0^l \rho A \phi_{\mathrm{n}}(x)^2 \, \mathrm{d}x + \sum_{j=1}^J m_j \phi_{\mathrm{n}}(x_j)^2} = \frac{K_{\mathrm{n}} + \sum_{i=1}^I k_i \phi_{\mathrm{n}}(x_i)^2}{M_{\mathrm{n}} + \sum_{j=1}^J m_j \phi_{\mathrm{n}}(x_j)^2}$$

$$= \frac{\omega_{\mathrm{n}}^2 + \dfrac{1}{M_{\mathrm{n}}} \sum_{i=1}^I k_i \phi_{\mathrm{n}}(x_i)^2}{1 + \dfrac{1}{M_{\mathrm{n}}} \sum_{j=1}^J m_j \phi_{\mathrm{n}}(x_j)^2} \tag{12.73}$$

where the energy contributions in other modes have been ignored and ω_{n} is the original natural frequency. This can be generalised for two-dimensional plate or shell structures in a similar manner. Depending upon the relative size of the terms, the new natural frequency may be greater or less than the original natural frequency. A more accurate procedure to analyse this problem is by the use of the receptance or mobility approaches, as described in Chapter 9. It is important to note that adding mass or stiffness changes the mode shapes as well as the natural frequencies. Although they can be described in terms of the original mode shapes (for example, using a component mode synthesis approach (CMS) or Lagrange multiplier techniques) they can be quite different if either the attached mass(es) or stiffness(es) are significant. Essentially they introduce additional boundary conditions that are shown in the qualitative diagrams in Figure 12.34 where it can be seen that there is a reduction in the beam displacement for higher values of both stiffness and mass and a movement of nodal points (N) towards the addition. For distributed mass and stiffness changes similar effects occur, but may be now observed over a more significant portion of the structure, as shown in Figure 12.35.

This form of analysis can be extended to two-dimensional problems, such as a plate that is stiffened by a beam. In general, if the fundamental natural frequency of the uncoupled stiffener is higher than the fundamental natural frequency of the plate then the latter will increase, and if it is less then it will decrease. Consider an isotropic uniform rectangular plate that is stiffened by a beam attached either along one edge or parallel to the edge. The beam provides a stiffening effect at excitation frequencies that is below the beam fundamental frequency, and a mass effect at excitation frequencies above the beam fundamental frequency.

Consider the plate stiffened along one edge, as shown in Figure 12.36. For a beam attached to a simply supported rectangular plate, one can formulate the KE and PE and derive an approximate formula for the natural frequency of the combined system. For simplicity it is assumed that the beam does not have significant torsional stiffness and inertia. On the basis

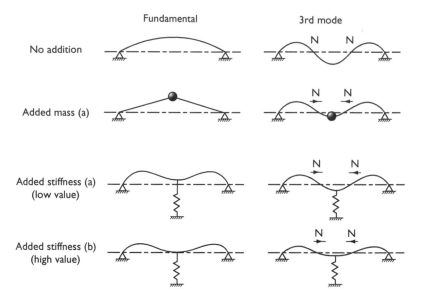

Fundamental
3rd mode
No addition
N
N
Added mass (a)
N
N
Added stiffness (a)
(low value)
N
N
Added stiffness (b)
(high value)
N
N

Figure 12.34 Illustration of the way in which the vibration of a simply supported beam is changed when a mass or a stiffness is attached.

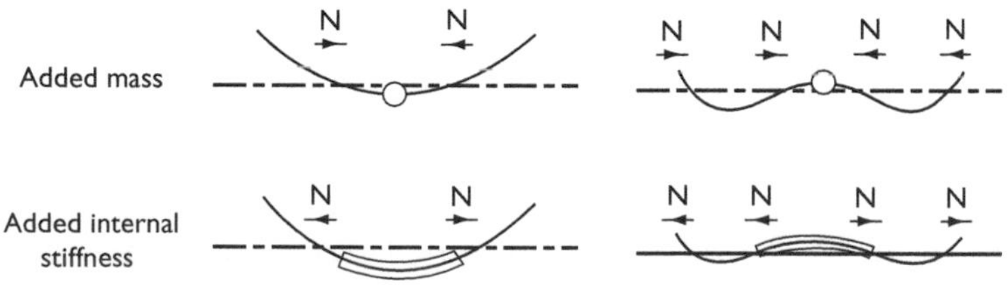

Free-free beam
Fundamental
3rd mode
No addition
N
N
N
N
N
N
Added mass
N
N
N
N
N
N
Added internal
stiffness
N
N
N
N
N
N

Figure 12.35 Illustration of the way in which the vibration of a simply supported beam is changed when a mass or a distributed stiffness is attached.

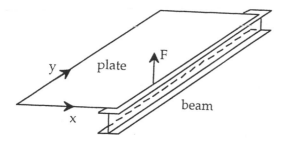

Figure 12.36 A plate stiffened by a beam along one edge.

of the assumptions of the single mode approximation, harmonic response, orthogonality of the modes and compatibility along the joint between the plate and the beam, the approximate formula involving the combined modal KE and PE can be applied, namely

$$\omega^2 = \frac{PE_{plate} + PE_{beam}}{KE_{ref(beam \mid plate)}} = \frac{\omega_p^2 M_p + 2\omega_b^2 M_b \sin^2\left(\frac{m\pi x_o}{a}\right)}{2 M_b \sin^2\left(\frac{m\pi x_o}{a}\right) + M_p} \tag{12.74}$$

which is expressed in terms of the masses of the plate and beam M_p and M_b and their uncoupled modal natural frequencies ω_p and ω_b, respectively and x_o is the position at which the beam is attached to the plate $(0 \le x_o \le a)$. In general, the beam can be attached anywhere but Figure 12.36 shows the attachment along one edge.

12.6.1.1 Cases (for a beam attached parallel to an edge)

(a) When the beam is attached along a nodal line of the plate mode, $\sin(m\pi x_o/a) = 0$ and $\omega^2 = \omega_p^2$; there is no change of mode shape due to the attached beam.

(b)
$$\omega^2 = \omega_p^2 \frac{\left[2\left(\frac{\omega_b}{\omega_p}\right)^2 M_b \sin^2\left(\frac{m\pi x_o}{a}\right) + M_p\right]}{\left[2 M_b \sin^2\left(\frac{m\pi x_o}{a}\right) + M_p\right]}$$

If $\omega_b/\omega_p > 1$ then $\omega^2 > \omega_p^2$. That is, if the beam mode natural frequency is higher than the original plate mode natural frequency then there is an increase in natural frequency. This means that the contribution from the beam's stiffness is more significant than that of its mass. If $\omega_b/\omega_p < 1$ then $\omega^2 < \omega_p^2$. That is, if the beam mode natural frequency is lower than the original plate mode natural frequency there is a decrease in

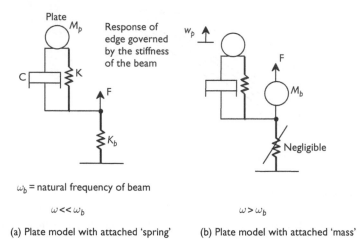

ω_b = natural frequency of beam

$\omega \ll \omega_b$ $\omega > \omega_b$

(a) Plate model with attached 'spring'   (b) Plate model with attached 'mass'

Figure 12.37 Representation of the addition of a beam to the modal parameter model of the plate with excitation at frequency ω also along the edge.

natural frequency. This means that the contribution from the beam's mass is more significant than that of its stiffness. Figure 12.37 shows the resulting dynamic representation for forced vibration when both excitation and stiffening beams are at the edge of the plate.

12.6.2 Reducing the number of significant responding modes

With broadband excitation, the natural frequencies may not be avoided and more than a single mode will normally dominate the response. Increasing stiffness will reduce the displacement or strain response if the loss factor is not decreased, because the mean square displacement response of a single mode, excited by a broadband excitation is proportional to powers of the modal stiffness, mass and damping, i.e. $\propto m^{-\frac{1}{2}}k^{-\frac{3}{2}}\eta^{-1}$ (see Newland 1993). Likewise, the mean square velocity and acceleration response are both proportional to these factors, i.e. $\propto m^{-\frac{3}{2}}k^{-\frac{1}{2}}\eta^{-1}$ and $\propto m^{-\frac{5}{2}}k^{-\frac{1}{2}}\eta^{-1}$ respectively, assuming broadband excitation.

Typically the strain is important for dynamic fatigue considerations. Increasing the overall stiffness will affect all modes, and lead to a reduction in the displacement response. A stiff structure has higher natural frequencies than a similar but more flexible structure, so has fewer modal natural frequencies in a given bandwidth, as natural frequency $\propto \sqrt{\text{stiffness/mass}}$. If modal stiffness is increased uniformly by a factor S, so that the natural frequencies are increased by $\sqrt{S}$, the number of modal frequencies in a given bandwidth are decreased by $1/\sqrt{S}$. This could be achieved, for example, by replacing a plate of thickness h by a sandwich plate where the original plate is divided into two equal faceplates separated by an incompressible lightweight core of thickness d. The thickness of the resulting sandwich plate will be $h + d$, and for the lower order bending modes $S = 3\,(d/h)^2$. Thus, far fewer modes will respond in a given bandwidth and there will be a reduced displacement response, but possibly greater sound radiation or response to acoustic excitation. This latter consequence is important for acoustic fatigue and the response of spacecraft structures. It is due to the reduced critical frequency, that is inversely proportional to the square root of the stiffness, resulting in higher radiation efficiency at lower frequencies, compared to the original plate of similar density but lower stiffness (see Fahy 1987).

When coupling different forms of structure it is possible to modify connection locations. The most noticeable examples of this are for well-defined structures such as periodic structures either in one, two or three dimensions. For example, effects due to periodic spacing can be avoided, as periodic structures exhibit well-defined pass and stop bands of vibration propagation (Langley *et al.* 1997). Conversely a designer could choose to use a periodic design, taking advantage of the stop bands to attenuate certain frequency components. A phenomenon is also observed, known as Anderson localisation, where disordering of a periodic structure results in a response that becomes localised to a small number of bays within a periodic structure rather than distributed. Hence, an alternative to reducing the number of modes globally is to choose a specific bandwidth in which vibration propagation is minimised.

12.6.3 Modification of spatial coupling and response

Locating point force sources at the nodal points of a mode results in a zero-generalised force on that mode. Mead (1998) describes the moving of nodes to source or receiver locations as 'nodalizing' the structure. This can also be achieved by the simple addition of secondary masses or by a structure that will change the response of the primary structure at the various locations. A simple analysis, in terms of mobilities, can then be formulated and extended to include rotational excitation (couples/moments), with rotational degrees of freedom, in order to predict the combined structural response, and likewise, multiple points of connections can be incorporated, but increasing complexity may result. These methods are viable if the excitation frequencies do not vary significantly. Another approach is to avoid coupling between different mode types of connecting structures. For example, a beam with asymmetric cross-section, will have coupled bending, warping and torsional modes that would couple to flexural modes of an attached plate, and in-plane vibration of a plate will couple to flexural vibration of a plate connected at 90°. When the excitation has a particular spatial variation, for example an incident plane acoustic wave, and the wavelength is long compared with the structure, then it is preferable to design the structure so that its modes have wavelengths that are short compared with the acoustic trace wavelength. For example, stiffeners with low torsional stiffness, but high bending stiffness, can be introduced to provide stiffened plate modes with successive bays in opposite phase leading to a lower coupling to the acoustic field. Likewise, there may be scope for altering the geometric orientation of the structure to the excitation field and hence reducing the response of the structure. Genetic algorithms may be used for searching for optimum structural modification from the dynamical point of view (Keane 1995). No simple or general guidelines exist for most cases, but the disruption of periodic geometry is generally desirable.

12.7 Material selection

It has already been shown that the structural response depends upon mass, stiffness and damping. For a single mode of vibration, modelled as a single-degree-of-freedom system of mass m and complex stiffness $k(1+j\eta)$, the modal displacement is given by

$$X = \frac{F}{k - \omega^2 m + jk\eta} \tag{12.75}$$

where F is the modal force. At the resonance frequency the response is given by

$$X = \frac{F}{jk\eta} \tag{12.76}$$

As the modal stiffness is proportional to the Young's modulus of the structure, the displacement response is inversely proportional to $E\eta$. Thus the material with the greatest $E\eta$ product gives the lowest displacement at resonance. Now the stress σ is proportional to the product of the displacement and Young's modulus. So, using equation (12.76) gives

$$\sigma \propto \frac{1}{\eta} \tag{12.77}$$

Thus, the material that has the greatest loss factor gives the lowest resonant stress. In cases of random excitation the rms response of the structure is the quantity of interest. The dependence of the responses on the material properties is different depending upon whether the input is a force source

Table 12.2 Some material properties

Material	Young's modulus E (N m^{-2})	Density ρ (kg m^{-3})	Poisson's ratio ν	Loss factor η
Mild steel	2e11	7.8e3	0.28	0.0001–0.0006
Aluminium alloy	7.1e10	2.7e3	0.33	0.0001–0.0006
Brass	10e10	8.5e3	0.36	<0.001
Copper	12.5e10	8.9e3	0.35	0.02
Glass	6e10	2.4e3	0.24	0.0006–0.002
Cork		1.2–2.4e3		0.13–0.17
Rubber	1e6–1e9	≈1e3	0.4–0.5	≈0.1
Plywood*	5.4e9	6e2		0.01
Perspex	5.6e9	1.2e3	0.4	0.03
Light concrete *	3.8e9	1.3e3		0.015
Brick*	1.6e10	2e3		0.015

Note
* Average values.

or a velocity source, as discussed in Section 12.2 on vibration sources. For excitation producing uniform force spectral density, the relationship is

$$X_{rms} \propto \frac{1}{E^{\frac{3}{4}} \rho^{\frac{1}{4}} \eta^{\frac{1}{2}}} \qquad (12.78)$$

and for excitation producing uniform displacement spectral density

$$X_{rms} \propto \frac{E^{\frac{1}{4}}}{\rho^{\frac{1}{4}} \eta^{\frac{1}{2}}} \qquad (12.79)$$

where ρ is the material density. The expressions in equations (12.78) and (12.79) can be thought of as figures of merit (Mead 1998). Typical material properties required for structural dynamics analysis are presented in Table 12.2.

References

Brennan, M.J. (1997) Vibration control using a tunable vibration neutraliser. *Proceedings of the Institution of Mechanical Engineers*, Vol. 211, Part C, 91–108.

Chow, L.C. and Pinnington, R.J. (1987) Practical methods of increasing structural damping in machinery, I: Squeeze-film damping with air. *Journal of Sound and Vibration*, 118(1), 123–129.

Crede, C.E. and Ruzicka, J.E. (1995) *Chapter 30 in Shock and Vibration Handbook Fourth Edition* (ed. C.M. Harris). Theory of Vibration Isolation (McGraw-Hill).

Cremer, L., Heckl, M. and Ungar, E.E. (1998) *Structure-Borne Sound* (2nd edn) (Berlin: Springer-Verlag).

Den Hartog, J.P. (1984) *Mechanical Vibrations* (Dover Publications, Inc.).

Ehrich, F.F. (1995) *Chapter 5 in Shock and Vibration Handbook Fourth Edition* (ed. C.M. Harris). Self-Excited Vibration (McGraw-Hill).

Everett Reed, F. (1995) *Chapter 6 in Shock and Vibration Handbook Fourth Edition* (ed. C.M. Harris). Dynamic Vibration Absorbers and Auxiliary Mass Dampers (McGraw-Hill).

Fahy, F.J. (1987) *Sound and Structural Vibration* (Academic Press).

Fahy, F.J. and Walker, J.G. (1998) *Fundamentals of Noise and Vibration* (E & FN Spon).

Fuller, C.R., Elliott, S.J. and Nelson, P.A. (1997) *The Active Control of Vibration* (Academic Press).

Holmes, R. and Box, S. (1992) On the use of squeeze-film dampers in rotor support structures. *Machine Vibration*, I, 71–79.

Keane, A.J. (1995) Passive vibration control via unusual geometries: The application of genetic algorithm optimization to structural design. *Journal of Sound and Vibration*, **185**, 441–453.

Langley, R.S., Smith, J.R.D. and Fahy, F.J. (1997) Statistical energy analysis of periodically stiffened damped plate structures. *Journal of Sound and Vibration*, **208**(3), 407–426.

Lyon, R.H. (1987) *Machinery Noise and Diagnostics* (Butterworths).

Mead, D.J. (1982) *Chapter 25 in Noise and Vibration* (eds R.G. White and J.G. Walker). Vibration Control (I) (Ellis Horwood).

Mead, D.J. (1998) *Passive Vibration Control* (John Wiley & Sons).

Nashif, A.D., Jones, D.I.G. and Henderson, J.P. (1985) *Vibration Damping* (John Wiley & Sons).

Nevzat Özgüven, H. and Houser, D.R. (1988) Mathematical models used in gear dynamics – A review. *Journal of Sound and Vibration*, **121**(3), 383–411.

Newland, D.E. (1993) *An introduction to random vibrations, spectral and wavelet analysis* (Longman Scientific & Technical).

Newton, R.E. (1995) *Chapter 31 in Shock and Vibration Handbook Fourth Edition* (ed. C.M. Harris). Theory of Shock Isolation (McGraw-Hill).

Petersson, B.A.T. and Gibbs, B.M. (1993) Use of the source descriptor concept in studies of multi-point and multi-directional vibration sources. *Journal of Sound and Vibration*, **168**, 157–176.

Petersson, B.A.T. and Gibbs, B.M. (2000) Towards a structure-borne sound source characterisation. *Applied Acoustics*, 61, 325–343.

Pinnington, R.J. (1987) Vibrational power transmission to a seating of a vibration isolated motor. *Journal of Sound and Vibration*, **118**(3), 515–530.

Pinnington, R.J. and White, R.G. (1981) Power flow through isolators to resonant and non-resonant beams. *Journal of Sound and Vibration*, **75**(2), 179–197.

Preumont, A. (1997) *Vibration Control of Active Structures* (Kluwer Academic Publishers).

Rao, S.S. (1995) *Mechanical Vibrations* (Addison Wesley).

TA Instruments (1997) DMA 3980 Dynamic Mechanical Analyser Operators Manual.

Tse, F.S., Morse, I.E. and Hinkle, R.T. (1978) *Mechanical Vibrations: Theory and Applications* (Allyn and Bacon, Inc.).

Ungar, E.E. (1992a) *Chapter 11 in Noise and Vibration Control Engineering* (eds L.L Beranek and I.L. Vér). Vibration Isolation (John Wiley & Sons).

Ungar, E.E. (1992b) *Chapter 12 in Noise and Vibration Control Engineering* (eds L.L. Beranek and I.L. Vér). Structural Damping (John Wiley & Sons).

Chapter 13

Vibration measurement techniques using laser technology: laser vibrometry and TV holography

N.A. Halliwell and J.N. Petzing

13.1 Introduction

The measurement of vibration of a solid surface is usually achieved with an accelerometer or some other form of surface-contacting transducer. There are, however, many cases of engineering interest where this approach is precluded. A prime example is that of loudspeaker vibration shown in Figure 13.1 which shows a loudspeaker in a 'rocking' mode at 3 kHz. In order to improve designs, vibration data are required both at a point (time-resolved) and over the whole surface. These data cannot be provided by contacting transducers. Other examples include very hot and rotating surfaces such as engine exhausts and rotating driveshafts, respectively. Since the advent of the laser in the early 1960s, optical metrology has provided engineering data for designers which, hitherto, would have been considered unobtainable. In this chapter we introduce the techniques of laser vibrometry and TV holography which are capable of providing remote time-resolved vibration data at a point on a surface in addition to both visualising and quantifying vibration behaviour over an area. We also address the difficult measurement problem of torsional vibration and show how this has been solved by laser technology. The basic physics of each technique is explained prior to descriptions of applications and a discussion of accuracy and limitations.

13.2 Laser vibrometry

Laser velocimetry is now a well-established technique for non-intrusive, time-resolved velocity measurements in fluid flows. The technique relies on detecting the velocity of seeding particles or tracers as they pass though a small volume in the flow. The tracer particles scatter light from the laser beams and the Doppler frequency shift in the light is detected to provide a measure of the tracer particle velocity. In this way, the flow is essentially sampled by the presence of tracer particles which are present in the volume and are specially chosen both to scatter sufficient light to

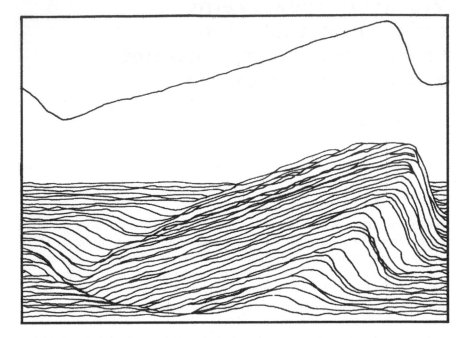

Figure 13.1 Vibration mode of a loudspeaker.

a photodetector and to follow the flow velocity fluctuations of interest. An obvious practical problem to be considered is the intermittent presence of tracer particles since a Doppler signal processing system which aims to provide time-resolved information must be able to accommodate periods of signal 'drop out' where either no tracers are present in the region or the vector addition of the scattered light from each particle produces a very low amplitude signal from the photodetector. Choice of tracer particles, and their spatial distribution in the flow, is a major issue in the accurate use of laser velocimetry and much of the research work published in establishing the technique has concerned the validity of statistical conclusions drawn from a particle measurement.

When laser velocimetry is used to take a velocity measurement of a solid body, the instrument or measurement system is usually referred to as a laser vibrometer. In this situation the tracer particles can be considered to be 'replaced' by scattering elements within a small surface area of the target body which is illuminated by the incident laser beam. The first major international conference on laser velocimetry was held in Copenhagen, Denmark in 1975 and when commercial instrumentation, designed for use in fluid flows, appeared in the marketplace, its use for solid surface velocity measurement was considered to be straightforward in comparison since the intermittency of the Doppler signal was assumed to be no longer a problem and in general, with appropriate collection optics, it was always possible (even from matt black surfaces!) to collect sufficient light to take a reliable measurement. In fact, when laser light is scattered from a solid surface, which is rough on the scale of the optical wavelength, a speckle pattern (Dainty 1984) is formed in space in front of the target. The speckle pattern (discussed in Section 13.3) consists of a random array of bright and dark 'cigar shaped' volumes. The photodetector may sample one or more speckles and this fact brings its own problems when laser vibrometry is used in practice. Rather than the intermittency of tracer particles presenting problems in the fluid flow case the permanent presence of scattering elements in the solid body case dictates speckle pattern dynamics which, as we shall see, can affect the ultimate performance of the instrument. It is this physical fact which primarily distinguishes the operation of laser vibrometry from its fluid flow counterpart.

13.2.1 Physical principles of laser vibrometry

The basic working principle of laser vibrometry is the detection of the Doppler shift in coherent light which is scattered from the moving surface and is best explained as an extension to the general fluid flow case shown in Figure 13.2. In Figure 13.2, when a tracer particle with a velocity vector U scatters light of wavelength λ in a direction K_2 from an incident laser beam in a direction K_1, where both K_2 and K_1 are unit vectors, the scattered light undergoes a Doppler shift in frequency f_D given by (Watrasiewicz and Rudd 1976)

$$f_D = (K_2 - K_1) \cdot U/\lambda \tag{13.1}$$

Laser vibrometry involves directing an incident laser beam at the target surface of interest and an instrument will collect light which is directly

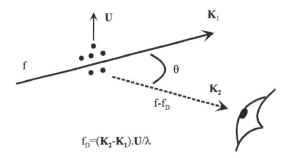

Figure 13.2 Doppler shift in scattered light (fluid flow).

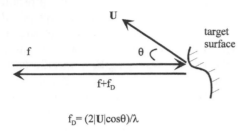

$$f_D = (2|U|\cos\theta)/\lambda$$

Figure 13.3 Backscatter geometry for laser vibrometry.

back-scattered off the target surface as shown in Figure 13.3. In Figure 13.3, the Doppler shift f_D is given by

$$f_D = |2K_1 \cdot U/\lambda| = \frac{2|U|\cos\theta}{\lambda} \tag{13.2}$$

since now $K_1 = -K_2$. In this way, by demodulating and tracking the changing value of f_D, it is possible to produce a time-resolved measurement of the component of solid surface velocity which is in the direction of the incident laser beam.

The scattered light beam from the target surface has a frequency of typically 10^{15} Hz which cannot be demodulated directly. The Doppler frequency f_D is detected using an optical interferometer by mixing the scattered beam coherently with a reference beam derived from the same laser source onto the surface of a photodetector. The photodetector responds to a time average of the *intensity* of the total light collected and this form of non-linear detection produces a heterodyne or 'beat' in the current output whose frequency is equal to the difference in frequency between the two beams (Watrasiewicz and Rudd 1976). This arrangement is shown schematically in Figure 13.4 in a Michelson interferometer geometry. Figure 13.4 shows that

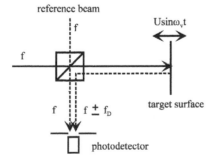

Figure 13.4 Reference beam heterodyne (Michelson interferometer).

the incident laser beam has a frequency f and, without loss of generality, the target surface is shown vibrating with a sinusoidal velocity $U \sin \omega_v t$ in a direction parallel with the incident laser beam, where ω_v is the angular vibration frequency.

The light intensity incident *at a point* on the photodetector surface $I(t)$, neglecting polarisation effects, is given by (Watrasiewicz and Rudd 1976)

$$I(t) = I_R + I_T + 2\sqrt{I_R I_T} \cos[2\pi f_D t + \phi_R - \phi_T] \qquad (13.3)$$

where I_R is the reference beam intensity, I_T is the target beam intensity, ϕ_R is the constant reference beam phase, ϕ_T is the target beam phase, and $f_D = (2U \sin \omega_v t)/\lambda$ is the 'instantaneous' Doppler frequency shift in Hz (strictly the phase change due to the Doppler effect at time t is $\int f_D dt$).

This arrangement is ambiguous in the measurement of the direction of the vibration velocity. When the Doppler frequency shift is non-zero, it is not possible to distinguish whether the target is moving towards or away from the detector. Figure 13.5 shows the spectrum of the photodetector output. Since the vibrometer provides a time-resolved measure of surface velocity by tracking the Doppler frequency, it is also clear that the signal will be discontinuous with this arrangement and thus generally not possible to decode properly. Vibration engineers require a measure of amplitude *and* phase of the target velocity and this is provided by frequency pre-shifting the reference beam.

Figure 13.6 shows a Michelson interferometer geometry where the reference beam has been frequency shifted by a constant amount f_R. In this case the photodetector output is given by

$$I(t) = I_R + I_T + 2\sqrt{I_R I_T} \cos[2\pi(f_R - f_D)t + \phi_R - \phi_T] \qquad (13.4)$$

With reference to Figure 13.6, the output spectrum of the photodetector now shows how the frequency shift provides a carrier frequency in the

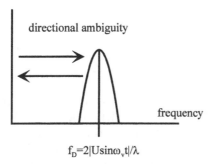

$f_D = 2|U \sin \omega_v t|/\lambda$

Figure 13.5 Photodetector output spectrum.

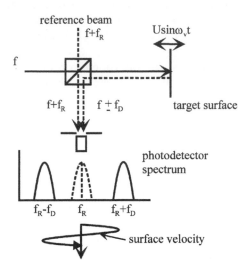

Figure 13.6 Frequency shifting (Michelson interferometer).

photodetector output which the target surface velocity frequency modulates. In this way, frequency demodulating and tracking the photodetector output signal provide a time-resolved measurement of target surface velocity.

Laser vibrometer systems assembled for laboratory use from bulk components, or purchased as a commercial instrument, all work on the physical principles so far described. They will differ in the choice of optical geometry, means of frequency shifting and/or direction ambiguity removal and electronic design of Doppler signal processor. In the next section practical considerations governing this choice are considered.

13.2.2   Frequency shifting

Various means of frequency shifting have been employed in laser vibrometer systems as the technology has developed; these include rotating scattering discs (Ballantyne *et al.* 1974), Kerr cells (Drain and Moss 1972), rotating diffraction gratings (Oldengarm *et al.* 1973), piezo-electric elements (Baker *et al.* 1990), fibre optic modulators (Lewin *et al.* 1985), Bragg cells (Buchave 1975) and laser diode current modulation (Laming *et al.* 1985).

An early, simple and robust frequency shifter in a vibrometer was provided by a rotating scattering disc (Halliwell 1979; Pickering and Halliwell 1986) and this is shown in Figure 13.7. The normal to the plane of the disc is at a slight angle θ to the incident laser beam so that scattering elements passing through the beam have a velocity component in the direction of the latter. The magnitude of the frequency shift is readily controlled by the speed

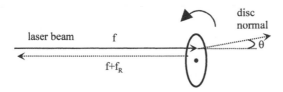

Figure 13.7 Frequency shifting by scattering disc.

of rotation and/or the radial position of the incident beam. In this way, the reference beam takes the form of a scattered light beam in a solid angle which is dictated by the limiting aperture (usually at the photodetector) in the light collection system. Shifts of typically 1 MHz and higher are easily provided. The disc is covered with retro-reflective tape to ensure adequate intensity in direct backscatter. Frequency shifting in this way provides a self-aligning property, useful in large-scale bulk component systems, which is not integral to other forms of frequency shifting. This is at the expense of a noise floor performance which, when measured on a stationary target, displays speckle noise peaks in the form of a periodogram (to be discussed in Section 13.3.2) with a fundamental at the rotation frequency. The minimum vibrational level measurable at these frequencies is increased. Commercial instrumentation which utilised this form of shifting provided a choice of disc speeds to avoid this problem.

Another means of frequency shifting is provided by a rotating diffraction grating (Oldengarm *et al.* 1973). This is shown in a Michelson interferometer in Figure 13.8. A small glass disc ($\cong 300$ mm diameter) is radially etched so that the incident laser beam is subject to periodic thickness variations as the disc is rotated. In this way periodic changes in optical path length on a spatial scale comparable with the laser wavelength provide a moving phase diffraction grating. The light is diffracted into distinct orders which are frequency shifted in proportion to the disc speed which is easily

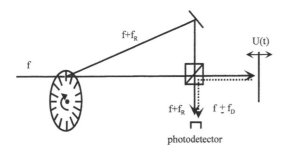

Figure 13.8 Frequency shifting by rotating diffraction grating.

controlled. Early commercial versions of this frequency shifter (Oldengarm 1977) provided shifts of about 1 MHz. Choice of optical geometry in the vibrometer dictates which orders are used. For a normal-to-surface vibration measurement, it is usually the first order which is chosen. The etched profile is designed to ensure adequate intensity in this order.

The noise floor of this vibrometer when measured on a stationary target surface is a periodogram with a fundamental frequency at the disc rotational speed. This is due to the machining process in manufacturing the disc. Consequently, as with the scattering disc, the minimum measurable vibration level at these frequencies is increased. The rotating grating provides a flexible tool in bulk component vibrometer systems. They are inherently fragile and experienced users treat them with care!

To date, the most commonly used form of frequency shifting in commercial vibrometers is the Bragg cell. This was employed in the first commercially available vibrometer (Buchave 1975). A Bragg cell design is shown in Figure 13.9. The incident laser beam traverses a transparent medium (usually a crystal) in which an ultrasonic wave is travelling normal to the beam direction. Density variations, which are on the same spatial scale as the optical wavelength, produce refractive index variations and, in this way, provide a moving phase diffraction grating. The incident beam is diffracted into distinct orders, each undergoing a frequency shift which is a direct multiple of the frequency of the ultrasonic wave (typically 40 MHz). Normally, the cell is designed so that the first order diffracted beam is used as the reference beam which emerges at an angle to the incident beam and may, therefore, require alignment compensation in some geometries. The solid-state nature of Bragg cells make them the first choice as frequency shifters in commercial instruments but their presence does add to the expense since they require additional electronic and optical components.

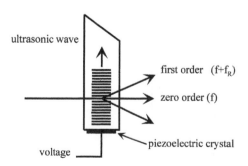

ultrasonic wave

first order $(f+f_R)$

zero order (f)

piezoelectric crystal

voltage

Figure 13.9 The Bragg cell.

As an alternative to direct frequency shifting, directional ambiguity in the Doppler signal can be removed by quadrature detection (Scruby and Drain 1990). This is so called because polarisation optics are used to provide two separate detection channels in which the reference beam in one channel is $\pi/2$ out of phase with the other. In this way, the sine–cosine relationship between the detector outputs provides for a deterministic response in their relative phase when the target surface changes direction. This technique is normally used in a Michelson interferometer geometry and one such method (Reibold and Molkenstruck 1981) is shown in Figure 13.10. The reference beam passes twice through a $\lambda/8$ retardation plate after reflection from a reference mirror to return as a circularly polarised beam before recombining with the signal beam at the beamsplitter. The combined beams are split by a Wollaston prism (polarising beamsplitter) whose axis is at 45° to the original polarisation direction of the laser. In this way, two signal channels are created where the reference beams are in quadrature; thus the signals from the photodetectors have the desired sine–cosine relationship. Quadrature detection has been incorporated into some commercially available vibrometer systems (Ometron 1997).

13.2.3 Choice of optical interferometer

The two most popular choices of optical interferometers are the Michelson and Mach-Zender. The Michelson is shown in Figures 13.7, 13.8 and 13.10. A Mach-Zender interferometer which utilises a Bragg cell as a frequency shifting element is shown in Figure 13.11. A polarising beamsplitter (PBS1) splits the incident beam into target and reference beams. The target beam is

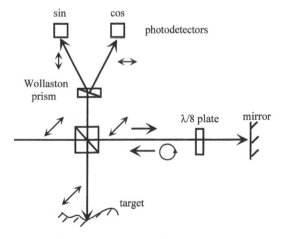

Figure 13.10 Quadrature detection. Source: Reibold and Molkenstruck (1981).

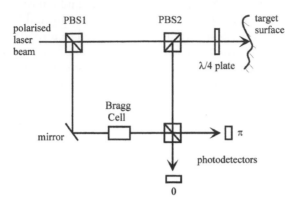

Figure 13.11 Mach-Zender interferometry design.

transmitted by a second polarising beamsplitter (PBS2). Assuming no change in polarisation after scattering from the target surface, two passes through the $\lambda/4$ retardation plate rotate the plane of polarisation through 90° so that the target beam is then reflected by PBS2. This offers the additional advantage that the laser source is isolated from the target light which avoids feedback problems with the laser. A final beamsplitter combines the target and reference beams after the latter has been frequency shifted by the Bragg cell. The beamsplitter produces two separate channels for detection which are π out of phase. This is an advantage, since taking the difference of the two detector outputs means the detection system is insensitive to stray coherent light and to intensity modulation of the reference beam. The latter is important when high ratios of reference beam to scattered light intensity are used to obtain increased sensitivity when the scattered light intensity is low. Modulation of the reference beam intensity can occur through power supply ripple to the laser and intermode beating effects (Watrasiewicz and Rudd 1976). The level is typically 1 per cent for a Helium Neon laser.

13.2.4   Doppler signal processing

The choice of a Doppler signal processing scheme is, in general, dictated by the characteristics of the Doppler signal itself and, in laser vibrometry applications, Doppler signals are continuous but can be subject to periods of very low amplitude due to speckle effects which are described in Section 13.3.1. These periods of low amplitude are analogous to 'drop-outs' in the fluid flow application of laser Doppler velocimetry.

The continuous nature of the Doppler signal in laser vibrometry lends itself to frequency tracking (Wilmshurst 1974) and is shown schematically in Figure 13.12. A voltage-controlled oscillator (VCO) is used to track the Doppler signal and is controlled via a feedback loop. A mixer at the input

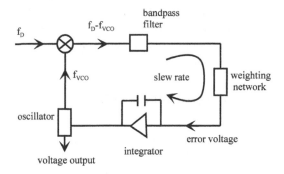

Figure 13.12 An Autodyne frequency tracker.

stage produces an 'error' signal between the Doppler and VCO frequencies which is bandpass filtered and weighted before being integrated and used to control the oscillator to drive the error to a minimum. The feedback loop has an associated 'slew rate' which limits the frequency response of the processor. With respect to the Doppler signal, the tracker acts effectively as a low pass filter which outputs the VCO voltage as a time-resolved voltage analogue of the changing frequency and hence vibration velocity. Some trackers incorporate sophisticated weighting networks which tailor the control of the VCO according to the signal-to-noise ratio of the Doppler signal. The simplest form of this network will hold the last value of the Doppler frequency being tracked should the amplitude of the signal fall below a pre-set value. In this way, the real Doppler signal effectively 'drops out' and consideration must be given to the statistics of what is therefore an essentially sampled output especially when high-frequency vibration information of the order of the drop-out period (typically several microseconds) is required.

Electronic designs of Doppler signal processors vary; each will compensate in some way for loss of Doppler signal amplitude in order to produce a time-resolved output. Commercially available laser vibrometer systems will only specify the performance of the processor in the absence of speckle effects where the Doppler signal amplitude is maintained at a high level. This is commensurate in practice with the target surface moving strictly parallel to the incident laser beam. Such 'ideal' signals can be demodulated with high resolution, accuracy and frequency response. Commercially available laser vibrometers specify low vibration velocity limits of typically fractions of a micrometre per second (bandwidth dependent) and frequency responses of MHz.

It should be remembered that the Doppler frequency is proportional to surface velocity and, as such, may in practice relate to high surface displacement amplitudes at low vibration frequencies or vice versa for the same amplitude value of Doppler frequency. At very low vibration frequencies, processors may operate in 'zero-crossing' or 'fringe counting' mode (Drain 1980). They then produce a time-resolved voltage output proportional to surface displacement rather than velocity. Commercial instruments specify displacement resolution limits of nanometres.

13.3   Solid surface scattering of laser light: Laser speckle

For a laser vibrometer to operate it is necessary to ensure that a sufficient intensity of light is backscattered along the path of the incident laser beam. With reference to equation (13.4), the target intensity I_T must be large enough to ensure that the amplitude of the AC term provides an adequate signal-to-noise ratio for the Doppler signal processor. For this reason, highly polished and reflective targets cause problems since they obey the law of reflection (angle of incidence = angle of reflection) with little diffuse scattering occurring. In laser vibrometry applications the target surface should scatter the incident light diffusely.

An ideal diffuse scattering surface obeys Lambert's cosine law of diffusion given by

$$W(\theta) = \frac{\Gamma}{\pi} W_i \cos \theta \tag{13.5}$$

where $W(\theta)$ is the scattered light power per unit solid angle in a direction θ relative to the normal to the surface, W_i is the incident light power and the constant Γ is unity for a white surface but is less for absorbing surfaces which are typically grey or black. Most target surfaces in engineering behave somewhere between the two extremes of specular reflection and ideal diffuse scattering. A large advantage can be obtained if the surface can be treated with retroreflective tape or paint. In this case light is always scattered back along the path of the incident beam almost independently of the surface orientation. Such surface treatment contains small spheres of high refractive index embedded in a medium and these return the light to the source over an angle of a few degrees. Reflections from sphere to sphere within the medium still cause a laser speckle effect which is introduced as follows.

When a laser beam is incident on a target whose surface is rough on the scale of the optical wavelength, the irradiated area behaves as a dense collection of individual scattering elements. Each element scatters the coherent light so that, at a point in space in front of the target, the contributions from each element add together constructively or destructively. In this way, if the intensity distribution is observed on a screen in front of the target, a laser speckle pattern is formed (Dainty 1984). The pattern consists of a random array of bright and dark areas (speckles); a typical example is shown in Figure 13.13 where each speckle represents a small area where it can be

Figure 13.13 A laser speckle pattern.

assumed that the resultant amplitude and phase from the vector sum of the scattered light from each element is constant. In practice, speckles form approximately ellipsoidal (cigar shaped) volumes in space and the picture shown in Figure 13.13 depicts a cross-sectional intensity map from each adjacent speckle volume.

The photodetector in a laser vibrometer samples the speckle pattern generated by the target surface. Speckle size is inversely proportional to the dimensions of the illuminated area and is dictated by the limiting aperture in the light collection and/or imaging system used (see Section 13.5.1). Most commercially available laser vibrometers are designed to focus the laser beam on to the target and operate by sampling a single bright speckle in the pattern. With reference to equation (13.4) this means that, in practice, the target intensity term I_T is maximised so that the signal-to-noise ratio is high. This situation is an optimum for laser vibrometer operation and is commensurate with instrument specifications quoted in commercial literature. The effect of the initial speckle, which is sampled by the detector, changing during a measurement is discussed in the next section.

13.3.1   Speckle noise in laser vibrometers

The generation of a laser speckle pattern was described above. It is the fact that the photodetector in a laser vibrometer samples a laser speckle pattern which distinguishes laser vibrometer operation from that of laser Doppler anemometry in the fluid flow case. One or more speckles may be sampled by the detector depending upon the optical geometry used and/or the focusing/collection optics used in a particular system or instrument. Most commercial instruments operate by sampling just one bright speckle which is heterodyned with a uniform frequency shifted reference wave. Initially the instrument is aligned with a stationary target so that a bright speckle is sampled by the photodetector. Since a speckle is a correlation region in space in which it can be reasonably assumed that the amplitude and phase of the light remains constant if, during a measurement, the sampled speckle does not change then the vibrometer operates as a classical interferometer where the *rate of change* of the speckle phase modulates the carrier frequency provided by mixing with the frequency shifted reference beam.

In this situation we rewrite equation (13.4) describing the intensity $I(t)$ at a point on the detector as

$$I(t) = I_R + I_p + 2\sqrt{I_R I_p} \cos[2\pi(f_R - f_D)t + \phi_R - \phi_p] \qquad (13.6)$$

where the suffix p now indicates that it is the pth speckle which is contributing to the detector output at this point on its surface. I_p is the intensity of the speckle and ϕ_p is the speckle phase.

In practice, the photodetector responds to an instantaneous sum of $I(t)$ over its active area A and we must therefore consider the spatial distributions of I_R and I_p and their respective phases. For convenience, we can assume the reference beam intensity and phase is uniform in space and we need only consider the speckle pattern distribution. The detector output current $i(t)$ is therefore given by

$$i(t) = \beta I_R A + \beta \int_A I_p(a)\, da + 2\beta \int_A \sqrt{I_R I_p(a)}$$
$$\cos[2\pi(f_R - f_D)t + \phi_R - \phi_p(a)]\, da \quad (13.7)$$

where β is a photoelectric constant, $I_p(a)$ is the speckle intensity distribution on the photodetector and $\phi_p(a)$ is the speckle phase distribution on the photodetector.

Examination of the photodetector output shows that when the speckle intensity and phase distributions change with time during a measurement, spurious information will be contributed when $i(t)$ is frequency demodulated. The major noise source is due to time dependence of the phase term ϕ_p contributing a frequency modulation $d\phi_p/dt$ which is indistinguishable from the change in Doppler frequency associated with the intended measurement. For a commercial instrument which detects just one speckle on the photodetector, it must be ensured that the same speckle remains on the photodetector during a measurement to avoid this problem. The contribution of speckle noise termed 'pseudo vibrations' has been examined in detail and readers are referred to Rothberg *et al.* (1989) for a full discussion.

In brief, if the surface moves strictly parallel to the incident laser beam during an intended normal to surface vibration measurement then the speckle pattern characteristics of intensity and phase can be assumed to be constant. This is because the speckles form approximately ellipsoidal volumes in space and the detector plane effectively samples at different positions along the major axis of each volume during this form of target motion. When correctly aligned on an initially stationary target surface, an instrument operating by collecting just one bright speckle will essentially 'stay with' this speckle during the measurement.

If, however, the target surface tilts, moves in-plane or rotates during a measurement, the speckle characteristics will change. When the scattering elements in the laser spot rotate locally due to surface tilt, the associated speckle pattern on the detector moves in sympathy and this is usually the dominant noise source. If a significant change in population of scattering elements occurs (perhaps due to in-plane target motion) then the speckle pattern will evolve or 'boil' in sympathy to produce noise. In practice, both mechanisms generate pseudo vibration signals in the laser vibrometer output; which one dominates depends upon the measurement attempted.

It is important to note that commercial instrument manufacturers quote performance for laser vibrometers when operating on a stationary target. In practice, this noise performance will only be obtained when the bright speckle detected during the alignment of the instrument on a stationary target does not change during the measurement. This point is given emphasis by considering the design of a laser vibrometer which utilises a rotating scattering disc as a frequency shifter (Halliwell 1979). The noise floor spectrum, for a stationary target and a tripod-mounted instrument is, as anticipated, a periodogram at a 5 Hz fundamental disc frequency generated by speckle noise as shown in Figure 13.14. This is comparable with the noise floor spectrum which would be produced by using a laser vibrometer with frequency shifting by a Bragg cell when the vibrometer was taking a measurement from a target which was rotating simply at 5 Hz. This emphasises the point that noise floors quoted by commercial instrument manufacturers do not apply to every measurement situation but rather depend on the speckle pattern dynamics in each case.

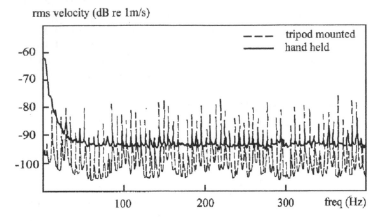

Figure 13.14 Vibrometer noise floor (5 Hz scattering disc).

Commercial instrument specifications of dynamic range, frequency response and accuracy are also measured when the speckle detected during alignment remains on the photodetector during the measurement. These specifications may change significantly if this situation is not maintained and speckle noise is produced. Figure 13.15 identifies practical situations where speckle noise is present and careful consideration is needed in interpreting the vibrometer data. Speckle noise is inherent in the vibrometer output if the target surface tilts, rotates or moves with a velocity component perpendicular to the incident laser beam. The noise peak heights in the spectrum of a laser vibrometer output on a rotating target can be of the order of $1\,\mathrm{mm\,s^{-1}}$ rms at the rotational frequency and higher order harmonics

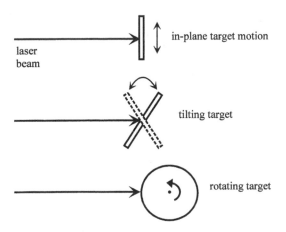

Figure 13.15 Target motions for speckle noise generation.

(Rothberg *et al.* 1989). Unfortunately, in practice, this noise is often linked to the vibration measurement frequency of interest. It is in this situation where engineering judgement is needed in interpreting the results.

It is interesting to note the effect on speckle noise of manually supporting a laser vibrometer. This is illustrated by Figure 13.14 where the laser beam from the hand-held instrument (Halliwell 1979) is incident on a stationary target. Operator body movement is known to be restricted to below 30 Hz (Griffin 1990) and this is shown clearly in the result. It can be seen that operator movement has the effect of causing the speckle pattern from the target to change so that the effect of the periodic repeat is weakened. This causes a reduction in the height of the speckle noise peaks and a corresponding slight increase in the *average* noise level.

13.3.2 Other limitations

The maximum working distance from the laser vibrometer to the target surface is dictated by the need to backscatter sufficient light intensity to the instrument to ensure adequate signal-to-noise ratio for the Doppler signal processor. The latter will often operate successfully at low signal-to-noise ratios whilst offering a reduced frequency response; there is a trade-off to be obtained here depending upon the measurement requirement. The intensity of the light collected by the photodetector depends upon the laser power, transmission/collection optics and the scattering properties of the surface as described earlier in Section 13.3. Commercially available instruments specify working distances of up to 30 m from untreated surfaces, with a laser beam power of less than one milliwatt, depending upon the choice of optical head for the vibrometer. This distance can be extended to 200 m if surface treatment is used. Civil engineering applications (Bougard and Ellis 1998) have seen laser vibrometers used at distances of more than 500 m. At these distances the speckle on the photodetector may begin to wander due to atmospheric turbulence effects and special speckle tracking systems are necessary to overcome this problem.

It is important to note that a laser vibrometer measures the 'instantaneous' rate of change in optical path length between the instrument and illuminated scattering elements on the target surface (see equation (13.3)). In practice, therefore, it is important to ensure that the vibrometer is effectively vibration isolated from the target surface over the frequency bandwidth of interest. An accelerometer fixed to the front of the instrument can be used to check the level of this noise source during a measurement.

Laser technology in the engineering workplace is still relatively new, and if its use is to become commonplace, then safety and user friendliness are prime concerns. With current codes of practice for safety this means that the output beam should be class II, which ensures safety for momentary accidental viewing of the beam directly or by specular reflection.

13.4   Measurement of angular vibration velocity

13.4.1   Introduction

Power transmission by machinery is most often achieved through rotating shafts and/or components. Designers of rotating machinery have to be crucially aware of the magnitude of angular oscillations and/or torsional resonances inherent in these systems in order to avoid such phenomena as fatigue failure of components, rapid bearing wear, gear hammer, fan belt slippage and excessive noise. A typical problem is the control of torsional resonances of a crankshaft in an internal combustion engine. Engine designers must avoid running the engine at 'critical' speeds where the piston firing frequency coincides with a torsional resonance of the crankshaft system. In higher power engines it is standard practice to fit torsional dampers to the crankshaft in order to suppress torsional oscillations over the working speed range of the engine. Damper failure can result in catastrophic fatigue failure of the crankshaft.

Until the advent of laser technology the measurement of torsional vibration was a particularly difficult measurement problem. Strain gauges with associated telemetry or slip ring systems were used but these are notoriously difficult to fix, calibrate and use successfully. For automotive engine test cell applications a slotted disc system was often used. The disc, with slots in its periphery, was fixed to the end of the engine crankshaft and a proximity transducer monitored the slot passing frequency as the disc rotated. Demodulation of the slot passing frequency produced a voltage analogue of the rotational speed, and hence torsional oscillation, within a limited frequency range dictated by the number of slots. In general, the measurement of angular or torsional vibration was problematic for contacting transducer technology and, of course, necessitated machinery downtime whilst arrangements were made for fitting and calibration, etc. The cost of machinery downtime could preclude a measurement being attempted even though the engineer had concluded that a level of torsional oscillation was the root cause of machine failure.

There was, therefore, a real need for a user-friendly instrument which could provide torsional vibration data in situ. The advent of laser technology solved this problem. The laser torsional-or rotational vibrometer, which is now used routinely for this form of measurement, is described below.

13.4.2   The laser torsional or rotational vibrometer

The original use of laser technology for the measurement of angular motion was the application of the cross beam or in-plane vibrometer to measure the tangential surface velocity of a shaft of circular cross section (Halliwell *et al.* 1983). Figure 13.16 shows how this instrument could be used when the solid

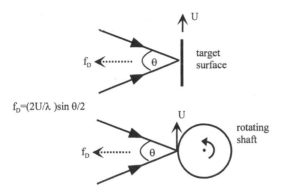

$f_D = (2U/\lambda)\sin\theta/2$

Figure 13.16 The cross beam laser velocimeter.

surface remained in the intersection region of the laser beams throughout the course of the measurement. This required the shaft or component to have a circular cross-section and/or the absence of gross solid body oscillations above about 0.5 mm in amplitude. However, the cross beam vibrometer, in measuring tangential surface velocity, also measures any solid body vibration velocity component which adds to the tangential velocity. It does not distinguish the true angular variation in speed. Successful measurements could, therefore, only be taken where it could be assumed that this effect was negligible.

These problems were overcome using a parallel beam optical geometry proposed by one of the authors (Halliwell *et al.* 1984; Halliwell and Eastwood 1985). Figure 13.17 shows two parallel laser beams incident on a solid body of arbitrary cross section rotating with an angular velocity $\omega(t)$ about an axis within the body. Without loss of generality, the point O defines an origin on the axis which itself is vibrating with a translational velocity $V(t)$. The two laser beams are incident on the body at points P_1 and P_2 and their direction is defined by the unit vector $\hat{s}$. The position

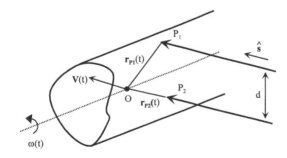

Figure 13.17 Laser rotational vibrometer: theory of operation.

vectors of the points P_1 and P_2 with respect to the origin are $r_{p_1}(t)$ and $r_{p_2}(t)$, respectively. We now consider the total velocity of the scattering elements $V_{p_1}(t)$ and $V_{p_2}(t)$ illuminated by the laser beams at P_1 and P_2 respectively which can be expressed as

$$V_{p_1}(t) = \omega(t) \times r_{p_1}(t) + V(t) \tag{13.8}$$

$$V_{p_2}(t) = \omega(t) \times r_{p_2}(t) + V(t) \tag{13.9}$$

With reference to equation (13.2), the Doppler shifts f_{DP_1} and f_{DP_2} in light directly backscattered from P_1 and P_2 respectively are given by

$$f_{DP_1} = \frac{2}{\lambda}\hat{s} \cdot V_{P_1}(t), \qquad f_{DP_2} = \frac{2}{\lambda}\hat{s} \cdot V_{P_2}(t) \tag{13.10}$$

Mixing this backscattered light on to a photodetector produces a beat frequency f_B in the output proportional to the difference $f_{DP_1} - f_{DP_2}$ (Watrasiewicz and Rudd 1976). Substitution for $V_{P_1}(t)$ and $V_{P_2}(t)$ from equations (13.8) and (13.9) respectively gives

$$f_B = \hat{s} \cdot [\omega(t) \times (r_{p_1}(t) - r_{p_2}(t)] = \omega(t) \cdot [(r_{p_1}(t) - r_{p_2}(t)) \times \hat{s}] \tag{13.11}$$

This result shows that the beat frequency is insensitive to the translational vibration velocity $V(t)$. The modulus of the vector product in equation (13.11) is the separation distance d of the laser beams. Substituting for the angular velocity vector and taking the scalar product (Halliwell 1996) show that

$$f_B = \left(\frac{4\pi}{\lambda}\right) Nd \sin\theta \tag{13.12}$$

where N is the rotational speed of the body, d is the laser beam separation distance and θ is the included angle between the rotational axis of the body and the plane defined by the incident laser beams. In this way, frequency demodulation of the photodetector signal provides a time-resolved voltage analogue of the speed of the rotating body $N(t)$, the ac part of which is the angular oscillation or torsional vibration.

The optical geometry proposed by Halliwell to produce the beat frequency is shown in Figure 13.18. This parallel beam geometry is now used in commercially available laser torsional or rotational vibrometers used for the measurement of angular velocity. It is important to note that equation (13.12) also shows that the measurement is not sensitive to the cross-sectional shape of the body. This fact allows a laser rotational vibrometer to work successfully on targets of arbitrary cross section. The beat frequency is simply dependent upon the rotational speed N and the laser beam separation d. The $\sin\theta$ term shows that the output of a laser rotational vibrometer

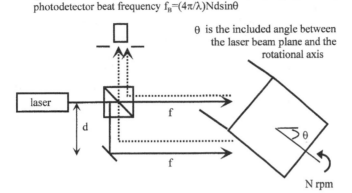

photodetector beat frequency $f_B = (4\pi/\lambda)N d\sin\theta$

θ is the included angle between the laser beam plane and the rotational axis

Figure 13.18 Rotational velocity measurements. Source: Halliwell *et al.* (1984).

is proportional to the component of angular velocity of the body which is resolved along the direction of the normal to the laser beam plane. Discussion of the effect of variations in θ is included in Section 13.4.3.

With reference to equation (13.12), successful operation of a vibrometer which uses direct optical heterodyning of scattered light in this way relies on a non-zero mean value of the rotational speed N. This means that the vibrometer is unable to measure target 'tilt' or angular oscillation without full rotations. A solution is to frequency shift one of the incident laser beams relative to its parallel counterpart (Halliwell *et al.* 1997). In this way, a stationary target still provides a 'carrier' beat frequency in the photodetector output which the subsequent angular velocity of the target frequency modulates.

Use of separate optical interferometers to provide each parallel beam in the laser rotational vibrometer has been adopted by commercial instrument manufacturers. These are frequency shifted to allow the vibrometer to measure tilt or angular oscillation without the need for full rotation of the target. Laser rotational vibrometers are commercially available with specifications that include speed ranges in excess of 10,000 rpm and frequency response in excess of 1 kHz. The error figure quoted for torsional vibration amplitude (AC rotational vibration) is typically less than 2 per cent of the full-scale reading.

13.4.3   Practical considerations and examples of use

The calibration in volts per rpm from a laser rotational vibrometer can usually be carried out in situ. It is not necessary to attempt to measure the beam separation d or the angle θ between the laser beam plane and the rotational axis. In most practical situations it is possible to rotate the target at several known speeds so that an in-situ calibration of DC volts per rpm is possible. If this is not the case, then a small disc can be used which is driven at known speeds and interposed in the laser beams in place of the target. It is straightforward to monitor the DC voltage from the instrument at three known speeds to produce a calibration factor giving negligible error. It is usual to use the vibrometer in conjunction with a real-time spectrum analyser which will integrate the output directly to produce vibration spectra in terms of torsional vibration or angular oscillation displacement referenced to degrees peak. If a real-time analyser is used in situ, a measure of the fundamental frequency of the periodogram produced by the speckle noise (see Section 13.3.1) can conveniently be noted to provide an alternative measure of the rotational speed of the target for calibration purposes. If a real-time analyser is not used, the vibrometer output can be tape-recorded for later analysis, but it must be remembered that the DC signal level is needed if an in-situ calibration has taken place.

In practice, it is often more convenient to tripod-mount the instrument so that the plane of the laser beams is incident at an arbitrary angle to the rotational axis of the target (for example, the end face of a rotating shaft). It should be noted, however, that the dependency of the beat frequency f_B on this angle, defined by equation (13.12), means that a local, time-varying tilt of the axis will modulate f_B and contribute a source of noise. Miles *et al.* (1999) examined this effect in detail and concluded that the error is negligible for values of $70° \leq \theta \leq 90°$. The ideal mode of operation for minimum error is for $\theta = 90°$ corresponding to the plane of the incident beams being at right angles to the local rotational axis. As previously noted, if the instrument is hand held, tilt of the operator will only affect the spectra obtained below a frequency of 30 Hz.

Laser rotational vibrometers provide a convenient means of measuring the revolution rate of a rotating target. However, the more usual measurand of interest is the angular oscillation or torsional vibration spectrum. In practice, the highest levels of torsional vibration can be anticipated on the crankshafts of large diesel engines or reciprocating compressors. These levels may be as high as a few degrees of peak displacement but, in Doppler frequency terms, these levels still represent a small modulation of the beat frequency f_B. Consequently, measuring a high level of torsional vibration does not usually present a problem and it is necessary to ensure simply that the levels are within the dynamic range of the Doppler signal processor.

The lowest level of torsional vibration which can be measured is limited by speckle dynamics as discussed in Section 13.3.1. A rotating (or tilting) target produces speckle noise in the spectrum which forms a periodogram at the fundamental rotation frequency. This is true whether a direct optical heterodyne or separate optical interferometers have been used to provide the beat frequency. Figure 13.19 shows a displacement spectrum obtained from a measurement of the speed of rotation of a DC motor in a tape

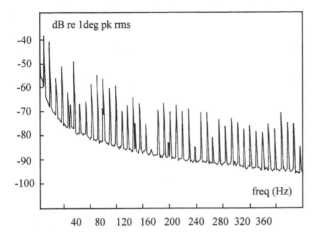

Figure 13.19 Periodic noise floor.

recorder. The spectrum exhibits the periodic nature of the speckle noise which completely masks the intended speed variation measurement. Fortunately, the insensitivity of the geometry to solid body translational oscillation means that the speckle periodicity can be removed by moving the laser beam plane during the course of a measurement (whilst avoiding tilt of the plane!) so that the speckle patterns incident on the photodetectors do not repeat with each revolution. Figure 13.20 shows an equivalent speed measurement of the tape recorder motor when the incident laser beam plane was moved from side to side at a frequency of approximately 1 Hz.

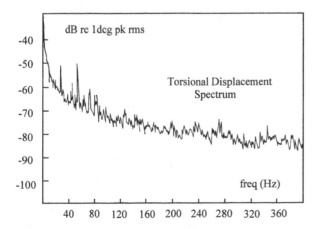

Figure 13.20 Speckle noise peak reduction.

The figure shows that the speckle noise peaks have been effectively removed and the 'wow' of the motor can be distinguished. For most mechanical engineering applications a level of $-60\,dB$ re1° peak corresponding to 10^{-3} degrees means that torsional vibrations are negligible and speckle noise periodicity can be ignored.

The range of possible rotational speeds found in practice presents a problem for Doppler signal processors. At a beam separation d of 10 mm, for a wavelength $\lambda = 0.6$ microns and a beam plane perpendicular to the rotational axis ($\theta = 90°$), a speed variation of 500–15,000 rpm produces a beat frequency range of 1.5–45 MHz. The theory analysing the Doppler shift in the backscattered light (Halliwell 1996) assumes the incident laser beams are infinitely thin and therefore, by an order of magnitude argument, a minimum beam separation of typically 10 mm is required. Commercial instruments usually have a fixed beam separation of this order and employ different speed ranges to address this problem. At higher rotational speeds use of a vibrometer so that the incident laser beam plane is at an angle to the rotational axis can reduce the mean beat frequency but attention must be paid to the problem of error caused by local tilt of the rotational axis during a measurement, as discussed earlier (Miles *et al.* 1999).

Very low values of rotational speed do not cause a problem for rotational vibrometers which utilise frequency shifting and separate interferometers to provide the parallel beam geometry. This is because frequency shifting inherently provides the necessary carrier frequency for a Doppler signal processor independent of the magnitude of the rotational speed. At high rotational speeds, however, care must be taken that the Doppler signal produced by a separate interferometer remains within the dynamic range of the signal processor: for example, with a laser beam of wavelength $\lambda = 0.6\ \mu m$, incident in a direction perpendicular to the rotational axis of a shaft, at a distance of 50 mm from the latter, a rotational speed of 5000 rpm produces a Doppler shift in the backscattered light of 87 MHz. This consideration may prevent the use of a vibrometer in situations where a measurement cannot be taken with the incident laser beams positioned on either side of, or close to, the rotational axis of the target.

13.5 TV holography

13.5.1 Introduction

TV holography, also often referred to as Electronic Speckle Pattern Interferometry (ESPI), is an optical metrology technique for obtaining spacewise maps of static deformation and vibration displacements in both amplitude and phase. During the 1970s and 1980s, the majority of use has been for qualitative analysis of surface deformations, but advances in laser technology and computing power during the 1990s have led to the use of

TV holography for quantitative analysis of both vibration amplitude and phase (Petzing and Tyrer 1998).

The development of TV holography can be traced to initial simultaneous work during the early 1970s (Leendertz 1970; Macovski *et al.* 1971). Whilst holographic interferometry (HI) was a recognised tool for deformation analysis by this time (and still is to this day), the inconvenience of wet-processing and other issues associated with HI created a demand for similar tools based on more convenient technology. The significant breakthrough came when it was recognised that the speckle pattern generated when object surfaces were illuminated with a laser, whilst regarded as a source of noise in HI, could be used as an information carrier in its own right. The use of speckle as a metrology tool was then enabled by using TV camera technology to record maps of speckle intensity, with speckle size being controlled by the viewing optics. The speckle was used to produce fringe patterns or interferograms which were contours of surface displacement.

The basic concepts and methods of TV holography developed during the 1970s still hold true today. Digital storage of speckle patterns allows the user to correlate patterns from before and after an object deformation, generating correlation fringe patterns describing surface displacement. Use of modern pulsed lasers sources, CCD-TV digital cameras and image processing technology allows the analysis of static, periodic dynamic and transient deformations. Furthermore, specific optical configurations of the interferometers allow the user to selectively analyse specific displacement components; out-of-plane, horizontal in-plane and vertical in-plane, with negligible cross-sensitivity to other components (Sirohi 1993; Rastogi 2001).

In the following sections, we introduce the basic principles of TV holography (ESPI), and the various correlation techniques and optical configurations which can be used to measure surface deformation and vibration.

13.5.2 TV holography and laser speckle

Speckle in TV holography serves the same information-carrying purpose as already described in Section 13.3 for laser vibrometry. In the case of TV holography, we are now interested in illuminating large surface areas rather than point illumination ideally required for the vibrometer. The speckle effect occurs when the surface is optically rough, such that the height variation is of the order of, or greater than, the wavelength of the laser light. The intensity of the scattered light is found to vary randomly with position (Dainty 1984).

If the illuminated surface is imaged onto the screen with a lens/aperture system as shown in Figure 13.21, then subjective speckle is formed. The subjective speckle size (diameter) σ_s is related to the effective numerical aperture (NA) of the lens by:

$$\sigma_s \approx \frac{0.6\lambda}{\text{NA}} \tag{13.13}$$

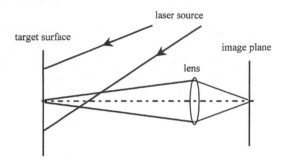

Figure 13.21 Subjective speckle.

However, it is more useful to relate the speckle size to the aperture ratio F of the lens and the magnification M at which the lens system is operating, such that:

$$\sigma_S \approx 1.2(1 + M)\lambda F \tag{13.14}$$

Equation (13.14) allows the use of the camera lens aperture to optimise the speckle size to the image plane pixel resolution and potentially maximise the benefits of the speckle statistics (Dainty 1984), although issues of object illumination intensity and low laser power may limit the aperture range used, in order to maintain good fringe contrast (Jones and Wykes 1989).

13.5.3   Wholefield speckle from object surfaces

It is important to consider the role of the object surface and material in generating a speckle pattern. As already highlighted in Section 13.3, most target surfaces will range between the extremes of specular reflection and ideal diffuse scattering. The brightness distribution of a speckle, and consequently the contrast and quality of the TV holography correlation fringes, depends upon whether or not it is fully developed. A fully developed pattern is only derived from interference of light that is all polarised in the same manner, and consequently the speckle field itself should ideally be polarised in the same manner. This implies that the object surface must maintain the polarised nature of the laser light, and such a surface would typically be composed of lightly abraded metal. For a fully developed speckle pattern, the most probable brightness value will be zero, and there will be more dark speckle than any other brightness (Dainty 1984).

By contrast, a surface which allows multiple scattering such as a matt white paint surface will tend to depolarise the laser light and consequently will not generate fully developed speckle. Multiple scattering in this case will be caused by light penetrating the initial surface before being reflected back, along with top-surface-reflected components. For such a partially developed speckle pattern, the most probable brightness value will be half the average brightness. Experimental work carried out on the surface finish of an object has established this relation (Moore and Tyrer 1990).

This discussion is important because there is generally very little control over the object material characteristics by the analyst. In many cases it may be possible to apply a process or coatings to the object surface in order to achieve fully developed speckle. Matt silver spray paint is a convenient material for both speckle and holographic interferometry, providing little depolarisation, good surface reflectivity and reduced flare, compared to bright metallic surfaces. Matt white paint is another popular finish, aiding the uniform illumination and scattering of the surface leading to improved fringe contrast, as highlighted above, but the paint finish will depolarise the returning object light, thus generating partially developed speckle patterns.

The use of shot peening, bead and sand blasting will yield good reflectivity and reduced flare, but may introduce changes to the surface of the object under study. There are certainly many cases where modification of the surface is undesirable, impracticable or even dangerous. In these cases the user will be forced to tolerate the partially developed speckle. However, the consequence of partially developed speckle patterns may be poor fringe contrast, in which case there may be a reduction of the signal-to-noise ratio during automatic data processing, leading to reduced data accuracy.

13.6 Theory of operation

13.6.1 Speckle interferometry

Speckle could be used just as a marker on an object surface, but as a basis for measurement it is more desirable to use the optical phase information within a speckle and the coherent combination of speckle fields. Such an approach is properly interferometric in concept and execution, and is usually termed 'speckle interferometry' or 'speckle correlation interferometry' as distinct from speckle photography.

Scattering laser light from an object surface will form a speckle pattern, with the speckles displaying a random distribution of intensity. In most practical forms of speckle interferometry, the set-up is arranged so that a second laser illumination wavefront (reference) is introduced, one possible configuration being that shown in Figure 13.22. The effect of a reference beam is to make the speckle intensity dependent on the phase difference between the reference and object wavefronts, and is often regarded as providing a phase reference. (In commercial instrumentation the reference wavefront is smooth and provided directly from the same laser source.)

With reference to Figure 13.22, speckles from both the target surface and the reference surface overlap and interfere at the image plane, producing a resultant speckle pattern. Each overlapping speckle behaves like a miniature

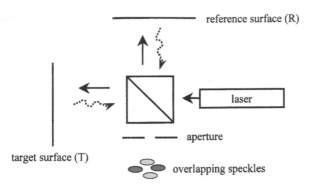

reference surface (R)

laser

aperture

target surface (T)

overlapping speckles

Figure 13.22 Interfering target and reference speckle.

Michelson interferometer and in this case a displacement (d) of the target surface produces a phase change $\Delta = 2kd$, where k is the wave number. When $2kd = 2n\pi$, where n is an integer, there is no change of speckle intensity, and conversely when $2kd \neq 2n\pi$, speckle intensity changes. The changes of speckle intensity are observed as correlation fringe patterns when the resultant speckle pattern (after object deformation) is subtracted from the resultant speckle pattern before object deformation.

At this stage it is helpful to introduce the two basic displacement-sensitive interferometers which form the basis for the speckle correlation displacement measurement techniques. This simplifies the analysis of speckle pattern correlation fringe formation because all interferometers are described by the same theory.

13.6.1.1   Optical arrangement for out-of-plane displacement interferometers

Figure 13.23 shows a schematic of an optical arrangement which is sensitive to displacement along the camera axis, and is commonly known as an 'out-of-plane' interferometer. Note that the reference beam is introduced directly to the image plane from the same laser source as the object illumination and after the lens/aperture system. The sensitivity of the interferometer is controlled by the angle θ between the camera axis and the illuminating source and is ideally reduced to zero. The out-of-plane displacement w which produces a phase change of $2n\pi$, where n is an integer, is given by:

$$w = \frac{n\lambda}{1 + \cos\theta} \tag{13.15}$$

where λ is the laser wavelength (Wykes 1982).

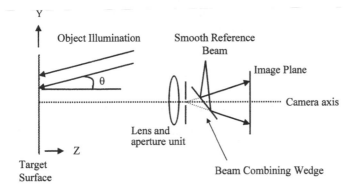

Figure 13.23 Out-of-plane TV holography optical configuration.

When the illumination/viewing angle is zero (source coincident with the camera axis), the interferometer sensitivity is maximised. In practice, for values of θ greater than 10°, in-plane displacement will begin to affect the intended measurement.

13.6.1.2 Optical arrangement for in-plane displacement interferometers

Figure 13.24 shows an optical geometry which is sensitive to displacement normal to the camera axis, and is commonly known as an in-plane interferometer. In this case, the reference beam also forms part of the object illumination geometry and no longer illuminates the image plane directly, the speckle patterns instead interfering at the object surface. The sensitivity of the interferometer is controlled by the subtended angles of the two illumination wavefronts with the camera axis. In the simplest case, if

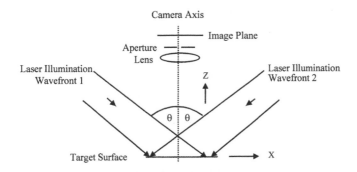

Figure 13.24 In-plane TV holography optical configuration.

the wavefronts are equal but opposite in direction (i.e. $\theta = -\theta$) as shown in Figure 13.24 then the in-plane displacement u which produces a phase change of $2n\pi$, where n is an integer, is given by:

$$u = \frac{n\lambda}{2\sin\theta} \tag{13.16}$$

13.6.2   Subtraction and addition correlation fringes

For simplicity, the theory of fringe formation will be introduced with respect to a normal to surface interferometer as shown in Figure 13.23. The target is illuminated with plane polarised monochromatic laser light. Each speckle in the image plane has a distinct phase and amplitude and the ensemble provide a unique description of the object surface at any point in space and time. The image plane of the interferometer receives a scattered wavefront $U_o(x, y)$ from the object, and a reference wavefront $U_r(x, y)$ introduced after the lens/aperture system. These wavefronts are described by:

$$U_o(x, y) = u_o(x, y)\exp i[\theta_o(x, y)] \tag{13.17}$$

$$U_r(x, y) = u_r(x, y)\exp i[\theta_r(x, y)] \tag{13.18}$$

where $u_o(x, y)$, $u_r(x, y)$ are the amplitudes and $\theta_o(x, y)$, $\theta_r(x, y)$ are the phase angles of overlapping points in the object and reference wavefronts, respectively. The total wavefront received at the image plane U_T is given by:

$$U_T = u_o(x, y)\exp i[\theta_o(x, y)] + u_r(x, y)\exp i[\theta_r(x, y)] \tag{13.19}$$

The intensity I of the combined wavefronts at the image plane is proportional to the square of the total wavefront amplitude, that is $I \propto U_T U_T^*$, where U_T^* is given by:

$$U_T^* = u_o(x, y)\exp -i[\theta_o(x, y)] + u_r(x, y)\exp -i[\theta_r(x, y)] \tag{13.20}$$

Using equations (13.19) and (13.20) we have

$$U_T U_T^* = u_o^2 + u_r^2 + 2u_o u_r \cos\psi \tag{13.21}$$

where $\psi = \theta_r(x, y) - \theta_o(x, y)$. Equation (13.21) can be further simplified to a form which is synonymous with speckle metrology and TV holography (Rastogi 2001) such that

$$I_A = I_o + I_r + 2\sqrt{I_o I_r}\cos\psi \tag{13.22}$$

Note that the descriptor I_A is used in this instance as a means of identifying the intensity distribution, attributable to the object in its initial state, and

I_o and I_r are the intensity of the object and reference beam wavefronts at the object and image plane, respectively.

If the object receives a static displacement, the optical path alters slightly, represented as a phase change of $\Delta(x, y)$ in the object wavefront only. This creates a second set of equations which describe the object in its new state after the target displacement, which include this change of phase:

$$U_o(x, y) = u_o(x, y) \exp \mathrm{i}[\theta_o(x, y) + \Delta(x, y)] \tag{13.23}$$

$$U_r(x, y) = u_r(x, y) \exp \mathrm{i}[\theta_r(x, y)] \tag{13.24}$$

Note that the reference wavefront does not change when the object surface is displaced.

Using similar mathematical manipulation to that employed in deriving equations (13.18)–(13.22), the total light intensity I_B at the image plane of the speckle displacement interferometer (after a static displacement) is given by:

$$I_B = I_o + I_r + 2\sqrt{I_o I_r} \cos(\psi + \Delta) \tag{13.25}$$

13.6.2.1 The formation of subtraction correlation fringes

Equations (13.22) and (13.25) describe the object in the two states; before and after displacement, but the formation of correlation fringes requires further processing of the data. Depending upon the type of recording media used at the image plane, and the laser source illuminating the object, it is possible to correlate the two intensity functions via a subtraction or an addition process.

Subtraction correlation fringes are generally formed when using a CCD camera and continuous wave (CW) laser illumination, the subtraction process being a function of hardware electronics or image-processor-based software algorithms. The change of intensity at the image plane denoted by $\delta I = I_A - I_B$ is achieved by storing an initial object state reference image, and subtracting subsequent images from this reference, such that

$$\delta I = 2\sqrt{I_o I_r}[\cos \psi - \cos(\psi + \Delta)] \tag{13.26}$$

This description may also be presented in TV holography texts after further trigonometric manipulation, as

$$\delta I = 4\sqrt{I_o I_r} \sin\left(\psi + \frac{\Delta}{2}\right) \sin\frac{\Delta}{2} \tag{13.27}$$

Examination of the relationship described in equation (13.27) leads to the conclusion that fringe intensity is at a maximum when $\Delta = (2n+1)\pi$, whereas minimum fringe intensity occurs when $\Delta = 2n\pi$. The zero order fringe pattern corresponding to the contour of zero displacement will normally be dark, although some commercially available subtraction hardware/software may invert the signal, thereby producing a white zero order (Cloud 1995).

13.6.2.2   The formation of addition correlation fringes

There are many cases of wholefield vibration analysis which ideally require strobed or pulsed laser sources. This type of laser illumination reduces interferometer sensitivity to environmental disturbances, allows the generation of quantitative data describing vibration amplitude and phase, and provides a means of measuring transient deformations (Petzing and Tyrer 1998). If a strobed or pulsed laser is used for object illumination whilst using a CCD camera, then the formation of addition correlation fringes is possible if two pulses are used and two images are recorded within a single camera frame at the image plane. The speckle map corresponding to the displaced target is superimposed onto the original speckle map of the object at the image plane. The addition process is attributable to the recording media rather than hardware or software manipulation used for subtraction fringes. With reference to equations (13.22) and (13.25), the change of intensity δI can be described in this case as $\delta I = I_A + I_B$, such that

$$\delta I = 2I_o + 2I_r + 2\sqrt{I_o I_r}[\cos\psi + \cos(\psi + \Delta)] \tag{13.28}$$

Using trigonometric identities this relationship may be further represented as:

$$\delta I = 2I_o + 2I_r + 4\sqrt{I_o I_r}\cos\left(\psi + \frac{\Delta}{2}\right)\cos\frac{\Delta}{2} \tag{13.29}$$

Consequently fringe intensity will be at a maximum when $\Delta = 2n\pi$, whilst minimum fringe intensity occurs when $\Delta = (2n+1)\pi$. Addition fringes are notorious for poor contrast which is caused by the first two background noise terms of equation (13.29), which are not a feature of equation (13.27) for the subtraction case.

13.6.3   Practical consideration in fringe production

Whilst production of correlation fringes has been examined in a theoretical sense, it is necessary to consider the practicalities of producing correlation fringe patterns from an out-of-plane or in-plane interferometer. Generally, the majority of TV holography applications using CW lasers will produce subtraction fringes, and the typical camera image processing design is shown in Figure 13.25. Fringe production requires some basic operations, which are outlined below.

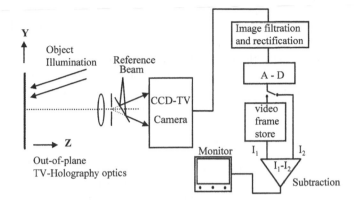

Figure 13.25 TV holography basic speckle processing.

1 The TV signal is taken into a filtration unit which allows simple bandpass filtering, to reduce speckle noise inherent to the TV holography technique. Total removal of speckle noise is not possible because the noise terms overlap the fringe data terms within the frequency domain, and fringe data will be eroded if all speckle noise is removed.

2 A reference image of the laser-illuminated object corresponding to I_A (equation (13.22)) is stored in the video frame store. This will occur when the object is at rest (non-displaced), or in a predetermined state.

3 All subsequent camera images corresponding to I_B (equation (13.25)) are subtracted from the reference image in 'real-time' (typically 25 Hz camera frame rate).

4 The subtraction correlation fringes of target surface displacement corresponding to δI (equation (13.29)) are displayed on a TV-monitor.

Hardware-based frame stores provide simple hands-off operation and only require a camera input and monitor output. These systems are typically suitable for analogue output CCD-TV cameras. It is now very common, however, to find the processing of subtraction correlation fringes performed in software/hardware on a PC- or workstation-based image processor. This entails a more complex system, but will follow the same basic four steps described above. The advantage of the computer-based system is that the data may eventually be processed to produce wholefield quantified maps of displacement amplitude and/or phase (to be discussed in Section 13.8). Computer-based image acquiring equipment will also be necessary if digital output CCD-TV cameras are used for the instrumentation. These cameras are becoming more common because they offer better signal-to-noise characteristics, improved image resolution and faster acquisition.

In order to maintain consistent interferometer sensitivity over the entire target object surface (either out-of-plane or in-plane) collimated illumination wavefronts should ideally be used in all cases of TV holography, although this condition is dependent upon experimental restrictions. Situations will often arise where the target object is large enough to make the use of collimating optics impracticable. Care must be taken, however, as the use of spherical wavefronts will lead to changes of fringe sensitivity and accuracy across an object surface (Wan Abdullah *et al.* 2001).

13.7   Basic static displacement analysis

TV holography can provide simple yet effective measurements of wholefield static deformations, just using the basic understanding of the fringe formation and fringe-displacement relationships (Section 13.6.2). Static surface displacement analysis using TV holography requires a single change of state of an object, from a non-displaced or reference/datum state to a displaced or further displaced state. Once this change of state has occurred, then it is expected that the object will remain at this state for some period of time, generally regarded as being considerably longer than the frame period of a CCD-TV camera (40 ms – CCIR standard in Europe). To complete an analysis of the static displacement, a speckle pattern is recorded at the non-displaced/datum state and stored in a hardware/software frame store (Section 13.6.3). All subsequent speckle patterns of the object surface in its displaced state are then subtracted from the reference, thereby generating correlation fringe patterns which describe the surface displacement.

As an illustrative example, the out-of-plane displacement of a 1 mm thick square stainless steel plate clamped around all edges, with a free surface measuring 200 mm × 200 mm, was measured. A static displacement of 3 μm was applied to the centre of the plate using a differential micrometer. The plate deformation was analysed using a commercial 'Vidispec' system, manufactured by Ealing Electro-Optics Ltd (laser wavelength $\lambda = 632.8$ nm). A reference image of the plate was stored before the displacement was applied, and once deformation had occurred, all subsequent TV images were subtracted from the stored reference.

Figure 13.26 shows the fringe pattern derived from the 3 μm displacement, in terms of a series of concentric subtraction correlation fringes, where each fringe represents a line of constant out-of-plane displacement. With reference to the figure, a simple quantitative analysis can be performed by taking a cross-section of the plate deformation along the line AA. By using the relation in equation (13.15) where $\theta = 0$, the out-of-plane displacement can be related to the fringe order and the number of fringes; this is shown plotted in Figure 13.27. Here the plate profile can be seen (exaggerated by scaling) and a peak displacement of 2.5 μm identified.

The interferometer will provide an absolute measurement and is therefore potentially more accurate than the micrometer reading. However, there

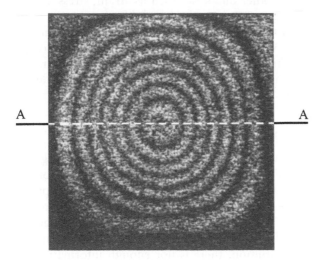

Figure 13.26 Static out-of-plane deformation of a square clamped plate.

is considerable uncertainty associated with the manual measurement of fringe position (specifically identifying fringe centres), and generally a ±1 fringe uncertainty on the results is often observed. Furthermore, the manual assessment of the subtraction correlation fringes is only suitable for simple cross-sections or peak assessments as shown here. Ideally, the analysis of the target surface should be represented as a mesh of data depicting displacement, with the information in a form suitable for engineering numerical

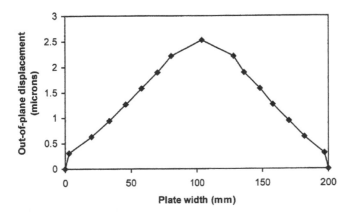

Figure 13.27 Cross section through Figure 13.26 showing a simple representation of the plate deformation.

manipulation to produce other quantities such as strain, stress and phase. This is possible, but the generation of wholefield quantitative measurement data using the full information content of the interferogram requires additional optics in the TV holography optical designs and the use of 'phase-stepping' techniques, which will be discussed in the next section. Further information on static analysis can be found in review articles and texts (Shchepinov and Pisarev 1995; Doval 2000).

13.8 Phase stepping and quantitative data

A fundamental problem with TV holography in its basic form using subtraction correlation fringes (Figures 13.25 and 13.26) is that it does not provide any directional sense to the fringe patterns and subsequent displacement measurement. The fringe pattern generated from the loaded square plate in the last section (Figure 13.26) will look identical to another fringe pattern generated when the load is applied in the opposite direction. In general, apart from engineering intuition, there is not enough information in one fringe pattern to determine automatically which direction the load was applied to the plate.

'Phase stepping' is the term used for the most common method of generation of wholefield quantitative TV holography data, which solves the problem of interferometer directional ambiguity. A simple analogy for the phase-stepping process is the generation of a set of fringe patterns which are described by simultaneous equations, with one of the unknowns being the optical phase change caused by the target surface displacement. Phase stepping involves introducing an additional controlled optical phase change into the object or reference beam of the interferometer. Normally a linear optical phase ramp is applied, introduced between camera frames in predetermined step sizes. Various algorithms exist for calculation using phase stepping (Robinson and Reid 1993), with a typical scheme involving steps of $0, 2\pi/3$ and $4\pi/3$ of optical phase introduced between the object and reference wavefronts of the interferometer.

The two-phase-step approach generates three subtraction correlation fringe patterns described by the three intensity maps I_1, I_2 and I_3 which take the form

$$I_{(k+1)} = 4\sqrt{I_o I_R} \left[\sin\left(\psi + \frac{\Delta}{2} + \frac{2k\pi}{3}\right) \sin\left(\frac{\Delta}{2} + \frac{2k\pi}{3}\right) \right], \quad k = 0, 1, 2$$

$$(13.30)$$

In terms of additional instrument hardware for an out-of-plane TV holography system, a mirror is typically mounted on a piezo-electric transducer (PZT) and introduced into the reference beam path, as shown in Figure 13.28. Application of a known voltage to the PZT transducer causes the mirror to move backwards or forwards in the order of fractions of

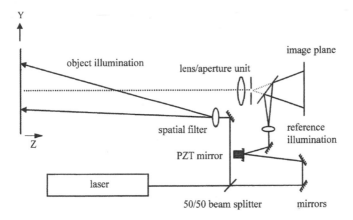

Figure 13.28 Out-of-plane TV holography optical configuration with phase stepping optics.

the laser wavelength. This change in reference beam path length causes the fringe pattern to move on the target. For a point on the target surface, in intensity terms the change from one black fringe to a second black fringe represents an optical phase change of 2π.

If a subtraction correlation fringe pattern describing a static plate displacement is formed on a TV-monitor, then the result of applying an optical phase shift (via the PZT mirror) will be to visually shift the fringes on the TV-monitor. An example of this can be seen in Figure 13.29, which depicts three sets of circular fringe patterns obtained from a square clamped plate, with a micrometer-induced deformation. The fringes change position (with respect to the centre of the image) each time a $2\pi/3$ phase change is introduced into the reference path of the interferometer.

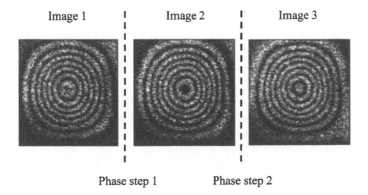

Figure 13.29 Three sets of correlation fringes with two phase steps.

Each of the three fringe patterns corresponding to equation (13.30) is acquired by an image-processing-based computer, which stores the data. Each pixel is now represented three times, once in each of the three sets of data (I_1, I_2 and I_3), and will have a different intensity value in each. We now solve the three simultaneous equations for the pixel phase value (Δ) which is given by

$$\Delta = \arctan\left(\frac{\sqrt{3}(I_3(x, y) - I_2(x, y))}{2I_1(x, y) - I_2(x, y) - I_3(x, y)}\right) \qquad (13.31)$$

In practice, this is achieved by reading across the images (x and y basis) and transferring the pixel intensity values (measured in grey scale values; black, 0–255, white) for each of the three images into equation (13.31) (Rastogi 2001). Equation (13.31) is used when using a three fringe pattern method. There are several other different forms of phase stepping which use one, three, four or many phase steps, and all have a slightly different phase calculation algorithm associated with them (Robinson and Reid 1993; Huntley 1998). The differences are caused by a compromise between data accuracy and speed of data acquisition. The choice of number of steps, step size and phase calculation algorithm will depend upon the target motion and the experimental environment. If the environment is active with vibration, temperature gradients, mass air-movement, etc. all contributing noise terms to the TV holography, then the acquisition time should be shortened as much as possible to reduce the influence of these noise sources. In these cases, alternative algorithms which are insensitive to phase step error can be employed (Huntley 1998).

The result of this calculation for all x-by-y pixels in the three sets of data is a grey scale modulo 2π image (0–255) of optical phase. This has been calculated from the fringe patterns shown in Figure 13.29 and is shown in Figure 13.30. If a cross section were produced then a saw-toothed profile would be generated due to the changing values of phase and the binary jumps or discontinuities between black and white (0 and 255). Consequently it is necessary to re-map the grey scale levels such that a continuous image of optical phase is produced in terms of the grey scales. This requires a third stage known as 'phase unwrapping'. This is a computer-based algorithm which identifies black/white discontinuities and calculates the cumulative optical phase across the whole of the image (Ghiglia and Pritt 2000). The result of this process on the data in Figure 13.30 is then finally re-mapped between 0 and 255 and is shown in Figure 13.31.

The final stage in generating the wholefield quantitative data is to calibrate the wholefield optical phase data with respect to the interferometer used to produce the original correlation fringes. For the out-of-plane interferometer corresponding to the clamped plate experiment, equation (13.15) is used. Numerical manipulation of the optical phase intensity values

Figure 13.30 Wrapped optical phase.

(0–255) is performed, generating a data set which depicts the wholefield out-of-plane displacement of the square clamped plate. These data can be visualised in many different forms, the mesh plot shown in Figure 13.32 being one of these. It should be noted that the phase-stepping techniques are applicable to out-of-plane and in-plane interferometers, using continuous wave or pulsed lasers. Phase stepping can also be applied to an in-plane TV holography interferometer. In this case, the PZT-based mirror can be introduced into either path of the interferometer, the important issue being to create a controlled change of optical phase between the two paths (Moore and Tyrer 1990). An example of an in-plane analysis can be seen in Figure 13.33, which shows the in-plane vector plot for the deformation of a human mandible under load. The purpose of the analysis was to map the deformation of the mandible and identify why mandibles tend to fracture in specific regions along the sides of the bone structure. This information is useful for orthodontic surgeons during facial reconstruction surgery.

Figure 13.31 Unwrapped continuous grey scale image of optical phase.

13.9 Dynamic displacement

Dynamic surface displacement analysis must take into account a constantly changing surface state. The data requirement will involve the measurement of vibration amplitude and vibration phase at points on the moving surface. If the vibration frequency is greater than the camera frequency (25 Hz for Europe and 30 Hz for the USA) then the analysis is known as 'time-averaged'. The analysis of time-averaged TV holography has many parallels with time-averaged holography and was developed during the mid to late 1970s (Løkberg and Høgmoen 1976a) and has been routinely used for qualitative analysis of target surface vibration over the last 20 years, especially in the automotive and aerospace industries (Williams 1993). The technique is very good at identifying modal patterns, clearly highlighting the nodes and anti-nodes of the vibration modes, with fringe density (number of fringes per unit length) again being a direct indication of displacement gradient and amplitude is useful for qualitative comparison with finite element modal analysis predictions.

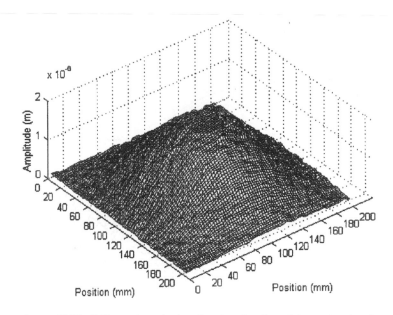

Figure 13.32 Calibrated mesh plot showing the plate deformation in microns.

13.9.1 Time-averaged TV holography

In time-averaged TV holography, vibrating target surfaces will have a surface displacement frequency considerably higher than the camera acquisition frequency, such that many vibration cycles will occur during one camera frame. In this time-averaged mode, with the camera shutter open (physically or electronic equivalent) the camera image plane records a continuous set of speckle patterns added together, which results in an average intensity developed over the frame period. When using a continuous wave laser as the illumination source, this initially leads to an addition process at the image plane, but once the averaging effect is included then an integration term is required to represent the combined effect. The integration of the speckle pattern on the image plane can be used to reduce speckle noise, and leads to the direct imaging of the object surface vibration nodes and anti-nodes.

For simplicity we will confine our discussion to a time-varying displacement amplitude $z(x, y)$ along the Z-axis (camera axis) corresponding to out-of-plane displacement. The analysis is simplified if the surface displacement D is assumed to vary sinusoidally with time, such that:

$$D(x, y, t) = z(x, y) \sin(\omega t) \tag{13.32}$$

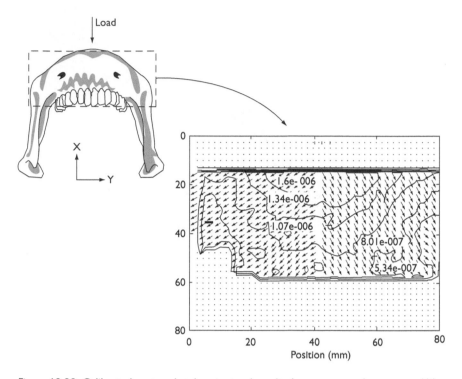

Figure 13.33 Calibrated vector plot showing in-plane displacement on a human mandible.

where ω is the vibration frequency $2\pi f$. At a given instant in time t, the intensity I of a point at the image plane is given by

$$I(t) = I_o + I_r + 2\sqrt{I_o I_r} \cos\left[\psi + \frac{4\pi}{\lambda}D(t)\right] \tag{13.33}$$

which corresponds to equation (13.25), where the optical phase Δ at a point is related to $D(t)$ by

$$\Delta = \frac{4\pi}{\lambda}D(t) \tag{13.34}$$

The function $D(t)$ represents the position of a given point on the object surface at time t. However, equation (13.34) only represents the image plane for a single moment in time, and the full frame time of the CCD-TV camera must be considered. Hence the average intensity I_τ over time τ is given by

$$I_\tau = I_o + I_r + \frac{1}{\tau}2\sqrt{I_o I_r}\int_0^\tau \cos\left[\psi + \frac{4\pi}{\lambda}D(t)\right]dt \tag{13.35}$$

Assuming that the vibration period is considerably smaller than the frame period $(2\pi/\omega\nu \ll \tau)$, the image plane intensity average can be reduced to

$$I_\tau = I_o + I_r + 2\sqrt{I_o I_r} J_0\left(\frac{4\pi}{\lambda}D\right)\cos\psi \tag{13.36}$$

where J_0 is the zero-order Bessel function (Rastogi 2001).

The correlation fringes observed during time-averaged TV holography describe the variation of the amplitude of vibration across the object surface, but include the influence of the zero-order Bessel function. TV holography hardware and software will typically cause the Bessel function in equation (13.36) to be squared, so that the correlation fringe contrast will vary as J_0^2. If the maxima points of the squared Bessel function are found and used to solve equation (13.36) (Løkberg 1984), the amplitude of vibration is found to be equivalent to $\lambda/4$ per fringe. One practical consequence of the Bessel function modulation of the fringe pattern is decreasing fringe visibility as the number of fringes increases. This tends to limit the maximum displacement amplitude which can be measured and depends on the number of fringes which can be identified. Typically, visual examination of time-averaged TV holography fringe patterns will be able to identify up to ten consecutive fringes, before poor contrast may make further analysis impossible.

It should also be noted that in the time-averaged TV holography, the vibration phase information is normally lost due to the averaging of the surface displacements over one or more vibration cycles, within one active camera frame. The sign of the time-varying displacement cannot be automatically calculated, although general intuition will aid the understanding of the vibration phase characteristics of an object surface. However, if additional optical elements are added to the TV holography systems, then it is possible to heterodyne the interferometer and use phase modulation techniques. This provides the flexibility to identify the phase of vibration of anti-nodes (Løkberg and Høgmoen 1976b; Rastogi 2001).

We now describe an example of time-averaged displacement measurement. The object examined was the plate described in Section 13.7. A fundamental mode (1,1) was excited at resonance at 200 Hz using a function generator and a loudspeaker. Figure 13.34 shows the time-averaged correlation fringe pattern obtained from the resonating plate in terms of a series of concentric fringes, where each fringe represents a contour of out-of-plane displacement modulated by the zero order Bessel function. A simple quantitative analysis can be performed by taking a cross-section of the plate deformation along the line BB. The out-of-plane amplitude is shown plotted in Figure 13.35, calculated on the basis of $\lambda/4$ displacement amplitude per fringe. The plate profile can be seen (exaggerated by scale), and a peak amplitude of 0.8 μm identified. The peak-to-peak amplitude is therefore

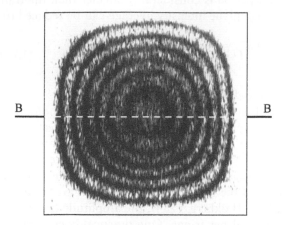

Figure 13.34 Time-averaged fringe pattern showing the resonant out-of-plane deformation of a square clamped plate.

approximately $1.6\,\mu m$ for this vibration anti-node. Notice that the fringe contrast decreases as fringe density increases, this being an influence of the modulating Bessel function, as previously discussed.

A further example of time-averaged TV holography applied to a three-dimensional object is shown in Figure 13.36. This analysis involved the identification of mode shapes, characteristics and frequencies for a church bell. The bell was excited using a frequency generator and 20 N shaker, with the input sinusoidal frequency set at specific values previously identified using an accelerometer-based frequency analysis. The result shown

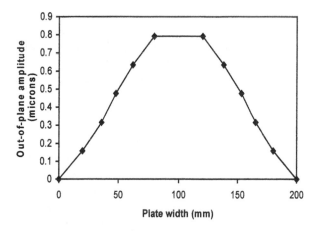

Figure 13.35 Cross section through Figure 13.34 showing a simple representation of plate deformation.

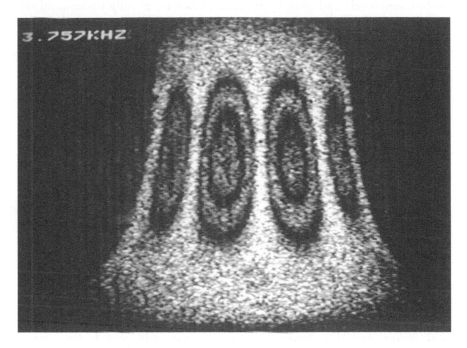

Figure 13.36 Time-averaged fringe pattern showing the mode shape of a bell at 3.757 kHz.

in Figure 13.36 shows the bell resonating with four longitudinally orientated anti-nodes at 3.757 kHz, each anti-node displaying an amplitude of approximately $\lambda/2$ or 0.3 microns.

It is important to note that a consequence of using a CCD-TV camera as the speckle recording instrument, is that three-dimensional objects are recorded on a two-dimensional image plane. This can lead to interferometer sensitivity errors as a function of target surface form or shape, and care should be taken when analysing three-dimensional objects, in order to avoid misinterpretation of the fringe patterns (Rastogi 2001). This issue is evident in Figure 13.36, because for an axisymmetric object such as the church bell, all anti-nodes should be of equal amplitude, yet the first and fourth anti-node in the image appear to have a lower fringe density. This is caused by the curved form of the bell, and the changing out-of-plane sensitivity of the interferometer.

13.9.2 Pulsed laser TV holography

Time-averaged fringe patterns are associated with the analysis of objects excited sinusoidally using continuous-wave-based speckle pattern interferometers, and are regarded as being very useful for quick analysis of target surface amplitude and modal structure. However, to fully understand the vibration amplitude and vibration phase with wholefield quantified meshes of data, it is necessary to use a strobed or pulsed laser. Vibration phase measurement is important in applications when it is necessary to distinguish travelling waves from standing waves, a distinction which is not possible with time-averaged TV holography.

A strobed laser is a continuous laser with a mechanical, electro-optic or acousto-optic chopper in the beam which can be synchronised to the TV camera or the vibration frequency. This produces long laser pulses (microseconds to milliseconds) but dramatically reduces the effective output power of the laser. The currently favoured option is a flash lamp or diode pumped pulsed laser (Nd:YAG). With a pulse length typically in the order of 10–15 ns, the laser strobes the motion of the object, such that any object displacement during the illumination period is negligible in practice.

The pulsed laser can be used in two forms. The commonly used single-pulse mode is synchronised to the CCD-TV camera such that one pulse illuminates the object every camera frame. Synchronisation of the laser pulse and camera frame initiation have to be linked to the object vibration phase, in order to provide a stationary fringe pattern. This may be via a direct link to the frequency generator used for experimentation, but this may not be possible in objects excited by means other than a frequency generator and shaker, and this approach does not necessarily provide the correct phase relationship between the generator and the TV holography system. A practical solution is to place an accelerometer on the target surface or use a laser vibrometer to provide vibration phase feedback from specific locations on the target surface. The signal from either of these transducers can then be used by pulse generators to correctly trigger synchronisation timing pulses for the laser and the camera, as shown in Figure 13.37.

Data are generated by storing a reference speckle pattern from the object, typically when the vibration amplitude is zero, and then subtracting all

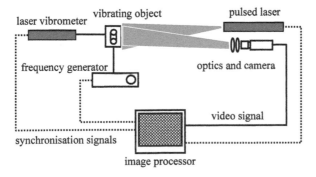

Figure 13.37 Pulsed laser TV holography with vibrating object synchronisation.

subsequent speckle patterns from the stored reference, as described by equation (13.26). The subtraction correlation method produces time-resolved fringes of good visibility and signal-to-noise ratio, at TV frame rates. This approach presents sinusoidal fringe patterns, removing the Bessel function modulation seen in the time-average approach. This is important because with time-averaged TV holography, the vibration phase information is usually lost due to the averaging of the speckle patterns at the image plane. The pulsed interferometer results now present fringe patterns of equivalent quality to static displacement analysis with the image plane intensity being the same as a continuous wave TV holography system (equation (13.26)).

In extreme environments, however, a pulse separation of milliseconds makes the TV holography interferometer susceptible to environmental disturbances, which can manifest as fringe movement in the order of $\pi/6$ (Moore and Perez-Lopez 1996). This problem can be solved by modifying the synchronisation and laser trigger signals so that two laser pulses are used to illuminate the object during every CCD-TV frame, which produces pulsed addition fringes as defined by equation (13.29). This approach has recently been successfully used for transient vibration analysis (Fernández et al. 1997). If the pulse separation is small, for example $100\,\mu s$, then instabilities caused by the environment are ignored because they generally occur at much lower frequencies.

Unfortunately, addition correlation fringes as described in Section 13.6.2.2 have poor contrast and visibility and require demanding image-processor-based manipulation to be usable. This problem has previously been addressed using filters, and algorithms designed to improve addition fringe contrast have been produced (Davila et al. 1994). The pulsed addition fringe approach is uncommon and this is also due, in part, to the lasers required for double pulsed analysis, which are optically complicated and expensive.

An example of a pulsed TV holography result is shown in Figure 13.38, which shows a resonant mode shape for a vented brake disc at 1.1 kHz. Brake discs and braking systems are prone to generating brake squeal, which is an irritating and unwanted side effect of the braking process. Brake system manufacturers are interested in understanding the vibration characteristics of the brake discs, with experimental data being used to validate modal analysis models. Whilst accelerometers and laser vibrometers provide excellent time-based analysis, TV holography has the benefit of providing complete spatial analysis of the vibrating brake disc.

Disc excitation was via a frequency generator and 20 N shaker, with specific input frequencies identified from previous accelerometer-based modal analysis. Laser and camera synchronisation were provided from an accelerometer attached to the brake disc surface. In siting the accelerometer (also applicable to a laser vibrometer), it is important to note that changing input frequencies will excite different mode shapes, causing changing

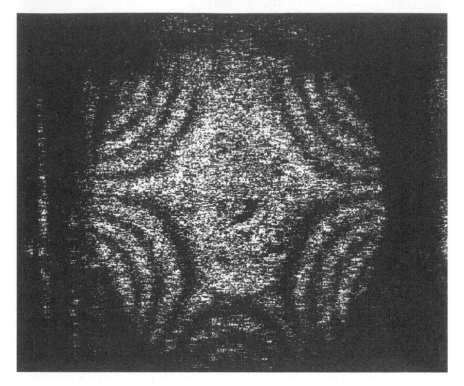

Figure 13.38 Pulsed fringe pattern showing the vibration mode of a brake disc.

positions of nodes and anti-nodes. If the accelerometer and a nodal region coincide, then the return signal will be poor and the TV holography system will not necessarily be correctly synchronised.

The interferograms which are generated using pulsed TV holography and subtraction correlation are suitable for phase stepping (Section 13.8). This is again an important part of the analysis because it allows the user to gain directional sense from the interferometer, eventually providing the correct visualisation of the vibration amplitude and vibration phase (Valera *et al.* 1992). This process has been applied to the data in Figure 13.38 using PZT-based phase-stepping optics described in Section 13.8 and detailed in Figure 13.28. A reference speckle image of the brake disc was acquired before excitation of a specific mode shape. Surface displacement amplitude was optimised by adjusting the laser pulses with respect to the synchronisation signal from the accelerometer. Stationary correlation fringes were formed by subtracting the deformed object speckle pattern from the reference, with phase stepping applied using the process described by equation (13.31).

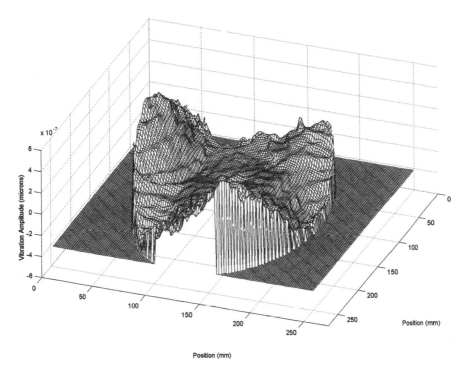

Figure 13.39 Brake disc vibration amplitude.

Optical phase calculation was completed using the two-step three-image routine described in Section 13.8, followed by optical phase unwrapping producing a wholefield mesh of surface displacement amplitude, with the data calibrated with respect to the out-of-plane interferometer sensitivity. The wholefield map of vibration amplitude is shown in Figure 13.39, identifying vibration nodes and anti-nodes.

Acknowledgements

N.A. Halliwell acknowledges permission from IOP Publishing Ltd to reproduce material from his contribution to their text 'The Handbook of Laser Technology' (to be published).

References

Baker, J.R., Laming, R.I., Wilmshurst, T.H. and Halliwell, N.A. (1990) A new high sensitivity laser vibrometer. *Optics and Laser Technology*, **22**(4), 241–244.

Ballantyne, A., Blackmore, C. and Rizzo, J.E. (1974) Frequency shifting for laser anemometers by scattering. *Optics and Laser Technology*, **6**, 170–173.

Bougard, A.J. and Ellis, B.R. (1998) Laser measurement of building vibration and displacement. *Proc. SPIE 3rd Int. Conf. on Vibration Measurements by Laser Techniques: Advances and Applications*, No. 3411, 254–265.

Buchave, P. (1975) Laser Doppler vibration measurement using variable frequency shift. *DISA Information Bulletin*, No. 18, 15–29.

Cloud, G. (1995) *Optical Methods of Engineering Analysis*. Cambridge University Press.

Dainty, C.J. (ed.) (1984) *Laser Speckle and Related Phenomena*. Spinger-Verlag.

Davila, A., Kerr, D. and Kauffman, G.H. (1994) Digital processing of electronic speckle pattern interferometry addition fringes. *Applied Optics*, 33(25), 5964–5969.

Doval, A.F. (2000) A systematic approach to TV holography. *Measurement Science and Technology*, 11, R1–R36.

Drain, L.E. (1980) *The Laser Doppler Technique*. Chichester, UK: John Wiley & Sons pp. 85–91.

Drain, L.E. and Moss, B.C. (1972) The frequency shifting of laser light by electro-optic techniques. *Opto-Electronics*, 4, 429.

Fernández, A., Moore, A.J., Pérez-López, C., Doval, A.F. and Blanco-García (1997) Study of transient deformations with pulsed TV-holography: Application to crack detection. *Applied Optics*, 36(10), 2058–2065.

Ghiglia, D.C. and Pritt, M.D. (2000) *Two Dimensional Phase Unwrapping: Theory, Algorithms and Software*. Wiley.

Griffin, M.J. (1990) *Handbook of Human Vibration*. Academic Press.

Halliwell, N.A. (1979) Laser Doppler measurement of vibration surfaces: A portable instrument, *Journal of Sound and Vibration*, 62(2), 312–315.

Halliwell, N.A. (1996) The laser torsional vibrometer: A step forward in rotating machinery diagnostics. *Journal of Sound and Vibration*, 190(3), 399–418.

Halliwell, N.A. and Eastwood, P.G. (1985) The laser torsional vibrometer. *Journal of Sound and Vibration*, 101(3), 446–449.

Halliwell, N.A., Pullen, L. and Baker, J. (1983) *Diesel Engine Health: Laser Diagnostics*. SAE Technical Paper Series 831324, International Off-highway Meeting and Exposition, Milwaukee, USA, also in Trans. SAE, 986–994.

Halliwell, N.A., Pickering, C.J.D. and Eastwood, P.G. (1984) The laser torsional vibrometer: A new instrument. *Journal of Sound and Vibration*, 93(4), 588–592.

Halliwell, N.A., Hocknell, A. and Rothberg, S.J. (1997) On the measurement of angular vibration displacements: A laser tiltmeter. *Journal of Sound and Vibration*, 208(3), 497–500.

Huntley, J.M. (1998) Automated fringe pattern analysis in experimental mechanics: A review. *Journal of Strain Analysis*, 33(2), 105–127.

Jones, R. and Wykes, C. (1989) *Holographic and Speckle Interferometry* (2nd edn), Cambridge Studies in Modern Optics 6. Cambridge University Press.

Laming, R.I., Gold, M.P., Payne, D.N. and Halliwell, N.A. (1985) Fibre-optic vibration probe. *Electronics Letters*, 22(3), 167–168.

Leendertz, J.A. (1970) Interferometric displacement measurement on scattering surfaces utilizing speckle effect. *Journal of Physics E*, 3, 213–218.

Lewin, A.C., Kersey, A.D. and Jackson, D.A. (1985) Non-contact surface vibration analysis using a monomode fibre optic interferometer incorporating an open air path. *Journal of Physics E: Scientific Instruments*, 18, 604–607.

Løkberg, O.J. (1984) ESPI – The ultimate holographic tool for vibration analysis? *Journal of the Acoustical Society of America*, 75(6), 1783–1791.

Løkberg, O.J. and Høgmoen, K. (1976a) Vibration phase mapping using electronic speckle pattern interferometry. *Applied Optics*, 15(11), 2701–2704.

Løkberg, O.J. and Høgmoen, K. (1976b), Use of modulated reference wave in electronic speckle pattern interferometry. *Journal of Physics E*, 9, 847–851.

Macovski, A., Ramsey, S.D. and Schaefer, L.F. (1971) Time lapse interferometry and contouring using television systems, *Applied Optics*, 10(12), 2722–2727.

Miles, T.J., Lucas, M., Halliwell, N.A. and Rothberg, S.J. (1999) Torsional and bending vibration measurements on rotors using laser technology. *Journal of Sound and Vibration*, 226(3), 441–467.

Moore, A.J. and Pérez-López, C. (1996) Fringe visibility enhancement and phase calculation in double-pulsed addition ESPI. *Journal of Modern Optics*, 9, 1829–1844.

Moore, A.J. and Tyrer, J.R. (1990) An electronic speckle pattern interferometer for complete in-plane displacement measurement. *Measurement Science Technology*, 1, 33–39.

Oldengarm, J. (1977) Development of rotating diffraction gratings and their use in laser anemometry. *Optics and Laser Technology*, April, 69–71.

Oldengarm, J., van Krieken, A.H. and Raternik, H.J. (1973) Laser Doppler velocimeter with optical frequency shifting. *Optics and Laser Technology*, 5(6), 249–252.

OMETRON Ltd (1997) *VPI 4000 User Manual*.

Petzing, J.N. and Tyrer, J.R. (1998) Recent developments and applications in electronic speckle pattern interferometry. *Journal of Strain Analysis*, 33(2), 153–170.

Pickering, C.J.D. and Halliwell, N.A. (1986) The laser vibrometer: A portable instrument. *Journal of Sound and Vibration*, 107(3), 471–485.

Rastogi, P.K. (2001) *Digital Speckle Pattern Interferometry and Related Techniques*. Wiley.

Reibold, R. and Molkenstruck, W. (1981) *Acoustica*, 49, 205.

Robinson, D.W. and Reid, G.T. (1993) *Interferogram Analysis: Digital Fringe Pattern Measurement Techniques*. IOP Publishing Ltd.

Rothberg, S.J., Baker, J.R. and Halliwell, N.A. (1989) Laser vibrometry: Pseudo vibrations. *Journal of Sound and Vibration*, 135(3), 516–522.

Scruby, C.B. and Drain, L.E. (1990) *Laser Ultrasonics*. Bristol, UK: Adam Hilger, pp. 116–122.

Shchepinov, V.P. and Pisarev, V.S. (1995) *Strain and Stress Analysis by Holographic and Speckle Interferometry*. Wiley.

Sirohi, R.S. (1993) *Speckle Metrology*. Marcel Dekker.

Valera, J.D., Doval, A.F. and Jones, J.D.C. (1992) Determination of vibration phase with electronic speckle pattern interferometry (EPSI). *Electronics Letters*, 28(25), 2292–2294.

Wan Abdullah, W.S., Petzing, J.N. and Tyrer, J.R. (2001) Wave front divergence: A source of error in quantified speckle shearing data. *Journal of Modern Optics*, 48(5), 757–772.

Watrasiewicz, B.M. and Rudd, M.J. (1976) *Laser Doppler Measurements*. London, UK: Butterworths.

Williams, D.C. (1993) *Optical Methods in Engineering Metrology*. Chapman & Hall.

Wilmshurst, T.H. (1974) An autodyne frequency tracker for use in laser doppler anemometry. *Journal of Physics E: Scientific Instruments*, 7, 924.

Wykes, C. (1982) Use of electronic speckle pattern interferometry (EPSI) in the measurement of static and dynamic surface displacements. *Optical Engineering*, 21(3), 400–406.

Index

Note: Bold type indicates graphical illustration.